MECHANICAL DRAWING

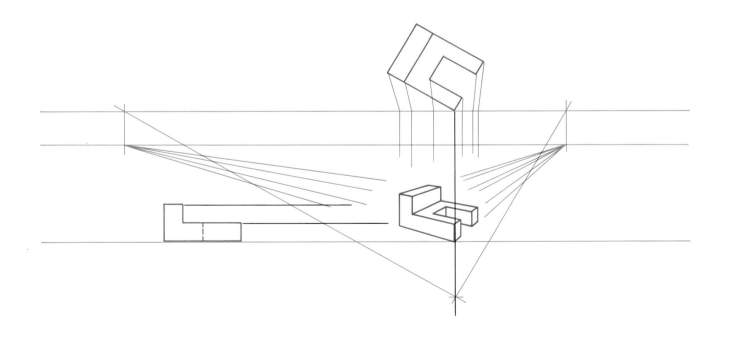

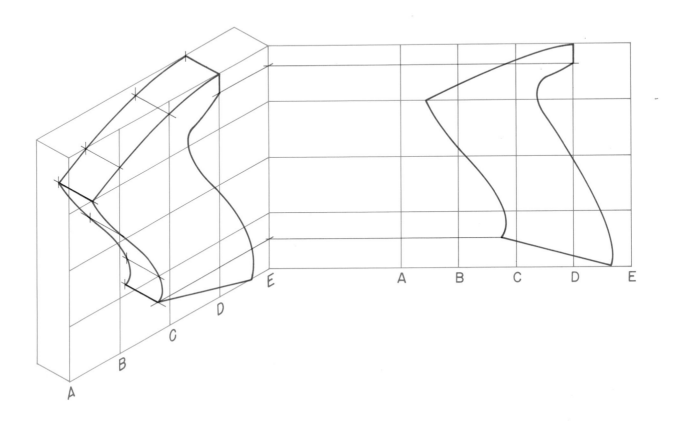

A · B · C · D · E

A · B · C · D · E

MECHANICAL DRAWING
8th edition

THOMAS E. FRENCH/CARL L. SVENSEN
JAY D. HELSEL/BYRON URBANICK

McGraw-Hill Book Company

New York St. Louis San Francisco Dallas Düsseldorf Johannesburg Kuala Lumpur
London Mexico Montreal New Delhi Panama Rio de Janeiro Singapore Sydney Toronto

Editor: Hal Lindquist   Designer: Peter Bender
Editing Supervisor: Paul Farrell   Production Supervisor: John Sabella

Library of Congress Cataloging in Publication Data
Main entry under title:

Mechanical drawing.
    SUMMARY: A textbook introducing the basic theory,
techniques, and uses of drafting for industrial arts
and vocational high school students.
    First-7th editions by T. E. French and C. L. Svensen;
1st-3d editions published under title: Mechanical
drawing for high schools.
    1. Mechanical drawing. [1. Mechanical drawing]
I. French, Thomas Ewing, 1871–1944. Mechanical drawing
for high schools.   II. Svensen, Carl Lars, 1884–1970.
[T353.M48 1974]   604'2   73-8419
ISBN 0-07-022310-6

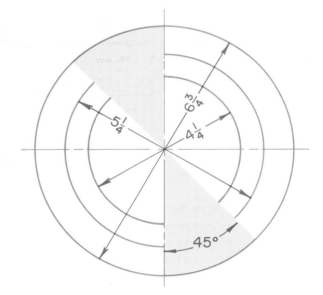

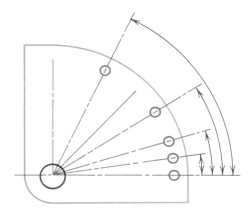

About the Authors

JAY D. HELSEL is an Associate Professor of Industrial Arts and Director of Campus Planning at California State College in Pennsylvania. He completed his undergraduate work in industrial arts at California State College and was awarded a master's degree from The Pennsylvania State University. He has done advanced graduate work at West Virginia University and at the University of Pittsburgh, where he is now working toward a doctoral degree in educational communications and technology. In addition, Mr. Helsel holds a certificate in airbrush techniques and technical illustration from the Pittsburgh Art Institute.

He has worked in industry and has taught drafting, metalworking, woodworking, and a variety of laboratory and professional courses at both the secondary and college levels. During the past fifteen years, he has worked as a freelance artist and illustrator. His work appears in a great variety of technical publications.

Mr. Helsel is coauthor of *Programmed Blueprint Reading, Drawing and Blueprint Reading Transparencies, Architectural Drafting Transparencies, Woodworking Transparencies, and Automotive Transparencies.* He is also the author of a series of *Mechanical Drawing Film Loops.*

BYRON URBANICK is Chairman of the Industrial Arts Department at Oak Park and River Forest High School in Oak Park, Illinois, where he has taught architectural and engineering graphics for sixteen years. He has also taught drafting and graphics at the Illinois Institute of Technology, Chicago State University, Northern Illinois University, Junior College of DuPage, and for the U.S. Naval Ordnance staff. His professional experience includes work for the architectural firm of Friedman, Alschuler and Sincere, Fairbanks Morse and Company, Starme Industrial Engineers, and Malco Tool and Manufacturing Company; he also has had his own design and drafting practice.

Mr. Urbanick studied at Wilson Junior College, the Illinois Institute of Technology, Illinois State University, and the University of Illinois; he earned his B.S. and M.S. degrees from Northern Illinois University. He is a member of the American Society for Engineering Education, the American Industrial Arts Association, the American Vocational Association, and the Illinois Technical Drawing Teacher's Association, where he served as secretary in 1963 and 1971.

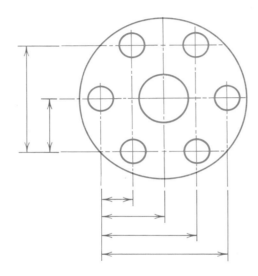

Contents

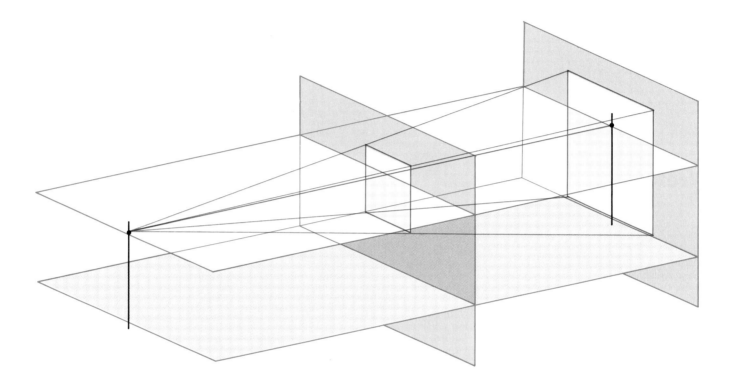

Preface

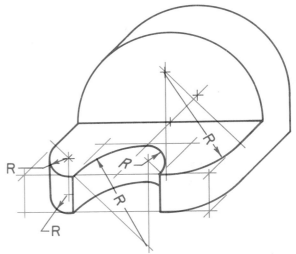

Thomas E. French and Carl L. Svensen wrote the first edition of *Mechanical Drawing* to help students learn to visualize in three dimensions, to develop and strengthen their technical imagination, to think precisely, to read and write the language of the industries, and to gain experience in making working drawings according to modern commercial practices. Those ideals were carefully upheld from the first through the seventh editions. Each edition was also updated to conform with changes in standards and with advancements in the technologies of industry.

In preparing the eighth edition, the original ideals have been maintained as the primary objectives, and many new concepts have been added. A substantial portion of the text has been printed in four colors to help clarify many of the subject areas. New chapters, such as "Systems for Graphic Communications," "Basic Descriptive Geometry," and "Drafting Media and Reproduction," have been added. Also, many of the chapters have been completely rewritten and reillustrated.

Since the book has been almost completely reillustrated, color is now used much more effectively as a teaching and learning device. The use of four colors not only enhances the appearance and aesthetic appeal of the book, but also helps to graphically show procedural steps and methods of projection in some of the more complex concepts.

Technical drawing is still the basic language of industry. Since technology is rapidly advancing the elements and concepts of industry, its language must keep pace accordingly. The eighth edition of *Mechanical Drawing* has been prepared to reflect such changes. Chapters such as "Architectural Drafting," "Aerospace Drafting," "Multiview Drawing," and others have been researched and rewritten in a modern manner designed to stimulate the student's imagination and to encourage problem solving and creative thought.

In this edition, the problems sections appear at the end of each chapter. This provides easy access to the problems and makes the related chapter readily available to the student for reference purposes. All problems throughout the book are entirely new. An adequate number of problems has been provided so that student assignments may be different for several classes over a number of years without duplication or overlap. In addition, problems are designed to provide for individual differences, special student interests, and maximum instructional value.

Basic elements of the Metric System have been introduced throughout the book in keeping with international trends. An introduction to the Metric System is offered in the Appendix, dual dimensioning is covered in Chapter 6, and various problems throughout the book require the use of the Metric System.

Basic mechanical drawing courses require the same fundamentals, regardless of the student's age or grade level. Therefore, the content and arrangement of the text and problems is practical and suitable for a great variety of educational programs.

We wish to offer sincere and special thanks to those users of previous editions whose suggestions have helped shape this edition. Their continued interest and help are greatly appreciated. Credit is also given to the many companies who contributed technical assistance as well as illustrations and photographs. Special thanks to Mrs. Gazella Mucy and Mrs. Eleanor Mars for secretarial help. Photographic research was done by Marie Baxter. Comments and suggestions are welcome and appreciated.

JAY D. HELSEL
BYRON URBANICK

Twentieth-Century Drafting

Chapter 1

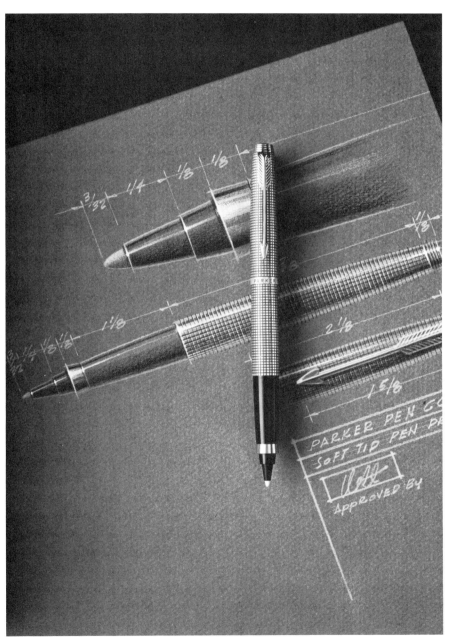

(THE PARKER PEN CO.)

Monuments of History. Drafting has played an important part in man's progress from the time of the ancient pyramids and the classical Parthenon (Fig. 1-1) to the geodesic domes and the skylabs of today (Fig. 1-2). Drafting has made it possible for man to develop and record a vast amount of knowledge that is useful to society. Without drafting, man's progress would have been much slower.

History of a Graphic Language. In ancient times men drew pictures to show others their ideas. Some of these drawings from ancient civilizations were crude sketches made on clay tablets. Many of them are still in existence. Early builders probably made detailed plans of their buildings on parchment or papyrus. These were less durable, and few have been preserved. The people of Mesopotamia used drafting materials as early as 2200 B.C. Figure 1-3 shows a statue of one of their kings, Gudea, with a drawing of a building on his lap.

Ancient Egyptian stonemasons made plans for the pyramids and other structures on papyrus, slabs of limestone (Fig. 1-4), and sometimes on wood. In the construction stage, they drew lines on the ground to locate the key positions of large stone blocks for temples and other large structures. However, it was the Romans who probably made the best mechanical drawings of the classical period. They drew highly detailed pictures and plans for their buildings, aqueducts (Fig. 1-5), temples, and roadways.

1

Fig. 1-1 The Parthenon—inaugurated 483 B.C., and the best known of all Greek temples. (*Greek National Tourist Office.*)

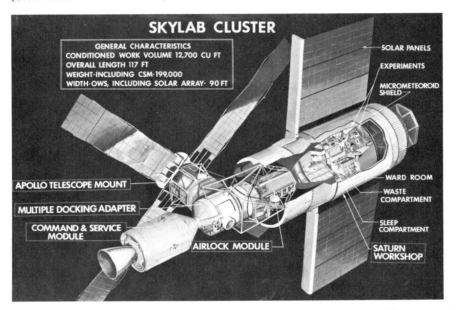

Fig. 1-2 Skylab—manned orbital scientific space station—an extension of the manned Mercury-Gemini-Apollo space programs. (*McDonnell Douglas, Inc.*)

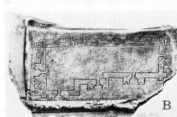

Fig. 1-3 At A, headless statue of Gudea, engineer and ruler (2000 B.C.). The inscription on the statue tells about the design and building of the temple of Ningirsu. At B, a plan view drawn on the stone tablet on Gudea's lap. (*From Ernest de Sarzec, Decouverte en Chaldee, 1891.*)

Fig. 1-4 The plan on limestone of the tomb of an Egyptian king. (*From Clark and Engelbach, Ancient Egyptian Masonry.*)

Fig. 1-5 Engineers of ancient Rome knew the importance of making drawings. The Pont du Gard, near Nimes, France, is a good example of the aqueducts they built. (*From T. Schreiver, Atlas of Classical Antiquities, 1895.*)

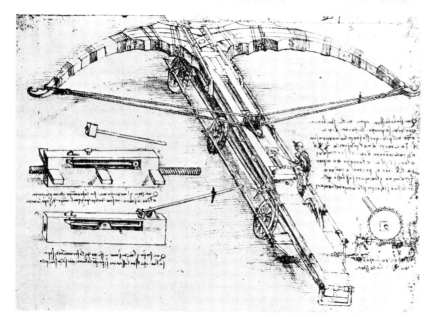

Fig. 1–6 Leonardo da Vinci's sketch of a giant crossbow. Notice the two suggestions showing the tripping action of the crossbow. Da Vinci often wrote his comments from right to left as a mirror image of ordinary writing. In this way he kept many of his ideas as private as possible. (*Lieb Museum.*)

Fig. 1–7 Leonardo da Vinci's sketch of a sprocket chain. Notice the many different views. (*Lieb Museum.*)

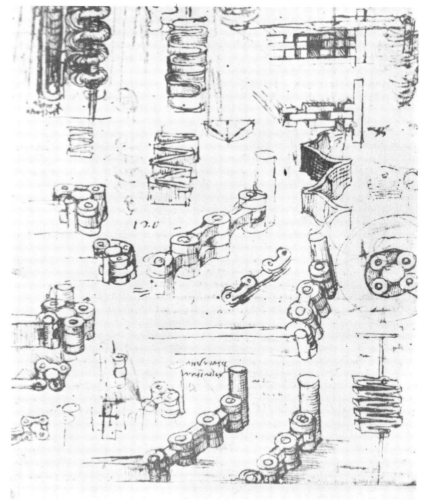

The methods used in drafting did not develop suddenly. Men struggled for centuries with the problem of drawing three-dimensional solid objects on flat surfaces. It was difficult to show accurately the dimensions of length, width, and height on drawings that had only two dimensions. A giant step forward was made by that genius of the fifteenth century Leonardo da Vinci (1452–1519). He was a master of art and technical drawing. His sketches were easy to understand (Fig. 1–6), and for years he taught others his method. After his death many Europeans continued to study da Vinci's work, although his studies were not published until 1651. Today there are 7,000 pages of his notes and sketches that reveal his genius as a designer, architect, and inventor (Fig. 1–7).

The eighteenth century ushered in the Industrial Revolution. This machine age was dependent upon sketches and drawings for new machinery and industrial structures. In this age, drafting was identified as the "language of industry," and standard

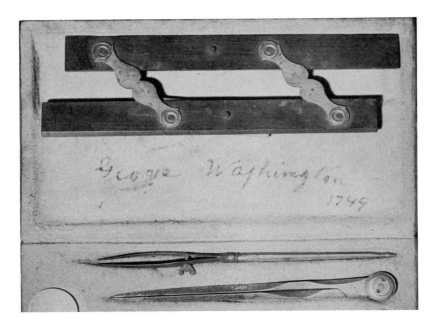

symbols were becoming a part of the universal language.

Americans have always made use of the graphic language, and many prominent early Americans used drawing instruments (Fig. 1–8) and made accurate technical drawings. George Washington and his officers employed drafting tools for maps and early buildings. Thomas Jefferson prepared house plans, as shown in Figs. 1–9 and 1–10. From these plans, Monticello, a beautiful 35-room mansion, was built in Virginia.

Fig. 1–8 George Washington's drawing instruments shown are parallel rulers, dividers, and a ruling pen. (*Mount Vernon Ladies Association, Mount Vernon, VA.*)

Fig. 1–9 Thomas Jefferson's plan for the first story of his home, Monticello. (*Massachusetts Historical Society.*)

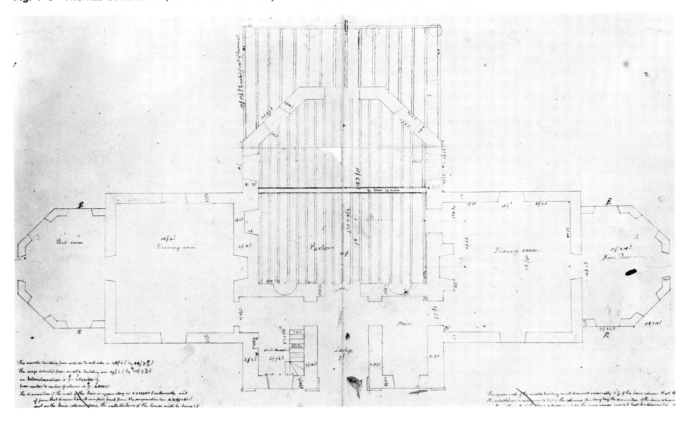

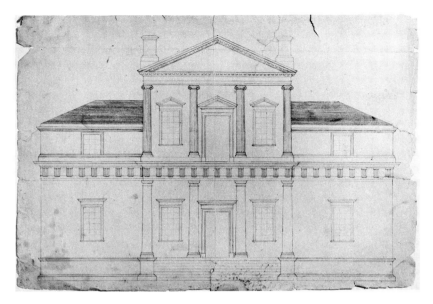

Fig. 1-10 An early drawing for Monticello by Jefferson. (*Massachusetts Historical Society.*)

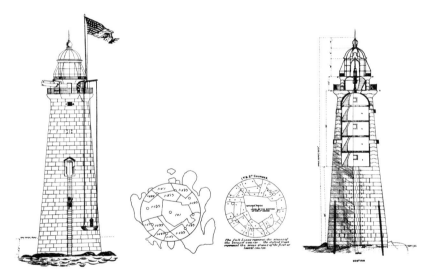

Fig. 1-11 Minot's Ledge Lighthouse in Boston Bay was designed by West Point graduates. (*U.S. Military Academy.*)

our technical progress. At the U.S. Military Academies, cadets have learned to describe and prepare plans for dams, waterways, bridges, lighthouses (Fig. 1-11), maps, machines, and buildings necessary for military operations.

American architecture and industry advanced rapidly from the 1880s to the present as the steel industry expanded and structures pressed skyward (Fig. 1-12). Plans for new machines and structures have developed with the birth of each new industry.

Today, drafting is the most precise method of graphically recording and communicating man's ideas. Without this means of communication it would be difficult for man to transfer technical ideas accurately within a reasonable length of time.

Fig. 1-12 Walt Disney World, FL, A shape begins to rise. Structural steel for the "A-framed" Tempo Bay Hotel begins to rise around the concrete elevator core. (*United States Steel.*)

Gaspard Monge (1746-1818), a Frenchman of Napoleon's time, discovered the principles and systems of graphics that we use today. His methods were considered a military secret for many years. Later, Claude Crozet, another French scholar, taught Monge's graphic-projection methods of drawing for the first time in this country. American teachers and engineers have added to Crozet's work and have further developed the graphic language.

West Point graduates, among whom were the first trained engineers of our country, have often contributed to

Technical Drawing: The Universal Language. Technical drawing is the universal language of industry. Drawings can be prepared by draftsmen and engineers in New York and used by technicians in Chicago. Drawings that are made correctly can be used not only in all parts of this country but also, with few additions or changes, in countries where other written and spoken languages are used.

Different kinds of industries have developed special symbols that are useful in their daily work and save time and money in making and reading drawings. The electronics industry, for example, has its own symbols for representing special parts, such as transistors, batteries, fuses, switches, and hundreds of other parts (Fig. 1-13). Architects use special simplified symbols for showing bricks, wood, stone, glass, and many other materials (Fig. 1-14). Welding drawings have special symbols for describing the shape, size, and type of weld to be made at each joint (Fig. 1-15). A special set of symbols is used in drawing maps in order to show the details of a large geographic area on a small sheet of paper (Fig. 1-16).

Drafting as a universal language involves more than technical drawings. Scale models and mock-ups (full-size models) are also used to show complex parts three dimensionally (length, width, and height). They may also be used with floor-plan drawings to determine the best possible equipment or furniture arrangement or to study the appearance and layout of a building or plant, as shown in Fig. 1-17. A mock-up is useful as a means of checking a design as it appears in three dimension in comparison with the two-dimensional technical drawing. Mock-ups are also used to check the assembly and rela-

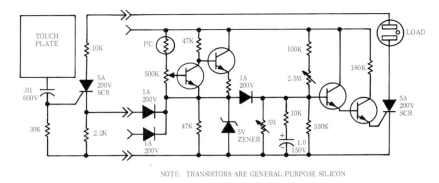

Fig. 1-13 An example of an electronic schematic diagram.

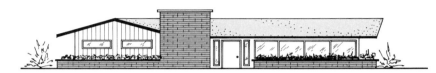

Fig. 1-14 This elevation drawing shows various materials through the use of architectural symbols.

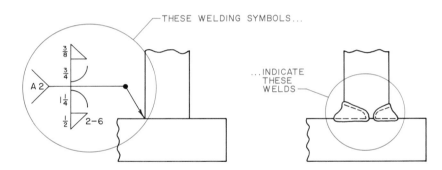

Fig. 1-15 Welding symbols are easier to draw than the finished shape of the welded joint.

tionship of moving parts of a product before it is mass produced.

Without technical drawing, designers and technicians would have a very difficult time communicating their ideas. But technical drawing is also very important to everyone living in an industrial-technical society. It is used by the average person as well as by the technician, scientist, and engineer. Nearly everyone reads road maps, makes sketches, and studies

house plans at one time or another. Graphs and charts showing business conditions and the weather appear daily in newspapers and magazines. Technical drawing and blueprint reading no longer belong to industry alone. They must be understood by the home owner, the driver, and the office worker as well as by the technician (Fig. 1-18).

Learning technical drawing, then, involves more than simply learning to

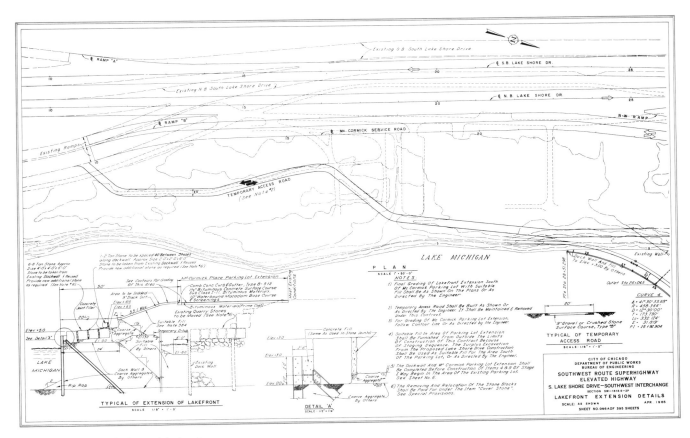

Fig. 1–16 Map of Chicago lakefront.

draw. It involves learning to read and write a new language, just as real a language as English or French.

Communication. Passing information from one person to another in a way in which it can be understood is *communication*. People communicate in many ways: writing and reading, speaking and listening, and sketching, instrument drawing, and blueprint reading are all means of communicating. Verbal (speaking and listening) and written (writing and reading) are probably the most widely used methods of communicating. However, in many cases they are not the best methods. Information and ideas in technical and scientific fields can very often be more

easily communicated through sketching or instrument drawing and blueprint reading.

Figure 1–19 is a written description of a V-block. It is difficult for the machinist to manufacture the part described unless he already knows

Fig. 1–17 A model of an electric power plant.

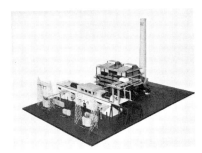

Fig. 1–18 A home owner studies a set of house plans.

what a V-block looks like. Even if he knows what it looks like, it would still take a great deal of his time to read and fully understand the details of how it is to be made. Figure 1–20 shows the same V-block with its sizes in one simple drawing. Only a quick glance is needed to understand the shape and sizes. This example is used to illustrate the saying, "A picture is worth a thousand words."

Four Levels of Graphic Communication. Graphic communication may be described at four distinct levels: (1) *Creative communication,* (2) *Technical communication,* (3) *Market communication,* and (4) *Construction communication.* Each level is important in the field of drafting, and each has its own special function to perform (Fig. 1–21).

Level one—creative communication. Graphic communication usually begins as an idea in the mind of a designer. The first sketches form the principal shape of the product or building. This initial work may be termed *the birth of an idea.* Creative communication is important to the individual who needs a quick means of expressing ideas and capturing them for further study.

Level two—technical communication. Original designs may occur while the designer is working alone. As the design process continues, technical assistance is needed. The complex nature of most projects demands the special talents of many professionals, including engineers, architects, and others. This inter-professional phase is necessary for refinement of the project design. It usually results in a series of drawings that fully explain the design (Fig. 1–22).

Level three—market communication. Before the design is totally refined, a presentation of the study is

THE V-BLOCK IS TO BE MADE OF CAST IRON AND MACHINED ON ALL SURFACES. THE OVERALL SIZES ARE TWO AND ONE-HALF INCHES HIGH, THREE INCHES WIDE, AND SIX INCHES LONG. A V-SHAPED CUT HAVING AN INCLUDED ANGLE OF 90° IS TO BE MADE THROUGH THE ENTIRE LENGTH OF THE BLOCK. THE CUT IS TO BE MADE WITH THE BLOCK RESTING ON THE THREE INCH BY SIX INCH SURFACE. THE V-CUT IS TO BEGIN ONE-QUARTER INCH FROM THE OUTSIDE EDGES. AT THE BOTTOM OF THE V-CUT THERE IS TO BE A RELIEF SLOT ONE-EIGHTH INCH WIDE BY ONE-EIGHTH INCH DEEP.

Fig. 1–19 Description of a V-block.

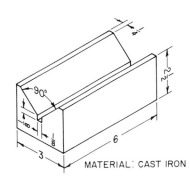

Fig. 1–20 Technical drawing of a V-block.

prepared for potential clients. This level includes client evaluation of style, form, and function of the project. An example would be the rendering of a residence designed for a client before final plans are prepared for construction (Fig. 1–23).

COMMUNICATIONS

Fig. 1–21 Four levels of graphic communication exercised by the designer and draftsman.

Level four—construction communication. This level deals with details necessary for manufacturing and construction. Graphic communication probably began with the need for manufacturing and construction drawings. These drawings must be

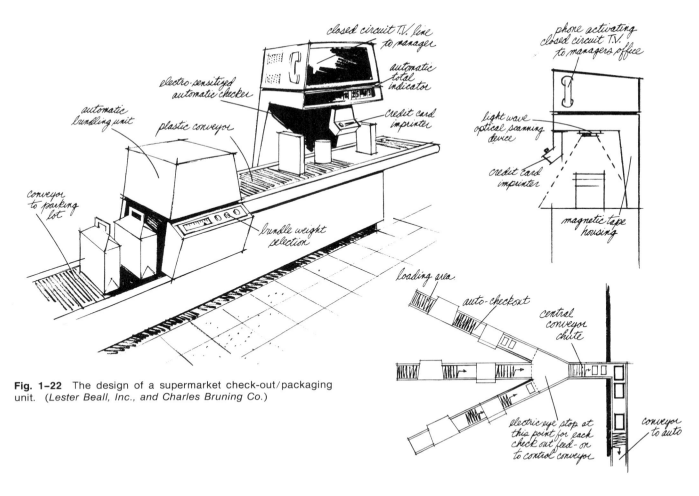

Closed circuit T.V. line to manager

electro-sensitized automatic checker

automatic total indicator

automatic bundling unit

plastic conveyor

credit card imprinter

conveyor to parking lot

bundle weight selection

phone activating closed circuit T.V. to managers office

light wave optical scanning device

credit card imprinter

magnetic tape housing

loading area

auto-checkout

central conveyor chute

electric eye stops at this point for each check out feed-on to control conveyor

conveyor to auto

Fig. 1-22 The design of a supermarket check-out/packaging unit. (*Lester Beall, Inc., and Charles Bruning Co.*)

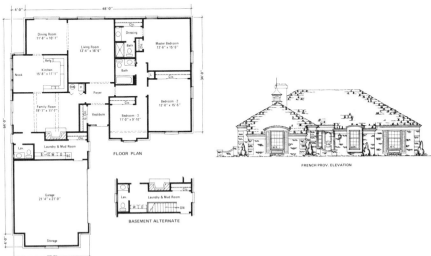

Fig. 1-23 Architectural rendering of a French Provincial residence for a client, with a floor plan and elevation and perspective shown on page 10. (*Inland Steel Co.*)

prepared with sufficient information so that cost estimators and shop superintendents can interpret the designer's project without having to guess about details or go back to the professional for clarification.

Types of Technical Drawings. The three types of technical drawing are *multiview, pictorial,* and *schematic.* Each is used for its own specific purpose, and each is equally as valuable as the other. One should not be used where another will serve the purpose better.

Multiview drawing. A *multiview drawing* (also called an *orthographic projection*) shows two or more views

Fig. 1–23 (*Cont.*) Architectural perspective rendering.

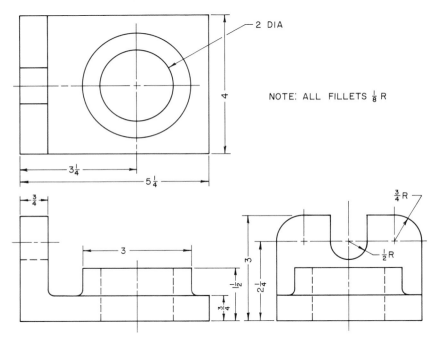

Fig. 1–24 A multiview drawing.

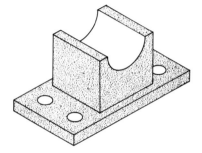

Fig. 1–25 A pictorial drawing shows more than one surface of an object in a single view.

or faces of an object (Fig. 1–24). This is accomplished by drawing views of the object as seen from different positions and arranging these views in a systematic relationship to one another. By doing so, we represent the exact shape and proportion of every part of a three-dimensional object on a single, flat sheet of paper. *Pictorial drawings and photographs show objects as they appear; multiview drawings show them as they really are.* Multiview drawings have as many views and as many dimensions (sizes) as is necessary for a complete understanding of the shape and size of the object. They are generally used for manufacture or construction of a product.

Pictorial drawing. A *pictorial drawing* shows two or three surfaces of an object in a single view (Fig. 1–25). It is used to show objects as they appear and not as they actually are. For this reason, technicians seldom work from pictorial representations (drawings). Assembled and exploded pictorial drawings are, however, used to show how the parts of machines, appliances, and other devices are to be put together. Figure 1–26 is an exploded pictorial drawing. Figure 1–27 shows an assembled pictorial drawing.

Schematic drawing. A *schematic drawing* shows the parts of a system by using various types of diagrams and a great number of graphic symbols (Fig. 1–28). Electricity-electronics, piping, and other industries that are concerned with showing components and their connecting elements, such as wires, printed circuits, or pipes, usually use schematic drawings.

Specialized Fields of Technical Drawing. The field of technical drawing is so large and so complex that draftsmen usually find it neces-

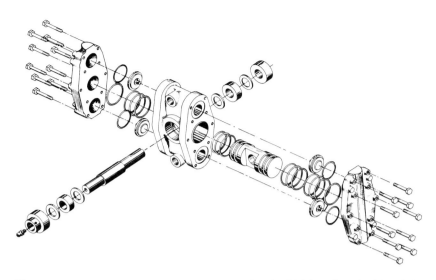

Fig. 1–26 An exploded pictorial drawing. (*Industrial Division, Standard Precision, Inc.*)

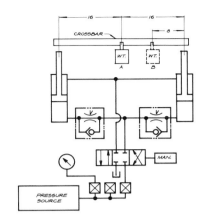

Fig. 1–28 Experimental circuit for synchronizing hydraulic cylinders.

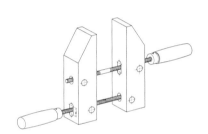

Fig. 1–27 An assembled pictorial drawing of a hand screw clamp.

sary to specialize in one type of drafting. These specialized fields are usually named for the type of manufacturing or construction in which they are done. For example, a draftsman working in the building industry is called an *architectural draftsman.* One who prepares technical drawings in the aircraft industry is an *aeronautical* or *aerospace draftsman,* and so on. Other specialized fields include *structural, machine, electrical* and *electronics, surface-development* (sheet metal), and *technical-illustration.*

Architectural drafting involves the same general principles as other types of technical drawing, except for cer-

tain special methods of representation and special symbols that are necessary because of the small scale that must be used to show a large building on a small sheet of paper. This type of drawing includes the representation and specifications of buildings and other structures of various kinds. Figure 1–23 shows a floor plan, an elevation, and a pictorial rendering of a residence. These are examples of architectural drawings.

Aeronautical or *aerospace draftsmen* prepare drawings for all types of aircraft: single-part details, subassemblies (drawings showing assembly of two or more parts), major assemblies, installation drawings, and layouts. The space program has opened this drafting field into a very large industry. Figure 1–29 is an example of an aerospace drawing.

Structural drafting is concerned with drawings made for the framework and supporting members of buildings and other kinds of large structures. In many cases architectural and structural draftsmen work closely together during the design of a building. The architectural draftsman prepares floor plans, elevations,

and other similar drawings, while the structural draftsman prepares drawings of the framework and other supporting members, such as columns, roof trusses, and floor members. Figure 1–30 shows a structural steel framework being erected. The drawings for this structure were prepared by a structural draftsman.

Machine draftsmen prepare detail and assembly drawings for manufacture of machine parts and machines. They are responsible for specifying and drawing all information necessary for manufacturing and assembling entirely from the drawings. No other information should be necessary. Figure 1–31 is an example of a machine drawing.

Electrical and electronics draftsmen prepare drawings for electrical and electronic machines, fixtures, apparatuses, and devices. Circuits and other special features are usually planned by an electrical design engineer, and rough sketches are submitted to the draftsman for the preparation of final drawings. Figure 1–13 is an example of an electronic schematic drawing.

A *sheet-metal* or *surface-development draftsman* prepares the layouts for such items as sheet-metal ducts

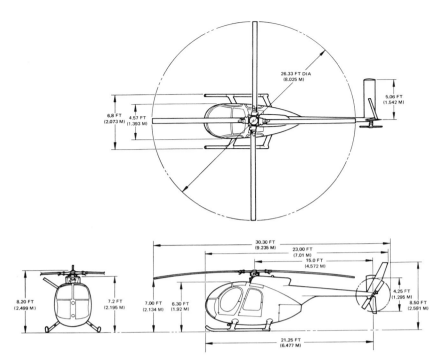

Fig. 1-29 A three-view drawing of a small jet helicopter. (*Hughes Tool Co.*)

shows one application of surface development.

Technical illustration has an important place in all phases of engineering and science. Technical illustrations vary from simple sketches to rather complex shaded drawings. A technical illustration, in general, means a pictorial drawing made to provide technical information by visible methods. These drawings are used extensively in technical literature, such as service, repair, and operation manuals. Exploded views such as the one shown in Figure 1–26 show how various parts are to be assembled. Technical illustrations are also used extensively in catalogs and other types of advertising materials.

Careers in Drafting. Drafting relates to technical fields in which associates, called *draftsmen,* translate ideas of architects, engineers, and designers into exact detailed instructions. The draftsman applies his knowledge of mathematics, science, and technology as well as common sense and experience in preparing technical drawings and detailed specifications (Fig. 1–33). These are some of the features that help to make a career in drafting both interesting and rewarding.

(pipes), boxes and cartons, automobile body parts, aircraft, and other products using sheet-type materials. The draftsman prepares a layout, called a *pattern,* which is later transferred to flat sheets of material and then cut, folded, rolled, or otherwise formed to provide the required shape. Sheet materials such as paper, cardboard, plastics, metal, fiberboard fabrics, and many others are used. Figure 1–32

Fig. 1-30 Structural steel is lifted into place successfully because the detail drawings were accurately prepared for the steel fabricators. (*Western Electric.*)

Fig. 1-31 An example of a machine drawing.

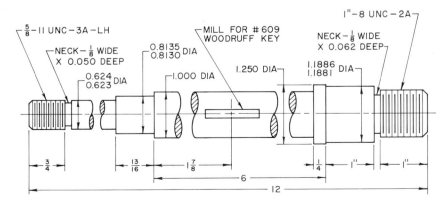

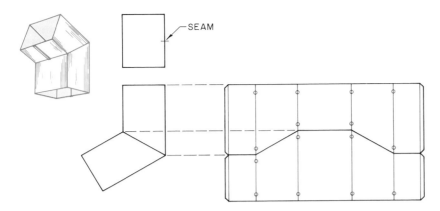

Fig. 1–32 Surface development.

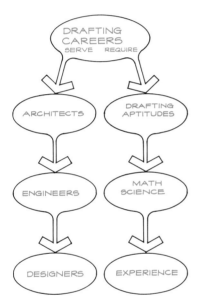

Fig. 1–33 Careers in drafting require technical skills that will complement engineering services.

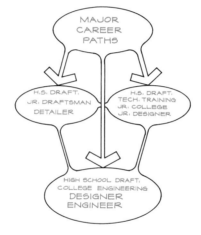

Fig. 1–34 Career decisions. Some ways to become professionally involved in graphic communication.

Three Roads to a Career in Drafting. While there may be many ways to become a draftsman, the three most common ones are described in the following paragraphs (Fig. 1–34).

High school followed by apprenticeship training. A good background in a sound high school drafting program may provide adequate qualifications for the person who wants to serve a drafting apprenticeship in an industrial drafting room. The broader the background from high school, the better the chances of being accepted as a trainee in an apprenticeship program.

In the training program the beginner may be called a *junior draftsman* or a *detailer*. He makes simple detail drawings under the supervision of an experienced detailer. He learns a great deal about his company's product and manufacturing specifications. As skills are acquired he will be charged with developing more complex drawings.

High school followed by technical training. There are excellent two-year courses available to young students who wish to work as drafting technicians. The technician is the engineer's assistant. Drafting technicians are classified as to the work they do or the level of responsibility they assume. The training may include courses in surveying, plan reading, mathematics, basic estimating, drafting, materials of industry, production techniques, and construction systems. This is the level at which the draftsman's title may change to senior detailer or junior designer. To work on design drawings the senior detailer must exercise some independent judgment. He may be required to use handbook references for calculations and strength of materials to be used. A general knowledge of manufacturing methods and procedures is necessary. The writing of technical reports is often a part of the job; these are usually submitted to the chief draftsman.

High school followed by college. Engineering students must carefully select high school courses which will complement their college major. Four years of mathematics and two or more years of science are recommended when taking the college preparatory course. The studies of the engineering student on the college level prepare the graduate engineer so that he can assist industry in making the properties of nature useful for man. All types of engineers employ technical drawing (drafting) in their work. Engineers can be identified according to the tasks they perform in the following ways: research, development, design, testing, planning production, construction operation, sales, service, and standards. The engineer is generally the top man in the engineering graphics department. Upon graduation from college, the beginner usually takes a position in industry with the title of engineer-in-training or engineer trainee. Upon completion of his training program he is usually given the title of engineer.

Career Qualifications. It is important for all draftsmen to have an *aptitude for detail.* The detail of the draftsman's work requires *accuracy, neatness,* and the *ability to achieve order.* He should be *logical, resourceful,* and *systematic about developing detailed communications* (Fig. 1–35).

A professional draftsman will constantly learn more about his chosen field of work and the problems which are a part of his industry. A strong mechanical aptitude with the desire and ability to solve problems are essential. These traits are generally complemented by an interest in mathematics and science.

The draftsman should be able to think and see in two or three dimensions and be familiar with geometric shapes. He should be capable of developing proper proportion of various forms at convenient scales. Size and shape description are essential to graphic interpretation. Technical sketching and skill in the use of instruments are required for the development of technical communications.

All draftsmen must be able to coordinate work and ideas with other members of the technical staff. Concentration and perseverance are factors which frequently help them understand problems and projects of their fellow workers. Teamwork is essential when something is to be designed and constructed from drawings.

There are several hundred definable skills needed in creative design which may be described as the manufacturing and the construction process of industry. These combine to form thousands of career opportunities

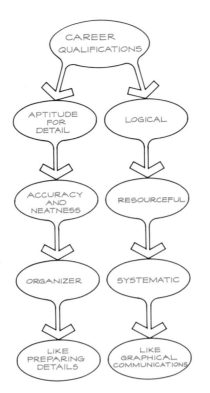

Fig. 1–35 Career qualifications. Some important traits for professional growth.

with unique qualifications. Specific information on careers in drafting and related fields may be obtained from the following organizations:

American Institute for Design
and Drafting
349 Price Road
Bartlesville, OK 74003

American Federation of Technical
Engineers
1126 16th Street, N.W.
Washington, DC 20036

High School Background Is Important. The importance of getting a good background in high school cannot be overemphasized. Upon completion of a successful high school drafting program the student should be able to:

1. Solve technical problems through the use of the tools and techniques of drafting

2. Prepare technical drawings that adequately communicate the design idea

3. Read charts and obtain data from technical references

4. Write technical reports that are clear, concise, accurate, and easily understood

In addition, a good background in mathematics, science, and English are essential. The technical and scientific principles of math and science linked with the communications (speaking, reading, and writing) aspects of English provide a strong basis for a high degree of success in a career in drafting.

Review

1. Name three materials on which ancient drawings were made.

2. Who was the famous designer, architect, and inventor of the fifteenth century that had a great deal to do with the development of technical drawing?

3. Drafting is often called the *language of* _____.

4. A full-size model is called a _____.

5. Passing information from one person to another in a way that can be understood is _____.

6. Name the four levels of graphic communication.

7. Name the three types of technical drawings.

8. Name three specialized fields of technical drawing.

9. What is the job name given to a person who prepares technical drawings?

10. There are basically three roads to a career in drafting. Name them.

Sketching | Chapter **2**

(SAMSONITE CORPORATION.)

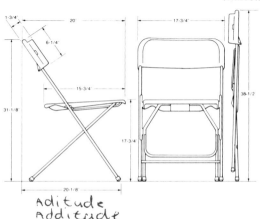

Aditude
Additude

Technical Discussions. Freehand sketching is the simplest form of drawing and one of the quickest ways to express ideas. A sketch can help simplify a technical discussion. Designers, draftsmen, technicians, engineers, and architects, when discussing technical problems, attempt to understand verbal descriptions used in the conversation. A freehand sketch will often explain verbal expressions which are not immediately clear or are too complicated to describe in words. While ideas are developing within the imaginative powers of the mind, sketches can capture the thoughts formed and hold them with simple lines for further study. The pencil is an instrument which aids clear thinking and creative communication. Figure 2–1 shows when sketching can become an important part of technical discussions.

Types of Sketches—Rough (Unrefined) and Refined. Any image drawn on paper freehand may be called a *sketch*. The technique for sketching may be rough and drawn quickly with jagged lines (unrefined). The rough sketch can quickly direct thoughts to the reason for the communication. Figure 2–2 shows rough sketches which were used to develop preliminary or early designs of a two-position automobile mirror. The sketches illustrate several possible design choices. The other technique, refined (neat and finished looking), may be carefully drawn, showing good

15

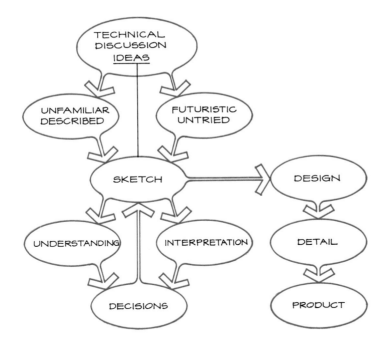

Fig. 2–1 Sketches can assist technical discussions.

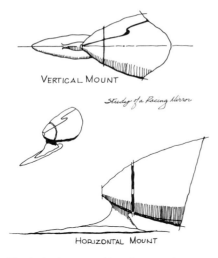

VERTICAL MOUNT

Study of a Racing Mirror

HORIZONTAL MOUNT

Fig. 2–2 A two-positional automobile racing mirror.

Fig. 2–3 Sketching—rough to refined.

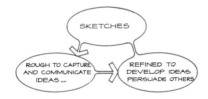

SKETCHES

ROUGH TO CAPTURE AND COMMUNICATE IDEAS ....

REFINED TO DEVELOP IDEAS PERSUADE OTHERS

proportion and excellent line values in an attempt to be more persuasive than an unrefined sketch. Many refined sketches are developed after a rough sketch has captured the general idea to be illustrated. Figure 2–3 shows the process normally followed when sketching is used to communicate. An excellent method of refinement or improvement is the *overlay* method.

The Overlay. Sketches are often drawn on a paper which can be seen through. This paper is called *translucent,* or *tracing, paper.* The best parts of a sketch which are to be kept may be quickly traced by putting a new piece of this paper on top of (overlay) the previous sketch (Fig. 2–4). When sketching on tracing paper, the refinement of ideas becomes a continuing process until a suitable design is achieved.

Two Uses for Overlays. The overlay has two important uses which may seem similar. The first purpose is the reshaping of an idea, such as refining the proportions of the structural features and altering the physical form. The secondary use of the overlay is the refinement or improvement of the sketch. These two kinds of refinement can happen at the same time (Fig. 2–5).

Importance of Proper Materials. To deliberately limit sketches to opaque paper, through which you cannot see, limits sketch development. The overlay technique is important. Sketching is generally more than a one-stage process when striving for refinement of the object or the drawing. No single sketch is as meaningful as its relationship to the sketch below and the sketch above in the refinement of an idea.

Reasons for Sketching. The reasons for technical sketching are many, but the following are the most important.

1. To persuade people who make decisions about a project that an idea has merit.

2. To develop a refined sketch for a client's response to a proposed solution to a problem (Fig. 2–6).

3. To interpret or define a complicated multiview projection detail with a simple pictorial sketch or enlargement.

4. To forward ideas of the design to the draftsman for the development of details.

5. To develop a series of studies for the refinement of a new product or machine part.

6. To analyze and develop computation necessary for production and to choose the best materials for a product.

7. To record a practical design change on a project which exists. The

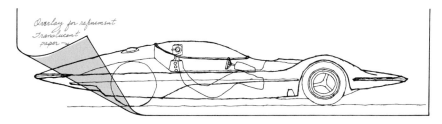

Fig. 2–4 The overlay can speed up the design process.

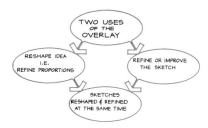

Fig. 2–5 Two uses for the overlay process.

Fig. 2–6 Sketching for the client. (*A. W. Wendell and Sons Designer–Contractor.*)

Fig. 2–7 Sketches are classified in three ways.

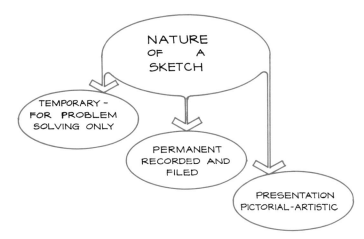

change may result from the need to repair a part which continually breaks, or from an easier and cheaper way to make parts. The change would permanently record the improvement.

8. To show that many interpretations or solutions to problems are developed by sketching.

9. To reduce time spent in drawing. To prepare a sketch on a pad of paper is quicker than to use all the equipment for a mechanical drawing.

Nature of a Sketch. A sketch is an important form of graphical (drawn) communication and may serve many levels of interest (Fig. 2–7). The person possessing a complete knowledge of mechanical drawing as a language will know how to execute quick, clear, and accurate sketches to complement his creative expressions.

Temporary. Many technical sketches have short lives; that is, they are prepared merely to resolve an immediate problem and then they are thrown away. Still other technical solutions are sketched and the sketch is kept for a longer period of time. Perhaps weeks or even months are required for some sketches to be studied and transformed into detailed mechanical drawings, but they too may be discarded eventually.

Permanent. Sometimes the engineering department or the management of a company will include a sketch as a part of a notice to other employees; thus, the sketch becomes an important record to keep. Some sketches therefore are recorded and filed as part of a company's permanent records.

Presentation. The sketch which is refined for presentation is generally a pictorial (picturelike) sketch used to convince a client or management that it should approve and accept the ideas presented. The pictorial sketch offers a three-dimensional view which can be understood easily by nontechnical people. Drawing techniques which appear glamorous, artistic, or eye-appealing are generally used in the sketch.

A Language of Visual Symbols. *The idea that progress begins on paper applies dramatically to the sketches* which can bring new ideas alive. When words cannot adequately describe new or futuristic forms, sketching takes over. The language of sketching has four basic visual symbols: a point, a line, a plane, and a texture or surface quality (Fig. 2–8). Any idea which takes form, no matter how simple, complicated, or spectacular, can be sketched by using these four ingredients. In learning to sketch, a person can work to achieve skills which may simplify the detail drafting of the item sketched.

Materials for Sketching. The advantages of sketching are that only a

Fig. 2–8 Elements of sketching are simple visual symbols.

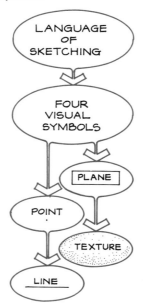

few materials are needed and that it can take place anywhere. You are well prepared if your equipment includes a pencil, an eraser, and a pad of paper. The person who believes he needs more equipment than that will generally find that he is not a great draftsman.

Paper. Plain paper may be used for sketching. Tracing paper is suggested if the refinement process seems necessary. Cross-section, also called graph or squared, paper is suggested for controlled proportions in sketching. Usually, graph paper is ruled in one-inch squares with a heavy rule. The one-inch squares are then subdivided into one-tenth, one-eighth, one-quarter, or one-half inch squares with a light rule. It is referred to as ten to the inch, eight to the inch, and so forth. Many specially ruled graph papers are available to aid in special kinds of drawing such as isometric and perspective drawings. These drawings are explained in Chapter 12.

Any convenient size of paper may be used, but standard letter size ($8\frac{1}{2}'' \times 11''$) is preferred for quick, small sketches. Working sketches can be held conveniently on stiff cardboard or on a clipboard. Cross-section paper may be placed under tracing paper to act as a guide for spacing lines.

Pencils and erasers. Most draftsmen prefer to include soft pencils (grades F, H, or HB), properly sharpened, along with erasers suitable for soft leads, such as a plastic or a kneaded rubber eraser.

Use a draftsman's pencil sharpener to remove the wood from the *plain* end of a pencil so that the grade mark (F, H, or HB) will always show. Sharpen the lead to a point on a sandpaper block, file, or in a lead pointer if using a lead holder. Do not forget to adjust the grade mark in the window of the lead holder if it has one.

Be careful to remove the sharp needle point by touching it gently on a piece of scrap paper. Then the pencil will not groove or tear the drawing paper.

Four types of points are convenient for sketching—sharp, near-sharp, near-dull, and dull (Fig. 2–9). The points should produce dense lines in the following variations:

Sharp-point—thin black line for center lines, dimension lines, and extension lines.

Near-sharp point—visible or object lines.

Near-dull point—cutting plane and border lines.

Dull-point—construction lines.

Lines drawn freehand have a natural look. They exhibit freedom of movement with slight variations in direction. Hold the pencil far enough from the point to allow easy movement of the fingers and yet allow adequate pressure on the point for dense black lines when necessary. Light construction lines should be made with very little pressure on the point and should be light enough that they need not be erased.

Straight Lines. Lines may be sketched (1) as a continuously sketched line, (2) as a line guided by a short dash at the start and end (place the pencil point on the starting dash, keep the eye on the ending dash, and draw toward it), (3) as a line made up of a series of strokes (touching or separated by very small spaces), or (4) as a line made up of a series of overlapping strokes (Fig. 2–10). Practice sketching straight lines to improve line technique before starting to draw objects. Horizontal lines are drawn from left to right and vertical lines from the top down (Fig. 2–11).

Slanted Lines or Specific Angles. Slanted or inclined lines can be

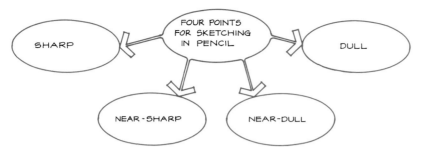

Fig. 2-9   Four convenient pencil points for sketching.

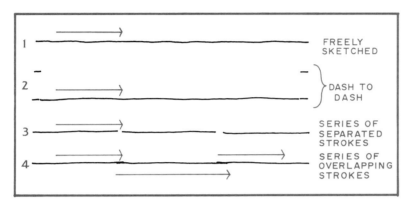

Fig. 2-10   Some ways of sketching straight lines.

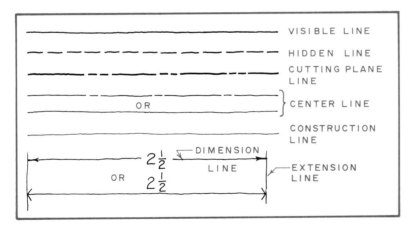

Fig. 2-11   Sketching horizontal and vertical lines and types of lines used in sketching.

sketched from left to right. It might be easiest to rotate the paper so as to sketch an inclined line the same way a horizontal line is sketched. When attempting to sketch a specific angle, draw a vertical line and a horizontal line to form a right angle. Divide the right angle into known equal spaces, such as 45° or 30°-60°, so as to calculate the desired angle. Note the direction of the inclined line drawn to form a desired angle (Fig. 2-12).

Circles and Arcs. Two methods of sketching circles can be used. In the first, draw very light horizontal and vertical lines and mark off the estimated radius (the distance from the center of the circle to the edge). Then

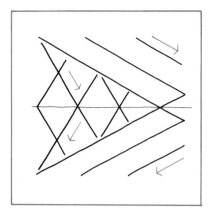

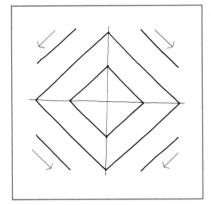

Fig. 2–12 Sketching slanted lines or specified angles.

draw a square in which the circle can be sketched (Fig. 2-13). In the second method, draw very light center lines with bisecting lines drawn through the center at convenient angles (Fig. 2-14). Proceed by marking off estimated radii on all lines, equidistant (same length) from the center (Fig. 2-15). The concave side of the curve is generally easier to form (Fig. 2-16).

Arcs (part of a circle), tangent arcs (parts of two circles which touch), and concentric circles (circles of different diameters which have the same center) may be sketched by using the same methods described to sketch circles (Fig. 2-17). Use light, straight, construction lines to block in the area of the arc.

For large circles, arcs, and ellipses, use a scrap of paper with the radius marked off along one edge. Put one end of the radius on the center of the circle and draw the arc by placing a pencil at the other end of the radius and rotating the scrap paper (Fig. 2-18). Another clever method is to use the hand as a compass. The little finger is used as a pivot at the center of the circle. The thumb and forefinger hold the pencil at the desired radius in a rigid position. The paper is carefully rotated underneath the hand and the circle is drawn (Fig.

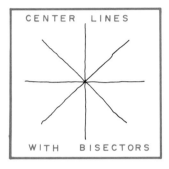

Fig. 2–14 Draw center lines for a circle with bisecting lines.

Fig. 2–15 Mark off the estimated radii on all lines before sketching circle.

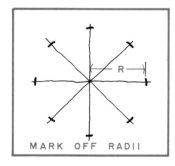

2-19). Similar results are obtained when two pencils are held in a rigid-cross position. Make the distance between the pencil points the desired radius. Put one pencil point on the center and hold it firmly in place while the paper is rotated to draw the circle (Fig. 2-20).

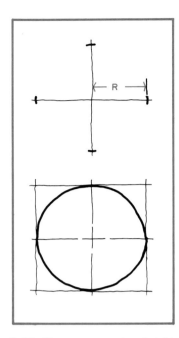

Fig. 2–13 Draw a square for sketching a circle.

Fig. 2–16 The bottom or concave side of the curve is easiest to form.

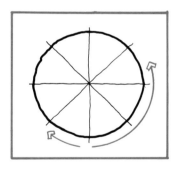

Proportions for Sketching. Since sketches are not usually made to scale (exact measure), it is important to keep sketches in proportion (similar size to the exact size). In preparing the layout, the larger overall dimensions (usually width) are examined and an estimated size made. The first

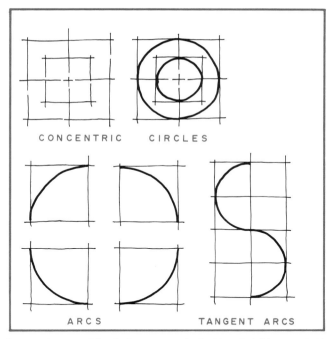

CONCENTRIC CIRCLES

ARCS   TANGENT ARCS

Fig. 2–17 Arcs and concentric circles are controlled by sketching squares.

Fig. 2–18 Large circles, arcs, and ellipses are easily sketched with the aid of a strip of scrap paper.

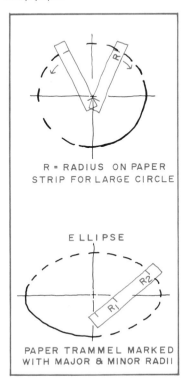

R = RADIUS ON PAPER STRIP FOR LARGE CIRCLE

ELLIPSE

PAPER TRAMMEL MARKED WITH MAJOR & MINOR RADII

Fig. 2–19 The hand as a compass, using the little finger as pivot point.

Rotate paper

important decision then is the proportion of the width to the height. Then as the front view using the width and height takes its form, the smaller details within the major framework are organized and compared to other proportions (Fig. 2–21).

It is important that the design draftsman, in sketching an object he is creating, have a good sense of related distances so that he can show the width, depth, and height of an object in their proportion to each other.

For example, the design draftsman plans a contemporary stereo cabinet to be four units wide. Each unit is about 15 in., for a total of 60 in. The height of the cabinet is two units. Each unit is about 15 in., for a total of 30 in. The depth of the cabinet is one unit. The one unit is 15 in. The overall front-view proportions are a width of 60 in. and a height of 30 in., or a two-to-one proportion, as the relation of width to height is compared. The overall side-view proportions are a height of 30 in. and a depth of 15 in., or a two-to-one proportion. The proportions are developed in two-to-one units in Fig. 2–22.

Technique in Developing Proportion. Good sketching technique requires that the design draftsman be able to divide a line in half by estimating by eye, or by *eyeballing*. The halves can be divided again to give fourths. Practice will train the eye to

Fig. 2–20 Two pencils can serve as a compass.

Rotate paper

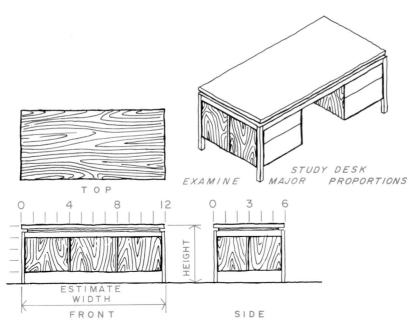

Fig. 2-21 Proportional study starts with overall width and height.

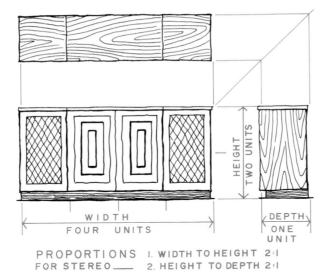

PROPORTIONS   1. WIDTH TO HEIGHT 2:1
FOR STEREO ___   2. HEIGHT TO DEPTH 2:1

Fig. 2-22 Sketching a contemporary stereo with proportional units.

work in at least two directions. Start by drawing a line of one unit and increasing it by one equal unit in width so that it is twice as long as the first. Develop the technique of adding an equal unit and of dividing a unit equally in half. Practice developing units on parallel horizontal lines and then at right angles or vertical so that comparisons can be used to improve this estimating technique (Fig. 2-23).

Design draftsmen may use a scrap paper or a rigid card as a straight edge when scales (measuring devices) are unavailable. The paper may be folded in half to determine the length of units on the edge marked, as in Fig. 2-24. In this way lines may be drawn or duplicated several times in different directions to help in maintaining proportions.

Figure 2-25 shows how to use diagonal lines to multiply proportional units. At A the square *ABCD* is sketched using diagonals. At B a horizontal centerline is sketched through the diagonal intersection and extended. At C the large diagonal lines *AL* and *BN* are sketched and the rectangle with a length three times the height is completed. The side view (at D) is sketched using diagonals, as at A. In Fig. 2-26 diagonal lines are used to reduce the height of the square by one-third. Square *ABCD* is sketched at A using diagonal lines. At B the centerline *MN* is sketched through the diagonal's intersection, locating one-half the height. Diagonal *AN* of rectangle *ABNM* is drawn at C. Point *Y,* the intersection of *AN* and *BD,* is one-third the height of the square. Rectangle *ABSR* is now one-third the height of *ABCD.* Figure 2-27 shows other methods of developing proportions for sketching.

Views Needed for a Sketch. In Chapter 5, complete details of multiview projection will be given. For sketching, some essentials of views and their placement are needed.

There are two types of drawings which are easily sketched. One is called *pictorial drawing,* meaning picturelike, in which the width, height, and depth of an object are shown in one view (Fig. 2-28). The other is called *multiview,* or *orthographic, projection* in which an object is usually shown in more than one view by drawing sides of the object and relating them to each other, as shown in Fig. 2-29.

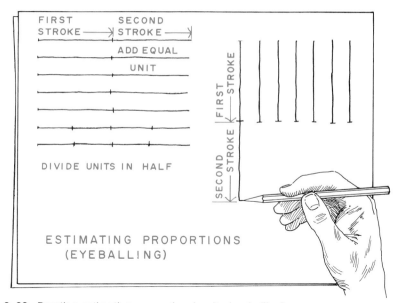

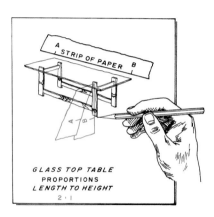

Fig. 2-24 Draftsmen develop techniques for blocking out sketches.

Fig. 2-23 Practice estimating proportional units (eyeballing).

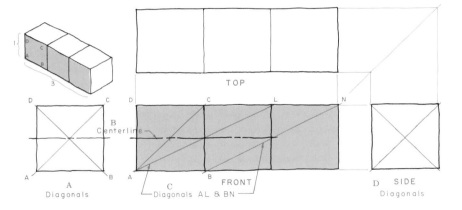

Fig. 2-25 The diagonal is used to multiply proportions.

Fig. 2-26 Diagonals are used to reduce proportional units.

Fig. 2-28 Comparing perspective and oblique pictorials.

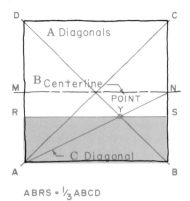

ABRS = 1/3 ABCD

Fig. 2-27 Study the ways in which the diagonal has divided the square.

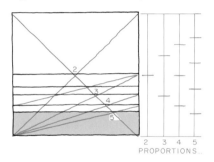

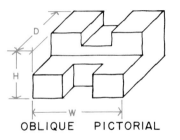

PERSPECTIVE PICTORIAL

OBLIQUE PICTORIAL

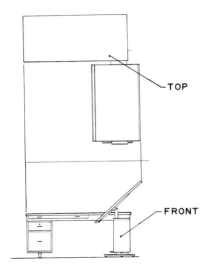

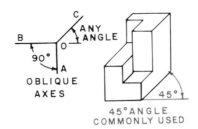

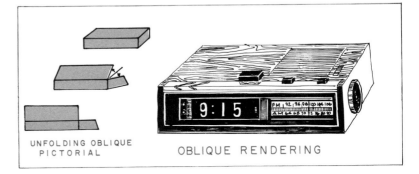

Fig. 2-31 Oblique drawing can be used with rendering techniques.

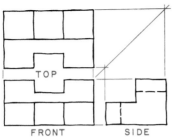

MULTIVIEW DRAWING

Fig. 2-29 Typical multiview drawings (two-view and three-view).

Pictorial Drawing. There are several different kinds of pictorial drawing, and they will be discussed in full in Chapter 12. Only two types, oblique and isometric, will be considered for sketching. Isometric and oblique sketching can help to develop visualization of objects which is necessary in preparing three-view drawings (multiview projection). The basic purpose of pictorials such as isometric is to communicate basic shapes to persons not trained to read the more specific detail of multiview drawing.

Oblique Sketching. One of the easiest pictorial drawings to sketch is the oblique sketch. *Oblique* means inclined. An oblique sketch has the depth or thickness portion of an object drawn at any convenient angle. An object has three dimensions, length (or width), height, and depth (or thickness). Each direction is called an *axis*. In oblique drawings, two of the axes are at right angles (90°) to each other and the third is drawn at any convenient angle to the other two (Fig. 2-30). Choose any side of an object to be the front view, representing length and height. Usually the side with the most detail is chosen. A digital clock radio (Fig. 2-31) is used as an example. The illustration shows it as a pictorial view. The dial side is chosen as the front view since it has the most detail and shows width and height. The front view is sketched just as it is seen, looking straight toward the clock radio with the dial in front.

Oblique Layout. A good oblique sketch is made by always following a good layout procedure.

A. Block in lightly the front face of the object with the estimated units for the width and height.

B. Sketch in lightly the receding lines at any convenient angle, say, approximately 30° with the horizontal. This receding angle can be chosen to give the proper emphasis to the top and side. So, if the third axis is drawn at 30° (a small angle compared to 60°), the side will show more clearly, but if 60° is selected, the top will be more clearly shown (Fig. 2-32).

C. The receding axis may have the same proportioned units as the front axis, or it may be reduced by any amount up to one-half. When the dimension chosen for the depth is exactly one-half the true dimension, it is called a *cabinet sketch*. The use of a full-depth dimension is called a *cavalier sketch* (Fig. 2-33).

D. Assuming that all the layout was drawn lightly, darken the final object lines.

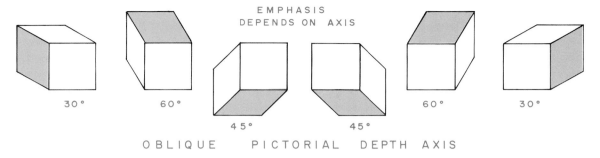

EMPHASIS
DEPENDS ON AXIS

30°   60°   45°   45°   60°   30°

OBLIQUE   PICTORIAL   DEPTH   AXIS

Fig. 2–32 Oblique drawing can change axis to build emphasis.

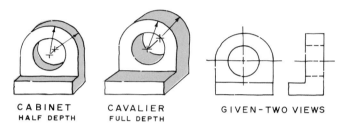

CABINET
HALF DEPTH

CAVALIER
FULL DEPTH

GIVEN-TWO VIEWS

Fig. 2–33 Variations in depth can change oblique appearance easily.

Fig. 2–34 Grid paper can assist in developing oblique pictorials.

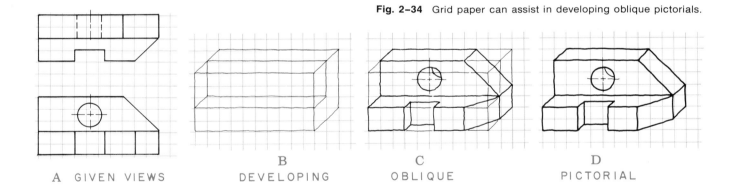

A GIVEN VIEWS

B
DEVELOPING

C
OBLIQUE

D
PICTORIAL

Oblique Sketching on Graph Paper. Graph paper is useful for oblique sketching since the front view of the oblique sketch is similar to the front view of a multiview sketch, as shown in Fig. 2–34 at A and B. If the oblique pictorial drawing is developed on graph paper from a multiview drawing sketched on graph paper, the dimensions are transferred from one line to the other simply by counting the square graph units.

A. Block in lightly the front face of the object by counting squares.

B. Sketch lightly the receding axis by drawing a line diagonally through the squares. The depth is established by using half as many squares, as shown on the side view of Fig. 2–34 at A.

C. Sketch in the required arcs and circles.

D. Assuming that all the layout was drawn lightly, darken the final object lines.

Oblique Circles. One of the important features of oblique sketching is that circles in the front view may be drawn in their true shape. The

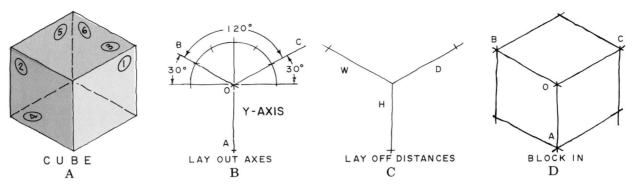

Fig. 2–35 Sketching isometric angles and layout of *Y* axis.

oblique circles drawn in the top or side view appear distorted and an ellipse must be drawn to represent them. Ellipses are not desirable for effective oblique pictorial sketching since the circular shapes of important features can be more easily sketched in the front view.

Isometric Axes. An isometric sketch is based on three lines called *axes,* which are used to show the three basic dimensions: width, height, and depth. A cube is an object with six equal square sides (Fig. 2–35). The isometric cube will portray three equal sides. *Isometric means equal measure* so that there are three equal angles at the *Y* axis at *B* (Fig. 2–35). The height *OA* is laid off on the vertical leg of the *Y* axis. The width *OB* is laid off to the left on a line 30° above the horizontal. The depth *OC* is laid off to the right on a line 30° above the horizontal. The 30° lines receding to the left and right may be located by estimating one-third of a right angle, as shown in Fig. 2–35. Lines parallel to the axes are called *isometric lines,* and the estimated distances are laid off on them only as shown for the cube at *C* (Fig. 2–35).

The sketched lines for isometric axes tend to become steeper than 30° if the layout is not carefully prepared. A better overall pictorial sketch will result when the angle is at 30° or a little less than 30°. Special isometric graph paper with 30° ruling allows good pictorial sketches to be made quickly and easily (Fig. 2–36). Figure 2–37 shows the steps in making an isometric sketch.

Nonisometric Lines on Isometric Sketches. Lines parallel to isometric axes are called *isometric lines.* Therefore lines that are not parallel to the isometric axes are *nonisometric lines* (Fig. 2–38).

Any object may be enclosed in a box, as suggested in Fig. 2–39. Note, however, that the objects in Figs. 2–40 and 2–41 have some lines which are not parallel to the isometric axes. Lines not parallel to the axes can be drawn by extending their ends to touch the blocked-in box. Points at the ends of slanted lines are located by estimating measurements along or parallel to isometric lines. With the ends of the nonisometric lines located, the line may be sketched from point to point. Lines which are parallel to each other will also show parallel on the sketch (Fig. 2–41). Note how the ends have been located on lines 1–2 and 1–3 in Fig. 2–41. Distances *a* and *b* are estimated and transferred from the figure at A to B. Any inclined line, plane, or specific angle must be found by locating two points of intersection on isometric lines.

Circles and Arcs on Isometric Sketches (Ellipses). To sketch a circle in an isometric view (Fig. 2–42), an isometric square is sketched first. The small-end arcs are sketched tangent to the square. The larger arcs are then sketched tangent at points *T* to complete the ellipse. Note that the major diameter (long axis) is longer

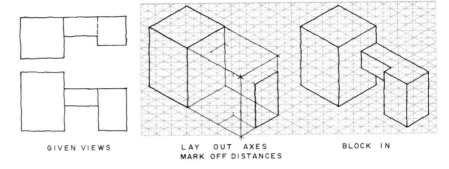

Fig. 2–36 Isometric grid paper can assist in developing quick sketches.

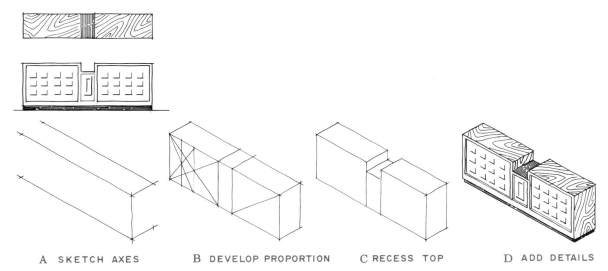

A SKETCH AXES
B DEVELOP PROPORTION
C RECESS TOP
D ADD DETAILS

Fig. 2–37 The steps in making an isometric sketch.

Fig. 2–38 Identifying nonisometric lines that form an inclined plane.

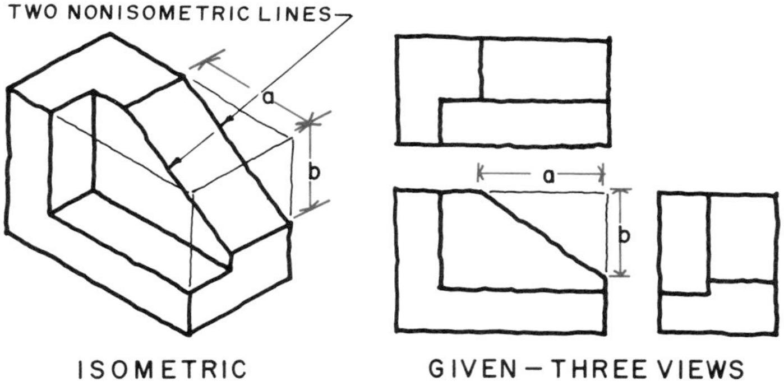

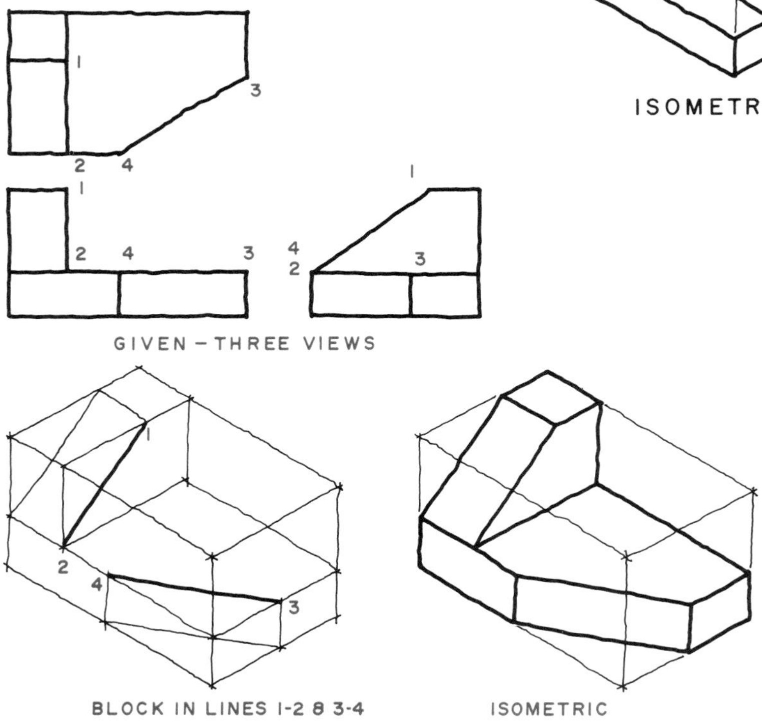

TWO NONISOMETRIC LINES
a
b
ISOMETRIC
a
b
GIVEN – THREE VIEWS

1
3
2 4
1
2 4 3
1
4
2
3
GIVEN – THREE VIEWS
1
2
4
3
BLOCK IN LINES 1-2 & 3-4
ISOMETRIC

Fig. 2–39 Development of two inclined planes in an isometric pictorial.

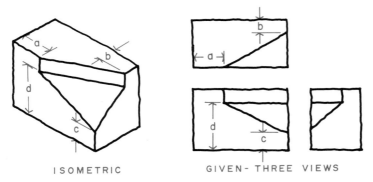

ISOMETRIC   GIVEN- THREE VIEWS

Fig. 2-40 Development of an oblique plane in an isometric drawing.

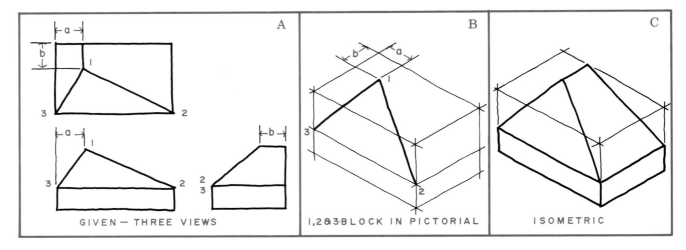

A

GIVEN — THREE VIEWS

B

1,2 &3 BLOCK IN PICTORIAL

C

ISOMETRIC

Fig. 2-41 Development of three inclined surfaces in isometric.

and the minor diameter is shorter than the true diameter of the circle. Figure 2-42 illustrates a typical ellipse for a top view only.

Circles on the three faces of an isometric cube are sketched in Fig. 2-43. Some methods of blocking in cylindrical shapes are illustrated in Fig. 2-44. Some methods of blocking in conical shapes are shown in Fig. 2-45.

Other Isometric Axes. Figure 2-46 shows a regular *Y*-axis condition. Other isometric axes may be drawn at any desired position as long as there

are 120° between the axes, as shown in Fig. 2-47. The reverse axis is illustrated for objects which are viewed better from the bottom. The position of the axes in Fig. 2-48 is effective for long pieces. In preparing a pictorial sketch which is to clearly describe the shape of an object, a decision must be made as to how to use the isometric axes on each object presented.

Sketching Curves and Arcs. Curves may be drawn in pictorial sketches by plotting points. To locate the curve, points (coordinates) are

Fig. 2-42 The isometric square with arcs tangent to form an ellipse.

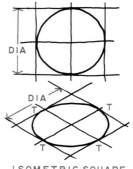

ISOMETRIC·SQUARE

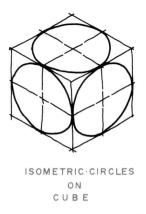

ISOMETRIC·CIRCLES
ON
CUBE

Fig. 2–43 Isometric circles placed on the front, top, and side of a cube.

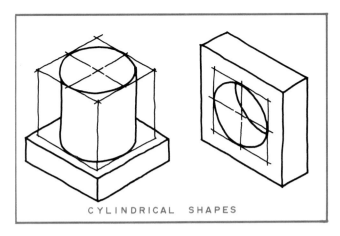

CYLINDRICAL SHAPES

Fig. 2–44 Isometric circles assist in describing cylindrical forms.

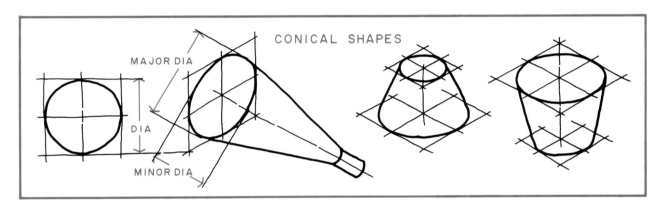

CONICAL SHAPES

MAJOR DIA

DIA

MINOR DIA

Fig. 2–45 Isometric circles on conical shapes.

Fig. 2–46 The regular isometric axes has three equal angles.

Fig. 2–47 The isometric axes is reversed to emphasize the bottom view.

Fig. 2–48 The isometric axes can emphasize length.

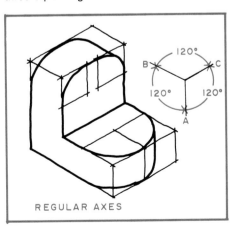

REGULAR AXES

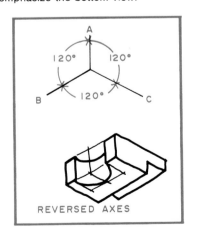

REVERSED AXES

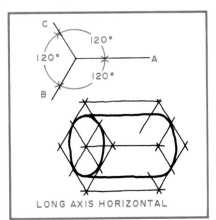

LONG AXIS HORIZONTAL

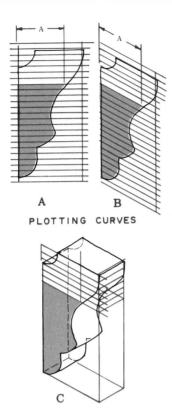

PLOTTING CURVES

C

Fig. 2–49 Irregular curves are plotted with a coordinated grid.

Fig. 2–50 Arcs are blocked in similar to isometric circles.

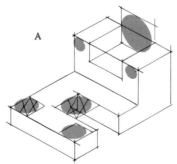

A

BLOCK IN ARCS

B

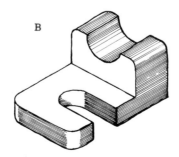

HEAVY IN ARCS

generally transferred from a multi-view sketch. In Fig. 2–49 at A, co-ordinates are plotted representing the width from the left edge of the front view and the height from the top edge of the front view. The series of lines at B drawn parallel to the vertical and horizontal axes locate coordinate points of intersection as shown. Similar coordinates are plotted on the pictorial at C, and the intersections serve as points for sketching the pictorial curve.

Arcs developed in an isometric view are shown in Fig. 2–50. The semicircular opening appears in the front view, with rounded corners. The object is blocked in lightly at A similar to plotting for isometric circles. However, only partial circles or arcs are necessary. Note that only the semicircular shape is darkened in at B, and the rounded corners require only a quarter of the full isometric circle which was plotted.

Multiview Study. A pictorial drawing shows how the object appears in a three-dimensional form. Three directions are suggested for viewing the residence pictured in Fig. 2–51. From in front, the width and height would show; this is the front view. From the side, the depth and height would show; this is the right-side view. From above, the depth and width would show; this is the top view.

A three-view pictorial does not completely describe the contemporary residence. There are *four* sides, or elevations, as they are called in architecture, but only *two* could be examined in the pictorial (Fig. 2–52). Needless to say, the floor plan could not be examined through the outside walls. There is a system for clearly defining the information needed to

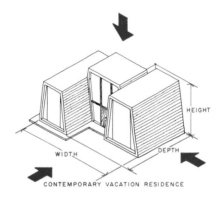

Fig. 2–51 A-frame residence in pictorial showing three-dimensional form.

Fig. 2–52 Two elevations can be projected from pictorial drawing.

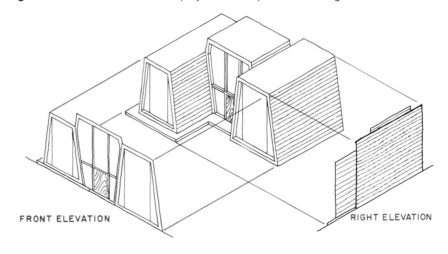

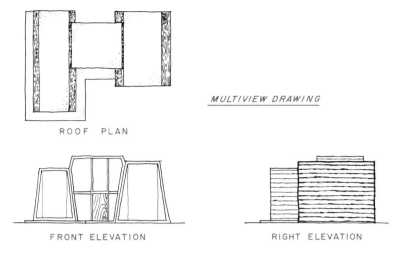

MULTIVIEW DRAWING

ROOF PLAN

FRONT ELEVATION

RIGHT ELEVATION

Fig. 2-53 Three-view drawing of A-frame residence.

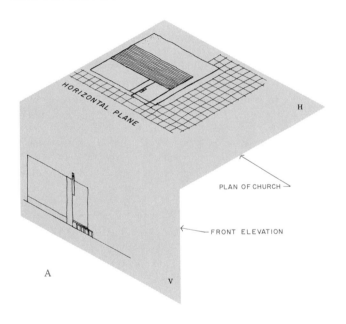

HORIZONTAL PLANE

H

PLAN OF CHURCH

FRONT ELEVATION

A

V

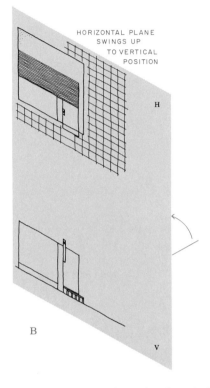

HORIZONTAL PLANE
SWINGS UP
TO VERTICAL
POSITION

H

B

V

Fig. 2-54 A. The horizontal and vertical planes hinged together reveal the roof plan and front elevation. B. The horizontal plane swings on a hinge.

develop this residence which would be more complete and accurate than the pictorial drawing.

Multiview projection (also called orthographic projection) is the system which arranges views in relation to each other (Fig. 2-53). Suppose a church is to be drawn from all sides. The draftsman-designer can now put his imagination to work. He looks

through a transparent plane set up in front of the church. The vertical plane (also called frontal plane) represents the paper, and the dashed lines are projection lines. The projection lines are perpendicular to the plane and bring a point on the object to the plane. When the points are connected on the plane they form the front view, or in architectural draw-

ing, the front elevation. This view will show the true width and height of the church form.

In Fig. 2-54 at A, a horizontal plane H is hinged at right angles to the vertical plane V. If the observer looks down through the plane H, he will see the top, or roof plan, of the church. Perpendiculars to this plane will give the horizontal projection, or top

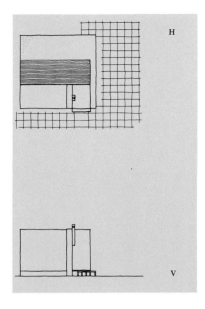

UNFOLDED PLANES

Fig. 2-55 Normal position of planes.

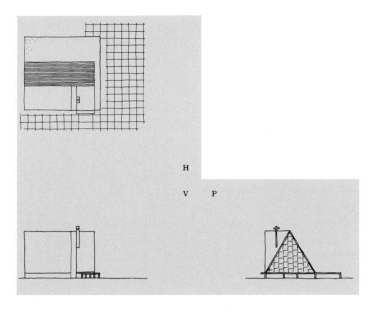

REGULAR POSITION OF THREE PRINCIPAL PLANES

Fig. 2-56 Three planes unfolded to see multiview projection.

view. These two planes represent the drawing paper. If, by using his imagination, the draftsman-designer can now swing the horizontal plane H up on the hinges until it lies in the extension of the vertical plane, as at B, the two views will appear in their proper relationship as they would be drawn on paper. Together the two views in Fig. 2-55 give the width W, the depth D, and the height H. The top view of any object is always drawn directly over the front view. The side view is always obtained in a similar manner by a profile plane P at right angles to both the vertical and the horizontal planes (Fig. 2-56).

Three planes have been mentioned, and these are considered the regular planes used to obtain the three regular views:

1. The front view on the vertical plane, indicated by the letter V.

2. The top view on the horizontal plane, indicated by the letter H.

3. The side view on the profile plane, indicated by the letter P.

The Glass Box. The church form can be thought of as being inside a transparent glass box, as shown in Fig. 2-57. When the glass box is fully opened up into one plane (Fig. 2-58), the views take their relative positions as they would be drawn on the paper. The church form has been projected to all planes and arranged with the entrance to the front.

Typical Studies. Most three-dimensional objects can be adequately described with three views. The top, front, and right-side views as ordinarily drawn are shown in Fig. 2-59 at A. When a left-side view is drawn, it is placed as shown at Fig. 2-59 at B. In this example, the left side at B is preferred over the right side at A because there are fewer hidden lines, and this will give better representation of the object.

Study the drawing in Fig. 2-60 to see how the theory of projection has been applied to five different shapes:

1. Three principal planes represented by A, B, and C.

2. Stepped planes represented by D, E, F, and G.

3. Inclined planes represented by R, S, and T.

4. Slanted planes represented by W, X, and Y.

Six Views vs. Three Views. The practical purpose of a sketch is to describe the geometric shape of something.

Six views are shown for the superb movie camera in Fig. 2-61. A study of the pictorial sketch and the six views will help determine which views are necessary. The top, front, and side views will give the best shape description and have the fewest hidden lines. Only necessary views

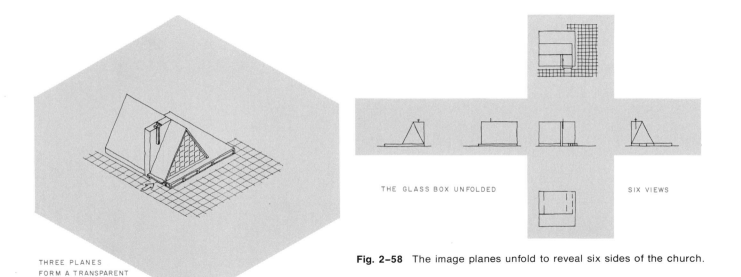

THREE PLANES
FORM A TRANSPARENT
GLASS BOX

Fig. 2–57 The church can be viewed through the image planes.

THE GLASS BOX UNFOLDED   SIX VIEWS

Fig. 2–58 The image planes unfold to reveal six sides of the church.

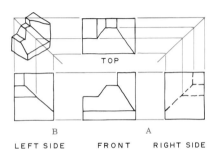

TOP

B   A
LEFT SIDE   FRONT   RIGHT SIDE

Fig. 2–59 Selecting the left side view for clarity.

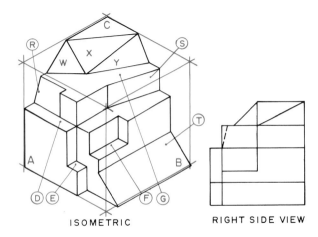

ISOMETRIC   RIGHT SIDE VIEW

Fig. 2–60 Describe and identify the planes in the right-side view.

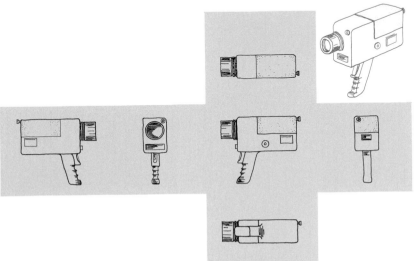

Fig. 2–61 The six views repeat geometric details.

should be sketched to describe the object.

Two-View Sketches. There are many objects which can be described by using only two views, as illustrated in Fig. 2-62. If two views are carefully selected so that they adequately describe the object, it can help to simplify the interpretation. Figure 2-63 illustrates 5 three-view drawings which could have been adequately described with two views. Which view is not necessary at A, B, C, D, and E?

One-View Sketch. If the object requires only one view to describe it, there are only two important descriptive dimensions for the shape—height and width. One-view drawings generally have a uniform depth or thickness which can be described with a note. A typical one-view drawing is illustrated in Fig. 2-64 at A. The stamping at B has a material thickness indicated as a note on the sketch. Many cylindrical objects can also be represented by single views with the proper diameters noted as in Fig. 2-65.

Line Interpretation. Sketched lines which are visible have three important roles:

1. The visible line forms the outline of the drawing.

2. The visible line shows how two surfaces meet or intersect, as shown in Fig. 2-66 at A.

3. The visible line shows the limit of a curved surface, as in Fig. 2-66 at B.

Sketched lines which are invisible (hidden) are used to describe the part of the object that normally cannot be seen. Correct practices in drawing hidden lines can be examined in Fig. 2-67. Invisible dashes are sketched approximately $\frac{1}{8}$ in. long with $\frac{1}{16}$-in.

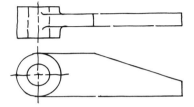
Fig. 2-62 A two-view sketch.

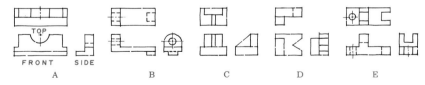
Fig. 2-63 Select the two required views.

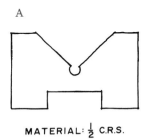
Fig. 2-64 Typical one-view drawings.

Fig. 2-65 A cylindrical object may require only one view.

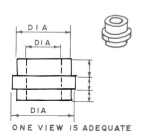
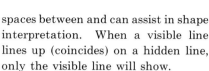
ONE VIEW IS ADEQUATE

Fig. 2-66 Visible line interpretation.

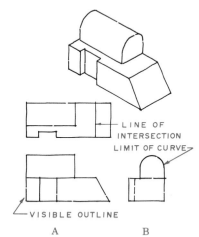

spaces between and can assist in shape interpretation. When a visible line lines up (coincides) on a hidden line, only the visible line will show.

Center lines are essential when working out sketches that have symmetrical features. An object is symmetrical when its shape is the same on both sides of a line drawn through the center. The center line (symbol ₵) can assist when developing propor-

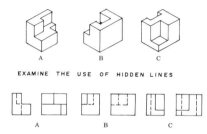

EXAMINE THE USE OF HIDDEN LINES

Fig. 2-67 Hidden lines are used to describe the unseen.

Fig. 2-68 Symmetrical features call for centerlines.

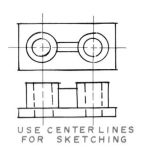

USE CENTER LINES FOR SKETCHING

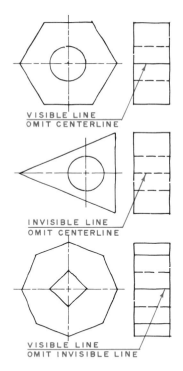

VISIBLE LINE
OMIT CENTERLINE

INVISIBLE LINE
OMIT CENTERLINE

VISIBLE LINE
OMIT INVISIBLE LINE

Fig. 2-69 Coinciding lines.

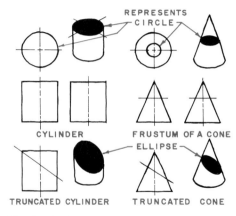

REPRESENTS CIRCLE

CYLINDER   FRUSTUM OF A CONE

ELLIPSE

TRUNCATED CYLINDER   TRUNCATED CONE

Fig. 2-70 Curved lines.

tions for cylinders and curved surfaces. The center line is illustrated in Fig. 2-68. The long dash of the center line can vary from $\frac{3}{4}$ to $1\frac{1}{2}$ in. or more in length, depending on the size of the object. The short dashes are usually $\frac{1}{8}$ in. long, with spaces on either side of about $\frac{1}{16}$ in. When preparing a sketch, make the center line thin to contrast well with hidden and visible lines. When center lines coincide with visible or invisible lines, the center line is omitted (Fig. 2-69).

Curved lines are used to describe the geometric shapes of cylinders, cones, and frustums of a cone. See typical shapes, Fig. 2-70.

Introduction to Dimensioning.
The sketch must adequately describe the object it represents. Generally the sketch is made before measurements are taken, since this gives a place to record the necessary dimen-

sions properly. There are two types of dimensions to specify (Fig. 2-71). Size dimensions at A are used to describe the overall geometric elements which give it form. Location dimensions (Fig. 2-71 at B) are used to describe or specify the relationship of these geometric elements to each other.

Definitions. A *dimension line* is used to show the direction of a dimension. Dimension lines are drawn with an arrowhead at each end and a break for the dimension numerals. *Arrowheads* are used at the ends of dimension lines to show where the dimension begins and ends. They generally stop at center lines or extension lines. *Extension lines* are thin lines used to extend the shape of the object. The extension line leaves a gap of $\frac{1}{16}$ to $\frac{1}{8}$ in. at the object and extends $\frac{1}{8}$ in. beyond the arrowhead on the dimension line. Center lines can

serve as extension lines. *Leaders* are thin lines drawn from a note or dimension to the place where it applies. A leader starts with a $\frac{1}{8}$ to $\frac{1}{4}$ in. horizontal dash and then angles off to the feature, usually at 30°, 45°, or 60°, and ends with an arrowhead.

Typical Multiview Dimensioning.
Dimensions can be used to describe linear distances, angles, and notes. A typical dimensioned sketch is shown in Fig. 2-72. See Chapter 6 for specific placement techniques.

Typical Oblique Dimensioning.
Oblique dimension lines are always parallel to the major axes, and lettering is always vertical. The ANSI approves the aligned and the unidirectional systems illustrated in Fig. 2-73. See Chapter 6 for specific placement techniques.

Example: Sketch of a Chair. The chair is shown in picture form in Fig. 2-74. Note that three views are necessary to adequately sketch all the important features. After careful observation, determine the approximate height, width, and depth. Block in major dimensions as shown. Large areas are first drawn in to the proper proportion, then the smaller features

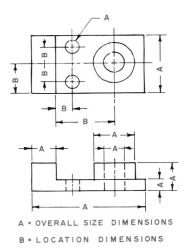

A = OVERALL SIZE DIMENSIONS

B = LOCATION DIMENSIONS

Fig. 2-71 Size dimensions at A, Location dimensions at B.

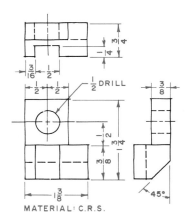

MATERIAL: C.R.S.

Fig. 2-72 Typical multiview dimensioning.

are added in their respective positions. Never attempt to complete one view at a time; block them all in and proceed to work out the corresponding details in each view. This will permit you to develop the relationship of the views of the object in correct relative proportion.

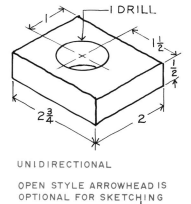

UNIDIRECTIONAL

OPEN STYLE ARROWHEAD IS OPTIONAL FOR SKETCHING

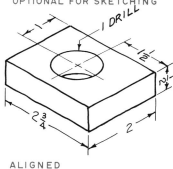

ALIGNED

Fig. 2-73 Dimensioning in pictorial.

Steps in making a sketch. The skills necessary in making good sketches will require careful practice. The steps in making a sketch are shown in Fig. 2-74.

1. Observe the chair at A.

2. Select the views necessary to describe the shapes involved.

3. Estimate the proportions carefully and mark off major distances for height, width, and depth in all three views at B.

4. Block in the enclosing rectangles as at C.

5. Locate the details in each of the views and block them in as at D.

6. Finish the sketch by darkening the object lines.

7. The preliminary blocking-in lines should be light enough so that they will not have to be erased. The dashes for hidden lines can be drawn effectively over preliminary light lines.

8. Add necessary dimensions and notes.

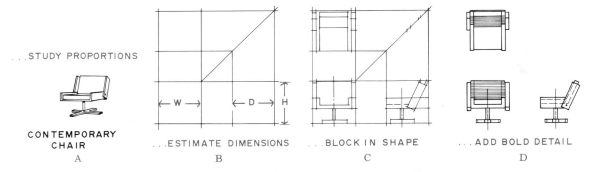

...STUDY PROPORTIONS

CONTEMPORARY CHAIR

A

...ESTIMATE DIMENSIONS

B

...BLOCK IN SHAPE

C

...ADD BOLD DETAIL

D

Fig. 2-74 The development of a multview drawing.

Review

1. What kind of paper is used in the overlay method?

2. Why is sketching considered a valuable skill by the draftsman, designer, and engineer who must participate in highly technical discussions?

3. Of all the reasons for sketching, which three seem most important?

4. Name the three characteristics which describe the nature of a sketch.

5. How does the nature of a sketch differ from the use of a sketch?

6. Why are the four visual symbols in the language of sketching essential elements in learning how to develop graphic communication?

7. What qualities would be good for the paper used for refining a sketch?

8. What is the term applied to the estimating of proportions, and why is estimating important to the draftsman and designer?

9. List four reasons why the pencil point used for freehand sketching can assist in giving the drawing a natural look.

10. Straight lines can be developed easily by remembering to draw vertical lines from the _____ down and horizontal lines from _____ to _____.

11. Name two advantages of pictorial sketching.

12. When preparing sketches of objects, what is the draftsman's first important decision?

13. What is another name for multiview drawing?

14. Name two basic types of pictorial sketching.

15. What are the three principal planes of projection, and what view appears on each plane?

16. When a visible line and a hidden line coincide, which line should be drawn? Which line if the lines are hidden and a centerline?

Problems for Chapter 2

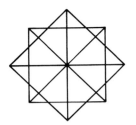

Fig. 2–75

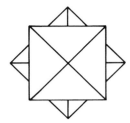

Fig. 2–76

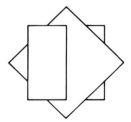

Fig. 2–77

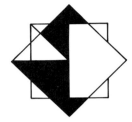

Fig. 2–78

Figs. 2–75 through 2–78 Sketch in 2-in. overlapping squares as creative visual studies.

Fig. 2–79

Fig. 2–80

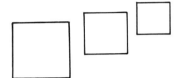

Fig. 2–81

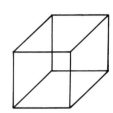

Fig. 2–82

Figs. 2–79 through 2–82 Sketch the squares overlapping, diminishing, and as a transparent cube. Sizes are about $1\frac{1}{2}$ in., $1\frac{1}{8}$ in., and $\frac{7}{8}$ in.

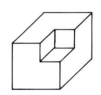

Fig. 2–83 Sketch a cube with the alteration as shown, and observe the optical illusion.

Fig. 2–84 Sketch a rectangular solid with the approximate proportions shown.

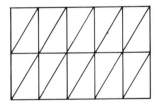

Fig. 2–85 Sketch the apparent two-dimensional rectangular form using six diagonals. How many times will Fig. 2–84 be found in Fig. 2–85?

Fig. 2–89

Fig. 2–90

Fig. 2–91

Fig. 2–92

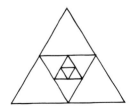

Fig. 2–86 Sketch a 3-in. equilateral triangle with diminishing triangles at midpoints. Note the proportions.

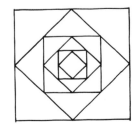

Fig. 2–87 Sketch a 3-in. square with diminishing squares at midpoints. Note the proportions.

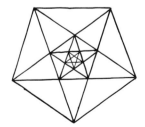

Fig. 2–88 Sketch a pentagon using 2-in. sides with diminishing five-pointed stars. Note the proportions.

Fig. 2–93

Figs. 2–89 through 2–93 Sketch the five basic solids.

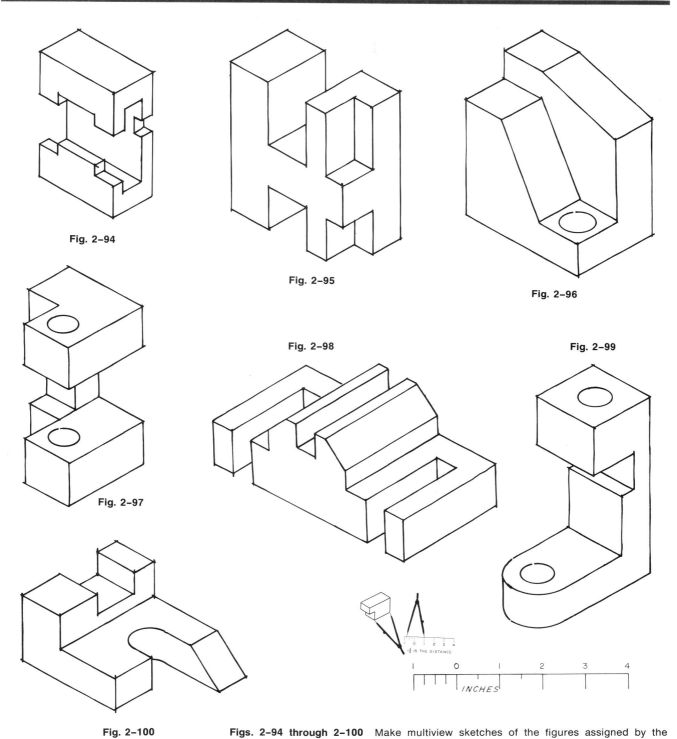

Fig. 2–94

Fig. 2–95

Fig. 2–96

Fig. 2–98

Fig. 2–99

Fig. 2–97

Fig. 2–100

IS THE DISTANCE

INCHES

Figs. 2–94 through 2–100 Make multiview sketches of the figures assigned by the instructor. Use cross-sectioned paper, or transfer dimensions from the scale with dividers as shown.

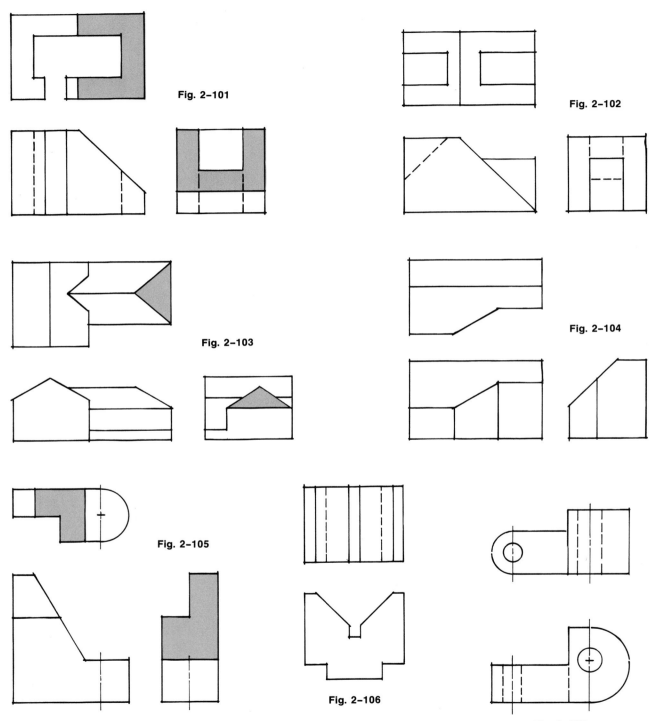

Fig. 2-101

Fig. 2-102

Fig. 2-103

Fig. 2-104

Fig. 2-105

Fig. 2-106

Fig. 2-107

Figs. 2-101 through 2-107 Make isometric or oblique pictorial sketches of the figures assigned by the instructor. Note proportions and estimate dimensions.

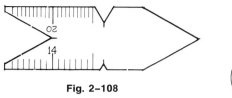

Fig. 2-108

Fig. 2-109

Fig. 2-110

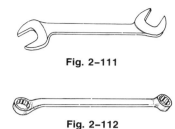

Fig. 2-111

Fig. 2-112

Fig. 2-113

Figs. 2-108 through 2-110. Make a sketch of the center gage, outside caliper, or insider caliper to approximate real sizes as assigned by the instructor.

Figs. 2-111 through 2-113 Make a sketch of the open-end wrench, socket wrench, or adjustable wrench as assigned by the instructor.

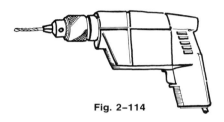

Fig. 2-114

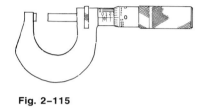

Fig. 2-115

Figs. 2-114 and 2-115 Make a pictorial sketch of the portable drill or micrometer as assigned by the instructor. Include shading.

Fig. 2-117 Prepare a multiview sketch of the digital-clock cabinet. Include dimensions and material specifications.

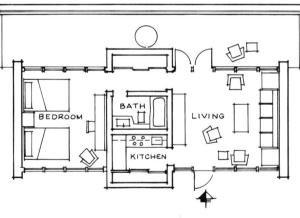

Fig. 2-116 Sketch the floor plan of the vacation home, and prepare an elevation of your own design to match.

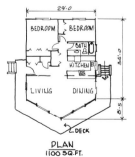

Fig. 2-118 Prepare a sketch of the ski lodge at $\frac{1}{4}$-in. scale. Add a side elevation as appropriate.

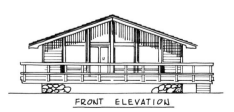

The Use and Care of Drafting Equipment

Chapter 3

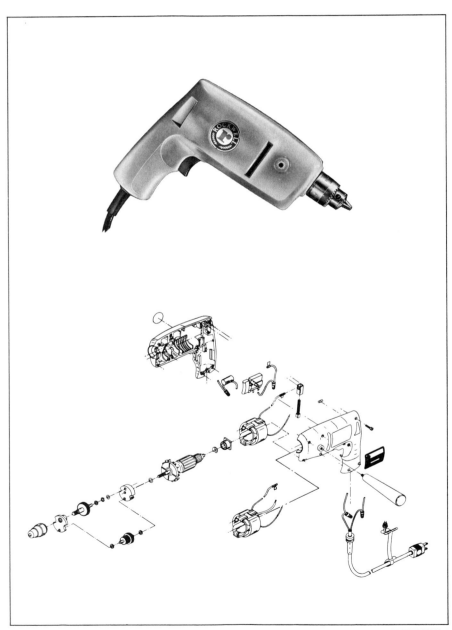

(ROCKWELL INTERNATIONAL.)

The Drafting Room. The necessary plans and directions for accomplishing engineering work of all kinds are prepared in the drafting room (Fig. 3–1). Here the designs are worked out and checked. Chapter 1 explained why drawing is really a language and why drawings are used in industrial, engineering, and scientific work. Sometimes designers or scientists make freehand sketches for preliminary study or to present ideas, but where accuracy is necessary, drawings are made with instruments. In learning to read and write this language we must learn what tools and instruments to use and how to use them skillfully, accurately, and quickly.

Basic Equipment. Figure 3–2 shows the basic drafting equipment often found in a student drafting kit. This equipment and other common items are named in the following list.

Drafting board

T-square, or parallel-ruling straight-edge, or drafting machine

Drawing sheets (paper, cloth, or film)

Drafting tape

Drafting pencils

Pencil sharpener

Eraser

Erasing shield

Triangles, 45° and 30°–60° (not required with drafting machine)

Architect's or engineer's scale

Irregular curve

Case instruments

Lettering instruments

Black drawing ink

Technical fountain pens

Brush or dust cloth

Protractor

Cleaning powder

Your instructor can list and specify the required items, sizes, and other necessary information needed for your course.

Drawing Tables and Desks. These items of equipment vary greatly in size and design (Fig. 3–3). They may have a fixed top to hold a separate, movable drawing board, a fixed top made for use as a drawing board, an adjustable top to hold a separate drawing board at a desired slope, or an adjustable top made for use as a drawing board.

They may be stand-up tables for use with high stools. They may be sit-down regular desk-high drawing tables. Or they may be various combinations of desks and tables for sit-down and stand-up use. The combination of a drafting table and desk used with a regular office chair provides a comfortable and efficient arrangement and is replacing the high drawing table in drafting rooms and engineering offices (Fig. 3–4).

Drawing Boards. The drawing sheet is attached to a *drawing board* (Fig. 3–5). Usual sizes for student or personal use are 9″ × 12″, 16″ × 21″, and 18″ × 24″. Larger sizes for engineering and architectural designs may have any necessary dimensions. Boards are generally made of soft pine or basswood and constructed so that they will stay flat and have the guiding edge (or edges) straight. Hardwood or metal (steel or aluminum)

Fig. 3–1 An industrial drafting room.

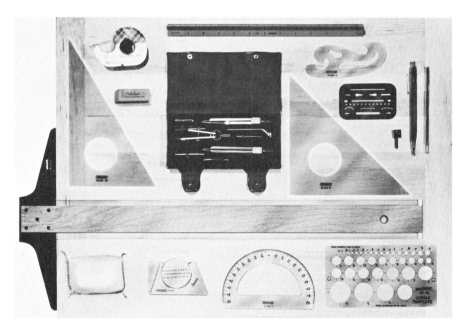

Fig. 3–2 Student drafting kit. (*Bruning Division, Addressograph Multigraph Corp.*)

strips are used at the ends on some boards to provide true guiding edges.

T-squares. T-squares (Fig. 3–6) are made of various materials, but most have plastic-edged wood blades or clear plastic blades and heads of wood or plastic. Stainless steel or hard aluminum blades with alloy heads are used where great accuracy is re-

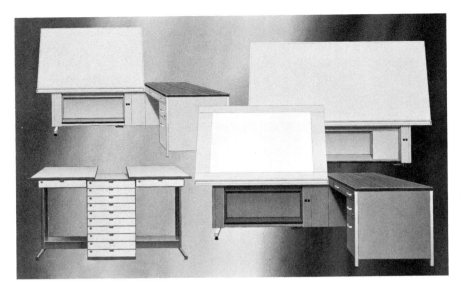

Fig. 3-3 Drafting tables are available in a variety of sizes and styles. (*Bruning Division, Addressograph Multigraph Corp.*)

Fig. 3-4 Modern combination desk and drafting table. (*Stacor Corp.*)

Fig. 3-5 Drawing boards are available in many types. (*Teledyne Post.*)

Fig. 3-6 T-squares are available in various styles and materials. (*Bruning Division, Addressograph Multigraph Corp.*)

quired. The blade (straightedge) must be straight and attached securely to the top surface of the T-square head.

The accuracy of the T-square can easily be checked, as shown in Fig. 3-7. First, draw a sharp line along the drawing edge of the T-square on a sheet of paper. Second, reverse the drawing sheet and align the drawing edge of the T-square with the other side of the line. If the drawing edge and the pencil line do not match, the T-square edge is not accurate.

Some T-squares are made with an adjustable head which allows the blade to be lined up with the drawing or set to any angle if the head has a protractor. The size of a T-square is the length of the blade measured from the contact surface of the head to the end of the blade.

Fig. 3-7 Check the T-square for accuracy.

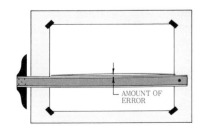

American National Standard Trimmed Sizes of Drawing Sheets. Two series of sizes are listed. One is based upon the standard-size letter paper, $8\frac{1}{2}'' \times 11''$, and the other upon a sheet $9'' \times 12''$. Multiples of these sizes are used for larger sheets, as follows:

Size	First series	Second series
A	$8\frac{1}{2}'' \times 11''$	$9'' \times 12''$
B	$11'' \times 17''$	$12'' \times 18''$
C	$17'' \times 22''$	$18'' \times 24''$
D	$22'' \times 34''$	$24'' \times 36''$
E	$34'' \times 44''$	$36'' \times 48''$

Drawing Sheets. Drawings are made on many different materials. Papers include white and tinted papers (cream and pale green). They are made in many thicknesses and qualities. For some purposes they are mounted on muslin or aluminum sheets. Most drawings are made directly in pencil on tracing paper, vellum, tracing cloth, glass cloth, or film. When such materials are used, copies can be made by blueprinting or other reproduction methods (Chapter 14).

Vellum is tracing paper which has been treated to increase its transparency. Tracing cloth is a finely woven cotton cloth treated to provide a good working surface and good transparency. Polyester films are widely used in industrial drafting rooms. They have great transparency, high strength, and exceptional durability. A matte surface is provided for drawing purposes and is suitable for both pencil and ink work.

Fastening the Drawing Sheet to the Board. The sheet may be held in place on the board by various methods (Fig. 3–8). Drafting tape may be placed across the corners of

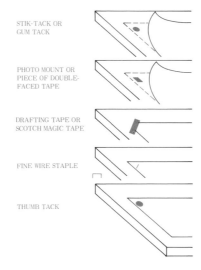

STIK-TACK OR GUM TACK

PHOTO MOUNT OR PIECE OF DOUBLE-FACED TAPE

DRAFTING TAPE OR SCOTCH MAGIC TAPE

FINE WIRE STAPLE

THUMB TACK

Fig. 3–8 The drawing sheet may be fastened to the board in a variety of ways.

the sheet and at other places, if needed. Stik-tacks (thin disks with adhesive on both sides) may be placed on the board and under the sheet. They may be removed and reused. Similar material may be obtained in squares or as tape. These methods leave the surface of the sheet entirely clear for the free movement of the T-square and triangles. They do not damage the corners or the edges of the sheet. They also can be used on composition or other hard-surface boards and are preferred by most draftsmen. Thumbtacks, if used, should be pushed straight down until the heads are in complete contact with the sheet. Small wire staples may be used but may damage the sheet. Thumbtacks and staples can be used only on soft pine or similar soft boards.

To fasten the paper, or other drawing sheet (Fig. 3–9), place it on the drawing board with the left edge an inch or so away from the left edge of the board. Left-handed students will work from the right-hand edge. The lower edge of the sheet should be placed about four inches up from the bottom of the board, or as much more as will provide a comfortable working position. Then line up the sheet with the T-square blade, as shown at A. Hold the sheet in position. Move the T-square down, as at B, keeping the head of the T-square against the edge of the board. Then fasten each corner of the sheet with drafting tape (or by other means). Drawing sheets $8\frac{1}{2}'' \times 11''$ and up to $12'' \times 18''$ may be held in place by fastening the two upper corners.

Drawing Pencils. Drawing pencils are available in the standard wood-case type and the mechanical type. There are presently four types of drawing-pencil leads. The first type is the conventional graphite lead used in pencils for over 200 years and made chiefly of graphite, clay, and resins. Drafting pencils of this type are normally made in 17 degrees of hardness (grades), as follows:

  6B (softest and blackest)
  5B (extremely soft)
  4B (extra soft)
  3B (very soft)
  2B (soft, plus)
  B (soft)
  HB (medium soft)
  F (intermediate, between soft and hard)
  H (medium hard)
  2H (hard)
  3H (hard, plus)
  4H (very hard)
  5H (extra hard)
  6H (extra hard, plus)
  7H (extremely hard)
  8H (extremely hard, plus)
  9H (hardest)

6B   F   9H

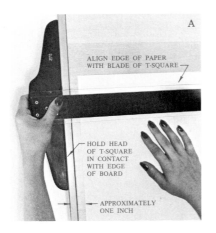
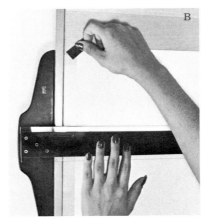

Fig. 3–9 Fastening the drawing sheet to the board.

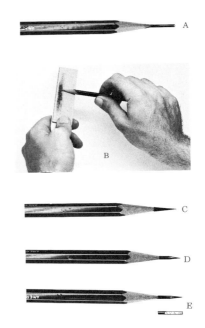

Fig. 3–10 Sharpening the pencil properly is important.

These pencils are made primarily for use on paper or vellum and for secondary use on cloth.

The right grade of pencil to use will depend upon the surface of the material on which the drawing is made, the kind of line required for blackness, or opacity, and the width of the line. For laying out views on fairly hard-surfaced drawing paper, the usual grades are 4H and 6H. For finished views on tracing paper or cloth for reproduction, the H or 2H pencil is favored. Grades sometimes used for sketching, lettering, arrowheads, symbols, border lines, and so forth, include HB, F, H, and 2H, according to the result desired on the surface of the material on which the drawing is made. Extremely hard and extremely soft leads are seldom used in ordinary drafting practices.

Since film has come into use for drawings, new types of pencil lead have been developed. Three types are described, based upon information furnished by the Joseph Dixon Crucible Company. The first type, developed for use on film, is called a *plastic pencil.* This type has a black crayon extruded lead formed by a "plasticizing" process; it is not fired. It has good microfilm reproduction charac-

teristics. The second type has a "combination" part-plastic and part-graphite fired lead. This type holds a point well, gives a good opaque line, does not smear easily, erases well, and has good microfilming qualities. Also, it can be used on paper or cloth as well as on film. The third type has an unfired combination lead of part plastic and part graphite. This type differs from the second type in that it is not fired but is extruded under pressure. It does not hold a point well, gives a fairly opaque line, erases well, does not smear easily, and has good microfilming qualities. It is made primarily for use on film. The three types of lead are made in a limited number of grades (five or six). The grades do not correspond to the grades used for the conventional leads. Different systems of letters and numbers are used by the pencil makers to tell the type of lead and the relative hardness of their products.

Sharpening the Pencil. Sharpen the pencil by cutting away the wood at a long slope, as at A in Fig. 3–10. Always sharpen the end opposite the grade mark. Be careful not to cut the lead. Leave it exposed for about $\frac{3}{8}$ to $\frac{1}{2}$ in. Then shape the lead to a

long conical point by rubbing it back and forth on a sandpaper pad (or on a fine file), as at B, while revolving it slowly to form the point, as at C or D. Some draftsmen prefer the flat, or chisel, point shown at E. Keep the sandpaper pad at hand so that the point can be kept sharp.

Mechanical sharpeners (Fig. 3–11) are made with special draftsman's cutters to remove the wood as shown. Special pointers are made for shaping the lead, as in Fig. 3–12. Such devices may be operated either by hand or electrically.

Mechanical pencils, also called *lead holders,* are widely used by draftsmen. They hold plain leads by means of a chuck which permits the exposed lead to be adjusted to any length. The lead is shaped as described for wood-cased leads; however, some refill pencils have a built-in sharpener that shapes the lead.

Pencil lines must be clean and dark enough to bring out the views using the standard lines of Fig. 3–13. You

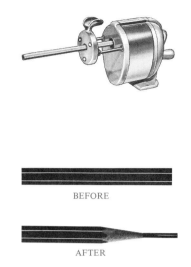

Fig. 3–11 A draftsman's pencil sharpener cuts the wood, not the lead.

Fig. 3–12 This lead pointer allows a choice of point shapes. (*Hunt Manufacturing Company.*)

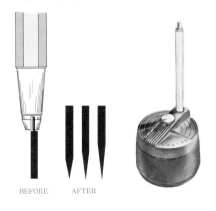

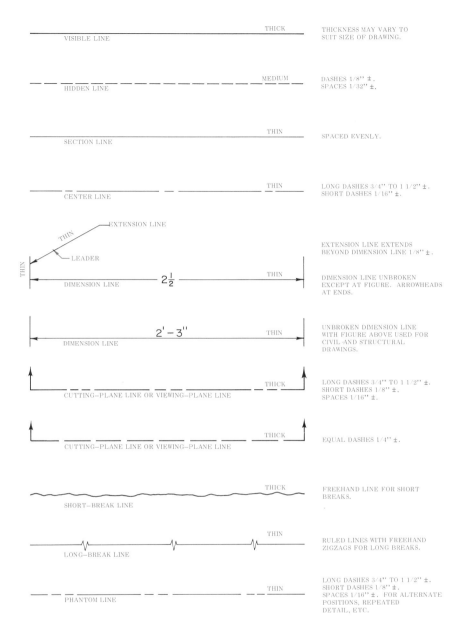

VISIBLE LINE — THICK — THICKNESS MAY VARY TO SUIT SIZE OF DRAWING.

HIDDEN LINE — MEDIUM — DASHES 1/8" ±. SPACES 1/32" ±.

SECTION LINE — THIN — SPACED EVENLY.

CENTER LINE — THIN — LONG DASHES 3/4" TO 1 1/2" ±. SHORT DASHES 1/16" ±.

EXTENSION LINE / LEADER / DIMENSION LINE $2\frac{1}{2}$ — THIN / THIN — EXTENSION LINE EXTENDS BEYOND DIMENSION LINE 1/8" ±. DIMENSION LINE UNBROKEN EXCEPT AT FIGURE. ARROWHEADS AT ENDS.

DIMENSION LINE $2'-3''$ — THIN — UNBROKEN DIMENSION LINE WITH FIGURE ABOVE USED FOR CIVIL AND STRUCTURAL DRAWINGS.

CUTTING–PLANE LINE OR VIEWING–PLANE LINE — THICK — LONG DASHES 3/4" TO 1 1/2" ±. SHORT DASHES 1/8" ±. SPACES 1/16" ±.

CUTTING–PLANE LINE OR VIEWING–PLANE LINE — THICK — EQUAL DASHES 1/4" ±.

SHORT–BREAK LINE — THICK — FREEHAND LINE FOR SHORT BREAKS.

LONG–BREAK LINE — THIN — RULED LINES WITH FREEHAND ZIGZAGS FOR LONG BREAKS.

PHANTOM LINE — THIN — LONG DASHES 3/4" TO 1 1/2" ±. SHORT DASHES 1/8" ±. SPACES 1/16" ±. FOR ALTERNATE POSITIONS, REPEATED DETAIL, ETC.

Fig. 3–13 The alphabet of lines for pencil drawing. (See Chapter 13 for ink lines)

can avoid using too much pressure and grooving the drawing surface if you use the proper grade of lead. Develop the habit of rotating the pencil between the thumb and forefinger as the line is being drawn. This will help you to keep a uniform line and will keep the point from wearing down unevenly. *Never sharpen a pencil over the drawing board. Wipe the lead with a cloth to remove the* *dust and the extra-fine or sharp needle point.* Care in such matters will do much to keep the drawing clean and bright. This is especially important when the original pencil drawing must be used to make copies.

Alphabet of Lines. The different lines, or line symbols, used on draw- ings represent a kind of graphical alphabet. The line symbols recom- mended by the American National Standards Institute are shown in Fig. 3-13 for pencil drawings. Three widths of lines—thick, medium, and thin—are generally used. Sometimes a medium-thick line may be used in place of the thick and medium lines,

Fig. 3–14 Erasing.

but the result is not so good. Drawings are much easier to read when there is good contrast between the different kinds of lines. Pencil lines should be uniformly sharp and black.

Erasers and Erasing Shields (Fig. 3–14). Soft erasers, such as the vinyl type, the Pink Pearl, or the Artgum, are used for cleaning soiled places or light pencil marks from drawings. Rub-kleen, Ruby, or Emerald erasers are good for most purposes to remove pencil or ink. Ink erasers contain grit and have to be used with extreme care, if used at all, to avoid injuring the drawing surface. Electric erasing machines (Fig. 3–15) are in common use in drafting rooms.

Fig. 3–15 An electric erasing machine saves time. (*Eugene Dietzgen Company.*)

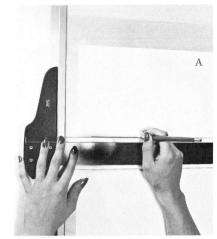

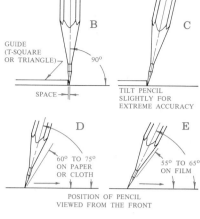

Fig. 3–16 Drawing a horizontal line.

Metal or plastic erasing shields have openings of various sizes and shapes. They are convenient for protecting lines which are not to be erased. Lines on paper or cloth are erased along the direction of the work. Lines on film are erased *at right angles to* the direction of the work. A vinyl-type or other eraser without grit should be used. Erasing on film is done very carefully by hand.

To Draw Horizontal Lines. The upper edge of the T-square blade is used as a guide for drawing horizontal lines (Fig. 3–16). Place the head of the T-square in contact with the left edge of the board with the left hand. Keep it in contact and move the T-square to the desired position and slide the left hand along the blade to hold it firmly against the drawing sheet. Hold the pencil about an inch from the point and inclined in the direction in which the line is being drawn (left to right for right-handers; the opposite for left-handers). While drawing the line, rotate the pencil slowly and slide the little finger along the blade of the T-square for better pencil control. On film it is necessary to keep the pencil at a constant angle

(55° to 65°) for the whole length of the line and to use less pressure than on paper or other material. The point of the lead should be kept a slight distance away from the corner between the guiding edge and the drawing surface, as shown, so that the draftsman can see where the line is being drawn and can also avoid a poor or smudged line. *Take care to keep the line parallel to the guiding edge.*

To Draw Vertical Lines. A triangle in combination with a T-square is used for drawing vertical lines (Fig. 3–17). Place the head of the T-square in contact with the left edge of the board. Keep it in contact and move the T-square to a position below the start of the vertical line. Place a triangle against the T-square blade and move it to the desired position. Keep the vertical edge of the triangle toward the left and draw upward. Incline the pencil at an angle in the direction in which the line is being drawn. Be sure to keep this angle constant when drawing on film. Keep the point of the lead far enough out from the guiding edge that you can see where the line is being drawn. Keep the line parallel to the guiding edge.

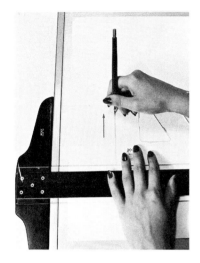

Fig. 3–17 Drawing a vertical line.

Fig. 3–18 Angles are measured in degrees, minutes, and seconds.

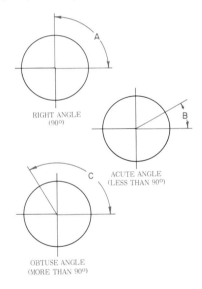

RIGHT ANGLE
(90°)

ACUTE ANGLE
(LESS THAN 90°)

OBTUSE ANGLE
(MORE THAN 90°)

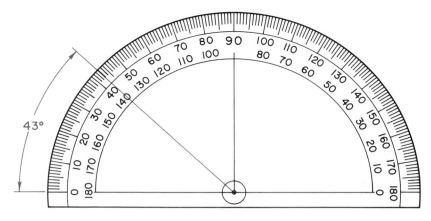

Fig. 3–19 A protractor is used to lay out, or measure, angles.

Fig. 3–20 The 45° triangle has angles of 45° and 90°.

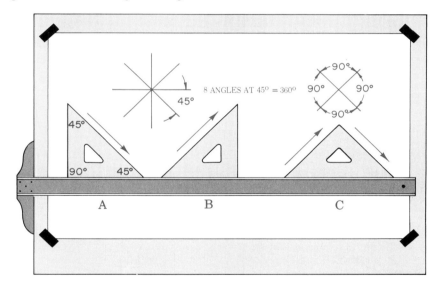

Angles. An angle is formed when two straight lines meet at a point. The point is called the *vertex,* and the lines are called the *sides* of the angle. The unit of measurement for angles is called a *degree* and is marked by the symbol °. If the circumference of a circle is divided into 360 parts, each part measures an angle of 1°. Then one-fourth of a circle would measure $^{360}/_4 = 90°$, as at A in Fig. 3–18. This is called a *right angle.* An *acute angle* (B) is less than 90°. An *obtuse angle* (C) is greater than 90°. For closer measurements, degrees are divided into 60 equal parts called *minutes* (') and minutes, into 60 equal parts called *seconds* ("). Notice that the size of an angle does not depend upon the length of the sides; therefore, the number of degrees in an angle will be the same regardless of the diameter of the circle.

An instrument used in measuring or laying out angles is called a *protractor.* A semicircular form is shown in Fig. 3–19, where an angle of 43° is measured. Drafting machines and adjustable-head T-squares have an adjustable protractor for laying out angles. Angles of 45°, 30°, and 60° can be drawn directly, and angles varying by 15° can be drawn with the

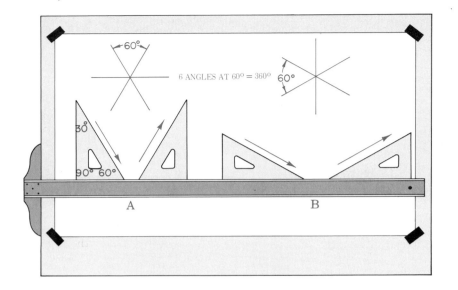

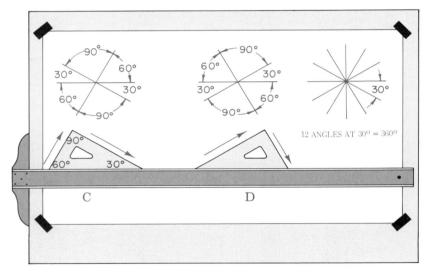

Fig. 3–21 The 30°-60° triangle has angles of 30°, 60°, and 90°.

off equal angles, 6 at 60° or 12 at 30°, about a center and for many constructions.

To Draw Lines at Angles Varying by 15°. The 45° and 30°–60° triangles, alone or together and in combination with the T-square, may be used to draw angles increasing by 15°, as 15°, 30°, 45°, 60°, 75°, and so forth, with the horizontal or vertical line. Some arrangements of the triangles for drawing angles of 15° and 75° are suggested at A, B, C, and D in Fig. 3–22. Methods of obtaining the different angles are shown in Fig. 3–23. In Space 0 the lines have been drawn for all the positions, and all the angles are 15°. By using the arrangements of Spaces B and M or Spaces F and I, two angles of 30° and two angles of 15° may be obtained. Try different combinations of triangles until you are familiar with the various arrangements. Any angle may be laid off with the protractor.

To Draw Parallel Lines. Parallel horizontal lines can be drawn with the T-square. Parallel vertical lines can be drawn by using a triangle in combination with the T-square. Parallel lines at regular angles may be drawn with the triangles, as suggested in Fig. 3–23. Parallel lines in any position may be drawn by using a triangle in combination with the T-square or another triangle, as shown in the steps of Fig. 3–24, or they may be drawn directly with a drafting machine.

To Draw a Line Parallel to a Given Line. Place a triangle against the T-square blade and move them together until one edge of the triangle matches the given line (Fig. 3–24, Space A). Hold the T-square firmly and slide the triangle along the blade until the desired position is reached. Then draw the parallel line (Space

triangles, as described in the following paragraphs.

To Draw Lines Inclined at 45°, 30°, and 60°. The 45° triangle (Fig. 3–20) has two angles of 45° and one of 90°. Slanted, or inclined, lines at 45° with horizontal or vertical lines are drawn with the triangle held against the T-square blade, or the horizontal straightedge, as shown at A, B, and C

in Fig. 3–20. The 45° triangle may be used to lay off eight equal angles of 45° about a center and for many constructions. The 30°–60° triangle (Fig. 3–21) has angles of 30°, 60°, and 90°. Inclined lines at 30° and 60° with horizontal or vertical lines are drawn with the triangle held against the T-square blade or horizontal straightedge, as shown in Fig. 3–21. The 30°–60° triangle may be used to lay

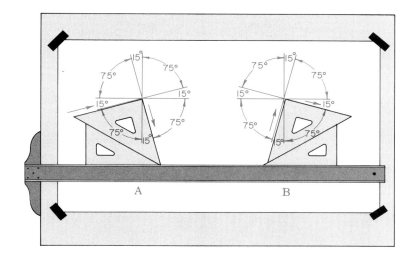

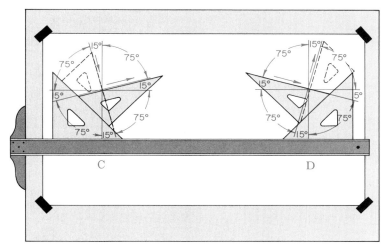

Fig. 3–22 Drawing lines at 15° and 75° using the two triangles.

be used instead of the 30°–60° triangle, as in Space C. A triangle may be used instead of the T-square, as in Space D.

Second method (Fig. 3–26). Place a triangle and the T-square together so that the hypotenuse of the triangle matches the given line, as in Space A. Turn the triangle about its right-angled corner, as in Space B, and slide it until it is in the desired position, Space C; draw the perpendicular. The 45° triangle may be used instead of the 30°–60° triangle. A triangle may be used instead of the T-square, as in Space D.

Drafting Machines. There are two types of drafting machines in general use. One type, shown in Fig. 3–27, uses an anchor and two arms to hold an adjustable protractor head with two scales, ordinarily at right angles. The arms allow the scales to be moved any place on the drawing parallel to the starting position. Another type, Fig. 3–28, uses a horizontal guide rail at the top of the board and a moving arm rail at right angles to the top rail. An adjustable protractor head and two scales, ordinarily at right angles, move up and down on the arm. The scales may be moved any place on the drawing parallel to the starting position. This type is convenient on large boards or on boards in a vertical position or at a steep angle.

A high percentage of the industrial drafting departments now use drafting machines. Many school drafting departments teach drawing with drafting machines. Drafting machines combine the functions of the T-square, triangles, scales, and protractor. Lines can be drawn to the exact lengths in the required places and at any angles by moving the scale ruling edge to the desired positions. This results in greater speed with less effort in making drawings. A com-

B). In Fig. 3–24 the hypotenuse of the triangle is used to draw the parallel line, but other edges or the 30°–60° triangle may be used. In Space C a triangle is used in place of the T-square. Parallel lines may also be drawn directly with a drafting machine.

To Draw Perpendicular Lines. Lines at right angles (90°) with each other are perpendicular. A vertical line is *perpendicular* to a horizontal line. Perpendicular lines may be

drawn using triangles and T-square or drawn directly with a drafting machine.

To Draw a Perpendicular to Any Line. Two methods can be used.

First method (Fig. 3–25). Place a triangle and the T-square together so that one edge of the triangle matches the given line, as in Space A. Hold the T-square firmly. Slide the triangle along the blade to the desired position, as in Space B, and draw the perpendicular. The 45° triangle may

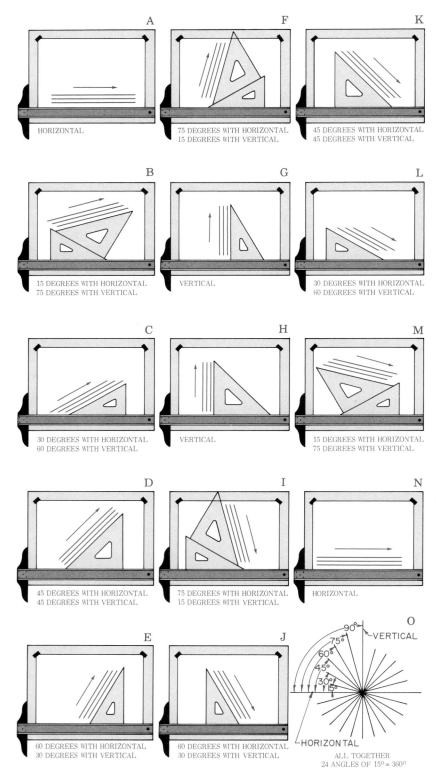

Fig. 3-23 Drawing lines with the T-square and triangles.

A — HORIZONTAL

F — 75 DEGREES WITH HORIZONTAL / 15 DEGREES WITH VERTICAL

K — 45 DEGREES WITH HORIZONTAL / 45 DEGREES WITH VERTICAL

B — 15 DEGREES WITH HORIZONTAL / 75 DEGREES WITH VERTICAL

G — VERTICAL

L — 30 DEGREES WITH HORIZONTAL / 60 DEGREES WITH VERTICAL

C — 30 DEGREES WITH HORIZONTAL / 60 DEGREES WITH VERTICAL

H — VERTICAL

M — 15 DEGREES WITH HORIZONTAL / 75 DEGREES WITH VERTICAL

D — 45 DEGREES WITH HORIZONTAL / 45 DEGREES WITH VERTICAL

I — 75 DEGREES WITH HORIZONTAL / 15 DEGREES WITH VERTICAL

N — HORIZONTAL

E — 60 DEGREES WITH HORIZONTAL / 30 DEGREES WITH VERTICAL

J — 60 DEGREES WITH HORIZONTAL / 30 DEGREES WITH VERTICAL

O — ALL TOGETHER / 24 ANGLES OF 15° = 360°

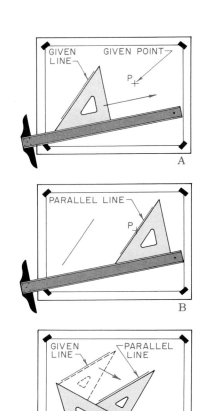

Fig. 3-24 Drawing parallel lines.

plete understanding of the possibilities, efficient use, and care of the drafting machine will reveal its value.

Parallel-ruling Straightedges. To many draftsmen, the use of parallel straightedges is especially convenient when working on large boards in a vertical or nearly vertical position. A guide cord, clamped to the ends of the straightedge, runs through a series of pulleys on the back of the board so that the straightedge may be moved up and down in parallel positions.

Scales. Scales are made in different shapes (Fig. 3-30) and of different materials, such as boxwood, white plastic on boxwood, plastic, and metal. They

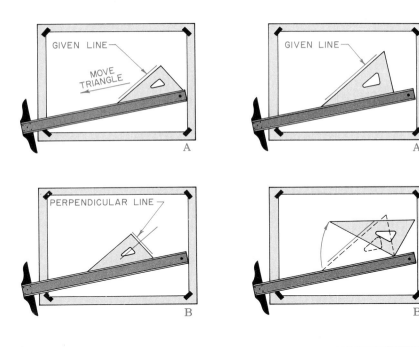

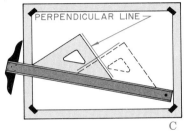

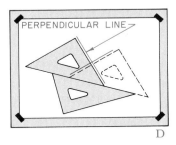

Fig. 3-25 Drawing a perpendicular line, first method.

Fig. 3-26 Drawing a perpendicular line, second method.

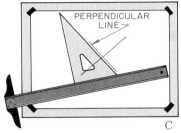

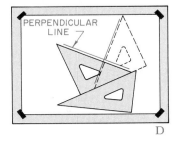

full size. Scales may be obtained *open divided* (Figs. 3–31 and 3–32), with only the end units subdivided; or they may be *full divided* (Figs. 3–33 and 3–34), with subdivisions over the entire length of the scale.

The Architect's Scale. The architect's scale (Fig. 3–31) is divided into proportional feet and inches. The triangular form shown is used a great deal in schools and in some drafting offices because it contains a variety of scales on a single stick. Many draftsmen prefer flat scales, especially when frequent changes of scale are not required. The symbol ' is used for feet and '' for inches. Thus three feet four and one-half inches is written $3'-4\frac{1}{2}''$. When all dimensions are in inches, standard practice is to omit the symbol.

The usual proportional scales are:

Full size,	$12'' = 1'-0''$
$\frac{1}{4}$ size,	$3'' = 1'-0''$
$\frac{1}{8}$ size,	$1\frac{1}{2}'' = 1'-0''$
$\frac{1}{12}$ size,	$1'' = 1'-0''$
$\frac{1}{16}$ size,	$\frac{3}{4}'' = 1'-0''$
$\frac{1}{24}$ size,	$\frac{1}{2}'' = 1'-0''$
$\frac{1}{32}$ size,	$\frac{3}{8}'' = 1'-0''$
$\frac{1}{48}$ size,	$\frac{1}{4}'' = 1'-0''$
$\frac{1}{64}$ size,	$\frac{3}{16}'' = 1'-0''$
$\frac{1}{96}$ size,	$\frac{1}{8}'' = 1'-0''$
$\frac{1}{128}$ size,	$\frac{3}{32}'' = 1'-0''$

These scales are used in drawing buildings and in making many mechanical, electrical, and other engineering drawings. They are much used as a general-purpose scale. The scale to which the views are made should be given on the drawing, either in the title or, if several parts are drawn to different scales, near the view, as:

Scale: $6'' = 1'-0''$
Scale: $3'' = 1'-0''$
Scale: $1\frac{1}{2}'' = 1'-0''$

are made with different divisions to meet the requirements for making different kinds of drawings. Commonly used scales include the architect's scale (Fig. 3–31), the mechanical engineer's scale (Fig. 3–32), and the civil engineer's scale (Fig. 3–34). These may be of any of the shapes shown in Fig. 3–30.

Scales are used for laying off distances and for making measurements, whether full size or in proportion to

Fig. 3–27 An arm-type drafting machine. (*Universal Drafting Machine Company.*)

Fig. 3–28 This track-type drafting machine is especially adapted for wide drawings, in addition to use for regular sizes of drawings. (*Teledyne Post.*)

Fig. 3–29 A parallel-ruling straightedge is another convenient device used to save time. (*Eugene Dietzgen Company.*)

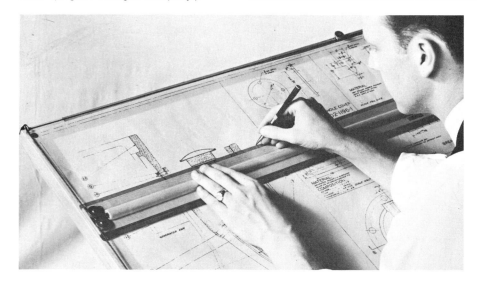

Fig. 3–30 Scales are made in various shapes.

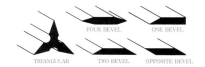

The Mechanical Engineer's Scale.

The conventional mechanical engineer's scale (Fig. 3–32) has inches and fractions of an inch divided to represent inches. The usual divisions are:

Full size—1 in. divided to read in 32nds.

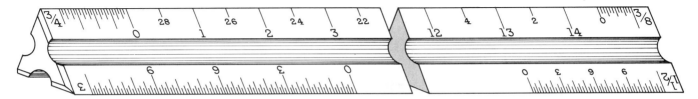

Fig. 3–31 Architect's scale, open divided. The triangular form has many proportional scales.

Fig. 3–32 Mechanical engineer's scale, open divided.

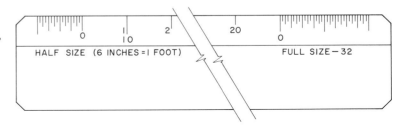

Half size—½ in. divided to read in 16ths.

Quarter size—¼ in. divided to read in 8ths.

Eighth size—⅛ in. divided to read in 4ths.

These scales are used for drawing parts of machines or where larger reductions are not required. The scale to which the views are drawn should be given on the drawing, either in the title or, if several parts are drawn to different scales, near the views, as: full size, half size or $1'' = 1''$, $\frac{1}{2}'' = 1''$.

The decimal-inch system uses a scale divided into decimals of an inch (Fig. 3–33). For full size, 1 in. is divided into 50 parts (each part of $\frac{1}{50} = 0.02''$) so that measurements can be readily made to hundredths of an inch by sight. Some usual divisions are:

Full size—1 in. divided into 50ths.

Half size—½ in. divided into 10ths.

Three-eighths size—⅜ in. divided into 10ths.

One-quarter size—¼ in. divided into 10ths.

The decimal-inch system has been used in the automotive industry for many years, and it is increasing in favor and use in other engineering fields.

Fig. 3–33 Decimal-inch scales are used in many industries.

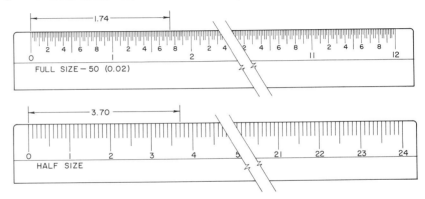

The Civil Engineer's Scale. The civil engineer's scale (Fig. 3–34) has inches divided into decimals. The usual divisions are:

10 parts to the inch
20 parts to the inch
30 parts to the inch
40 parts to the inch
50 parts to the inch
60 parts to the inch

With this scale 1 in. may be used to represent feet, rods, miles, and so forth, or to represent quantities, time, or other units. The divisions may represent single units or multiples of 10, 100, and so on. Thus the 20 parts to an inch scale may represent 20, 200, or 2000 units. This scale is used for civil engineering work, such as maps, roads, and other public projects. It is also used where decimal divisions are required, such as plotting data, drawing graphic charts, and so forth.

The scale used should be given on the drawing or other presentation as:

Scale: $1'' = 100$ feet
Scale: $1'' = 500$ miles
Scale: $1'' = 200$ pounds

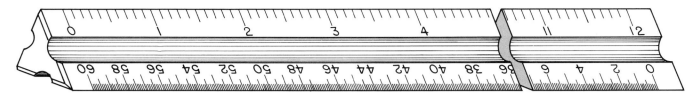

Fig. 3–34 Civil engineer's scale, chain divided.

For some purposes a graphic scale is placed on a map, drawing, or chart, as shown below.

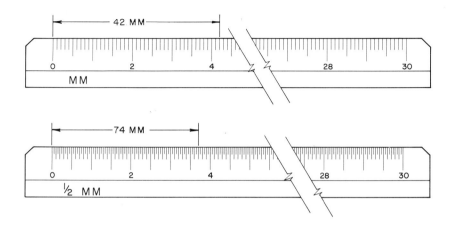

Fig. 3–35 Metric scales. Nearly all foreign countries use the metric system.

Metric Scales. Metric scales (Fig. 3–35) are divided into centimeters (cm), with centimeters divided into millimeters (mm) (10 divisions) or into half millimeters (20 divisions). Some scales are made with metric divisions on one edge and inch divisions on an opposite edge. The length of a 30-cm scale approximates the length of a 12-in. scale: $1'' = 2.54$ cm, and $12 \times 2.54 = 30.48$ cm. See the Appendix for a table of metric units and equivalents.

The *metric system* is the international system of measurement presently in use by 90 percent of the world's population. The United States and Canada are the last two major countries to convert from the U.S. Customary system to the metric system. Metric measurements are not entirely new to United States industry, however. By an Act of Congress in 1866 the metric system was made legal but not mandatory. As a result, it has long served as the standard of measure for the sciences, medicine, pharmaceuticals, and some hardware industries.

Many companies that deal in international trade have adopted a dual-dimensioning system expressed in both millimeters (metric) and inches (U.S. Customary). In order to standardize measurement, it appears as though the United States and Canada will convert entirely to the metric system within the next several years.

Full-size Drawings. When the object is not too large for the paper, it may be drawn full size, using the inches and fractions scale. *To make a measurement,* put the scale on the paper in the direction to be measured. Make a short, light dash opposite the zero on the scale (Fig. 3–36) and another dash opposite the division at the desired distance. Do not make a dot or punch a hole in the surface of the drawing sheet. Figure 3–37 shows a full-size distance of $1\frac{7}{16}''$ laid off.

Fig. 3–36 To make a measurement of $2\frac{5}{8}''$ at a scale of $3'' = 1'$-0''.

Fig. 3–37 Making a measurement ($1\frac{7}{16}''$) with the full-size scale.

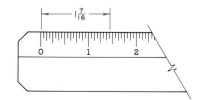

Drawing to Scale. If the object is large or has little detail, it may be drawn in a *reduced proportion*. The first reduction is to the scale of $6'' = 1'$-$0''$, commonly called *half size*. A full-size scale can be used to draw a half size by considering each half inch to represent 1 in. and each 12-in. scale to represent a 24-in. scale. This is illustrated in Fig. 3–38, where $3\frac{5}{8}''$ is laid off by three half inches and five-eighths of the next half inch. Always think full size. If one is available, a scale divided and marked for half size (Fig. 3–32) is more convenient to use.

If smaller views are required, the next reduction that may be used is the scale of $3'' = 1'$-$0''$, called *quarter size*. Find this scale on the architect's scale and examine it. The actual length of 3 in. represents 1 ft divided into 12 parts, each representing 1 in. and further divided into eighths. Learn to think of the 12 parts as representing real inches. An example of laying off the distance of $1'$-$3\frac{1}{2}''$ is shown in Fig. 3–39. Notice the position of the zero mark, placed so that inches are measured in one direction from it and feet in the other direction, as shown in the figure.

A regular mechanical engineer's scale may be used with a scale of $\frac{1}{4}'' = 1''$. For other reductions the scales mentioned in connection with Figs. 3–31 and 3–32 are used. For small parts, enlarged scales may be used, such as $24'' = 1'$-$0''$ for double-size views. Very small parts may be drawn four or eight times size or for some purposes 10, 20, or more times full size. The views of large parts and projects must be drawn to a small scale.

Case Instruments. The two patterns of instruments in general use are shown in Fig. 3–40, the flat pattern at A and the square pattern at B. Some

draftsmen prefer one pattern to the other. The usual set (Fig. 3–41) includes compasses with pen part, pencil part, and lengthening bar, dividers, bow pen, bow pencil, bow dividers, and one or two ruling pens.

Large bow sets (Fig. 3–42) are favored by some users of drafting instruments. They are known as *master*, or *giant*, bows and are made in several patterns. Large bows (6 in. or more) are capable of drawing circles up to 13 in. in diameter or, with lengthening bars, up to 40 in. in diameter. Large bow sets provide the convenience of using one instrument in place of the regular compasses, dividers,

and small bow instruments. Large bow instruments provide a means of securely holding the radius at any required distance up to their largest capacity.

The Dividers. The dividing of lines and the transferring of distances are done with dividers. The dividers are held in the right hand and adjusted as illustrated in Space A, Fig. 3–43.

To divide a line into three equal parts, adjust the points of the dividers until they appear to be about one-third the length of the line, and place one point on one end of the line and the other point on the line (Space B).

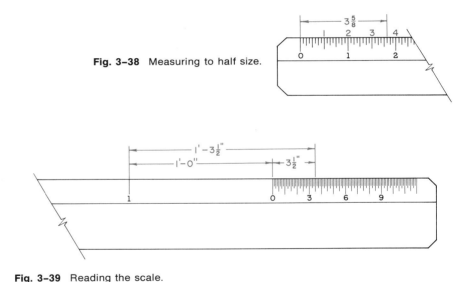

Fig. 3–38 Measuring to half size.

Fig. 3–39 Reading the scale.

Fig. 3–40 The two patterns of drawing instruments.

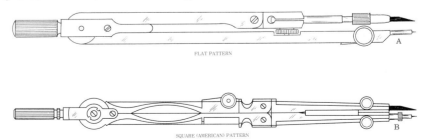

FLAT PATTERN

SQUARE (AMERICAN) PATTERN

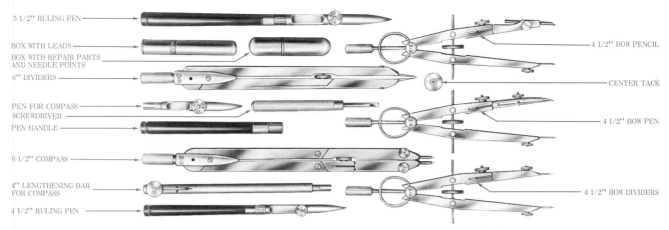

Fig. 3–41 A three-bow set of drawing instruments. (*Keuffel & Esser Co.*)

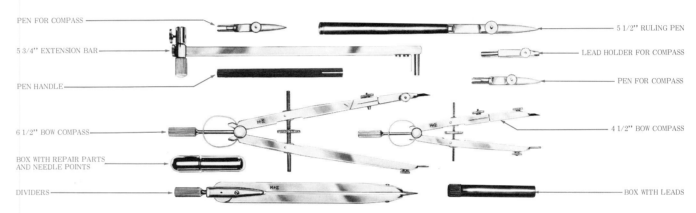

Fig. 3–42 A large-bow set of drawing instruments. (*Keuffel & Esser Co.*)

Turn the dividers about the point that rests on the line, as in Space C, then in the alternate direction, as in Space D. If the last point falls short of the end of the line, increase the distance between the points of the dividers by an amount estimated to be about one-third the distance *mn* and start at the beginning of the line again. Several trials may be necessary. If the last point overruns the end of the line, decrease the distance between the points by one-third the extra distance. For four, five, or more spaces, proceed as described, except that the correction will be one-fourth, one-

fifth, and so forth, of the overrun or underrun. An arc or the circumference of a circle is divided in the same way by using the distance between the points of the dividers as a chord.

The Compasses. Views on drawings are composed of straight lines and curved lines. Most of the curved lines are circles or parts of circles (arcs) and are drawn with the compasses (Fig. 3–44). The legs of the compasses may be left straight for radii under 2 in. For larger radii the legs should be adjusted perpendicular to the paper (Fig. 3–45). A lengthening bar (Fig.

3–46) may be inserted when a very large radius is needed (over 8 in. usually).

The compasses should be prepared for use by sharpening the lead as in Fig. 3–47 and allowing it to extend about $\frac{3}{8}$ in. A long bevel on the outside of the lead will keep a sharp edge when the radius is increased. Then adjust the shouldered end of the needle point until it extends a very little beyond the lead point, as illustrated in Fig. 3–47. The pressure used on the lead in the compasses cannot be as much as on the pencil. Therefore, it is desirable to use lead which is one or two degrees softer in the compasses in order to produce the same line weight.

When ready to use the compasses, locate the center of the required arc or circle by two intersecting (crossing) lines and lay off the radius by a short, light dash (Fig. 3–44 in Space A). The compasses are used entirely with the right hand. They are opened by pinching them between the thumb and second finger (Space A) and are set to the proper radius by placing the needle point at the center and adjusting the pencil leg with the first and second fingers (Space B). When the radius is set, raise the fingers to the handle (Space C) and revolve the compasses by twirling the handle between the thumb and finger. Start the arc near the lower side and revolve clockwise (Space D), inclining the compasses slightly in the direction of the line. *Do not force* the needle point way into the paper. Use only enough pressure to hold the point in place. Long radii are obtained by using a lengthening bar in the compasses to extend the pencil leg. For extra-long radii, beam compasses are used (Fig. 3–52).

Bow Instruments. A set of bow instruments (Fig. 3–48) consists of the bow pencil (A), the bow dividers (B), and the bow pen (C). Any of them may have the center-wheel adjustment, as at A and C, or the side-wheel adjustment, as at B. They may be of the hook-spring type, as at A, or the fork-spring type, as at B and C. The usual size is about four inches high.

The bow dividers are used for taking off (transferring) small distances, for marking off a series of small distances, and for dividing a line into small spaces. The bow pencil is used for drawing small circles. The choice of center- or side-wheel adjustment is a matter of personal preference. The

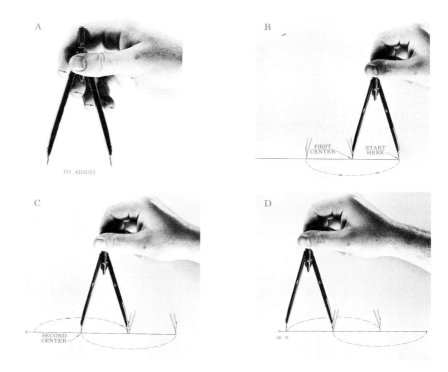

Fig. 3–43 The dividers are used to divide and to transfer distances.

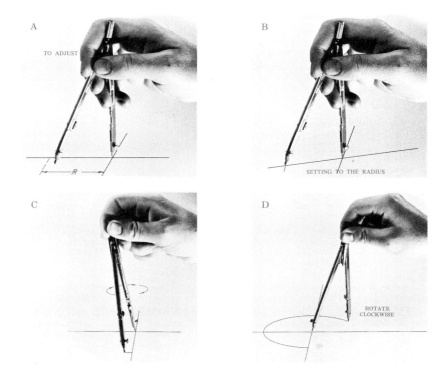

Fig. 3–44 The compasses are used to draw circles and arcs.

lead for the bow pencil is sharpened and adjusted as at A in Fig. 3-49. The inside bevel holds an edge for small circles and arcs, as indicated at B. For larger radii the outside bevel at C is better. Some draftsmen prefer a conical center point or an off-center point, as at D, E, and F. The bow instruments are convenient and accurate for small distances or radii (less than $1\frac{1}{4}$ in.), as they hold small distances better than the large instruments. Large adjustments can be made quickly with the side-wheel bows by pressing the fork and spinning the adjusting nut. Some center-wheel bows are made for large, rapid adjustments by holding one leg in each hand and either pushing to close or pulling to open. Small adjustments are made by the adjusting nut on both the side-wheel and the center-wheel bows. The bow pencil (Fig. 3-50) is used with one hand. Set the radius as in Space A. Start the circle near the lower part of the vertical center line (Space B). Revolve clockwise, as in Space C.

Drop-spring Bow Compass. The drop-spring bow (Fig. 3-51) is convenient for drawing very small circles, especially where the same size has to be repeated many times, as for rivets and similar conditions. The marking point (pencil or pen) is attached to a tube which slides on a pin. In use the pin remains stationary while the pencil point revolves about it. The radius is set by the adjusting-spring screw. The marking point is held up while the pin is placed on the center, then dropped and revolved. Circles drawn will all be the same size, as set.

Beam Compass. Beam compasses (Fig. 3-52) are used for drawing arcs or circles with large radii. They consist of a bar (beam) upon which mov-

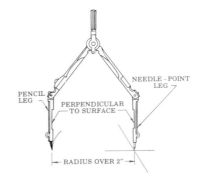

Fig. 3-45 Adjusting the compass for large circles.

Fig. 3-46 The lengthening bar is used in compasses for large radii.

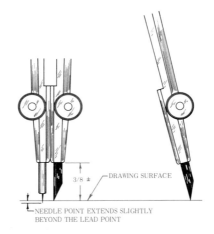

Fig. 3-47 Adjusting the point and shaping the lead of the compass.

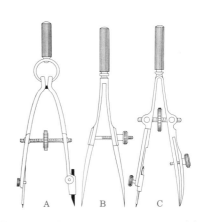

Fig. 3-48 Bow instruments are used for drawing small circles and arcs.

Fig. 3-49 Adjusting the lead for the bow pencil.

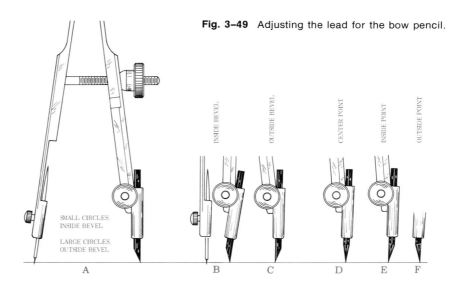

SETTING TO
THE RADIUS

A

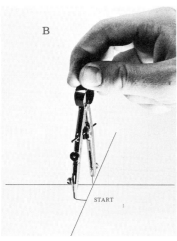

B

START

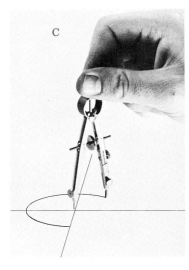

C

Fig. 3–50 Adjusting the radius for the bow pencil.

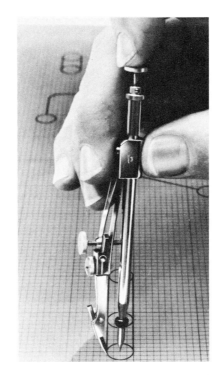

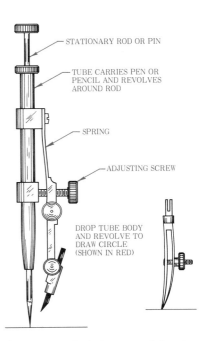

STATIONARY ROD OR PIN

TUBE CARRIES PEN OR PENCIL AND REVOLVES AROUND ROD

SPRING

ADJUSTING SCREW

DROP TUBE BODY AND REVOLVE TO DRAW CIRCLE (SHOWN IN RED)

Fig. 3–51 The drop-spring bow is used for drawing very small circles, especially where there are many to be drawn. *(Keuffel & Co.)*

able holders for a pencil part and a needle part may be positioned along the bar and fixed at any desired distance apart. The pencil part is interchangeable with a pen part. The pencil point may be replaced with a needle point in order to use the beam compasses as dividers or to set off long distances. The usual bar is about 13 in. long, but a coupling may be used to add extra length to permit drawing circles of almost any desired size.

Irregular Curves. Irregular, or French, curves (Fig. 3–53) are used for drawing many noncircular curves (involutes, spirals, ellipses, and so forth), for curves on graphic charts, and for plotting motions, forces, and some engineering and scientific graphs. French curves are made of sheet plastic and in a very large vari-

ety of forms, a few of which are illustrated. Sets are made for ellipses, parabolas, hyperbolas, and many special purposes.

To use a curve (Fig. 3–54), locate points through which the curve is to pass. This is done by construction, from tests, by computation, or otherwise. Then establish the path of the curve by drawing a light line, freehand, through the points, or adjust as necessary for a smooth curve for experimental or other data. Next, fit the irregular curve by trial against a part of the curved line and draw a portion of the line. Move the irregular curve to match the next portion, and so forth. Each new position should fit enough of the part just drawn (overlap) to make sure of continuing a smooth line. It is very important to notice whether the radius

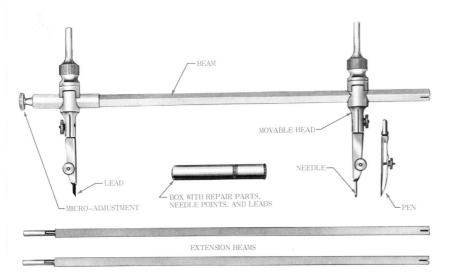

Fig. 3–52 Beam compasses are used for large radii.

Fig. 3–53 Some irregular or French curves. They are made in a great variety of forms. (*Teledyne Post.*)

of the curved line is increasing or decreasing and to place the irregular curve in the same way. Do not attempt to draw too much of the curve with one position. If the curved line is symmetrical about an axis, the position of the axis may be marked on the irregular curve with a pencil for one side and then reversed to match and draw the other side. Various flexible or adjustable curves are made that are useful for certain special kinds of work (Fig. 3–55 and 3–56). Other special curves are available, but for ellipses and many other purposes, templates are preferred.

Templates. Templates (Fig. 3–57) are an important part of the equipment for engineers and professional draftsmen, since they save a great deal of time in drawing shapes of details, such as bolt heads, nuts, electrical symbols, architectural symbols, plumbing symbols, outlines of tools and equipment, and many other outlines which are used often or repeated. Regular and special templates are made for drawing all kinds of symbols and many details for electronics, electrical and mechanical engineering, civil engineering, architecture, geometry, mathematics, and so forth.

Fig. 3–54 Steps in drawing a smooth curve.

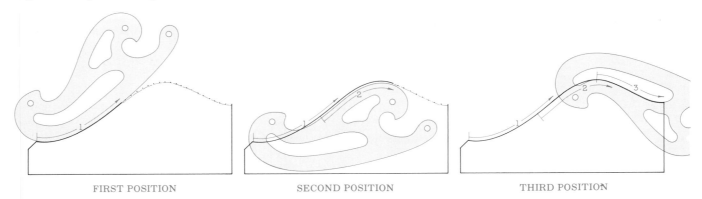

FIRST POSITION   SECOND POSITION   THIRD POSITION

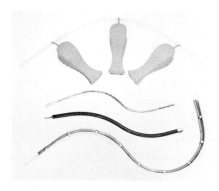

Fig. 3–55 Adjustable (flexible) curves are also used for drawing smooth, irregular curves. (*Bruning Division, Addressograph Multigraph Corp.*)

Fig. 3–56 Irregular curves may be drawn with spline held in position with lead weights. (*Keuffel & Esser Co.*)

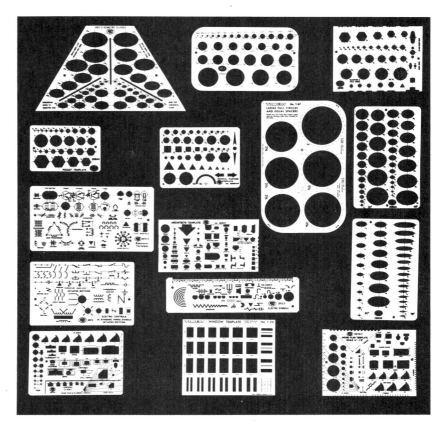

Fig. 3–57 Templates are made for all possible uses and save a lot of time. (*Teledyne Post.*)

Review

1. Explain how the accuracy of a T-square may be checked.

2. Name three kinds of drawing sheets.

3. Drafting pencils are made in 17 degrees of hardness from _____ (softest and blackest) to ___ (hardest).

4. The shape of a drafting pencil point may be _____ or _____.

5. How many widths, or thicknesses, of lines are generally used in drafting?

6. Right-handed draftsmen should draw horizontal lines in which direction? Left-handers?

7. In which direction should vertical lines be drawn?

8. Using the 45° and 30°–60° triangles with the T-square, angles in _____-degree intervals can be drawn.

9. What piece of drafting equipment may be used in place of a T-square and triangles?

10. Name three types of scales (measuring instruments) used in drafting.

11. The metric system is in common use in all countries except _____ and _____.

12. Using both English and metric dimensions on a drawing is called _____.

13. Another name for a French curve is _____.

14. How would a scale of one-half inch to the inch be designated on a drawing?

Problems for Chapter 3

Figs. 3–58 through 3–69 ASSIGNMENT: Make an instrument drawing of the figure assigned by your instructor. NOTE: Some dimensions have been deleted intentionally or placed incorrectly for instructional purposes. Include all center lines. Do not add dimensions unless instructed to do so.

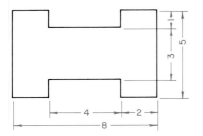

Fig. 3–58 Sheet-metal pattern.

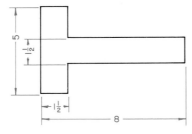

Fig. 3–59 Template for letter T.

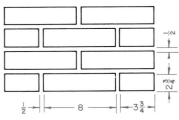

Fig. 3–60 Brick pattern.

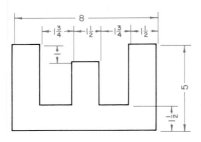

Fig. 3–61 Template for letter E.

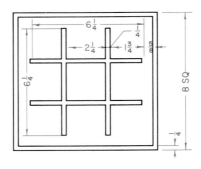

Fig. 3–62 Tic-tac-toe board.

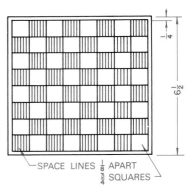

Fig. 3–63 Checker board.

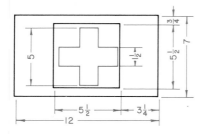

Fig. 3–64 First aid station sign.

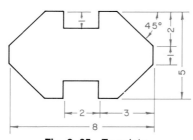

Fig. 3–65 Template.

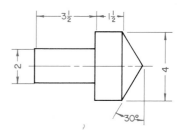

Fig. 3–66 Pivot.

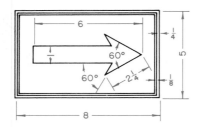

Fig. 3–67 Direction sign.

Fig. 3–68 International danger road sign.

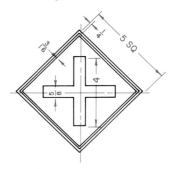

Fig. 3–69 Highway warning sign.

Figs. 3–70 through 3–81 ASSIGNMENT: Make an instrument drawing of the figure assigned by your instructor. NOTE: Some dimensions have been deleted intentionally or placed incorrectly for instructional purposes. Include all center lines. Do not add dimensions unless instructed to do so.

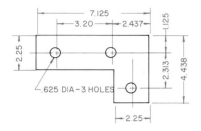

Fig. 3–70 Angle bracket.

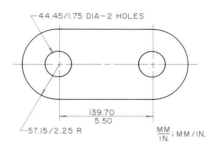

Fig. 3–71 Gasket.

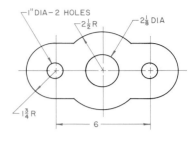

Fig. 3–72 Armature support.

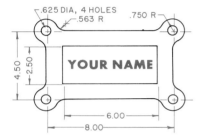

Fig. 3–73 Identification plate.

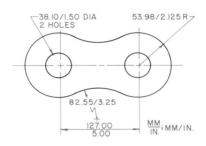

Fig. 3–74 Bicycle chain link.

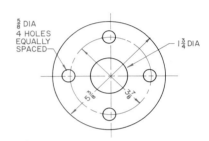

Fig. 3–75 Round gasket.

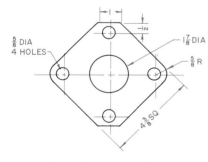

Fig. 3–76 Bronze shim.

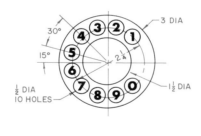

Fig. 3–77 Telephone dial.

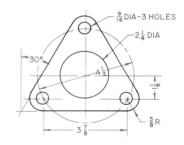

Fig. 3–78 Base plate.

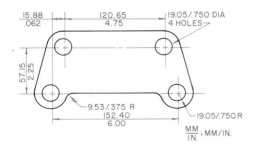

Fig. 3–79 Cover plate.

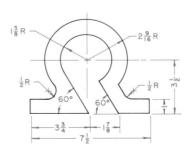

Fig. 3–80 Housing.

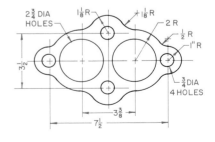

Fig. 3–81 Carburetor gasket.

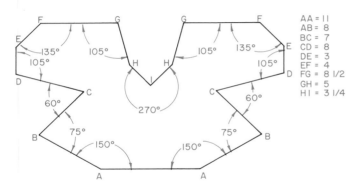

Fig. 3–82 Irregular polygon. Construct the irregular polygon as shown. Use a scale of ½″ = 1″. Begin by drawing line *AA* near the bottom of the sheet and centered horizontally. The length of each line is given at the right of the figure above. All angles may be drawn with the T-square and a combination of triangles.

AA = 11
AB = 8
BC = 7
CD = 8
DE = 3
EF = 4
FG = 8 1/2
GH = 5
HI = 3 1/4

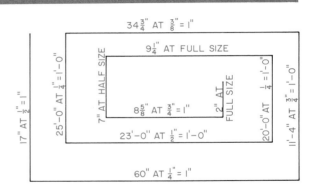

Fig. 3–83 Scale drawing. Draw a figure similar to the one shown above. Draw lines to the lengths and at the scales indicated.

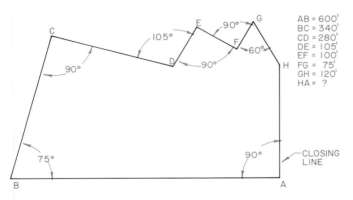

AB = 600′
BC = 340′
CD = 280′
DE = 105′
EF = 100′
FG = 75′
GH = 120′
HA = ?

Fig. 3–84 Civil engineer's scale. Draw a figure similar to the one shown above. Use a scale of 1″ = 40′-0″. Measure the length of the closing line to the nearest tenth of a foot and note it on your drawing.

Fig. 3–85 Measuring practice. Measure the lengths of lines *A* through *N* at full-size, ¾″ = 1″, ½″ = 1″, 1″ = 40′-0″, 1″ = 1′-0″, etc., as assigned by your instructor.

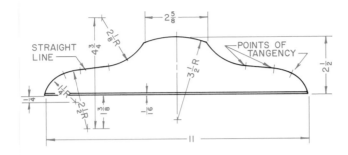

Fig. 3–86 T-square head. Draw to a scale of ¾″ = 1″. Be sure to locate points of tangency.

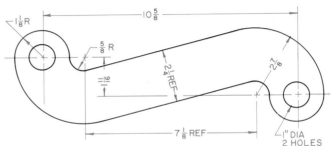

Fig. 3–87 Offset bracket. Locate all center points before beginning to draw circles and arcs.

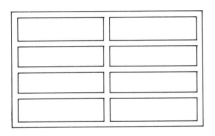

Fig. 3–88 Perforated sheet-metal design. Make drawing full size. Begin with a 5″ × 8″ rectangle in the center of the drawing space. Make all ribs ¼″ wide.

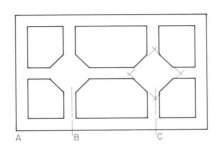

Fig. 3–89 Grill plate. Scale: full size or as assigned. Make all ribs ½″ wide. The distance from A to B is 2¼″; B to C is 3½″. The diamond shapes are 1½″ square.

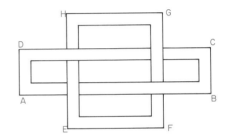

Fig. 3–90 Inlay. Scale: full size, half size, or as assigned. Rectangle *EFGH* is 8″ wide × 10″ high. All ribs are 1″ wide. Each rectangle is centered on the other.

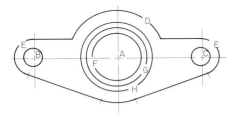

Fig. 3–91 Bearing bracket. Scale: full size or as assigned. The distance between the centers of circles B and C is 7″. The radius of arc D is 2″. The radius of E is ¾″. The diameter of F is 2″; G is 2½″; H is 3″. Be sure to locate carefully all centers of circles and points of tangency.

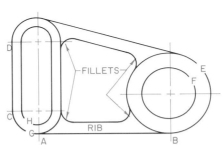

Fig. 3–92 Slide support. Scale: as assigned. Distance between center lines A and B is 11″; between C and D is 6″. Diameter of circle E is 7″; circle F is 4½″ Radius of fillets is 1″. Thickness of ribs is 1″. Radius of G is 2″; radius of H is 1¼″.

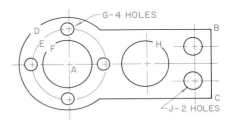

Fig. 3–93 Base gasket. Scale: as assigned. The distance between center point A and line BC is 4″. Line BC is 2″ long. Center point A to center of hole H is 2⅛″. Line BC to center of holes labeled J is ½″. Arc D is 1⅜″ R. Hole circle E is 2″ DIA. Circle F is 1⅜″ DIA. Four holes labeled G are ⅜″ DIA. Hole H is 1⁵⁄₁₆″ DIA. Two holes labeled J are ½″ DIA and are 1″ apart.

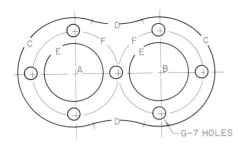

Fig. 3–94 Head gasket. Scale: as assigned. Use metric scale. The distance between center points A and B is 45.50 mm. Radius of arc C is 30 mm. Radius of arc D is 43 mm. Diameter of hole E is 32 mm. Diameter of circle F is 45.50 mm. Holes labeled G are 7 mm DIA.

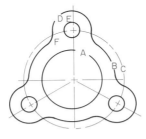

Fig. 3–95 Gasket. Scale: full size (decimal inches). Hole A = 3.125″. Arc B = 2.250″R. Hole circle C = 5.250″. Arc D = 1.00″R. Hole E = 0.750″ DIA, 3 holes. Arc F = 0.750″R. Locate all tangent points before darkening lines.

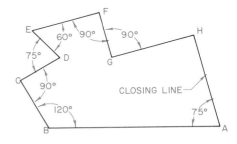

Fig. 3–96 Metric measurement. Scale: as assigned. Draw horizontal line *AB* 180 mm long. Work clockwise around the layout. *Remember!* Angular dimensions are the same in both the English and metric systems. *BC* = 60 mm. *CD* = 48 mm. *DE* = 42 mm. *EF* = 74 mm. *FG* = 50 mm. *GH* = 90 mm. Measure the closing line and mark it on your drawing. Also, measure and label the angle at *H*.

Geometry for Technical Drawing

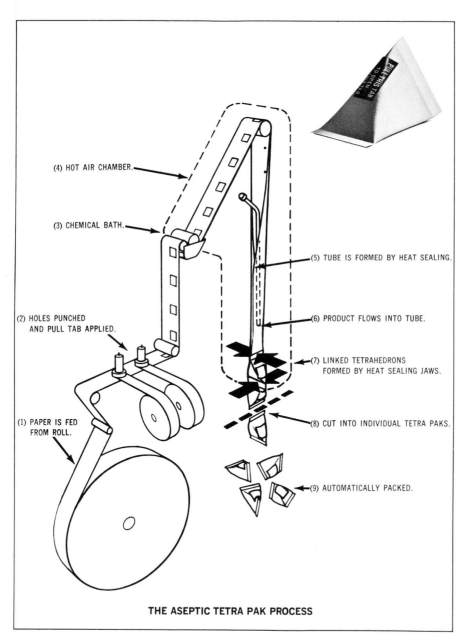

(4) HOT AIR CHAMBER.

(3) CHEMICAL BATH.

(5) TUBE IS FORMED BY HEAT SEALING.

(2) HOLES PUNCHED AND PULL TAB APPLIED.

(6) PRODUCT FLOWS INTO TUBE.

(7) LINKED TETRAHEDRONS FORMED BY HEAT SEALING JAWS.

(1) PAPER IS FED FROM ROLL.

(8) CUT INTO INDIVIDUAL TETRA PAKS.

(9) AUTOMATICALLY PACKED.

THE ASEPTIC TETRA PAK PROCESS

(BROOK HILL FARMS DIVISION, HUNT-WESSON FOODS.)

Importance of Geometry. Geometry has always been important to mankind. It was used in ancient times for surveying land and establishing right-angle (90°) corners for buildings and other structures. The Egyptians used men called *rope stretchers* (Fig. 4–1) for this purpose. A rope with marks or knots having 12 equal spaces was used. It was divided into 3-, 4-, and 5-space sections, as shown in Fig. 4–2. A square (right-angle) corner was established by stretching the rope and driving pegs into the ground at the 3-, 4-, and 5-space marks, as shown in Fig. 4–1. This was one early, practical use of geometry, and it is one of the most important concepts in drafting geometry even today.

The use of the 3-4-5 triangle for establishing a right angle was proved by the mathematician Pythagoras in the sixth century B.C. This proof is known as the Pythagorean theorem. The proof is easy to obtain graphically or mathematically. Stated in words: The sum of the square of each adjacent side of a right triangle is equal to the square of the hypotenuse. Stated in numbers: $3^2 + 4^2 = 5^2$.

$$
\begin{aligned}
3 \times 3 &= 9 \\
4 \times 4 &= 16 \\
\overline{5 \times 5} &= \overline{25} \\
25 &= 25
\end{aligned}
$$

This method works equally well for right triangles with the same propor-

Fig. 4-1 The Egyptian rope stretchers.

Fig. 4-2 Rope used by rope stretchers.

tions. For example, in a right triangle with sides of 6, 8, and 10 units, the sum of the squares of the adjacent sides will be equal to the square of the hypotenuse. The units may be inches, fractions of an inch, or units of the metric system—such as millimeters or centimeters.

The graphic proof of the Pythagorean theorem can be found in Fig. 4–3, the "dictionary of drafting geometry." It is further explained in Fig. 4–21.

Geometry is the study of the size and shape of things. It involves accurate measurements. It also involves the relationship of straight and curved lines in drawing figures. Some geometric figures used in drafting include circles, squares, triangles, hexagons, and octagons. Many other shapes and lines are shown in Fig. 4–3. Study the "dictionary of drafting geometry" (Fig. 4–3) before going on to the geometric constructions on the following pages. Be sure to refer back to the "dictionary of drafting geometry" as often as a review of geometric terms and figures is needed.

Geometric constructions consist of a number of individual lines and points placed in proper relationship to one another. Extreme accuracy in the measurement of lines and angles and the careful location of points is necessary to produce a precise geometric construction.

The geometric construction principles described in this chapter are essential to draftsmen, surveyors, engineers, architects, scientists, mathematicians, and designers. Geometric construction is important for making technical drawings and for solving technical problems by diagrams. Therefore, the student should become familiar with the more commonly used constructions explained in this chapter.

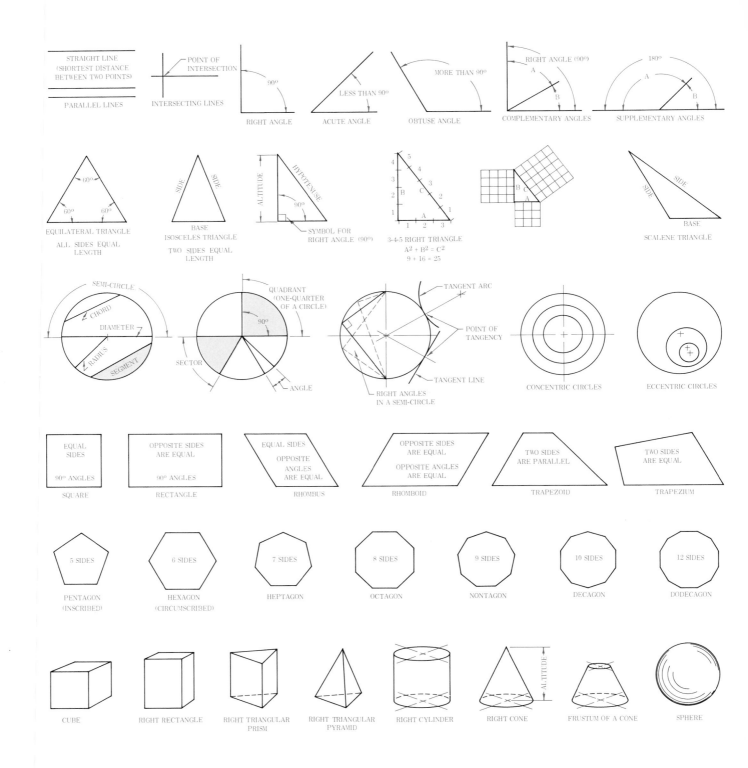

Fig. 4-3 Dictionary of drafting geometry.

Problem: Bisect line *AB* or arc *AB*. NOTE: "Bisect" means to divide into two equal parts.

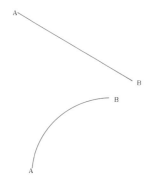

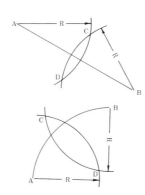

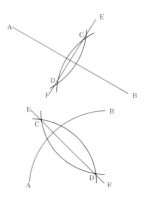

Given line *AB* or arc *AB*.

With *A* and *B* as centers and any radius *R* greater than one-half of *AB*, draw arcs to intersect at *C* and *D*. NOTE: "Intersect" means to cut across.

Draw line *EF* through intersections *C* and *D*.

Fig. 4-4 To bisect a straight line or arc.

Problem: Divide line *AB* into eight equal parts.

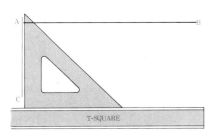

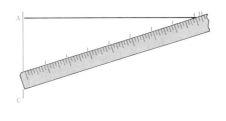

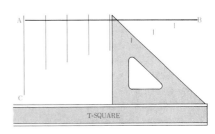

Draw a line of indefinite length at *A* perpendicular to line *AB*. NOTE: "Perpendicular" means at right angles, 90°.

Place scale with zero at point *B* and adjust along line *AC* until any eight equal divisions are included between points *B* and *C*. (In this case, eight inches.) Mark the divisions.

Draw lines parallel to *AC* through the division marks to intersect line *AB*.

Fig. 4-5 To divide a straight line into any number of equal parts (first method).

Problem: Divide line *AB* into five equal parts.

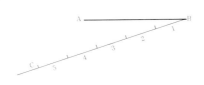

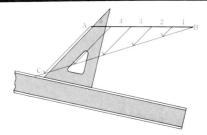

Draw line *BC* from point *B* at any convenient angle and of indefinite length.

Use dividers or a scale to step off five equal spaces on line *BC* beginning at point *B*.

Draw a line connecting points *A* and *C*. Draw lines through each point on *BC* parallel to line *AC* to intersect line *AB*.

Fig. 4-6 To divide a straight line into any number of equal parts (second method).

Problem: Erect a perpendicular at point *O* on line *AB*.

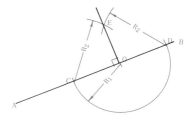

Given line *AB* and point *O*.

With *O* as the center and any convenient radius R_1, construct an arc cutting line *AB*, locating points *C* and *D*.

With *C* and *D* as centers and any radius R_2 greater than *OC*, draw arcs intersecting at *E*. Draw a line connecting points *E* and *O* to form the perpendicular.

Fig. 4–7   To erect a perpendicular to a given line at a given point on the line (first method).

Problem: Erect a perpendicular at *O* near the end of line *AB*.

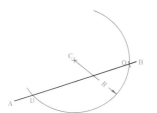

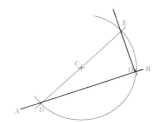

Given line *AB* and point *O*.

From any point *C* above line *AB*, construct an arc using *CO* as the radius and passing through line *AB* to locate point *D*.

Draw a line through points *D* and *C*, extending it through the arc to locate point *E*. Connect points *E* and *O* to form the perpendicular.

Fig. 4–8   To erect a perpendicular to a given line at a given point on the line (second method).

Problem: Erect a perpendicular to line *AB* through point *O*.

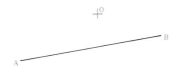

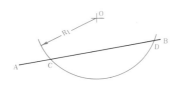

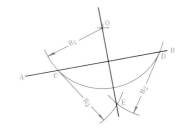

Given line *AB* and point *O*.

With *O* as the center, draw an arc with radius R_1 long enough to intersect line *AB* to locate points *C* and *D*.

With *C* and *D* as centers and radius R_2 greater than one-half of *CD*, draw intersecting arcs to locate point *E*. A line drawn through points *O* and *E* is the perpendicular.

Fig. 4–9   To erect a perpendicular to a given line through a point outside the line (first method).

Problem: Erect a perpendicular to line *AB* through point *O*.

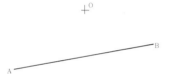

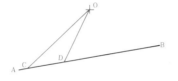

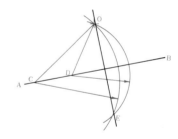

Given line *AB* and point *O*.

Draw lines from point *O* to any two points on line *AB*, locating points *C* and *D*.

With *C* and *D* as centers and *CO* and *DO* as radii, draw arcs to intersect, locating point *E*. Connect points *O* and *E* to form the perpendicular.

Fig. 4–10 To erect a perpendicular to a given line through a point outside the line (second method).

Problem: Erect a perpendicular to line *AB* through point *O*.

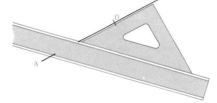

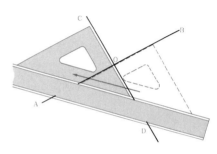

Given line *AB* and point *O*.

Place the T-square and triangle as shown.

Slide the triangle along the T-square until the edge aligns with point *O* on line *AB*. Draw the perpendicular *CD*.

Fig. 4–11 To erect a perpendicular to a given line through a point outside the line (third method).

Problem: Draw a line parallel to line *AB* through point *P*.

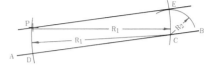

Given line *AB* and point *P*.

With point *P* as the center and any convenient radius R_1, construct an arc cutting line *AB* to locate point *C*. With point *C* as the center and the same radius R_1, construct an arc through point *P* and line *AB* to locate point *D*.

With *C* as the center and radius R_2 equal to chord *PD*, construct an arc to locate point *E*. Draw a line through points *P* and *E* which is parallel to *AB*.

Fig. 4–12 To construct a line parallel to a given line (first method).

Problem: Construct a line parallel to *AB* at a required distance from *AB*.

Given line *AB*.

Construct two arcs with centers at any location on line *AB* and with a radius *R* equal to the required distance between the parallel lines.

Draw the parallel line *CD* tangent to the arcs.

Fig. 4–13 To construct a line parallel to a given line (second method).

Problem: Construct a line parallel to *AB* through point *P*.

Given line *AB* and point *P*.

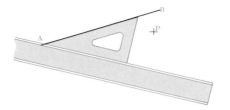

Place the T-square and triangle as shown.

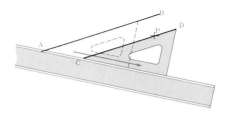

Slide the triangle until the edge aligns with point *P*. Draw the parallel line *CD*.

Fig. 4–14 To construct a line parallel to a given line (third method).

Problem: Construct a line to bisect angle *AOB*.

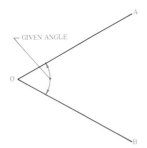

Given angle *AOB*.

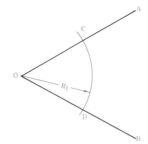

With point *O* as the center and any convenient radius R_1, draw an arc to intersect with *AO* and *OB* at *C* and *D*.

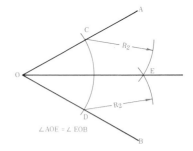

With *C* and *D* as centers and any radius R_2 greater than one-half of arc *CD*, draw arcs to intersect, locating point *E*. Draw a line through points *O* and *E* to bisect angle *AOB*.

Fig. 4–15 To bisect an angle.

Problem: Construct angle *AOB* in a new location.

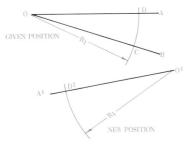

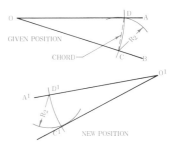

Given angle *AOB*.

Draw one side O^1A^1 in the new position. With *O* and O^1 as centers and any convenient radius R_1, construct arcs to cut *BO* and *AO* at *C* and *D* and A^1O^1 at D^1.

With D^1 as the center and radius R_2 equal to chord *DC*, draw an arc to locate point C^1 at the intersection of the two arcs. Draw a line through points O^1 and C^1 to complete the angle.

Fig. 4–16 To copy a given angle.

Problem: Construct an isosceles triangle. NOTE: An "isosceles" triangle has two equal sides.

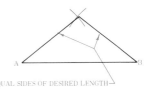

Given base line *AB*.

With points *A* and *B* as centers and a radius *R* equal to the desired length of the sides, construct intersecting arcs to locate the vertex (top point) of the triangle.

Draw lines through point *A* and the vertex and through point *B* and the vertex to complete the triangle.

Fig. 4–17 To draw an isosceles triangle.

Problem: Construct an equilateral triangle.

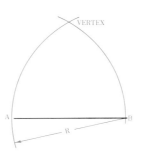

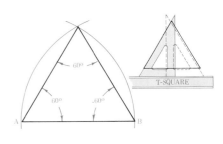

Given base line *AB*.

With points *A* and *B* as centers and a radius *R* equal to the length of line *AB*, draw intersecting arcs to locate the vertex.

Draw lines through point *A* and the vertex and through point *B* and the vertex to complete the triangle. NOTE: An equilateral triangle may also be constructed by drawing 60° lines through the ends of the base line with the 30°–60° triangle, as shown to the right.

Fig. 4–18 To draw an equilateral triangle.

Problem: Construct a right triangle with two sides given. NOTE: A right triangle has one right angle (90°).

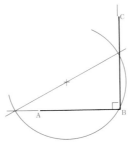

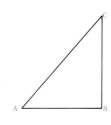

Given sides *AB* and *BC*.

Draw side *AB* in the desired position. Erect a perpendicular to *AB* at *B* equal to *BC*. NOTE: Use the method of Fig. 4–7 or the T-square and triangle method of Fig. 4–11 to construct the perpendicular.

Draw a line connecting points *A* and *C* to complete the right triangle. NOTE: Line *AC* is called the *hypotenuse*. It is always the side opposite the 90° angle.

Fig. 4–19 To draw a right triangle with two sides given.

Problem: Construct a right triangle with the hypotenuse and one side given.

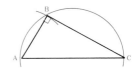

Given hypotenuse *AC* and side *AB*.

Draw the hypotenuse in the desired location. Draw a semicircle on *AC* using ½*AC* as the radius.

With point *A* as the center and a radius equal to side *AB*, draw an arc to cut the semicircle at *B*. Draw *AB* and then draw a line to connect *B* and *C* to complete the triangle.

Fig. 4–20 To draw a right triangle with the hypotenuse and one side given.

Problem: Construct a right triangle on a base line three units long.

Given base line *AB* three units long.

With *A* and *B* as centers and radii four and five units long, draw intersecting arcs to locate point *C*.

Draw lines *AC* and *BC* to complete the triangle.

Fig. 4–21 To draw a right triangle by the 3–4–5 method.

Problem: Construct a triangle with three sides given.

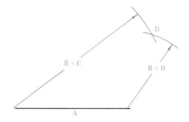

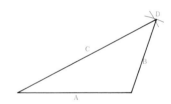

Given triangle sides *A*, *B*, and *C*.

Draw base line *A* in the desired location. Construct arcs from the ends of line *A* with radii equal to lines *B* and *C* to locate point *D*.

Connect the ends of line *A* with point *D* to complete the triangle.

Fig. 4-22 To draw a triangle with three sides given.

Problem: Draw a square within a circle with corners that touch the circle.

Problem: Draw a square outside the circle with the circle tangent to the midpoints of the sides of the square.

Problem: Construct a square.

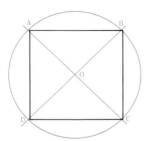

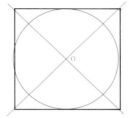

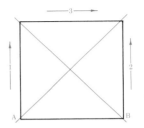

Given a circle with center point *O*. Construct 45° diagonals through the center point *O* to locate points *A*, *B*, *C*, and *D*. Connect points *A* and *B*, *B* and *C*, *C* and *D*, and *D* and *A* to complete the square.

Given a circle with center point *O*. Construct 45° diagonals through the center point *O*. Draw sides tangent to the circle, intersecting at the 45° diagonals to complete the square.

Given the length of the side *AB*. Construct 45° diagonals from ends of line *AB*. Complete the square by drawing the sides in the order shown by the numbered arrows.

Fig. 4-23 To draw a square within a circle.

Fig. 4-24 To draw a square about a circle.

Fig. 4-25 To draw a square.

Problem: Construct a regular pentagon.

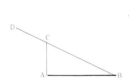

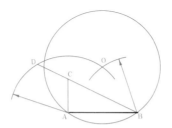

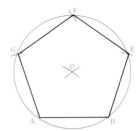

Given line *AB*, construct a perpendicular line *AC* equal to one-half of *AB*. Draw line *BC* and extend it to make line *CD* equal to *AC*.

With radius *AD* and points *A* and *B* as centers, draw intersecting arcs to locate point *O*. With the same radius and *O* as the center, draw a circle.

Step off *AB* as a chord to locate points *E*, *F*, and *G*. Connect the points to complete the pentagon.

Fig. 4-26 To construct a regular pentagon given one side.

Problem: Draw a regular pentagon in a given circle.

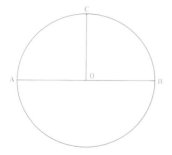 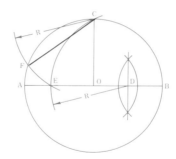 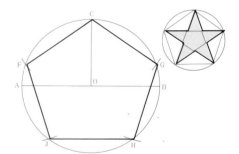

Given a circle with diameter *AB* and radius *OC*, draw a pentagon within the circle with points touching the circle.

Bisect radius *OB* to locate point *D*. With *D* as center and radius *DC*, draw an arc to locate point *E*. With *C* as center and radius *CE*, draw an arc to locate point *F*. Chord *CF* is one side of the pentagon.

Step off chord *CF* around the circle to locate points *G*, *H*, and *J*. Draw the chords to complete the pentagon. NOTE: Another method for constructing a pentagon in a circle is to use dividers and locate the points by trial-and-error, as shown to the right.

Fig. 4–27 To draw a regular pentagon in a given circle.

Problem: Construct a hexagon.

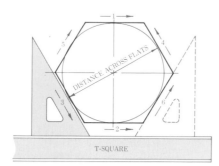

Problem: Construct a hexagon.

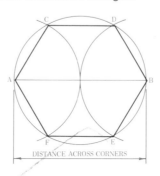

Problem: Construct a hexagon.

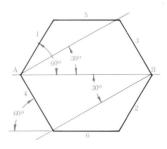

Given the distance across the flats of a regular hexagon. Draw center lines and a circle with the diameter equal to the distance across the flats. With the T-square and 30°–60° triangle, draw the tangents in the order indicated.

Given the distance *AB* across the corners. Draw a circle with *AB* as the diameter. With *A* and *B* as centers and the same radius, draw arcs to intersect the circle at points *C*, *D*, *E*, and *F*. Connect the points to complete the hexagon.

From points *A* and *B* draw lines of indefinite length at 30° to line *AB*. With the T-square and 30°–60° triangle, draw the sides of the hexagon in the order indicated.

Fig. 4–28 To draw a regular hexagon given the distance across the flats.

Fig. 4–29 To draw a regular hexagon given the distance across the corners (first method).

Fig. 4–30 To draw a regular hexagon given the distance across the corners (second method).

Problem: Construct an octagon outside a circle which touches the midpoints of each of its sides.

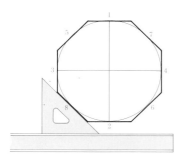

Given the distance across the flats, draw center lines and a circle with the diameter equal to the distance across the flats. With the T-square and 45° triangle, draw lines tangent to the circle in the order indicated to complete the octagon.

Fig. 4–31 To draw a regular octagon about a circle.

Problem: Construct an octagon within a circle so the corners touch the circle.

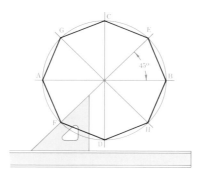

Given the distance across the corners, draw center lines *AB* and *CD* and a circle with the diameter equal to the distance across the corners. With the T-square and 45° triangle, draw diagonals *EF* and *GH.* Connect the points to complete the octagon.

Fig. 4–32 To draw a regular octagon within a circle.

Problem: Construct an octagon within a square.

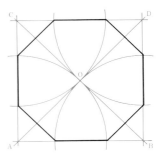

Given the distance across the flats, construct a square having the sides equal to *AB.* Draw diagonals *AD* and *BC* with their intersection at *O.* With *A, B, C,* and *D* as centers and radius *R = AO,* draw arcs to cut the sides of the square. Connect the points to complete the octagon.

Fig. 4–33 To draw a regular octagon in a square.

Problem: Construct a circle through points *A, B,* and *C.*

Given points *A, B,* and *C.* Draw lines *AB* and *BC.*

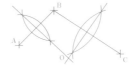

Draw perpendicular bisectors of *AB* and *BC* to intersect at point *O.*

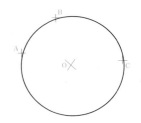

Draw the required circle with point *O* as the center, and radius *R = OA = OB = OC.*

Fig. 4–34 To construct a circle through any three points not in a straight line.

Problem: Draw a tangent line through point P on the circle.

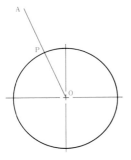 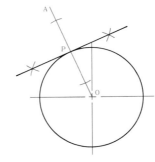

Problem: Draw a tangent line through point P on the circle.

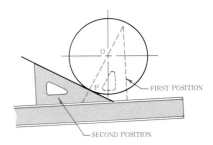

Given circle with center point O and tangent point P. Draw radial line OA to extend beyond the circle through point P.

Erect a perpendicular to OA at P. The perpendicular is the tangent line.

Given a circle with center point O and tangent point P. Place a T-square and triangle so that the hypotenuse of the triangle passes through points P and O (first position). Hold the T-square, turn the triangle to the second position at point P, and draw the tangent line.

Fig. 4–35 To draw a tangent to a circle at a given point P on the circle (first method).

Fig. 4–36 To draw a tangent to a circle at a given point P on the circle (second method).

Problem: Draw a line from point P tangent to the circle.

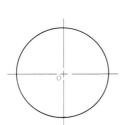 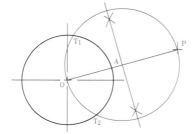 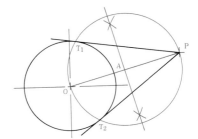

Given a circle with center point O and point P outside the circle.

Draw line OP and bisect it to locate point A. Draw a circle with center A and radius $R = AP = AO$ to locate tangent points T_1 and T_2.

Draw PT_1 and PT_2. These lines are tangent to the circle.

Fig. 4–37 To draw a tangent to a circle from a point outside the circle.

Problem: Construct an arc tangent to two straight lines.

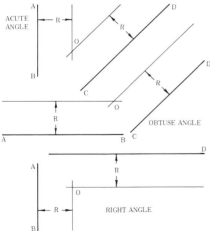

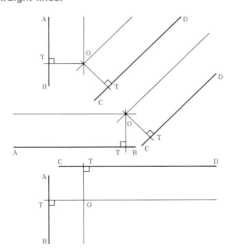

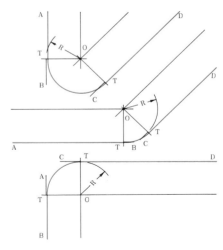

Given lines *AB* and *CD*. Draw lines parallel to *AB* and *CD* at a distance *R* from them on the inside of the angle. The intersection *O* will be the center of the required arc.

Draw perpendiculars from *O* to *AB* and *CD* to locate the points of tangency *T*.

With *O* as the center and radius *R*, draw the required arc.

Fig. 4–38 To construct an arc tangent to two straight lines at an acute angle, an obtuse angle, and a right angle.

Problem: Draw a reverse, or ogee, curve between two straight lines.

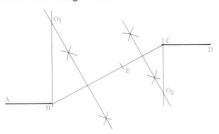

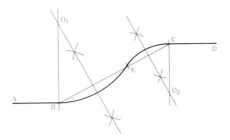

Given lines *AB* and *CD*, draw line *BC*. Select a point *E* on line *BC* through which the curve is to pass.

Draw perpendicular bisectors of *BE* and *EC*. Draw perpendiculars to *AB* at *B* and to *CD* at *C* to intersect the bisectors of *BE* and *EC* at O_1 and O_2, respectively.

Draw one arc with center O_1 and radius O_1E and the other with center O_2 and radius O_2E to complete the required curve.

Fig. 4–39 To draw a reverse, or ogee, curve.

Problem: Construct an arc tangent to two given arcs.

RADIUS OF TANGENT ARC

  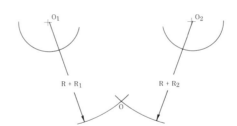 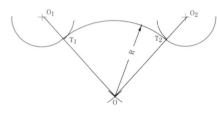

Given two arcs having radii R_1 and R_2 (radii may be equal or unequal) and radius R of the tangent arc.

Draw an arc with center O_1 and radius $= R + R_1$. Draw an arc with center O_2 and radius $= R + R_2$. The intersection at O is the required tangent arc center.

Draw lines O_1O and O_2O to locate tangent points T_1 and T_2. With point O as the center and radius R, draw the required arc.

Fig. 4–40 To draw an arc of a given radius tangent to two given arcs.

Problem: Draw a tangent line to the exterior of two circles.

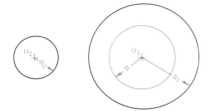 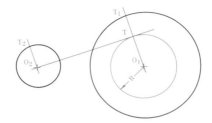 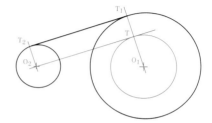

Given two circles with centers O_1 and O_2 and radii R_1 and R_2. With $R_1 - R_2$ as the radius R and point O_1 as the center, draw a circle.

From center point O_2 draw a tangent O_2T to the circle of radius R. Draw radius O_1T and extend it to locate tangent point T_1. Draw O_2T_2 parallel to O_1T_1.

Draw the required tangent T_1T_2 parallel to TO_2.

Fig. 4–41 To draw an exterior common tangent to two circles of unequal radii.

Problem: Draw a tangent line to the interior of two circles.

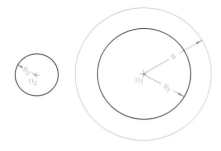 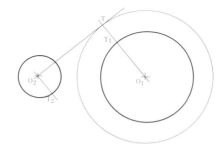 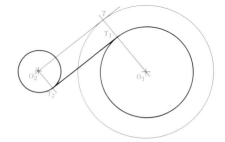

Given two circles with centers O_1 and O_2 and radii R_1 and R_2. With $R_1 + R_2$ as the radius R and point O_1 as the center, draw a circle.

From center point O_1, draw a tangent O_1T to the circle of radius R. Draw radius O_1T to locate tangent point T_1. Draw O_2T_2 parallel to O_1T.

Draw the required tangent T_1T_2 parallel to TO_2.

Fig. 4–42 To draw an interior common tangent to two circles of unequal radii.

Problem: Construct an arc tangent to line *AB* and arc *CD*.

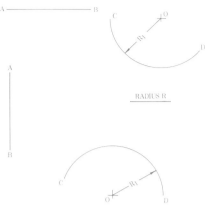

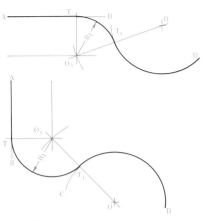

Given line *AB*, arc *CD*, and radius *R*.

Draw a parallel to *AB*, at distance *R*, toward arc *CD*. Use radius $R_1 + R$, locating O_1. A perpendicular O_1 to *AB* locates tangent point *T*.

Draw a line from *O* to O_1 to locate tangent point T_1 on *CD*. With point O_1 as the center and radius *R*, draw the tangent arc.

Fig. 4–43 To draw an arc of given radius tangent to an arc and a straight line.

Problem: Construct an ellipse by the pin-and-string method.

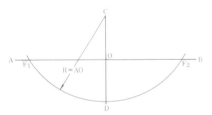

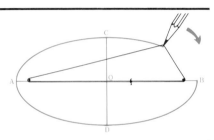

Given major axis *AB* and minor axis *CD* intersecting at *O*. With *C* as center and radius $R = AO$, an arc locates F_1 and F_2.

Place pins at points F_1, *C*, and F_2. Tie a string around the three pins and remove pin *C*.

Insert the point of a pencil in the loop and draw the ellipse. Keep the string taut when moving the pencil.

Fig. 4–44 To draw an ellipse by the pin-and-string method.

Problem: Construct an ellipse by the trammel method.

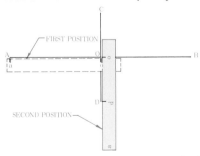

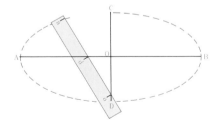

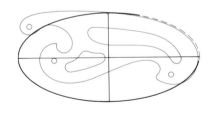

Given major axis *AB* and minor axis *CD* intersecting at point *O*. Cut a strip of paper or plastic (called a *trammel*). Mark off distances *AO* and *OD* on the trammel.

On the trammel, move point *o* along line *CD* (minor axis) and point *d* along line *AB* (major axis) and mark points at *a*.

Use a French curve or flexible curve to connect the points to draw the ellipse.

Fig. 4–45 To draw an ellipse by the trammel method.

Problem: Draw an ellipse using axes *AB* and *CD*.

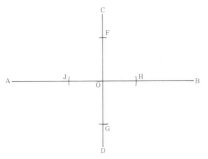

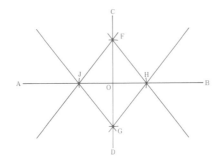

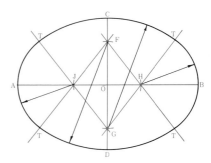

Given major axis *AB* and minor axis *CD* intersecting at point *O*. Lay off *OF* and *OG*, each equal to *AB* minus *CD*. Lay off *OJ* and *OH*, each equal to three-fourths of *OF*.

Draw and extend lines *GJ*, *GH*, *FJ*, and *FH*.

Draw arcs with centers *J* and *H* and radius *JA* and *HB* to complete the ellipse. The points of tangency are marked *T*.

Fig. 4–46 To draw an approximate ellipse when the minor axis is at least two-thirds the size of the major axis.

Problem: Draw an ellipse using axes *AB* and *CD*.

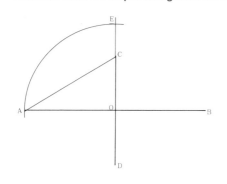

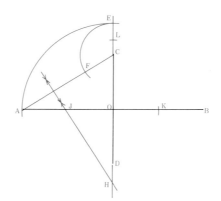

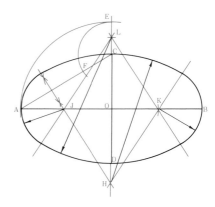

Given major axis *AB* and minor axis *CD* intersecting at point *O*. Draw line *AC*. Draw an arc with point *O* as the center and radius *OA* and extend line *CD* to locate point *E*.

Draw an arc with point *C* as the center and radius *CE* to locate point *F*. Draw the perpendicular bisector of *AF* to locate points *J* and *H*. Locate points *L* and *K*. *OL* = *OH* and *OK* = *OJ*.

Draw arcs with *J* and *K* as centers and radii *JA* and *KB*. Draw arcs with *H* and *L* as centers and radii *HC* and *LD* to complete the ellipse.

Fig. 4–47 To draw an approximate ellipse when the minor axis is less than two-thirds of the major axis.

Problem: Reduce or enlarge the drawing of the sailboat shown at A.

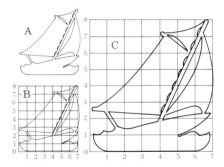

Lay out a grid over the drawing. Use squares of an appropriate size. Draw a larger or smaller grid on a separate sheet of paper. The size of the grid depends upon the desired amount of enlargement or reduction. Use dots to mark key points on the enlarged or reduced grid corresponding to points on the original drawing. Connect the points to complete the new drawing.

Fig. 4–48 To reduce or enlarge a drawing.

Problem: Change the proportion of the drawing shown at A.

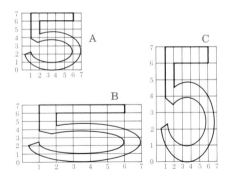

Draw a grid over the original drawing. Draw a grid on a separate sheet of paper in the desired proportion, as at B or C. Use dots to mark key points on the original drawing. Connect the points to complete the new drawing.

Fig. 4–49 To change the proportion of a drawing.

Problem: Enlarge or reduce the original rectangle.

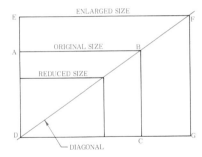

Draw a diagonal through corners D and B. Measure the desired width or height along line DC or DA (example: DG). Draw a perpendicular from that point (G) to the diagonal. Draw a line perpendicular to DE intersecting at point F. Reductions are made in the same way.

Fig. 4–50 To enlarge or reduce a square or rectangular area.

Review

1. What does bisect mean?

2. What is another name for perpendicular?

3. The radius of a circle is what part of the diameter?

4. What name is given to the distance around a circle?

5. Describe an isosceles triangle.

6. Describe an equilateral triangle.

7. What is an angle greater than 90° called?

8. What is an angle less than 90° called?

9. What is another name for a reverse curve?

10. Describe a right triangle.

11. A pentagon has how many sides?

12. Describe a perpendicular bisector.

13. The shape of an ellipse is determined by two axes. Name them.

14. Name three methods that may be used for drawing an ellipse.

Problems for Chapter 4

The problems for drafting geometry are designed to help the student develop accuracy in the use of instruments and to familiarize the student with the basic constructions that occur most frequently in drafting. Figures 4-51 through 4-70 involve the basic elements as described in the chapter; Figs. 4-71 through 4-85 are practical applications of the basic constructions. A complete understanding of the first group will increase speed and accuracy in the second group and in all types of drawing.

Suggestions

1. Make all construction (layout) lines thin and light.
2. Work accurately.
3. Do not erase construction lines.
4. Locate points by two short intersecting lines.
5. Show the exact length of lines by two short intersecting lines.

The first group of problems (Figs. 4-51 through 4-70) are prepared to be drawn within a rectangle that should be drawn four times the size shown in the book. Use dividers to pick up the dimensions from the illustration, and step off each measurement four times. This will result in a rectangle approximately $6\frac{3}{8}'' \times 4\frac{1}{2}''$. The number of constructions per drawing sheet will depend on the size of sheet used. The red reference lines may be used to locate points and lines within the layout and may show up as construction lines on the finished drawing.

Nearly all the Problems for Chapter 3 may be used as Problems for Chapter 4 and vice versa.

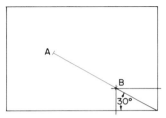

Fig. 4–51 Bisect line *AB*.

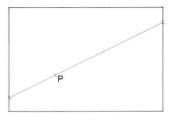

Fig. 4–52 Construct a perpendicular at point *P*.

Fig. 4–53 Divide line *AB* into line five equal parts.

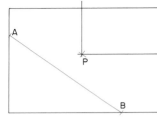

Fig. 4–54 Construct line *CD* parallel to *AB* and equal in length to *AB* through point *P*.

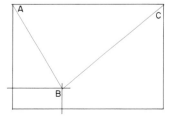

Fig. 4–55 Bisect angle *ABC*.

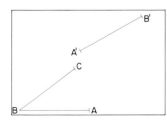

Fig. 4–56 Copy angle *ABC* in a new location, beginning with *A'B'*.

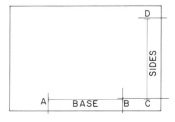

Fig. 4–57 Construct an isosceles triangle on base *AB* with sides equal to *CD*.

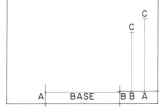

Fig. 4–58 Construct a triangle on base *AB* with sides equal to *BC* and *AC*.

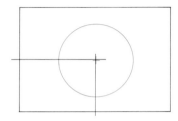

Fig. 4–59 Construct a square within the circle with corners touching the circle.

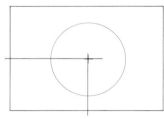

Fig. 4–60 Construct a regular pentagon within the circle.

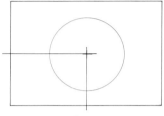

Fig. 4–61 Construct a regular hexagon around the circle.

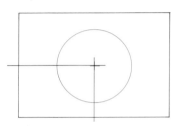

Fig. 4–62 Construct a regular octagon around the circle.

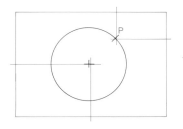

Fig. 4–63 Construct a tangent line through point *P* on the circle.

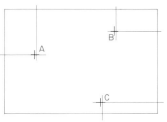

Fig. 4–64 Construct a circle through points *A, B,* and *C.*

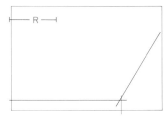

Fig. 4–65 Construct an arc having a radius *R* tangent to the two lines.

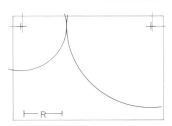

Fig. 4–66 Construct an arc having a radius *R* tangent to two given arcs.

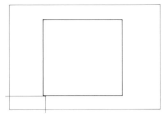

Fig. 4–67 Construct a regular octagon within the square.

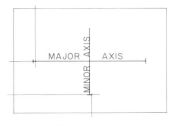

Fig. 4–68 Construct an ellipse on the major and minor axes. Use method assigned by the instructor.

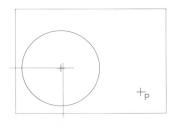

Fig. 4–69 Construct a line from point *P* tangent to the circle.

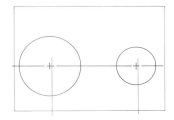

Fig. 4–70 Construct a tangent line to the exterior of the two circles.

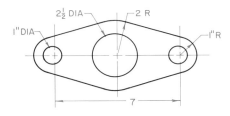

Fig. 4–71 Draw the view of the gasket full size or as assigned. Mark all points of tangency. Do not dimension.

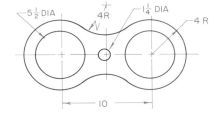

Fig. 4–72 Pipe support. Scale: as assigned. Locate and mark all centers and all points of tangency. Do not dimension.

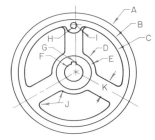

Fig. 4–73 Handwheel. Scale: as assigned. $A = 7''$ DIA, $B = 6\frac{1}{8}''$ DIA, $C = 2\frac{3}{4}''$ R, $D = 1\frac{1}{4}''$ R, $E = 2''$ DIA, $F = 1''$ DIA, G (keyway) $= \frac{3}{16}''$ wide $\times \frac{3}{32}''$ deep, $H = \frac{3}{8}''$ DIA, $I = \frac{3}{8}''$ R, $J = \frac{3}{16}''$ R, $K = 1''$.

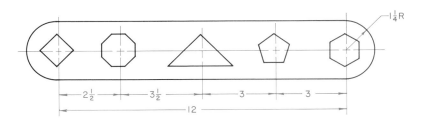

Fig. 4–74 Combination wrench. Scale: as assigned. Square: 1″; octagon: $1\frac{3}{8}''$ across flats; isosceles triangle: $2\frac{3}{4}''$ base, 2″ sides; pentagon: inscribed within $1\frac{3}{8}''$-diameter circle; hexagon: $1\frac{1}{4}''$ across flats. Use the method of your choice for constructing geometric shapes. Do not erase construction lines.

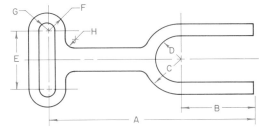

Fig. 4–75 Adjustable fork. Scale: as assigned. $A = 220$ mm, $B = 80$ mm, $C = 40$ mm, $D = 26$ mm, $E = 64$ mm, $F = 20$ mm, $G = 8$ mm, $H = 10$ mm.

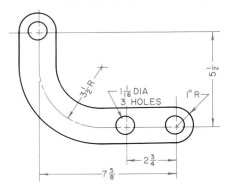

Fig. 4-76 Rod support. Scale: as assigned.

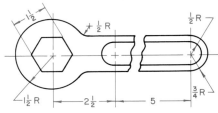

Fig. 4-77 Rocker arm. Scale: full size or as assigned.

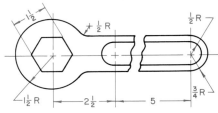

Fig. 4-78 Hex wrench. Scale: as assigned. Mark all tangent points. Do not erase construction lines.

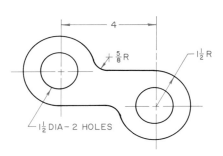

Fig. 4-79 Offset link. Scale: full size or as assigned. Locate and mark all points of tangency.

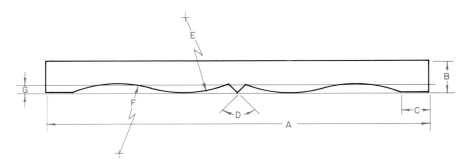

Fig. 4-80 Valance board. Scale: 1″ = 1′-0″ or as assigned. A = 8′-0″, B = 0′-8″, C = 0′-7″, D = 90°, E = 2′-6″, F = 2′-6″, G = 0′-2″. Locate and mark points of tangency. Do not erase construction lines.

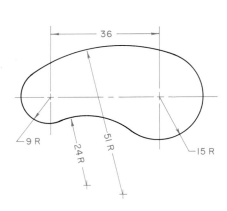

Fig. 4-81 Kidney-shaped table top. Scale: full size or as assigned.

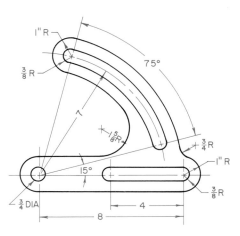

Fig. 4-82 Adjustable table support. Scale: as assigned.

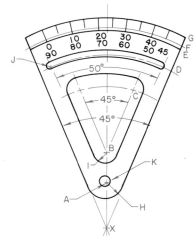

Fig. 4-83 Tilt scale. Scale: as assigned. AB = 1¾, AX = 2⅝, AC = 5½, AD = 7¼, AE = 8½, AF = 8¾, AG = 9¼, H = 1″ R, I = ⅝ R, J = 3/16 R, K = ½ DIA.

Multiview Drawing | Chapter 5

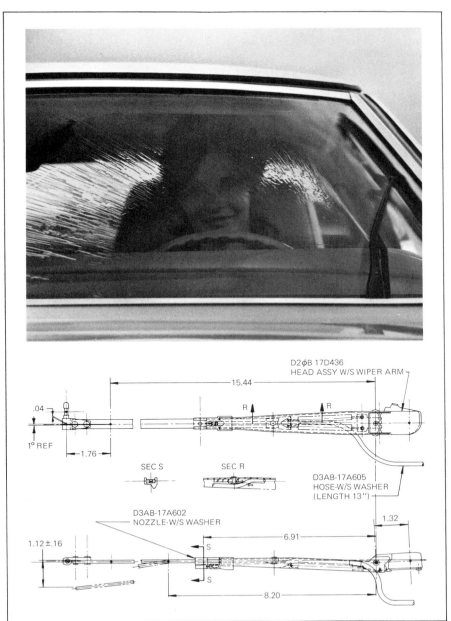

D2φB 17D436
HEAD ASSY W/S WIPER ARM

15.44

.04

1° REF

1.76

SEC S   SEC R

D3AB-17A605
HOSE-W/S WASHER
(LENGTH 13")

D3AB-17A602
NOZZLE-W/S WASHER

1.32

1.12 ±.16

6.91

8.20

(FORD MOTOR COMPANY.)

Introduction to Multiview Drawing. The role of technical drawing as a means of communicating ideas was described in Chapter 1. Man communicates by oral and written language and by graphic methods. One of the graphic methods is technical drawing. It is a universal language used and understood in all countries. Some of the similarities and differences in technical drawing between countries will be described in this chapter. The similarities are many; the differences, few. In all cases, where accurate visual understanding is necessary, technical drawing is the most exact method used.

Technical drawing involves two things: (1) *visualization* and (2) *implementation*. Visualization is the ability to see clearly what a machine, device, or other object looks like in the mind's eye. Implementation is the graphic representation (drawing) of the object which has been visualized. In other words, the designer, engineer, or draftsman first visualizes the object and then interprets it graphically by means of a technical drawing. In this way his idea is recorded in a form that can be used as a means of communication.

A technical drawing, properly prepared, will give a more accurate and clearer description of an object than a photograph or written analysis. In addition, a photograph cannot be made of something which does not exist. Technical drawings made according to standard principles result

89

in *views* which give an exact visual description of an object (Fig. 5–1).

Multiview Drawing. A photograph of a V-block is shown in Fig. 5–2. It shows the object as it appears to the eye. Notice that three sides of the V-block are shown in a single view. In Figure 5–3 the same photograph is shown with the three sides labeled according to their relative positions. If all sides could be shown in a single photograph, it would also include a left-side view, a rear view, and a bottom view. Nearly all objects have six sides.

Since an object cannot be photographed before it has been constructed, it becomes necessary to use another type of graphic representation. One possibility is to prepare a *pictorial drawing,* as shown in Fig. 5–4. It shows, just as a photograph, the general appearance of the object, but it does not show the exact forms and relations of the parts of the object. It shows the V-block as it appears, not as it really is. For example, the holes in the base appear as ellipses, not as true circles.

The problem, then, is to represent an object on a sheet of paper in a manner which will describe its exact shape and proportions. This is done by drawing views of the object as seen from different positions and by arranging these views systematically.

In order to accurately describe the shape of each view, one must imagine himself in a position directly in front of the object, then above it, and finally at the right side of it. This is where the ability to visualize is important. Figure 5–5 shows the exact shape of the V-block when viewed from the front. The dashed lines are used to show the outline of details behind the front surface (hidden details). Notice that the width and height of the object show on this view.

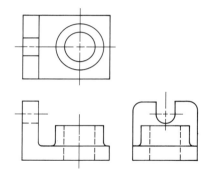

Fig. 5–1 This three-view drawing gives an accurate description of the object.

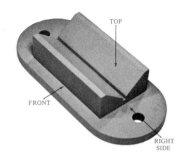

Fig. 5–3 Photograph of a V-block with front, top, and side views labeled.

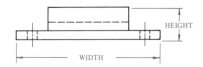

Fig. 5–5 Front view of V-block.

Figure 5–6 is a top view of the V-block. It shows the width and depth. Since the view is taken directly from above, the exact shape of the top is shown. Notice that the holes are true circles and that the rounded ends of the base are true radii. In the photograph and in the pictorial drawing, these appeared as elliptical shapes.

Figure 5–7 is a right-side view of the V-block. It shows the depth and height. Notice that the shape of the

Fig. 5–2 Photograph of a V-block.

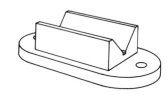

Fig. 5–4 Pictorial drawing of V-block.

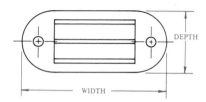

Fig. 5–6 Top view of V-block.

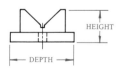

Fig. 5–7 Right-side view of V-block.

V appears to be balanced on the drawing, while it appears distorted on the photograph and on the pictorial drawing.

The relationship of views. In order for everyone to read and understand technical drawings consisting of various views, the views must be placed in proper relationship to one another. Figure 5–8 shows the V-block and how its three normal views have been revolved into their proper places. Notice that the top

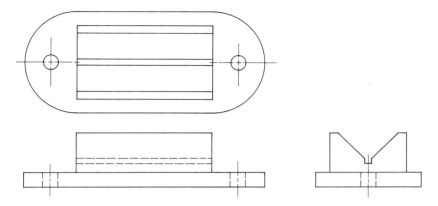

Fig. 5-8 The relationship of the three normal views of the V-block.

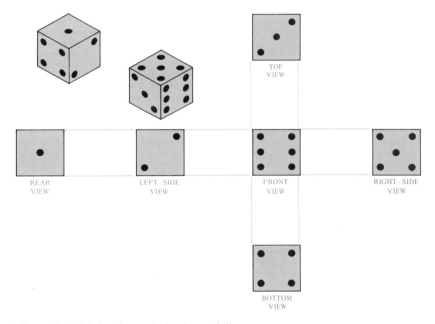

Fig. 5-9 Pictorial drawing and six views of dice.

view is directly above the front view and that the right-side view is directly to the right of the front view. Each of the normal views is where it logically belongs. When the normal views are placed in proper relationship to one another, a *multiview* drawing results. *Multiview drawing is the exact representation of two or more views of an object on a flat surface (plane).* These three views will usually give a complete description of the object. Figure 5-8 is an example of

a multiview drawing with views in their proper relationship.

Other views. It was stated earlier that most objects have six sides or six views. In most cases two or three views will completely describe the shape and size of all parts of an object. However, in some cases it may be necessary to show views other than the front, top, and right side. In Fig. 5-9 dice are shown pictorially in the upper left-hand corner. Since the detail is different on each of the six

sides, six views are required for a complete graphic description. The six views are shown in their proper locations in the lower part of Fig. 5-9. Only in very unusual cases will six views be necessary.

Orthographic Projection. It was stated earlier that multiview drawing is the exact representation of two or more views of an object on a flat surface (plane). The development of these views is based upon the principles of orthographic projection. *Ortho-* means "straight or at right angles," and *-graphic* means "written or drawn." *Projection* comes from two old Latin words, "pro," meaning "forward," and *"jacere,"* meaning "to throw." Thus, orthographic projection literally means "thrown forward, drawn at right angles." *Orthographic projection is the method of representing the exact form of an object in two or more views on planes, generally at right angles to each other, by perpendiculars from the object to the planes.*

Angles of projection. Orthographic projection involves the use of three planes. They include the vertical plane, the horizontal plane, and the profile plane. These are shown in Fig. 5-10. In technical drawing, a plane is an imaginary flat surface, having no thickness, upon which a view of an object is projected and drawn. Notice that the vertical and horizontal planes divide space into four quadrants (quarters of a circle). In orthographic projection, quadrants are usually called angles. Thus, we get the names *first-angle projection* and *third-angle projection.* First-angle projection is used in European countries. Third-angle projection is used in the United States and Canada. Second- and fourth-angle projections are not used in technical drawing.

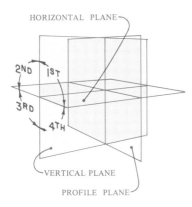

Fig. 5–10 The three planes used in orthographic projection.

Fig. 5–11 The position of the three planes used in first-angle projection.

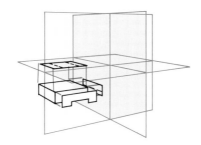

Fig. 5–13 The position of the three planes used in third-angle projection.

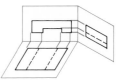

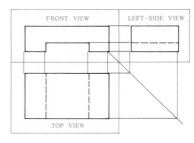

Fig. 5–12 Three views in first-angle projection.

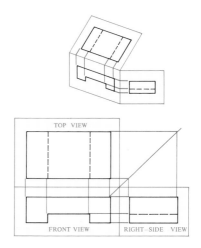

Fig. 5–14 Three views in third-angle projection.

First-angle projection. Figure 5–11 shows an object within the three planes of the first quadrant for developing a three-view drawing in first-angle projection. The front view is projected to the vertical plane. The top view is projected to the horizontal plane. The left-side view is projected to the profile plane. When the horizontal and profile planes are rotated into a single plane, the front view is above the top view, and the left-side view is to the right of the front view (Fig. 5–12).

Third-angle projection. Third-angle projection uses the same basic principles as first-angle. The main difference is in the relative positions of the three planes. Figure 5–13 shows the same object placed within the third quadrant for developing a three-view drawing in third-angle projection. In this case the front view is projected to the vertical plane. The top view is projected to the horizontal plane. The right-side view is projected to the profile plane. When the horizontal and profile planes are rotated into a single plane, the top view is above the front view, and the right-side view is to the right side of the front view (Fig. 5–14).

The Glass Box. The three views of an object have been developed by using imaginary transparent planes and projecting the views onto them. Yet it was mentioned earlier that most objects have six sides, and therefore, six views may result. In order to understand the theory of projecting all six views, an imaginary glass or other transparent box will be used.

Figure 5–15 shows the glass box partially opened with the six views labeled. When fully opened up into one plane (Fig. 5–16), the views take their relative positions as they would be drawn on paper. These views are arranged according to accepted practice for the six views. Notice that the rear view is located to the left of the left-side view.

Also notice that some views give the same information contained in other views. They may also be, for practical purposes, mirror images of one another. Thus, it is not necessary to show all six views for a complete description of the object. The three normal views, as ordinarily drawn, are shown in Fig. 5–17.

Hidden Lines. Since it is necessary to describe every part of an object, everything must be represented, whether it can be seen or not. Inte-

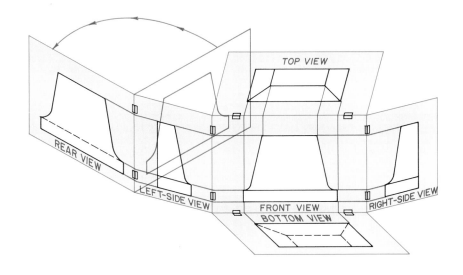

Fig. 5–15 Opening the glass box.

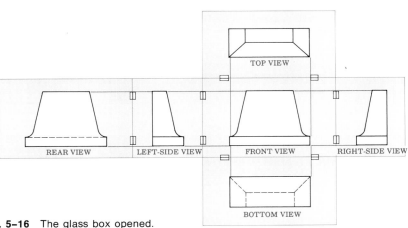

Fig. 5–16 The glass box opened.

Fig. 5–17 The front, top, and right-side views.

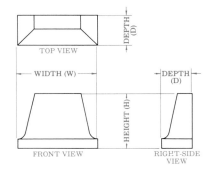

Fig. 5–18 Hidden lines.

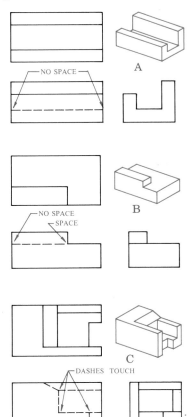

rior features are projected in the same way as for exterior features. The only difference is that parts which cannot be seen in the views are represented by hidden lines composed of short dashes (Fig. 5–18). Notice that the first dash of a hidden line touches the line at which it starts (Fig. 5–18 at A). If a hidden line is a continuation of a visible line, space is left between the visible line and the first dash of the hidden line, as at B. If the hidden lines show corners, the dashes touch at the corners, as at C.

Dashes for hidden arcs (Fig. 5–19 at A) start and end at the tangent points. When a hidden arc is tangent to a visible line, a space is left, as at B. When a hidden line and a visible line project at the same place, show the visible line (Fig. 5–19 at C). When a center line and a hidden line project at the same place (Fig. 5–20 at A), draw the hidden line. When a hidden line crosses a visible line (Fig. 5–20 at B), do not cross the visible line with a dash. When hidden lines cross (Fig. 5–20 at C), the nearest hidden line has the "right of way" and spaces are left as shown.

Center Lines. Center lines are used to locate views and dimensions. (See the alphabet of lines, Fig. 3–13.) Primary center lines, marked *P* in Fig. 5-21, locate the center on symmetrical views where one part is a mirror image of another part. Primary center lines are also used as major locating lines to aid in constructing the views and as base lines for dimensioning. Secondary center lines, marked *S* in Fig. 5-21, are axes for details of a part or construction. Primary center lines are, therefore, the first lines to be drawn, and the views are developed from them. Note that center lines represent the axes of cylinders in the side view and that the centers of circles or arcs are located first so that measurements can be made from them to locate the lines on the various views. A hidden line is shown in preference to a center line in Fig. 5–22.

Curved Surfaces. The fact that some curved surfaces, such as cylinders and cones, do not show as curves in all views is illustrated in Fig. 5–23. A cylinder with its axis (center line) perpendicular to a plane will show as a circle on that plane and as a rectangle on the other two planes. Three views of a cylinder when placed

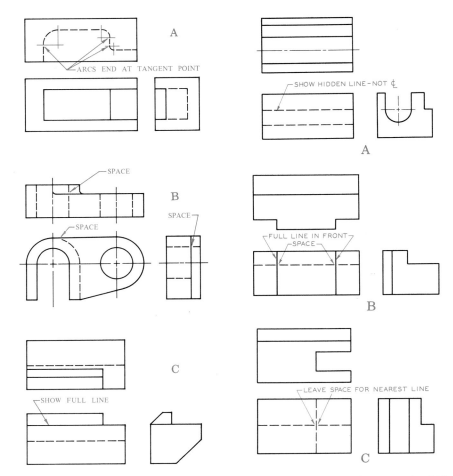

Fig. 5–19 Hidden arcs.

Fig. 5–20 Technique of representing hidden and visible lines.

Fig. 5–21 Center lines.

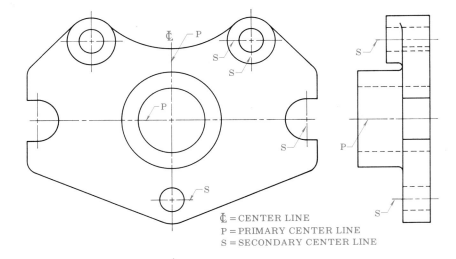

ℂ = CENTER LINE
P = PRIMARY CENTER LINE
S = SECONDARY CENTER LINE

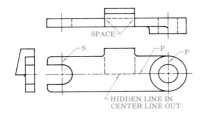

Fig. 5-22 Center lines and hidden lines.

in different positions are shown at B, C, and D. The holes may be considered as negative cylinders. A cone appears as a circle in one view and as a triangle in the others, as shown at E. For a frustum of a cone one view appears as two circles, as at F. In the top view the conical surface is represented by the space between the two circles.

Cylinders, cones, and frustums of cones have single curved surfaces and are represented by circles in one view and straight lines in the other. The

handles in Fig. 5-24 at A have double curved surfaces which are represented by curves in both views. The ball handle has spherical ends, and both views of the ends are circles because a sphere appears as a circle when viewed in any direction. The slotted link in Fig. 5-24 at B and C is an example of tangent plane and curved surfaces. Since the rounded ends are tangent to the sides of the link and the ends of the slot are tangent to the sides, the surfaces are smooth and there is *no* line of separation.

What Views to Draw. As already mentioned, six views are usually not needed to describe most objects. Three views are sufficient. The six views explain the theory of making drawings, but it is not necessary to draw them in order to tell which views are needed. The general characteristics of an object indicate the views required to describe its shape. Three

properly selected views will describe most shapes, but sometimes there are features that will be more clearly described by using more views or parts of extra views.

Most pieces have a characteristic view by which they can be recognized. This is the first view to consider and generally the first view to draw. Next, consider the normal position of the part when it is in use. This is often desirable but not always necessary. For example, tall parts such as vertical shafts are more readily drawn in a horizontal position; then, views with the fewest hidden lines are easiest to read and take much less time to lay out and draw.

The practical purpose for drawing views is to describe the shape of something. More views than necessary are a waste of time in making the drawing, as well as a waste of time in reading, because all views have to be read (the reader assumes they are neces-

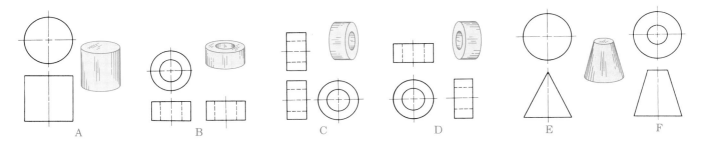

Fig. 5-23 Curved surfaces. Cylinders and cones.

Fig. 5-24 Curved surfaces.

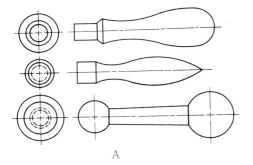

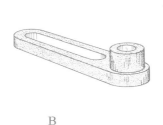

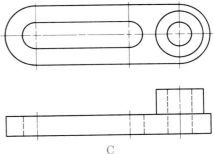

sary and has to be sure). Turned parts, such as the handles shown at A in Fig. 5–24, and sheet material, plywood, plate material, and parts of uniform thickness, such as the latch and the stamping in Fig. 5–25 at A and B, can often be described in one view. For the handles in Fig. 5–24 at A give the diameter; for the latch or the stamping in Fig. 5–25 at A and B give the thickness. Parts such as the bushing shown at D and the sleeve shown at E could be, and often are, shown in one view by dimensions for the diameters marked DIA, as indicated. The two views of the bushing at C are not necessary, as shown at D.

There are many things which can be described in two views, as indicated in Fig. 5–26. When two views are used, they must be carefully selected so as to be sure to describe the shape of the object. For the parts at A and B in Fig. 5–26 there would be no question; for the part at A the top view and either front or side view would be sufficient. For the placer cone at C a third view would add nothing to the description. There should be no question about the selection of views for the guide at D or for the wedge cam at E. Figure 5–27 shows some objects which can be described in two views. The top and front views at A and B are the same. Since the side views are necessary, the front and side views would be sufficient. At C and D the top views are necessary, so the top and front views would be sufficient. At E and F the front views are necessary so the front and top views or the front and side views would be sufficient. Some things, such as the angle in Fig. 5–28, require three views, since it is necessary to describe shapes in each of the views.

Six views are shown for the sliding base in Fig. 5–29. A study of the picture and the views will show that the top, front, and right-side views will give the best shape description and have the fewest hidden lines. The six views are shown here simply to illustrate and explain the selection of views. Only the necessary views would be drawn in practice.

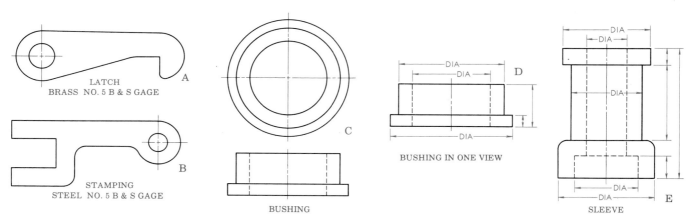

Fig. 5–25 One-view drawings.

Fig. 5–26 Two-view drawings.

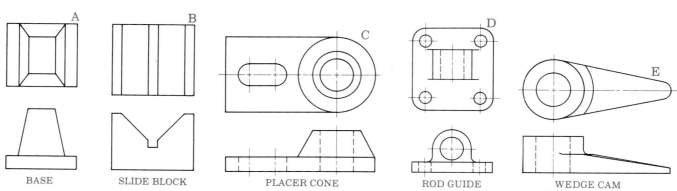

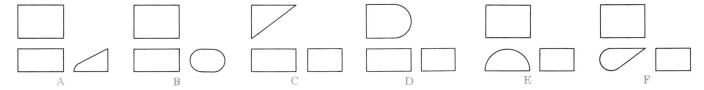

Fig. 5-27 Selection of two views.

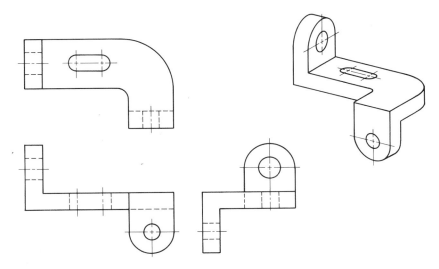

Fig. 5-28 Three-view drawings.

Fig. 5-29 Choice of views.

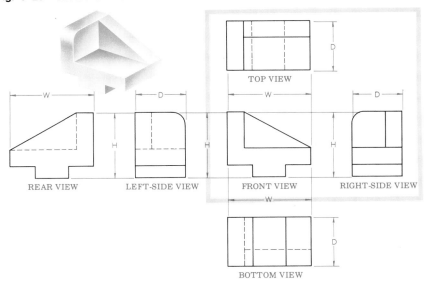

REAR VIEW   LEFT-SIDE VIEW   FRONT VIEW   RIGHT-SIDE VIEW

TOP VIEW

BOTTOM VIEW

Second Position of the Side View.
The proportions of an object or the size of the sheet sometimes makes it desirable to show a side view in the second position, or directly across from the top view, as in Fig. 5-30. This position is obtained by revolving the side plane about its intersection with the top plane.

Visualization and careful thought about a "mind's-eye picture" of an object will help decide which views should be drawn to describe its shape.

Placing Views. When drawings are made from the actual part, or for parts yet to be made, it is necessary to be able to place the views so that they will go into the space available on the size of sheet used. This is done by considering the space necessary for each view and comparing the total space required. The size of the drawing sheet selected should provide for views that will give a clear description of the part. The working space may be as suggested in Fig. 5-31 or as specified by the instructor. The method of working out the positions of the views is the same for any space.

A working space of $10\frac{1}{2}'' \times 7''$ is used to explain how to place the views of the slide stop in Fig. 5-32 at A. The overall dimensions are: width = $W = 5\frac{1}{4}''$; depth = $D = 1\frac{3}{4}''$; height = $H = 3''$.

In Fig. 5-32 at A, a pictorial drawing of a slide stop is shown with its overall width, height, and depth dimensions added. Some simple arithmetic is needed to properly place the three normal views. It may also be helpful

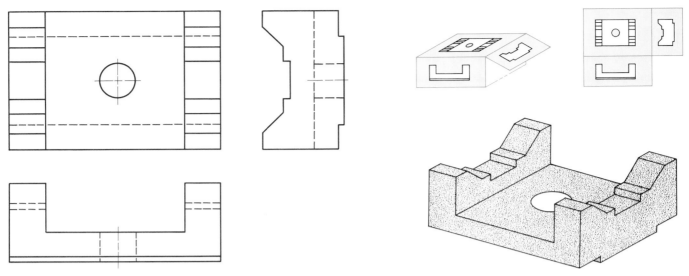

Fig. 5–30 Second position of the side view.

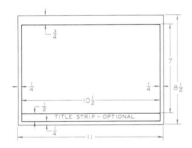

Fig. 5–31 Sheet layout.

to make a rough layout on scrap paper, as shown at B. This layout need not be made to scale. The calculations may be made on a separate scrap of paper or simply on the side of the sketch sheet.

The width, depth, and height dimensions are given in red on the sketch (Fig. 5–32 at B). The dimensions in blue indicate the spacing at the top, bottom, side, and between views. Use the following procedure to determine spacing. Refer to Fig. 5–32.

A. Add the width, $5\frac{1}{4}$ in., and the depth, $1\frac{3}{4}$ in.: $5\frac{1}{4} + 1\frac{3}{4} = 7$. Subtract 7 in. from the width of the drawing space, $10\frac{1}{2}$ in.: $10\frac{1}{2} - 7 = 3\frac{1}{2}$.

The remaining $3\frac{1}{2}$ in. is the amount left for the space at the left, right, and between views. If a space of about 1 in. is used between the front and side views, it will allow $1\frac{1}{4}$ in. on the left and right sides. These spaces may be more or less, depending upon the shapes of the views, the space available, and the space needed for dimensions and notes when added.

B. Next, add the height, 3 in., and the depth $1\frac{3}{4}$ in.: $3 + 1\frac{3}{4} = 4\frac{3}{4}$. Subtract $4\frac{3}{4}$ from the height of the drawing space, 7 in.: $7 - 4\frac{3}{4} = 2\frac{1}{4}$. The remaining $2\frac{1}{4}$ in. is the amount left for the space at the bottom, top, and between views. If a space of $\frac{3}{4}$ in. is used between the front and top views, it will allow $1\frac{1}{2}$ in. for spaces above and below the views. These could be $\frac{3}{4}$ in. each, but a better visual balance will result if $\frac{7}{8}$ in. is used below and $\frac{5}{8}$ in. above.

Once all calculations are made, proceed with the layout on the final drawing sheet, as shown at C in Fig. 5–32. Notice that the views are blocked in with light construction lines until all details have been

added. Figure 5–32 at D shows all necessary visible, hidden, and center lines darkened.

Figures 5–33 and 5–34 show the same procedure being used for a two-view drawing. Whether the views are arranged as in Fig. 5–33 at D or as in Fig. 5–34 at D, and regardless of the number of views, the basic procedure does not change.

Locating Measurements. After lines have been drawn to locate the views, it is necessary to make measurements for details and to draw the views (Fig. 5–35). Measurements made on one view can be transferred to another to save the time of making them again and to ensure accuracy and correctness. Distances in the three directions, width W, height H, and depth D, can be easily transferred, as illustrated in Fig. 5–35.

A. Width, W (horizontal), measurements made on the front view can be located on the top view by drawing up from the front view. In like manner, measurements can be projected down from the top view to the front view.

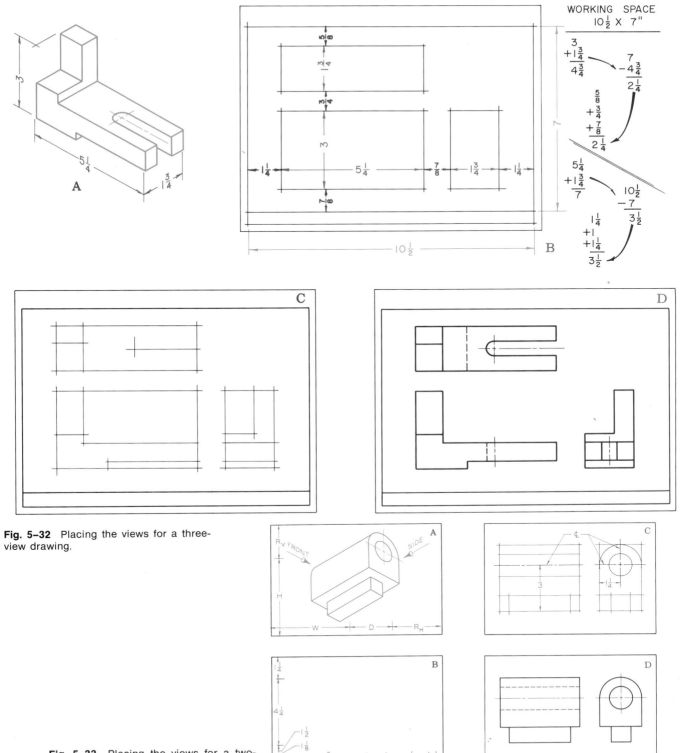

WORKING SPACE
10½ X 7"

$$\frac{3}{+1\frac{3}{4}}{4\frac{3}{4}} \qquad \frac{7}{-4\frac{3}{4}}{2\frac{1}{4}}$$

$$\frac{\frac{5}{8}}{+\frac{3}{4}}{+\frac{7}{8}}{2\frac{1}{4}}$$

$$\frac{5\frac{1}{4}}{+1\frac{3}{4}}{7} \qquad \frac{10\frac{1}{2}}{-7}{3\frac{1}{2}}$$

$$\frac{1\frac{1}{4}}{+1}{+1\frac{1}{4}}{3\frac{1}{2}}$$

A

B

C

D

Fig. 5–32 Placing the views for a three-view drawing.

A

B

C

D

Fig. 5–33 Placing the views for a two-view drawing.

MULTIVIEW DRAWING   99

B. Height, *H* (vertical), measurements on the front view can be located on the side view by drawing a light line across to the side view. Measurements can also be projected to the front view from the side view.

C. Depth, *D,* measurements show as vertical distances in the top view and as horizontal distances in the side view. Such measurements can be taken from the top view to the side view by drawing arcs from a center *O* (at C), by using a 45° triangle through *O* (at D), by using the dividers (at E), or by using the scale as shown at F.

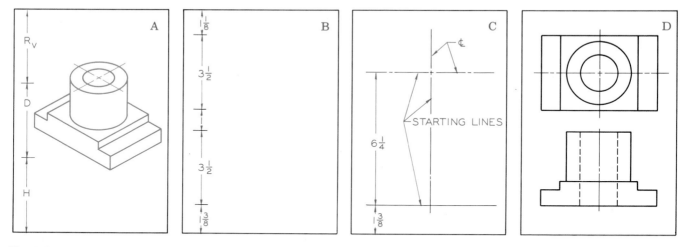

Fig. 5-34 Placing the views for a two-view drawing.

Fig. 5-35 Locating measurements.

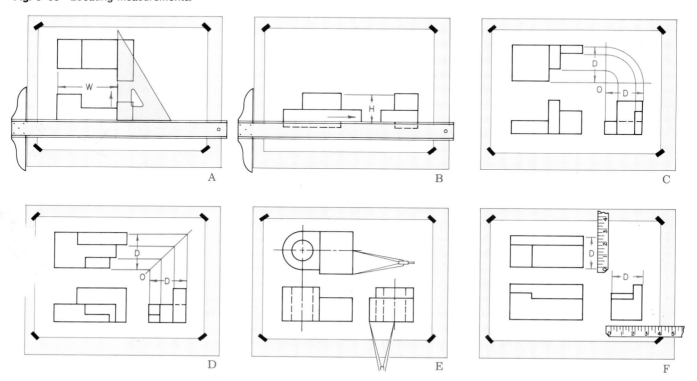

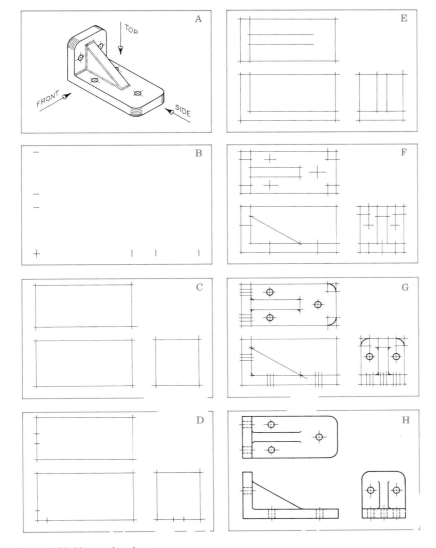

Fig. 5-36 Making a drawing.

3. Locate the views (at B).

4. Block in the views with light, thin lines (at C).

5. Lay off the principal measurements (at D).

6. Draw the principal lines (at E).

7. Lay off the measurements for details (at F; centers for arcs, circles, and triangular ribs).

8. Draw the circles and arcs (at G).

9. Draw any additional lines needed to complete the views.

10. Brighten the lines where necessary to make them sharp and black and of the proper thickness (at H).

Review

1. Describe multiview drawing.

2. Most objects have six sides or six views. Name them.

3. There are four angles of projection. Which two are used in technical drawing? Which one is used in the United States?

4. What type of line is used to represent interior details not visible on the outside of an object?

5. Name the two types of center lines.

6. How many views will usually be needed to completely describe an object having uniform thickness throughout?

7. How many views does a cylindrical shaped object usually require?

8. If a center line and a hidden line fall in the same place, which takes preference?

9. What determines the amount of space needed between views?

10. Which type of drawing gives the most accurate shape description of an object?

To Make a Drawing. A systematic method of working should be followed to ensure accuracy and understanding (Fig. 5-36). All views should be carried along together. Do not attempt to finish one view before starting the others. Use a hard pencil (4H or 6H) and light, thin lines for preliminary lines. Use a soft pencil (F, HB, or H) for final lines. The grade of pencil depends, to a certain extent, upon the surface of the paper, cloth, or film used. The following order of working is suggested:

1. Consider the characteristic view (at A; the front view).

2. Determine the number of views (at A; three views needed).

Problems for Chapter 5

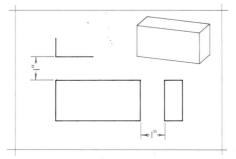

Fig. 5–37 Sanding block. A pictorial and two views of a sanding block are given. Draw full size the two views shown and complete the third (top) view. Do not draw the pictorial view. The block is $\frac{3}{4}$" thick, $1\frac{3}{4}$" wide, and $3\frac{1}{2}$" long.

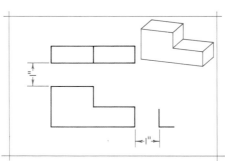

Fig. 5–38 Step block. Scale: full size. Draw the front and top views, but not the pictorial view. Complete the right-side view in its proper location. The step block is $\frac{3}{4}$" thick, $1\frac{3}{4}$" wide, and $3\frac{1}{2}$" long. The notch is $\frac{7}{8}$" × $1\frac{3}{4}$".

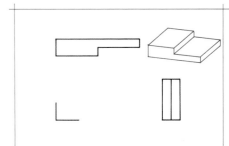

Fig. 5–39 Half lap. Scale: full size. Draw the top and right-side views, but not the pictorial. Complete the front view in its proper shape and location. The half lap is $\frac{3}{4}$" thick, $1\frac{3}{4}$" wide, and $3\frac{1}{2}$" long. The notch is $\frac{3}{8}$" × $1\frac{3}{4}$".

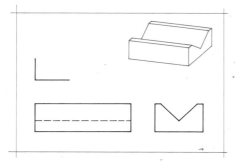

Fig. 5–40 V-block. Scale: full size. Draw the front and right-side views as shown. Complete the top view in its proper shape and location. Do not draw the pictorial view. The overall sizes are $1\frac{1}{4}$" high, 2" wide, and 4" long. The V-cut has a 90° included angle and is $\frac{3}{4}$" deep.

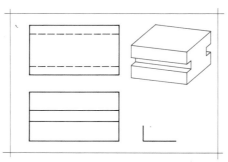

Fig. 5–41 Slide. Scale: full size. Draw the front and top views as shown. Complete the right-side view in its proper location. Do not draw the pictorial view. The overall sizes are $2\frac{1}{8}$" square and $3\frac{3}{4}$" long. The slots are $\frac{3}{8}$" deep and $\frac{1}{2}$" wide.

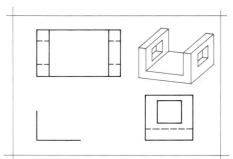

Fig. 5–42 Rod support. Scale: full size. Draw the top and right-side views, but not the pictorial. Complete the front view. The overall sizes are 2" square by $3\frac{1}{2}$" long. Bottom and ends are $\frac{1}{2}$" thick. The holes are 1" square and are centered on the upper portions.

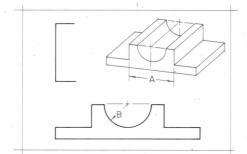

Fig. 5–43 Cradle. Scale: full size. Draw the front view, but not the pictorial. Complete the top view in the proper shape and location. Height = 2", width = $2\frac{1}{2}$", length = 6". Base = $\frac{1}{2}$" thick. A = 3", B = 1" R.

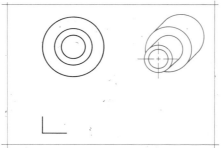

Fig. 5–44 Spacer. Scale: full size or as assigned. Draw the top view, but not the pictorial. Complete the front view. Base = $2\frac{1}{2}$" DIA × 1" high. Top = $1\frac{1}{2}$" DIA × $\frac{3}{4}$" high. Hole = 1" DIA. A vertical sheet will permit a larger scale.

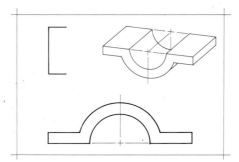

Fig. 5–45 Strap. Scale: full size. Draw the front view as shown. Complete the top view in the proper shape and location. Do not draw the pictorial view. Material is $\frac{1}{2}$" thick × 2" wide. The inside radius is $1\frac{1}{4}$". The overall length is 6".

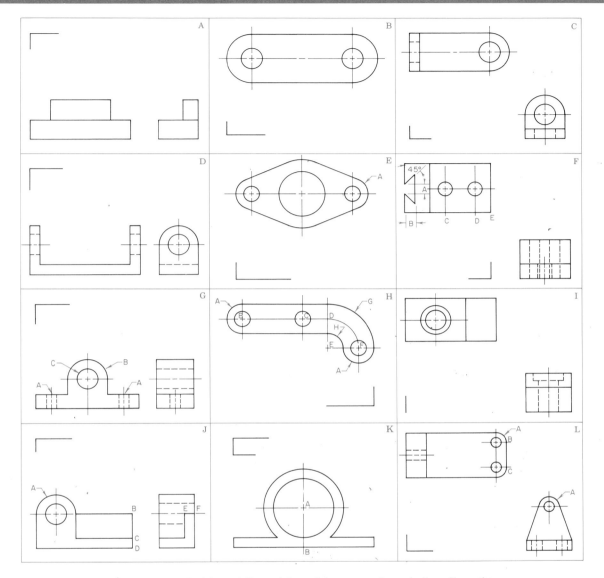

Fig. 5–46 Two- and three-view problems. Problems A through L each have one view missing. Draw the view or views given and complete the remaining view in the proper shape and location. Scale: full size or as assigned. Do not dimension unless instructed to do so.

A STOP: $L = 5$, $W = 2$, $H = 2$, base = $1 \times 2 \times 5$, top = $\frac{3}{4} \times 1 \times 3$.

B LINK: $L = 7\frac{1}{2}$, $W = 2\frac{1}{2}$, Thk = $1\frac{1}{4}$, holes = $1\frac{1}{16}$ DIA.

C ANGLE BRACKET: $L = 5$, $W = 2$, $H = 2\frac{1}{4}$, Matl thk = $\frac{1}{2}$, holes = 1" DIA.

D SADDLE: $L = 5\frac{1}{2}$, $W = 2$, $H = 2\frac{1}{2}$, Matl thk = $\frac{1}{2}$, hole = 1" DIA.

E SPACER: $L = 6\frac{1}{2}$, $W = 3\frac{1}{2}$, thk = 1", holes = $2\frac{3}{8}$ DIA, $\frac{3}{4}$ DIA, $A = \frac{3}{4}$ R.

F DOVETAIL SLIDE: $L = 4\frac{1}{4}$, $W = 2\frac{1}{2}$, $H = 2$, base thk = $\frac{3}{4}$, upright thk = $1\frac{1}{4}$, holes = $\frac{5}{8}$ DIA, $A = \frac{1}{2}$, $B = \frac{1}{2}$, $CD = 1\frac{1}{2}$, $DE = \frac{3}{4}$.

G ROD GUIDE: $L = 5\frac{1}{8}$, $W = 1\frac{7}{8}$, $H = 2\frac{1}{2}$, $C = 1$" DIA, $A = \frac{1}{2}$ DIA $3\frac{5}{8}$ apart, base thk = $\frac{3}{4}$, $B = 1$" R.

H HINGE PLATE: $A = \frac{3}{4}$ R, $BC = 3$, $CD = 1\frac{1}{4}$, $DE = 1\frac{1}{2}$, $EF = 1\frac{1}{2}$, $G = 2\frac{1}{4}$ R, $H = \frac{3}{4}$ R, holes $\frac{3}{4}$ DIA, thk = 1".

I OFFSET LUG: $L = 4\frac{1}{2}$, $W = 2\frac{1}{4}$, $H = 2$, notch = $\frac{3}{4} \times 1\frac{1}{2}$, hole = 1" DIA, Cbore = $1\frac{1}{2}$ DIA $\times$ $\frac{3}{8}$ deep.

J PIN HOLDER: $L = 4\frac{3}{4}$, $W = 1\frac{3}{4}$, $H = 2\frac{3}{4}$, hole = 1" DIA, $A = 1$" R, $BC = 1\frac{1}{4}$, $BD = 1\frac{3}{4}$, $EF = \frac{1}{2}$.

K RING: base = $\frac{1}{2} \times \frac{7}{8} \times 7$, ring = 4 OD, 3 ID, $AB = 2$.

L BRACKET: $L = 5$, $W = 2\frac{1}{4}$, $H = 2\frac{3}{4}$, base thk = $\frac{1}{2}$, upright = 1", $A = \frac{1}{2}$ R, $BC = 1\frac{1}{4}$, holes = $\frac{1}{2}$ DIA.

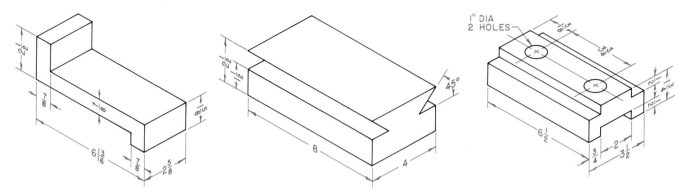

Fig. 5–47 Stop.

Fig. 5–48 Dovetail slide.

Fig. 5–49 Slide.

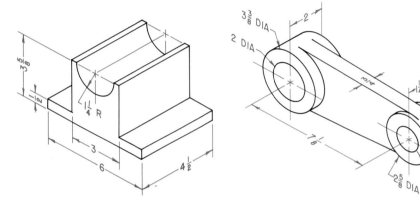

Fig. 5–50 Cradle block.

Fig. 5–51 Pivot arm.

Fig. 5–52 Base.

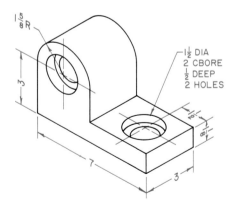

Fig. 5–53 Shaft support.

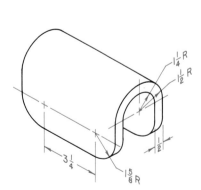

Fig. 5–54 Edge protector.

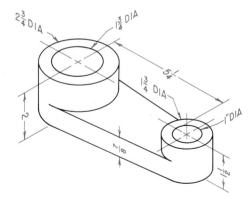

Fig. 5–55 Swivel arm.

ASSIGNMENT: Make two- or three-view drawings of objects on this page with instruments. Scale: as assigned. Do not dimension unless instructed to do so. Include all center lines.

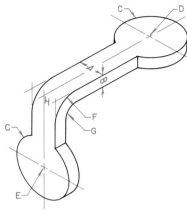
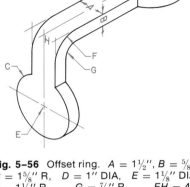

Fig. 5–56 Offset ring. $A = 1\frac{1}{2}''$, $B = \frac{5}{8}''$, $C = 1\frac{5}{8}''$ R, $D = 1''$ DIA, $E = 1\frac{1}{8}''$ DIA, $F = 1\frac{1}{2}''$ R, $G = \frac{7}{8}''$ R, $EH = 4''$, $HD = 6\frac{3}{4}''$.

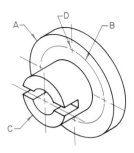

Fig. 5–57 Socket. Scale: as assigned. All dimensions are in millimeters (mm). $A = 50.50$ mm DIA $\times$ 7 mm thick, $B = 38.0$ mm, $C = 25.25$ mm DIA $\times$ 17 mm long with 13-mm-DIA hole through, Slots = 4.50 mm wide $\times$ 8.0 mm deep, $D = 6$ mm DIA, 4 holes equally spaced.

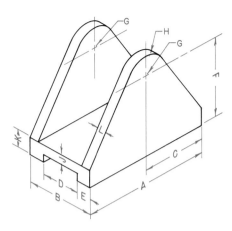

Fig. 5–58 Shaft guide. Scale: as assigned. $A = 7\frac{1}{4}''$, $B = 3\frac{7}{8}''$, $C = 3\frac{5}{8}''$, $D = 2\frac{1}{4}''$, $E = \frac{13}{16}''$, $F = 4\frac{1}{2}''$, $G = 1''$ DIA, 2 holes, $H = 1\frac{1}{4}''$ R, $J = \frac{1}{2}''$, $K = 1''$, $L = \frac{1}{2}''$.

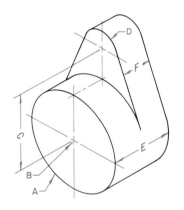

Fig. 5–59 Cam. $A = 2\frac{5}{8}''$ DIA, $B = 1\frac{1}{4}''$ DIA, $1\frac{3}{4}''$ Cbore, $\frac{1}{4}''$ deep, both ends, $C = 2\frac{1}{8}''$, $D = \frac{9}{16}''$ R, $E = 1\frac{3}{4}''$, $F = \frac{7}{8}''$.

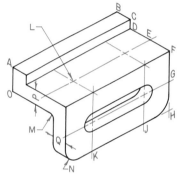

Fig. 5–60 Adjustable stop. Scale: as assigned. $AB = 10''$, $BC = 1\frac{3}{8}''$, $CD = \frac{3}{4}''$, $DE = 1\frac{7}{8}''$, $EF = 2''$, $FG = 2\frac{1}{2}''$, $GH = 2\frac{1}{2}''$, $FH = 5''$, $HJ = 2\frac{1}{2}''$, $JK = 5''$, $L = 1''$ DIA, 2 holes, $M = \frac{1}{2}''$ R, $N = 1\frac{1}{4}''$ R, $AO = 2''$, $P = 1\frac{1}{4}''$, $Q = 1\frac{1}{4}''$, Slot = $1\frac{1}{2}''$ wide.

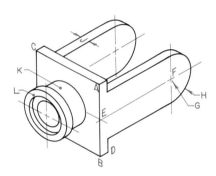

Fig. 5–61 Camera swivel base. Scale: as assigned. $AB = 40$ mm, $AC = 38$ mm, $BD = 5.5$ mm, $BE = 20$ mm, $EF = 45$ mm, $H = 12$ mm R, $G = 10$ mm DIA, $J = 6$ mm. Boss: $K = 24$ mm DIA $\times$ 9 mm long, $L = 28$ mm DIA $\times$ 4 mm long, hole = 12 mm DIA $\times$ 14 mm deep, Cbore = 18 mm DIA $\times$ 3 mm deep.

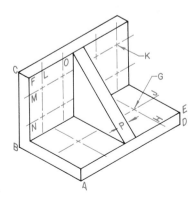

Fig. 5–63 Pipe support. Scale: as assigned. $AB = 8''$, $BC = 4''$, $AD = \frac{3}{4}''$, $E = 3\frac{7}{8}''$, $F = 2\frac{3}{4}''$ DIA $\times$ 2\frac{1}{4}''$ long, $G = 1\frac{1}{8}''$ DIA hole through, slots = $1''$ wide.

Fig. 5–62 Angle plate. Scale: as assigned. $AB = 6''$, $BC = 6\frac{1}{2}''$, $AD = 9\frac{3}{4}''$, $DE = 1''$, $CF = 1''$, $G = \frac{3}{4}''$ DIA, 2 holes, $EH = 2''$, $EJ = 2\frac{1}{2}''$, $FL = 1\frac{1}{8}''$, $LO = 2''$, $FM = 1\frac{1}{2}''$, $MN = 2\frac{1}{2}''$, $P = 1\frac{1}{8}''$, $K = \frac{1}{2}''$ DIA, 8 holes.

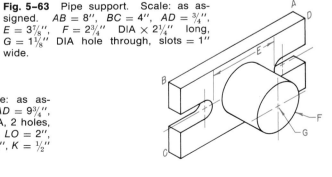

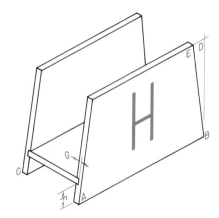

Fig. 5–64 Letter holder. Scale: full size or as assigned. Draw all necessary views. All material is $\frac{3}{16}''$ thick (plastic or wood). *AB* is 4″, *AC* is 2″, *BD* is 3″, *DE* is $\frac{3}{8}''$, *F* is $\frac{3}{8}''$, *G* is $\frac{3}{32}''$. Add a design to the front view. See the Appendix for the layout of block-style lettering. Dimension only if instructed to do so.

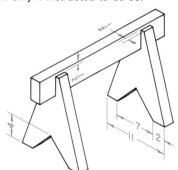

Fig. 5–67 Mini saw horse. Scale: $\frac{1}{4}'' = 1''$ or as assigned. Top rail is 2″ × 4″ × 24″. Legs are cut from 2 × 12, $14\frac{1}{2}''$ long. Dimension only if instructed to.

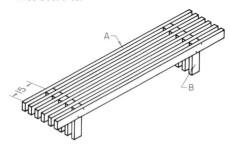

Fig. 5–69 Garden bench. Scale: 1″ = 1′-0″ or as assigned. *A* is 2 × 4 × 6′-0″, *B* is 2 × 4 × 1′-4″. Draw front, top, and right-side views of the bench. Use six or more top rails.

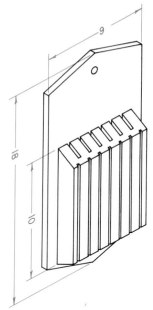

Fig. 5–65 Knife rack. Scale: half size or as assigned. Draw all necessary views. Back is $\frac{1}{2}'' \times 9'' \times 18''$. Front is $1\frac{1}{2}'' \times 7'' \times 10''$ with 30°-angle bevels on each end. Slots for knife blades are $\frac{1}{8}''$ wide × 1″ deep. Grooves on front are $\frac{1}{8}''$ wide × $\frac{1}{8}''$ deep. Estimate all sizes not given. Redesign as desired.

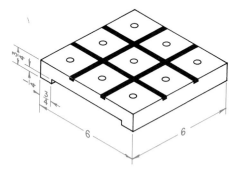

Fig. 5–70 Tic-tac-toe board. Scale: full size or as assigned. Material: hardwood with black plexiglas inlays. Inlays are $\frac{1}{8}'' \times \frac{1}{4}''$ and are located 2″ OC. Holes are $\frac{5}{16}''$ DIA × $\frac{1}{4}''$ deep and are centered within squares. Game board is designed to use marbles.

Fig. 5–66 Desk-top book rack. Scale: as assigned. Draw all necessary views. Material is laminated wood, plastic, or aluminum. *AB* is 8″, *BC* is $9\frac{1}{2}''$, *CD* is 2″, *DE* is $\frac{3}{8}''$, *EF* is 6″, *G* is $1\frac{1}{4}''$ R. All bends are 90°.

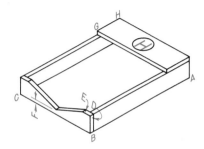

Fig. 5–68 Note-paper box. Scale: full size or as assigned. All stock is $\frac{1}{4}''$ thick. *AB* is $6\frac{5}{8}''$, *BC* is $4\frac{5}{8}''$, *BD* is 1″, *DE* is $\frac{1}{2}''$, *F* is $\frac{1}{4}''$, *GH* is $1\frac{1}{2}''$. Draw all necessary views. Do not dimension unless instructed to do so. Initial inlay is optional. Redesign as desired.

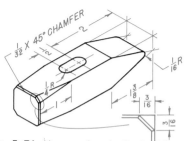

Fig. 5–71 Hammer head. Scale: 2″ = 1″ or as assigned. Draw all necessary views. Overall sizes are $\frac{7}{8}''$ square × $3\frac{1}{2}''$ long. Do not dimension unless instructed to do so.

Dimensioning | Chapter 6

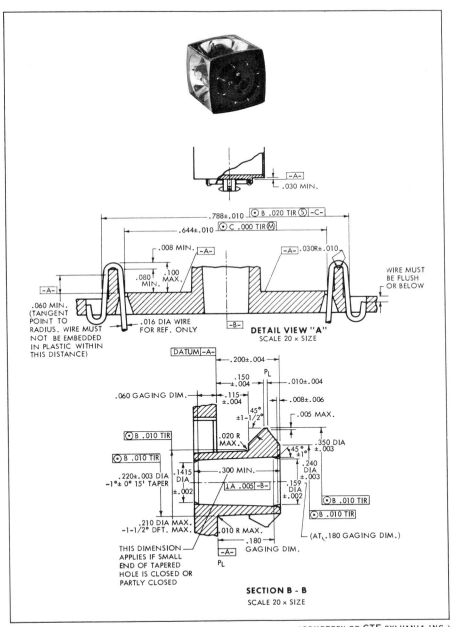

.030 MIN.

.788±.010

⊙ B .020 TIR Ⓢ -C-

.644±.010

⊙ C .000 TIR Ⓜ

.008 MIN.

-A-

-A- .030R±.010

.080 MIN.

.100 MAX.

WIRE MUST BE FLUSH OR BELOW

-A-

.060 MIN. (TANGENT POINT TO RADIUS. WIRE MUST NOT BE EMBEDDED IN PLASTIC WITHIN THIS DISTANCE)

.016 DIA WIRE FOR REF. ONLY

-B-

DETAIL VIEW "A"
SCALE 20 × SIZE

DATUM -A-

.200±.004

.150 ±.004

P L

.010±.004

.115 ±.004

.008±.006

.060 GAGING DIM.

45° ±1-1/2°

.005 MAX.

⊙ B .010 TIR

.020 R MAX.

.350 DIA ±.003

⊙ B .010 TIR

45° ±1°

.240 DIA ±.003

.220±.003 DIA -1°± 0° 15' TAPER

.1415 DIA ±.002

.300 MIN.

⌴A .005 -B-

.159 DIA ±.002

⊙ B .010 TIR

.210 DIA MAX. -1-1/2° DFT. MAX.

.010 R MAX.

⊙ B .010 TIR

THIS DIMENSION APPLIES IF SMALL END OF TAPERED HOLE IS CLOSED OR PARTLY CLOSED

.180 GAGING DIM.

(AT .180 GAGING DIM.)

-A-

P L

SECTION B - B
SCALE 20 × SIZE

(COURTESY OF **GTE** SYLVANIA INC.)

Size Description. There are two elements necessary in describing an object. One is its shape; the other, its size. In most chapters of this book, methods of describing shape are studied. In this chapter, methods of describing the size of an object will be shown and explained. The importance of developing a clear understanding of the rules and principles of accurate size description cannot be overemphasized. A machinist can manufacture a part only if he is given complete and accurate sizes on the drawing.

Another name for size description is *dimensioning.* It includes numerical values of measurement in inches and fractions, or feet and inches, or inches and decimals, or millimeters (metric system), as well as notes and symbols to specify the kind of finish, materials, and other information needed to make the part. When such information is added to shape description, the result is a complete *working drawing.*

Size description is an essential part of a working drawing. For some purposes it is enough to specify nominal and ordinary sizes in common fractions of an inch. Sometimes a note is added stating that dimensions are to be plus or minus a specified amount, as $\frac{1}{64}$ in. or $\frac{1}{32}$ in.; for large castings it might be that $\frac{1}{16}$ in. or more would be sufficiently close. Such a note may be placed on the drawing with the views, or it may be placed in the title block, usually in a space provided for this purpose.

When accurate dimensions are required, they are given in decimals to hundredths, thousandths, or ten-thousandths of an inch.

Dimensioning. The views on drawings describe the shape. Whether the views are drawn actual size, or to a proportional scale, it would not be practical to try to obtain the measurements by applying a scale. It would take too much time, and it would not be possible to make measurements of the accuracy necessary for interchangeable manufacture where large numbers of pieces must fit in place with mating parts (parts which fit together) and where there are other requirements.

This size information is provided by adding a system of lines, symbols, and numerical values, called *dimensioning*. To ensure accuracy and efficiency in use, size information is arranged on the drawing in a definite manner.

Lines and Symbols. Lines and symbols are used on drawings to show where the dimensions apply, as shown in Fig. 6–1. These lines and symbols are recognized by the men who use the drawings. Professional and trade associations, engineering societies, and certain industries have agreed upon the symbols to be used on drawings. Standards from the American National Standards Institute (ANSI), the Society of Automotive Engineers (SAE), and the Military Standards are sources of latest information on drawings and symbols.

To make a correct drawing, the draftsman must be familiar with these symbols, and he must know the principles of dimensioning. He must also be acquainted with the shop processes that will be used in building or making the product he is drawing. Symbols

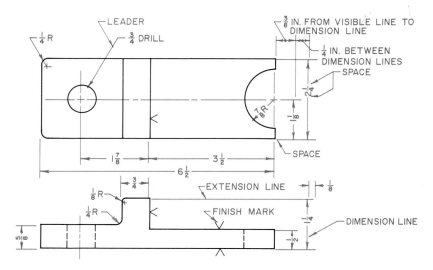

Fig. 6–1 Dimensioning requires the use of lines and symbols.

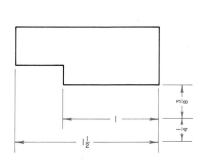

Fig. 6–2 Dimension lines must be spaced to provide clearness.

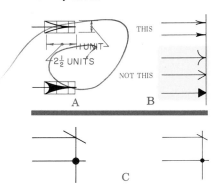

Fig. 6–3 Arrowheads.

are sometimes used to indicate such processes.

Dimension Lines. A dimension line is a thin line used to show the direction of a measurement (where it begins and where it ends) or to show the size of an angle (Fig. 6–2). The dimension line should have a "break" for the dimension numerals. To avoid crowding the numerals, dimension lines should be $\frac{3}{8}$ in. or more from the lines of the drawing and $\frac{1}{4}$ in. or more from each other.

Arrowheads. Arrowheads are used at the ends of dimension lines to show

where the dimension begins and ends (Fig. 6–2). They are also used at the end of a leader (Fig. 6–7) to show where a note or dimension applies to a drawing. Shapes of arrowheads are shown enlarged at A in Fig. 6–3 and reduced to actual size at B. They may be "open" or made solid, as shown. Arrowheads should be carefully drawn and of uniform size on a given drawing; however, a small space may require some variation.

In addition, some industries have adopted other techniques for indicating the termination point at the end of a dimension line or leader. Figure 6–3 at C shows examples of these.

While these methods also serve the purpose, the arrowheads shown at A and B are preferred.

Extension Lines. Extension lines are thin lines used to extend lines of the views and to indicate points or surfaces for which dimensions are given (Fig. 6-4). Since extension lines are not part of the views, they should not touch the outline. Extension lines start with a visible space, about $\frac{1}{32}$ to $\frac{1}{16}$ in. gap, and extend about $\frac{1}{8}$ in. beyond the last dimension line.

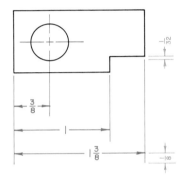

Fig. 6-4 Extension (witness) lines. A center line may be used as an extension line.

Numerals and Notes. Numerals and notes must be made carefully so that they will be easy to read. Do not make them so large, however, that they overbalance the drawing. In general, make numerals about $\frac{1}{8}$ in. high and fractions about $\frac{1}{4}$ in. high, with fraction numerals about $\frac{3}{32}$ in. high. Always make the fraction bar (division line) in line with the dimension, never at an angle.

Light guidelines for figures and fractions may be drawn quickly and easily with a lettering triangle or an Ames lettering instrument (Fig. 6-5).

When drawings are made to be reduced photographically for use at a smaller size or when used for micro-

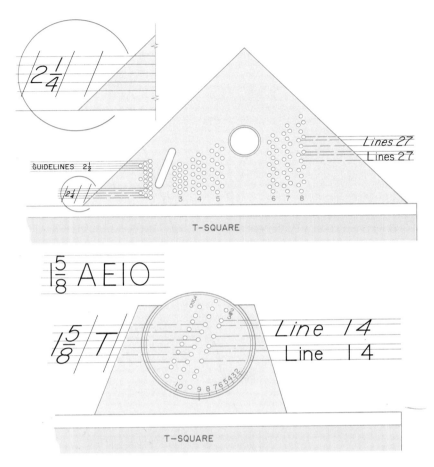

Fig. 6-5 Guidelines for letters, whole numbers, and fractions.

filming, the numerals must be made larger and with heavier strokes so that they will be clear when reduced. Capitals, either vertical or inclined, are preferred for lettering on most drawings.

The Finish Mark. The finish mark, or surface texture symbol, is used to indicate that a surface is to be machined (finished), as shown in Fig. 6-6. The old symbol form, $\mathcal{f}$, is still in use to some extent but is being replaced. The symbol V is now in general use. The point of the V is placed in contact with the edge view of the surface to be finished. Modifications of this symbol are used to

provide for an indication that allowance for machining is required, that a certain surface condition is required, and for other conditions described in ANSI B46.1, *Surface Texture.*

Fig. 6-6 The finish mark tells which surfaces are to be machined.

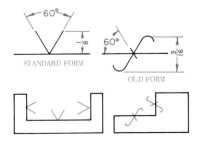

Leaders. Leaders are thin lines drawn from a note or dimension to the place where it applies (Fig. 6–7). Leaders are drawn at an angle to the horizontal; 60° is preferred, but 45°, 30°, or other angles may be used. A leader starts with a dash, or short horizontal line (about ⅛ in. is preferred, but it may be longer if needed), and generally ends with an arrowhead (a dot or other symbol may be used for special identification).

A number of leaders close together are best drawn parallel; a leader to be drawn to a circle or arc should be drawn in a radial direction. Don't cross leaders; don't draw long leaders; don't draw leaders horizontally, vertically, at a small angle, or parallel to dimension, extension, or section lines.

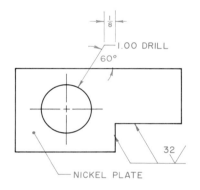

Fig. 6–7 Leaders point to the place where a note or dimension applies.

Scale of a Drawing. Scales used in making drawings are described in Chap. 3. The scale used should be given in or near the title. If a drawing has views of more than one part and different scales are used, the scale should be given close to the views. Usual scales are stated as full size, full, 1 = 1, 1.00 = 1.00; half size, ½ = 1, or 0.50 = 1.00; and so forth, for quarter or eighth size. If enlarged views are used, the scale would be shown as twice size, two times full size, 2 = 1, or 2.00 = 1.00, etc.

The scales generally used on metric drawings are based on divisions of 10. Scales such as 1 to 2, 1 to 5, and 1 to 10 are examples.

Units and Parts of Units. Dimensions on drawings are given in units of measurement, such as feet and inches, feet and decimals of a foot, inches and fractions of an inch, inches and decimals of an inch, and, on metric drawings, millimeters (mm).

Common fractions, such as ½, ¼, ⅛, etc., used with whole numbers have been and are used as generally suitable where particular accuracy is not required—say, not closer than ±1/64 in. When all dimensions are in inches, the inch symbol (″) is omitted. Standard practice, when feet and inches are used, is to show the symbol for feet but *not* for inches, as: 7′–5, 7′–0, etc.

Decimals are used where accuracy is required. For parts which must fit accurately, the dimensions are given in decimals, and the workman is required to work within specified limits. Such dimensions are used between finished surfaces, center dis-

tances, and places which must be held in a definite relationship to each other. Decimals to two places are used where limits of ±0.01 are sufficiently close (Fig. 6–8 at A, B, and C). For two-place decimals fiftieths are preferred, such as 0.02, 0.04, 0.24, etc., rather than 0.03, 0.05, etc. Such decimals can be divided by two and result in two-place decimals when used to get the radius from a diameter or for other purposes. The decimal point should be clear and placed on the bottom guideline in a space about the width of a zero. Decimals to three or more places are used where limits of less than ±0.01 are required, as in Fig. 6–9 at A and B.

Decimal dimensioning is used in many industries. It is indicated as the preferred method in drafting standards and is coming into general use for all drawings.

Millimeters (mm) are used for giving dimensions on drawings where the metric system of measurements is adopted.

In addition, a dual-dimensioning system is used in industries involved in international trade. This system

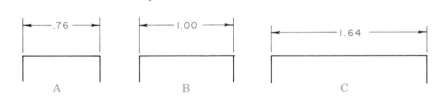

Fig. 6–8 Decimal dimensions: two places.

Fig. 6–9 Decimal dimensions: three places.

uses both the decimal inch and the millimeter (Fig. 6–10).

Placing Dimensions for Reading.
There are two methods in use: the *aligned system* and the *unidirectional system*.

1. The *aligned system* of dimensioning (Fig. 6–11) has the dimensions placed in line with the dimension lines. Horizontal dimensions always read from the bottom of the sheet. Vertical dimensions read from the right-hand side of the sheet. Inclined dimensions read in line with the inclined dimension line but should be kept outside the area indicated by the shading in Fig. 6–12 if possible.

2. The *unidirectional system* of dimensioning (Fig. 6–13) has all the dimensions placed to read from the bottom of the sheet, no matter where they occur. Automotive and aircraft companies have brought this system into general use. It is accepted practice and is being rapidly adopted by other industries, replacing the old aligned system which was in universal use for many years.

Special typing and dimensioning typewriters for use with drawings make it possible to save a great deal of time by using the unidirectional system. Both systems are in use in industry. Both systems are used in this book, since the student should become familiar with them. The one to use will be determined by company practice. On both systems, notes and dimensions with leaders should read from the bottom of the drawing.

Theory of Dimensioning. There are two basic kinds of dimensions: (1) *size dimensions* and (2) *location dimensions*. The theory of dimensioning considers any object as being made up of a number of geometrical shapes, such as prisms, cylinders, pyramids,

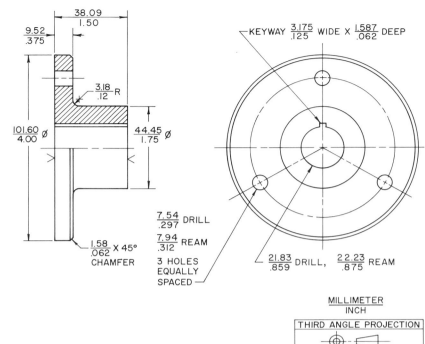

Fig. 6–10   Typical dual-dimensioned drawing.

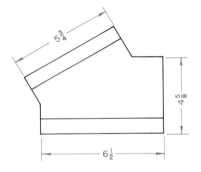

Fig. 6–11   Dimensioning: aligned system.

Fig. 6–13   Dimensioning: unidirectional system.

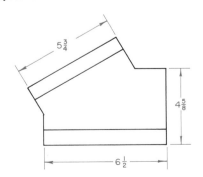

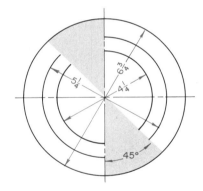

Fig. 6–12   Avoid placing dimensions in the shaded area.

cones, and so forth, or of parts of such shapes. This is illustrated in Fig. 6–14, where the bearing is separated into simple parts. A hole or hollow part can be considered as having the same outlines as one of these shapes. Such open spaces in an object may be considered as negative (not solid) shapes. It then becomes a matter of dimensioning a number of simple

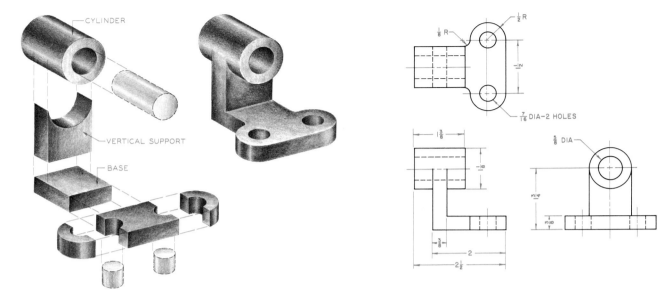

Fig. 6–14 Parts can usually be broken down into basic geometric shapes for dimensioning.

shapes. When the size of each simple piece is defined and the relative positions are given, the description is complete. Size dimensions are used to define the simple pieces and location dimensions to give relative positions. When a number of pieces are assembled, each piece is first considered separately and then in relation to the other pieces. In this way the size description of a complete machine, a piece of furniture, or a building is simply a matter of following an orderly procedure, as for a single part.

This analysis is applied in the aircraft and some other industries where the weights of parts are calculated by figuring volumes of parts as solid. From these solids the volumes of holes and hollow or open spaces (negative or minus shapes) are subtracted. The result is then multiplied by the weight per cubic inch of the material to obtain the total weight.

Size Dimensions. The first shape is the prism. For a rectangular prism (Fig. 6–15), the width (*W*), the height

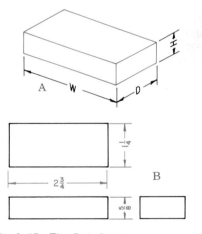

Fig. 6–15 The first shape.

Fig. 6–17 The first rule applied.

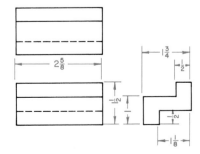

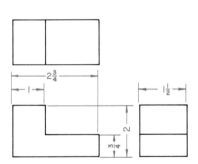

Fig. 6–16 The first rule applied.

(*H*), and the depth (*D*) will be required. Such an elementary shape may appear in a great many ways, a few of which are shown in Figs. 6–16 and 6–17. Flat pieces of irregular shape are dimensioned in a similar way (Figs. 6–18 and 6–19). The rule may be stated thus:

For any flat piece, give the thickness in the edge view and all other dimensions in the outline view.

The outline view is the one which shows the shape of the flat surface or surfaces. The front views in Figs. 6–18 and 6–19 are the outline views.

Prisms are illustrated in Fig. 6–20. A square prism requires two dimensions, a hexagonal or an octagonal prism may use two dimensions, a triangular prism may use three dimensions, and so on for other regular or irregular prisms.

The second shape is the cylinder, which requires two dimensions: the diameter and the length (Fig. 6–21). Three cylinders are dimensioned in Fig. 6–22, one of which is the hole. A washer or other hollow cylinder may be thought of as two cylinders of the same length (Fig. 6–23). The rule is as follows:

For cylindrical pieces, give the diameter and the length on the same view.

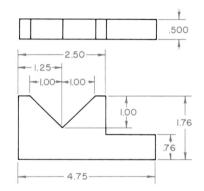

Fig. 6–18 An irregular flat shape.

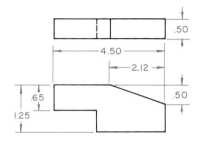

Fig. 6–19 An irregular flat shape.

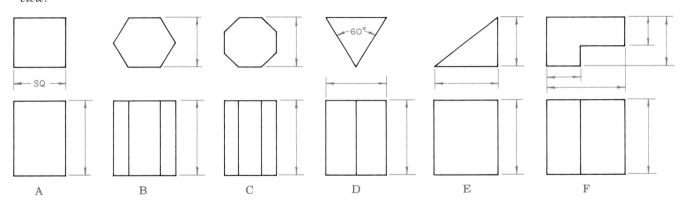

Fig. 6–20 Dimensioning prisms.

Fig. 6–21 Dimensioning a cylinder: the second shape.

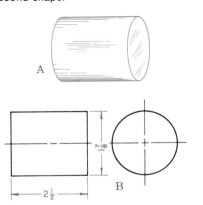

Fig. 6–22 The second rule applied.

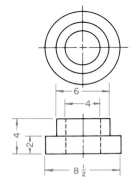

Fig. 6–23 The second rule applied.

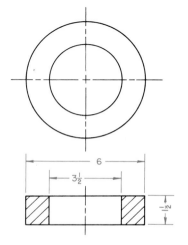

When the circular view of a cylinder is not shown, the abbreviation DIA is placed with the diameter dimension (Fig. 6–24). When parts of cylinders occur (Fig. 6–25 at A, B, and C), they are dimensioned in the view where the curves show by giving the radius dimension followed by the abbreviation R.

Notes are generally used to specify the sizes of holes. Such a note is usually placed on the outline view, especially when the method of forming the hole is specified (Fig. 6–25 at A). These notes show the operations necessary to form or complete the hole, such as drilling, punching, reaming, lapping, tapping, countersinking, spot facing, and so forth (Fig. 6–46). Either a dimension or a note may be used when a hole is to be formed by boring. When a hole in a casting is to be formed by a core, the word "core" is used in a note or with the dimension.

When parts of cylinders occur, such as fillets (Fig. 6–25 at B), rounds (Fig. 6–25 at C), and rounded corners, they are dimensioned in the views where the curves show (Fig. 6–25).

Other shapes include the cone, the pyramid, and the sphere. The cone, the frustum, the square pyramid, and the sphere may be dimensioned in one view (Fig. 6–26). To dimension rectangular or other pyramids and parts of pyramids, two views are required.

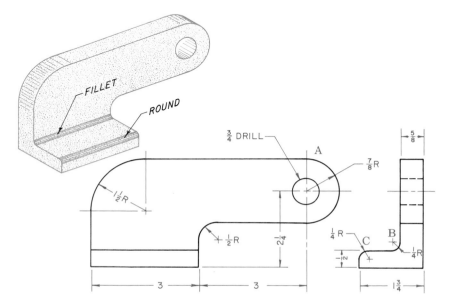

Fig. 6–25 Fillets, rounds, and radii.

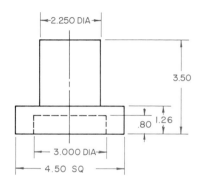

Fig. 6–24 Use of DIA on a single view.

Fig. 6–26 Dimensioning some elementary shapes.

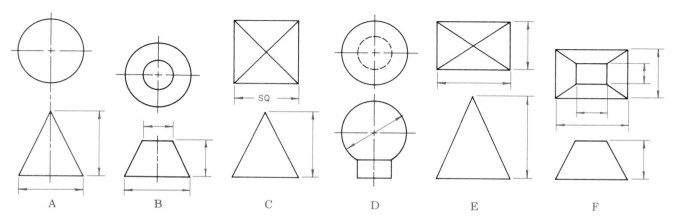

A   B   C   D   E   F

Location Dimensions. Location dimensions are used to specify the relative positions of the elementary shapes. Finished surfaces and center lines, or axes, are important for fixing the positions of parts by location dimensions. In general, location dimensions are necessary in three mutually perpendicular directions (up and down, crossways, and forward and backward).

Prisms are located by surfaces, surfaces and axes, or axes (center lines).

Cylinders are located by axes and bases. Three location dimensions are required. Location dimensions also locate holes, surfaces, and other features. The relative importance of the various surfaces and axes must be studied together so that the parts will go together as accurately as necessary. A knowledge of engineering practice in manufacture, assembly, and use is necessary if the draftsman is to do a good job of including the correct dimensions and notes on his drawing. Finished surfaces and center lines, or axes, are used to define positions with location dimensions. Two general rules will serve as a basis for showing location dimensions:

Prism forms are located by the axes and the surfaces (Fig. 6–27). Three dimensions are required.

Cylinder forms are located by the axis and the base (Fig. 6–28). Three dimensions are required.

Combinations of prisms and cylinders are shown in Figs. 6–29 and 6–30. The dimensions at L (Fig. 6–30) are location dimensions.

Datum Dimension. Datums are points, lines, and surfaces which are assumed to be exact. Such datums are used for purposes of computation or reference, and location dimensions are given from them. When positions are located from datums, the different features of a part are all located from the datum.

Two surfaces, two center lines, or a surface and a center line are typical datums. In Fig. 6–31 at A, two surface datums are used; at B, two center lines; and at C, a surface and a center line. A datum must be evident and accessible while the part is being made. Mating parts are parts which have contact and which must fit together; therefore, they should have the same datums.

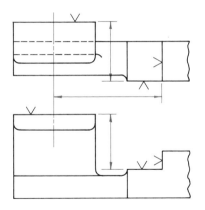

Fig. 6–27 Locating dimensions for prisms.

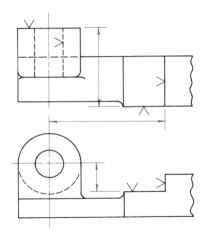

Fig. 6–28 Locating dimensions for prisms and cylinders.

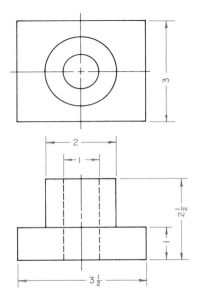

Fig. 6–29 First and second shapes.

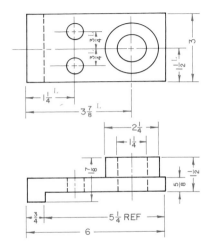

Fig. 6–30 First and second shapes.

General Rules. When adding dimensions to drawings, draftsmen follow certain practices which represent good form to the extent that they have the force of rules.

1. Dimension lines should be spaced about $\frac{1}{4}$ in. apart and about $\frac{3}{8}$ in. from the view outline (Fig. 6–2).

2. If the aligned system is used, dimensions must read in line with the dimension line and from the lower or right-hand side of the sheet (Fig. 6–11).

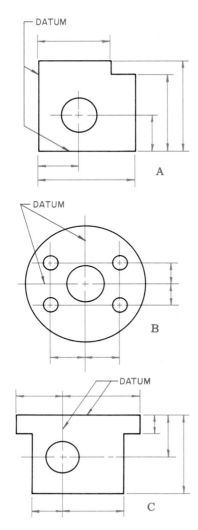

Fig. 6-31 Datum dimensioning.

3. If the unidirectional system is used, all dimensions must read from the bottom of the sheet (Fig. 6-13).

4. On machine drawings, detail dimensions up to 72 in. should be given in inches. Above this, feet and inches are generally used, except for gear drawings, bore of cylinders, length of wheel bases, and so forth. Aircraft and automotive drawings use inches or millimeters.

5. When all the dimensions are in inches or millimeters the symbol is generally omitted.

6. On architectural and structural drawings, dimensions of 12 in. and over are given in feet and inches. Millimeters may also be used.

7. Sheet-metal drawings are usually dimensioned in inches or millimeters.

8. Furniture and cabinet drawings are usually dimensioned in inches or millimeters.

9. Feet and inches are designated thus: 7′-3 or 7 ft 3 in. Where the dimension is in even feet, it is indicated thus: 7′-0.

10. The same dimension is not repeated on different views.

11. Dimensions not required for making a piece should not be given. This is especially important for interchangeable manufacture where limits are used. Figure 6-32 at A has "not required" dimensions. These have been omitted in Fig. 6-32 at B.

12. Overall dimensions should be placed outside the smaller dimensions (Figs. 6-32 and 6-33). With the overall dimension given, one of the smaller distances should not be dimensioned (Fig. 6-32 at B) unless it is needed for reference, and then it should be indicated by adding REF, as in Fig. 6-33.

13. On circular end parts the center-to-center dimension is given instead of an overall dimension (Fig. 6-34).

14. When it is necessary to place a dimension within a sectioned area, leave a clear space for the number (Fig. 6-35).

15. American National Standard practice is to avoid placing dimensions in the area indicated by shading in Fig. 6-36 when the aligned system is used.

16. Dimensions should be given from center lines, finished surfaces, or datums where necessary.

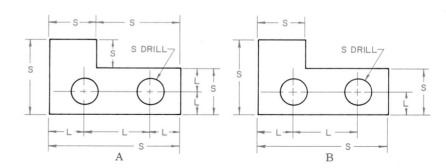

Fig. 6-32 Omit unnecessary dimensions.

Fig. 6-33 A dimension for reference should be indicated by REF.

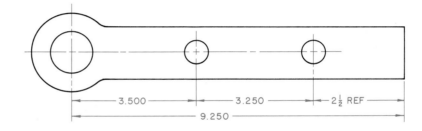

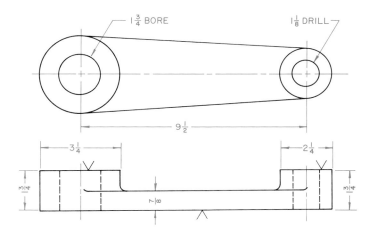

Fig. 6-34 Center-to-center dimensions.

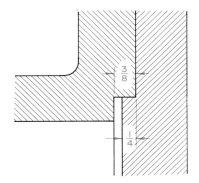

Fig. 6-35 Dimensions within a sectioned area.

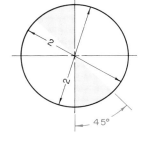

Fig. 6-36 Avoid shaded area with aligned dimensions.

17. Never use a center line or a line of the drawing as a dimension line.

18. Never have a dimension line as a continuation of a line of a view.

19. Never place a dimension where it is crossed by a line.

20. Always give the diameter of a circle, not the radius. The abbreviation DIA is used after the dimension, except when it is obviously a diameter.

21. The radius of an arc should always be given with the abbreviation R placed after the dimension.

22. In general, dimensions should not be placed inside the view outlines.

23. Extension lines should be drawn so that they do not cross each other, or cross dimension lines, if it can be avoided without making the drawing more complicated.

24. Do not dimension to hidden lines if it can be avoided.

25. It must be remembered that there are no absolutely hard-and-fast rules, nor any practice, not subject to possible change or modification under special conditions or requirements of a particular industry. When there is a variation of any rule, there must always be a reason which can be completely justified.

Standard Details. The shape of a part, the methods of manufacture, and the purpose for which the part is to be used generally indicate the kind and accuracy of the dimensions that must be given. A knowledge of manufacturing methods, patternmaking, foundry, machine-shop procedures, forging, welding, and so forth, is very useful when you are selecting and placing dimensions. In most cases such knowledge is essential. It is also important to consider whether only one part is to be made or whether quantity-production methods are to be used. In addition, there are purchased parts, identified by name or brand, that require few, if any, dimensions. Some companies have their own standard parts for use in different machines or constructions, and these are dimensioned according to use and production methods.

There are, however, certain more or less standard details or conditions for which methods of dimensioning may be suggested.

Angles and Chamfers. Angles are usually dimensioned in degrees (°), minutes('), and seconds ('') (Fig. 6–37 at A). The abbreviation DEG may be used instead of the symbol ° when only degrees are indicated. Angular tolerance is generally bilateral (plus or minus), as $\pm\frac{1}{2}$° for degrees and $\pm5'$ for minutes (Fig. 6–37 at B). Angular tolerance is stated either on the drawing or in a space provided in the title block. Angular measurements on structural drawings are given by run and rise, using 12'' for the horizontal side of the triangle (Fig. 6–38 at A). A similar method is used for slopes, as at B and C, where one side of the triangle is made equal to 1.

Two usual methods of dimensioning chamfers are shown in Fig. 6–39 at A and B.

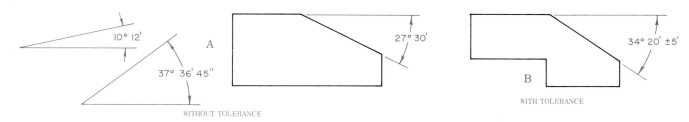

Fig. 6–37 Dimensioning an angle.

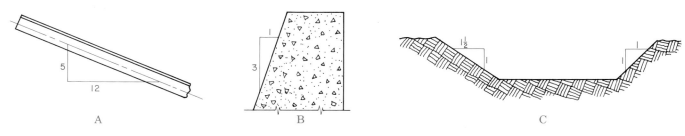

Fig. 6–38 Dimensioning angles of slopes.

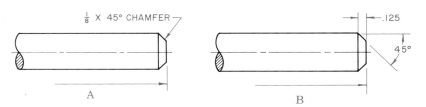

Fig. 6–39 Dimensioning chamfers.

Fig. 6–40 Dimensioning tapers.

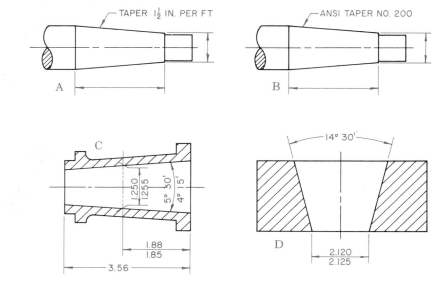

Tapers. Tapers may be specified by giving the length, one diameter, and the taper as in Fig. 6–40 at A. Another method is shown at B, where one diameter or width, the length, and the American National Standard, or other standard, taper number, are given. For a close fit the taper is dimensioned as at C, where the diameter is given at a located gage line. At D one diameter and the angle are given.

Dimensioning Curves. A curve composed of arcs of circles is dimensioned by the radii with centers located by points of tangency (Fig. 6–41 at A and B). Noncircular or irregular curves (Fig. 6–42) may be dimensioned as at A or from datum lines, as at B. A regular curve may be described and dimensioned by showing the construction or naming the curve, as at C, giving the basic dimensions.

Dimensioning a Detail Drawing. The drawing of a separate part with the dimensions, notes, and information for making the part is called a *detail drawing.* The order of dimensioning is as follows: The views of a drawing should be completed before starting to add any of the dimensions or notes. Then consider the actual shape of the part and the characteristic views. Now draw the extension lines, all of them including the lengthening of any center lines, if needed. Consider the size dimensions and the related location dimensions. Put on the dimension lines, leaders, and ar-

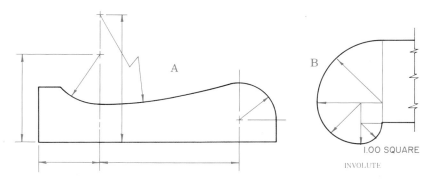

Fig. 6–41 Dimensioning curves composed of circular arcs.

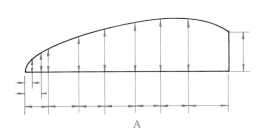

Fig. 6–42 Dimensioning noncircular curves.

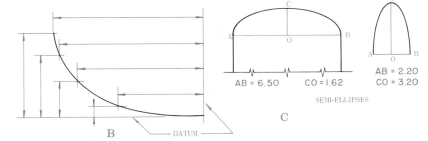

rowheads. After considering any changes, put in the dimensions and add any required notes.

Dimensioning an Assembly Drawing. When the parts of a machine are shown together in their relative positions, the drawing is called an *assembly drawing.* The rules and methods of dimensioning apply where a complete description of size is required.

Drawings of complete machines, constructions, and so forth, are made for different uses and have to be dimensioned to serve the purpose for which they were designed.

1. If the drawing is merely to show the appearance or arrangement of parts, the dimensions may be left off.

2. If it is desired to tell the space required, give overall dimensions.

3. If it is necessary to locate parts in relation to each other without giving all the detail dimensions, it is usual to give center-to-center distances and the dimensions needed for putting the machine or construction together or erecting it in position. For purposes indicated at 1, 2, and 3, photodrawings may be made of the completed machine with necessary information added.

4. In some industries, assembly drawings may be completely dimensioned either with or without extra part views (Chap. 11). Such drawings serve the purpose of both detail and assembly drawings. These are often referred to as *composite drawings.*

For furniture and cabinetwork sometimes only the major dimensions are given, such as length, height, and sizes of stock. The details of joints are left to the cabinetmaker or the standard practice of the company, especially where machinery is used and construction details are standardized.

Notes for Dimensions. Some of the operations for which information is given by notes are drilled holes (Fig. 6–43), reamed holes (Fig. 6–44), and counterboring or spot facing (Fig. 6–45). The use of notes for specifying these and other dimensions and operations is indicated in Fig. 6–46. In this figure, A is for a drilled hole, B is for a hole to be drilled and reamed,

Fig. 6–43 Drilling a hole.

Fig. 6–44 Reaming a hole.

Fig. 6–45 Counterboring to a specified depth. Spot facing is generally used to provide smooth spots.

C and D specify counterbore, E specifies countersink for a No. 24 flat-head screw, F, G, and H are for countersunk and counterdrilled holes, I specifies a spot face to provide for a nut, and J specifies a smooth-surface spot face.

When a hole is to be made in a piece after assembly with its mating piece, the note should read as previously stated but with the addition of the words "at assembly." Because such a hole is located when it is made during assembly with its mating part, no dimensions are required for its location. Other dimensions with machining operations are suggested in Fig. 6–47.

Abbreviations. Many of the abbreviations used as a part of dimensioning are familiar, such as the few examples from American National Standards which follow:

Allowance	ALLOW
Alloy	ALY
Aluminum	AL
Babbitt	BAB
Bevel	BEV
Cast iron	CI
Center line	CL or ℄
Chamfer	CHAM
Cold-rolled steel	CRS
Countersink	CSK
Degree	(°) DEG
Diameter	DIA
Dimension	DIM.
Inch	IN.
Key	K
Keyseat	KST
Keyway	KWY
Left	L
Left hand	LH
Limit	LIM
Material	MATL
Maximum	MAX
Millimeter	MM
National	NATL
Not to scale	NTS
Outside diameter	OD
Pattern	PATT

Fig. 6–46 Dimensions for holes.

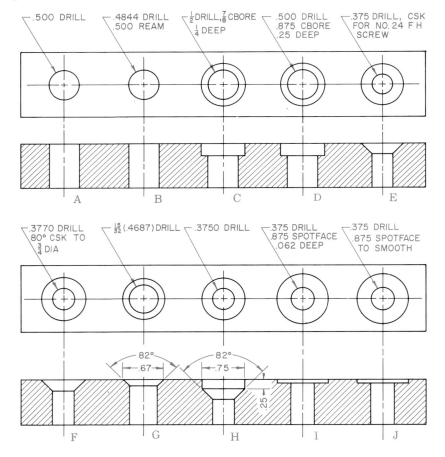

Radial	RAD
Radius	R
Reference	REF
Require	REQ
Revise	REV
Right hand	RH
Screw	SCR
Spherical	SPHER
Spot faced	SF
Square	SQ
Stock	STK
Surface	SUR
Tabulate	TAB
Thread	THD
Tolerance	TOL
United States Gage	USG
United States Standard	USS
Wrought Iron	WI

Refer also to the latest *American National Standard Abbreviations for Use on Drawings,* ANSI Z32.13.

Interchangeable Manufacturing. When large quantities of parts are made to be assembled with other parts, as on an assembly line, it is necessary to make the parts so that any part will fit into place without further machine or hand work. This requires specified allowances for size so that mating parts will fit together. Mating parts are parts which have contact and which must fit together to meet the requirements for which they are to be used. A rod or shaft and the hole in which it turns are mating parts. The diameter of the rod would be limited, and so would the diameter of the hole.

Limit Dimensioning. Since absolute accuracy cannot be expected, a workman is required to keep within a fixed limit of accuracy. The number of hundredths, thousandths, or ten-thousandths of an inch that are allowed as a variation from absolute measurements is called the *tolerance.* The tolerance may be specified by a note on the drawing or in a space provided in the title block, as: "Dimension Tolerance 0.01 Unless Otherwise Specified." Limiting dimensions, or limits to specify the maximum and minimum dimensions permitted, are used to show the necessary degree of accuracy. This is illustrated at A in Fig. 6–48. Note that the maximum limiting dimension is placed above the dimension line for the shaft (external dimension) and that the minimum limiting dimension for the hole in the ring is placed above the dimension line.

At B and C in Fig. 6–48 the basic sizes are given, and the tolerance specified, plus or minus, is shown. Consecutive dimensions are shown at B, where the dimension designated by *X* could have some variation. This dimension would not be given unless required for reference, in which case it would be followed by the abbreviation REF. Progressive dimensions are shown at C, where they are all given from a single surface (sometimes called *base-line dimensioning*).

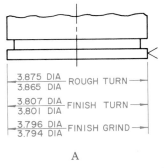

Fig. 6–47 Operations with limits specified.

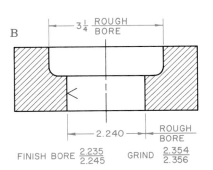

Fig. 6–48 Limit dimensions.

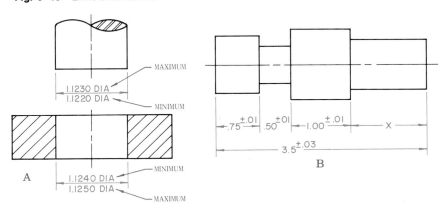

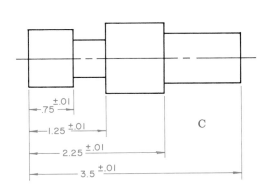

Accurate or limiting dimensions should not be called for unless necessary, for they greatly increase the cost of manufacturing the part. The detail drawing in Fig. 6–49 has limits for only two dimensions; all others are nominal dimensions, with variations permissible according to the purpose for which the part is to be used. In this case the general note calls for a tolerance of ±0.010 in.

Precision or Exactness. For precise information on dimensioning for accuracy of measurements and positions, the latest edition of ANSI Y14.5 should be studied and used as a reference. The following paragraphs are extracted from *American National Standard Drafting Practices* with the permission of the publisher, The American Society of Mechanical Engineers, 345 East 47th St., New York, NY, 10017.

Expressing Size and Position. *Definitions relating to size.* The following five terms have been defined to provide a common interpretation in respect to their use in this book.

Size. Size is a designation of magnitude. When a value is assigned to a dimension it is referred to hereinafter as the size of that dimension. *Note: It is recognized that the words "dimension" and "size" are both used to convey the meaning of magnitude.*

Nominal size. The nominal size is the designation which is used for the purpose of general identification. Example: ½ in. pipe.

Basic size. The basic size is that size from which the limits of size are derived by the application of allowances and tolerances.

Design size. The design size is that size from which the limits of size are derived by the application of tolerances. When there is no allowance the design size equals the basic size.

Actual size. An actual size is a measured size.

Limits of size. The limits of size (commonly referred to simply as "limits") are the applicable maximum and minimum sizes.

Position. Dimensions that establish position generally require more analysis than dimensions that only state sizes. Either linear or angular expressions may locate features with respect to one another (point-to-point), or from a datum. Point-to-point distances may be adequate for describing simple parts. Dimensions from a datum may be necessary if a part with more than one critical dimension must mate with another part.

Locating round holes. Figs. 6–50 through 6–55 illustrate the positioning of round holes by giving distances, or distances and directions, to the hole centers. These methods can also be used to locate round pins and other features of symmetrical contour. Allowable variations for any of the positioning dimensions illustrated may be specified by giving a tolerance with each distance or angle, by stating limits of dimensions or angles, or by true position expressions.

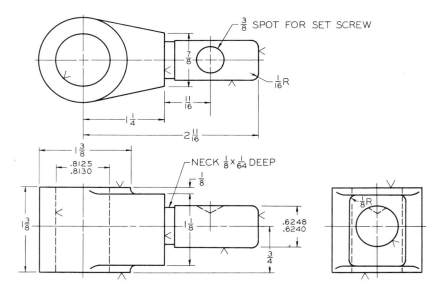

Fig. 6–49 A detail drawing with limits.

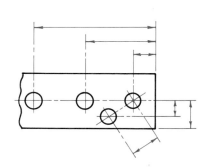

Fig. 6–50 Locating by linear distances.

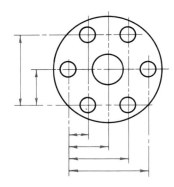

Fig. 6–51 Locating holes by rectangular coordinates. (*ANSI.*)

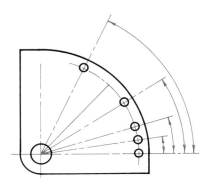

Fig. 6-52 Locating holes on a circle by polar coordinates. (*ANSI.*)

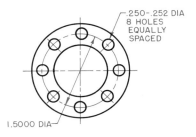

Fig. 6-53 Locating holes on a circle by radius or diameter and "equally spaced." (*ANSI.*)

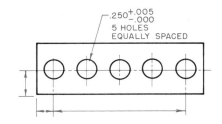

Fig. 6-54 "Equally spaced" holes in a line. (*ANSI.*)

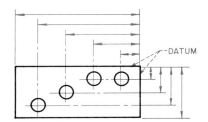

Fig. 6-55 Dimensions for datum lines. (*ANSI.*)

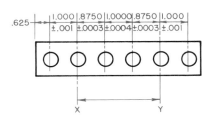

Fig. 6-56 Point-to-point, or chain, dimensioning. (*ANSI.*)

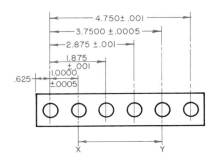

Fig. 6-57 Datum dimensioning. (*ANSI.*)

Tolerance. A tolerance represents the total amount by which a given dimension may vary. A tolerance should be expressed in the same form as its dimension; the tolerance of a decimal dimension should be expressed by a decimal to the same number of places, and the tolerance of a dimension written as a common fraction should be expressed as a common fraction. An exception to this is a close tolerance on an angle which may be expressed by a decimal representing a linear distance.

In a "chain" of dimensions with tolerances, overall variations in position that may occur are equal to the sums of the tolerances on the intermediate distances. The datum dimensioning method of Fig. 6-55 avoids overall accumulations, but the tolerance on the distance between two features equals the sum of the toler-

ances on two dimensions from the datum. Where the distance between two points must be controlled closely, the distance between the two points should be dimensioned directly, with a tolerance. Figure 6-56 illustrates a series of "chain" dimensions where tolerances accumulate between points X and Y; datum dimensions in Fig. 6-57 show the same accumulation with larger tolerances; Fig. 6-58 shows how to avoid accumulation without the use of extremely small tolerances.

Unilateral tolerance system. A unilateral system of tolerances allows variations in only one direction from a design size. This way of stating a tolerance is often helpful where a critical size is approached as material is removed during manufacture. See Fig. 6-59. For example, close-fitting holes and shafts are often given unilateral tolerances.

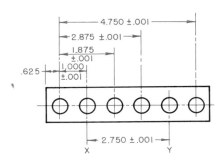

Fig. 6-58 Dimensioning to prevent tolerance accumulation between *X* and *Y*. (*ANSI.*)

Fig. 6-59 Giving a tolerance by a plus figure and a minus figure. (*ANSI.*)

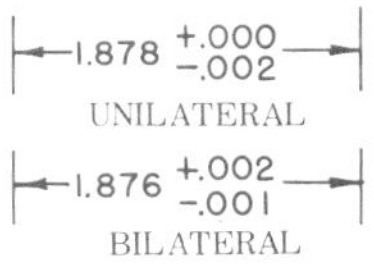

$$1.878 \begin{array}{c} +.000 \\ -.002 \end{array}$$
UNILATERAL

$$1.876 \begin{array}{c} +.002 \\ -.001 \end{array}$$
BILATERAL

Bilateral tolerance system. A bilateral system of tolerances allows variations in both directions from a design size. Bilateral variations are generally given with locating dimensions, or with any dimension that can be allowed to vary in either direction. See Fig. 6–59.

Limit System. A limit system indicates only the largest and smallest permissible dimensions. See Figs. 6–60 and 6–61. The tolerance is the difference between the limits.

Expressing allowable variations. Various expressions are used to state the amounts of variation permitted for the dimensions indicated on drawings. The expressions recommended in this book are listed and described as follows:

1. Two tolerance numerals are specified, one plus and one minus. This form of expression is necessary if the plus variation differs from the minus variation. It may be used in preference to 2 where the plus and minus variations are equal. See Fig. 6–59. NOTE: Two variations in the same direction should never be specified.

2. A combined plus and minus sign is followed by a single tolerance numeral. This method is very generally followed if the plus variation is equal to the minus variation. See Fig. 6–62.

3. The maximum and minimum limits of size are specified. The numerals should be arranged in one of two ways, but not both on the same drawing.

a. The high limit is always placed above the low limit where dimensions are given directly, and the low limit always precedes the high limit where dimensions are given in note form. See Fig. 6–60.

b. For location dimensions given directly (not by note), the high-limit numeral (maximum dimen-

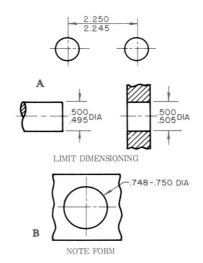

LIMIT DIMENSIONING

NOTE FORM

Fig. 6–60 Specifying limits: first method. (*ANSI.*)

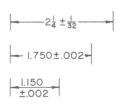

Fig. 6–62 Using a combined plus and minus sign. (*ANSI.*)

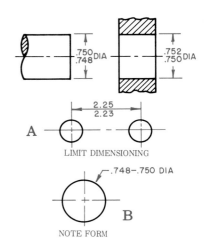

LIMIT DIMENSIONING

NOTE FORM

Fig. 6–61 Specifying limits: second method. (*ANSI.*)

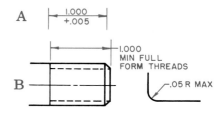

Fig. 6–63 Expressing a single tolerance or limit. (*ANSI.*)

sion) is placed above and the low-limit numeral (minimum dimension) is placed below. For size dimensions given directly, the numeral representing the maximum material condition is placed above and the numeral representing the minimum material condition is placed below. Where the limits are given in note form, the numeral that otherwise would be above shall precede the other. See Fig. 6–61.

4. It is not always necessary to state both limits.

a. A unilateral variation is sometimes expressed without stating that the variation in the other

direction is zero. See Fig. 6–63 at A.

b. MIN or MAX is often placed after a numeral where the other limit is not important. Depths of holes, lengths of threads, chamfers, etc., are often limited in this way. See Fig. 6–63 at B.

5. Other expressions allowing variations are described under Positional Tolerances and Positional Tolerances at Maximum Material Condition.

6. The above recommendations refer to linear tolerances, but the forms of expression are also used for angular tolerances. Angular tolerances may be in degrees, minutes and seconds, or in decimals.

Placing Tolerance and Limit Numerals. A numeral indicating a tolerance should be placed to the right of the dimension numeral, and in line with it. An accepted alternative method places the tolerance numeral below the dimension numeral, with the dimension line between. Figure 6–64 shows both arrangements.

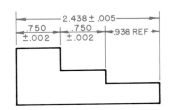

Fig. 6–64 Placing tolerance and limit numerals. Note the reference dimensions. (*ANSI.*)

Dimensioning for Fits. For interchangeable manufacture, the tolerances on dimensions must be such that an acceptable fit will result from assembly of parts having any combination of actual sizes that are within the tolerances. See Fig. 6–65.

Figure 6–66 shows a method of dimensioning mating parts that must fit one another, when they do not need to be interchangeable. The size of one part need not be held to a close tolerance, because it is to be modified at assembly to the size that is necessary for the desired fit.

"Hole basis" and "shaft basis." To specify the dimensions and tolerances of an internal and an external cylindrical surface so that they will fit together as desired, it is necessary to begin calculations by assuming either the minimum hole size or the maximum shaft size.

A basic hole system is a system of fits in which the design size of the hole is the basic size and the allowance is applied to the shaft.

A basic shaft system is a system of fits in which the design size of the

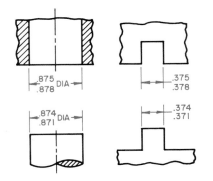

Fig. 6–65 Indicating dimensions of surfaces that are to fit closely. (*ANSI.*)

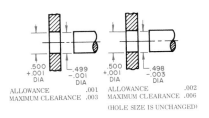

Fig. 6–67 "Basic hole" fits. (*ANSI.*)

shaft is the basic size and the allowance is applied to the hole.

NOTE: For further information on limits and fits, see ANSI B4.1–1967 or subsequent revision thereof.

"Basic hole" system. Limits for a fit in the basic hole system are determined by (1) specifying the minimum hole size, (2) determining the maximum shaft size by subtracting the desired allowance (minimum clearance) from the minimum hole size for a clearance fit, or adding the desired allowance (maximum interference) for an interference fit, and (3) adjusting the hole and shaft tolerances to obtain the desired maximum clearance or minimum interference. See Fig. 6–67. Tooling economies can often be realized by calculating from the basic hole size, providing the size selected can be produced by a standard tool (reamer, broach, etc.) or gaged with a standard plug gage.

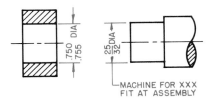

Fig. 6–66 Dimensioning noninterchangeable parts that are to fit closely. (*ANSI.*)

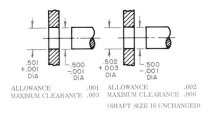

Fig. 6–68 "Basic shaft" fits. (*ANSI.*)

"Basic shaft" system. Limits for a fit in the basic shaft system are determined by (1) specifying the maximum shaft size, (2) determining the minimum hole size by adding the desired allowance (minimum clearance) to the maximum shaft size for a clearance fit, or subtracting for an interference fit, and (3) adjusting hole and shaft tolerances to obtain the desired maximum clearance or minimum interference. See Fig. 6–68. The "basic shaft" method is recommended only if there is a particular reason for it, for example, where a standard size of shafting can be used.

Geometric Tolerancing. Geometric tolerancing means specifying the maximum allowable variation in form (shape and size) and position (location). That is, it is the maximum allowable variation that can be tolerated in a system of interchangeable parts. Tolerances of position refer to the location of holes, slots, tabs, dovetails, etc. Tolerances of form refer to such characteristics as flatness, straightness, roundness, parallelism, perpendicularity, angularity, etc.

Specifying Tolerances of Position and Form. Tolerances of position and form may be specified in note form or in symbol form (Fig. 6-69). While either method is acceptable, most industries require that all draftsmen within their organization be consistent in using one or the other.

Geometric Characteristic Symbols. The symbols denoting geometric characteristics are shown in Fig. 6-70. These symbols represent a draftsman's shorthand for specifying form and position tolerances on drawings. These symbols may be used alone or they may be used in conjunction with other symbols and notes to give complete information on the shape or location of details.

Datum. A datum is a point, line, plane, cylinder, etc., assumed to be exact for purposes of computation, from which the location or geometric relationship (form) of features of a part may be established. In other words, a datum is a reference, determined to be true and accurate, from which measurements may be made.

Datum Reference Letters. Datum features requiring identification are assigned a reference letter. Any letter of the alphabet or combination of letters may be used except I, O, and Q. These letters tend to be confusing and may be misread. Single letters are used unless more are needed than are contained in the single-letter series. If more are needed, AA through AZ may be used. Figure 6-71 shows a

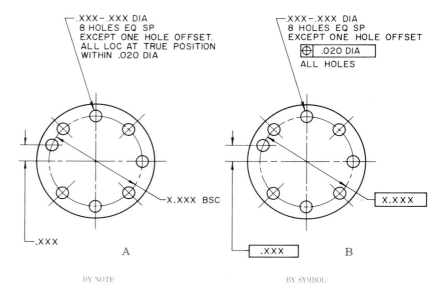

Fig. 6-69 Tolerances may be specified in a note or by a symbol.

GEOMETRIC CHARACTERISTIC SYMBOLS		
	CHARACTERISTIC	SYMBOL
FOR SINGLE FEATURE	FLATNESS	
	STRAIGHTNESS	
	ROUNDNESS (CIRCULARITY)	○
	CYLINDRICITY	
	PROFILE OF ANY LINE	
	PROFILE OF ANY SURFACE	
FOR RELATED FEATURES	PARALLELISM	∥
	PERPENDICULARITY (SQUARENESS)	⊥
	ANGULARITY	
	RUNOUT	
POSITIONAL TOLERANCES	TRUE POSITION	⊕
	CONCENTRICITY	◎
	SYMMETRY	

(FORM TOLERANCES covers the single feature and related features rows)

Fig. 6-70 Geometric characteristic symbols. *(ANSI.)*

Fig. 6-71 Datum-identifying symbol.

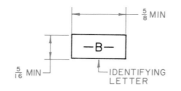

reference letter as it appears in a datum identifying symbol. The box containing the letter is drawn approximately $\frac{5}{16}$ in. high, and the length may vary according to its application on the drawing.

Symbols for MMC and RFS. The symbols Ⓜ and Ⓢ are used to designate *Maximum Material Condition* and *Regardless of Feature Size.* MMC exists when the feature contains the maximum amount of material, e.g., minimum hole diameter and maximum shaft diameter. RFS means tolerance of position or form must be met irrespective of where the feature lies within its size tolerance. The symbols are restricted to use as modifiers in feature control symbols.

Feature Control Symbols. The feature control symbol consists of a frame containing the geometric characteristic symbol followed by the permissible tolerance, and in some cases by the modifier Ⓜ or Ⓢ. A vertical line separates the symbol from the tolerance, as shown in Fig. 6–72. The datum-identifying symbol may also be added, as shown in Fig. 6–73. Refer to ANSI Y14.5–1966, *Dimensioning and Tolerancing for Engineering Drawings,* for a detailed description of terms, symbols, and their applications.

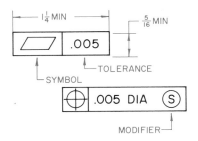

Fig. 6–72 Feature-control symbols.

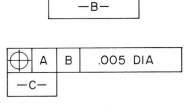

Fig. 6–73 Combined feature-control and datum-identifying symbols.

Surface Texture. In addition to size, it is sometimes necessary to indicate the condition of the surfaces of material. Absolute smoothness is not possible. Surfaces have irregularities, and standards have been developed for establishing classifications for roughness, waviness, and lay, as have symbols for indicating such conditions.

The subject is contained in the American National Standard ANSI B46.1–1962, to which reference is made for study and information. An idea of surface texture is indicated by the paragraphs which follow, extracted from *Surface Texture* (ANSI B46.1–1962) with the permission of the publisher, The American Society of Mechanical Engineers, United Engineering Center, 345 East 47th St., New York, NY, 10017.

Surfaces, in general, are very complex in character. This standard deals only with the height, width, and direction of surface irregularities, since these are of practical importance in specific applications.

Classification of Terms and Ratings Related to Surfaces. The terms and ratings in this standard relate to surfaces produced by such means as machining, abrading, extruding, casting, molding, forging, rolling, coating, plating, blasting, burnishing, etc.

Surface texture. Repetitive or random deviations from the nominal surface which form the pattern of the surface. Surface texture includes roughness, waviness, lay, and flaws.

Profile. The profile is the contour of a surface in a plane perpendicular to the surface, unless some other angle is specified.

Measured profile. The measured profile is a representation of the profile obtained by instrumental or other means. See Fig. 6–74.

Microinch. A microinch is one millionth of an inch (0.000001 in.). For written specifications or reference to surface roughness requirements, microinches may be abbreviated μin.

Roughness. Roughness consists of the finer irregularities in the surface texture, usually including those irregularities which result from the inherent action of the production process. These are considered to include trav-

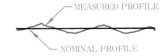

Fig. 6–74 An enlarged profile shows that a surface is not as it appears. (*ANSI.*)

erse feed marks and other irregularities within the limits of the roughness-width cutoff. See Fig. 6–75.

Roughness height. For the purpose of this standard, roughness height is rated as the arithmetical average deviation expressed in microinches measured normal to the center line. The preferred series of roughness height values is given in Table 1.

Fig. 6–75 Relation of symbols to surface characteristics. Refer to Fig. 6-78. (*ANSI.*)

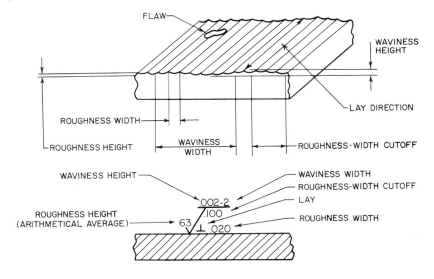

Table 1 Preferred Series Roughness Height Values (Microinches)				
	5	20	80	320
	6	25	100	400
1	8	32	125	500
2	10	40	160	600
3	13	50	200	800
4	16	63	250	1000

Table 3 Preferred Series Waviness Height Values (Inches)					
0.00002	0.00008	0.0003	0.001	0.005	0.015
0.00003	0.0001	0.0005	0.002	0.008	0.020
0.00005	0.0002	0.0008	0.003	0.010	0.030

Roughness width. Roughness width is the distance parallel to the nominal surface between successive peaks or ridges which constitute the predominant pattern of the roughness. Roughness width is rated in inches.

Roughness-width cutoff. The greatest spacing of repetitive surface irregularities to be included in the measurement of average roughness height. Roughness-width cutoff is rated in inches. Standard values are given in Table 2. Roughness-width cutoff must always be greater than the roughness width in order to obtain the total roughness height rating.

Table 2 Standard Roughness-Width Cutoff Values (Inches)					
When no value is specified, the value 0.030 is assumed					
0.003	0.010	0.030	0.100	0.300	1.000

Waviness. Waviness is the usually widely spaced component of surface texture and is generally of wider spacing than the roughness-width cutoff. Waviness may result from such factors as machine or work deflections, vibration, chatter, heat treatment or warping strains. Roughness may be considered as superimposed on a "wavy" surface.

Waviness height. Waviness height is rated in inches as the peak to valley distance. The preferred series of maximum waviness height is given in Table 3.

Waviness width. Waviness width is rated in inches as the spacing of successive wave peaks or successive wave valleys. When specified, the values shall be the maximum permissible.

Lay. Lay is the direction of the predominant surface pattern, ordinarily determined by the production method used. Lay symbols are shown in Fig. 6–79.

Flaws. Flaws are irregularities which occur at one place or at relatively infrequent or widely varying intervals in a surface. Flaws include such defects as cracks, blow holes, checks, ridges, scratches, etc. Unless otherwise specified, the effect of flaws shall not be included in the roughness height measurements.

Contact area. Contact area is the area of the surface required to effect contact with its mating surface. Unless otherwise specified, contact area shall be distributed over the surface with approximate uniformity. Contact area shall be specified as shown in Fig. 6–78 at E.

Designation of Surface Characteristics. Where no surface control is specified, it is to be assumed that the surface produced by the operation will be satisfactory. If the surface is critical, the quality of surface desired should be indicated.

Surface symbol. The symbol used to designate surface irregularities is the check mark with horizontal extension, as shown in Fig. 6–76. The point of the symbol shall be on the line indicating the surface, on the extension line, or on a leader pointing to the surface. The long leg and extension shall be to the right as the drawing is read. Where only roughness height is indicated, it shall be permissible to omit the horizontal extension. For typical applications of the symbol on a drawing, see Fig. 6–77.

Where the symbol is used with a dimension, it affects all surfaces defined by the dimension. Areas of transition, such as chamfers and fillets, shall conform with the roughest adjacent finished area unless otherwise indicated.

Surface-roughness symbols, unless otherwise specified, shall apply to the completed surface. Drawings or specifications for plated or coated parts shall definitely indicate whether the surface-roughness symbols apply before plating, whether the surface-roughness symbols apply after plating, or whether the surface-roughness symbols apply both before and after plating.

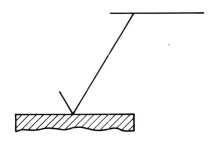

Fig. 6–76 The surface symbol. (*ANSI.*)

Fig. 6–77 The surface symbol on a drawing. (*ANSI.*)

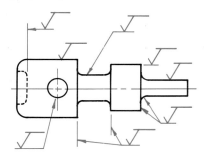

Application of symbols and ratings. Figure 6–78 illustrates the specification of roughness, waviness, and lay by insertion of the ratings in appropriate portions of the symbol. Only those ratings required to specify adequately the desired surface shall be shown in the symbol.

Symbols indicating direction of lay. Symbols for lay are shown in Fig. 6–79.

Roughness and waviness ratings, unless otherwise specified, shall apply in a direction which gives the maximum reading; this is normally across the lay.

This is the end of the material extracted (and adjusted) from ANSI *Surface Texture,* ANSI B46.1–1962. It is intended to suggest the consideration which may be given to surface quality. For further information and study, the complete standard should be used.

Fig. 6–78 Applications of symbols and ratings. *(ANSI.)*

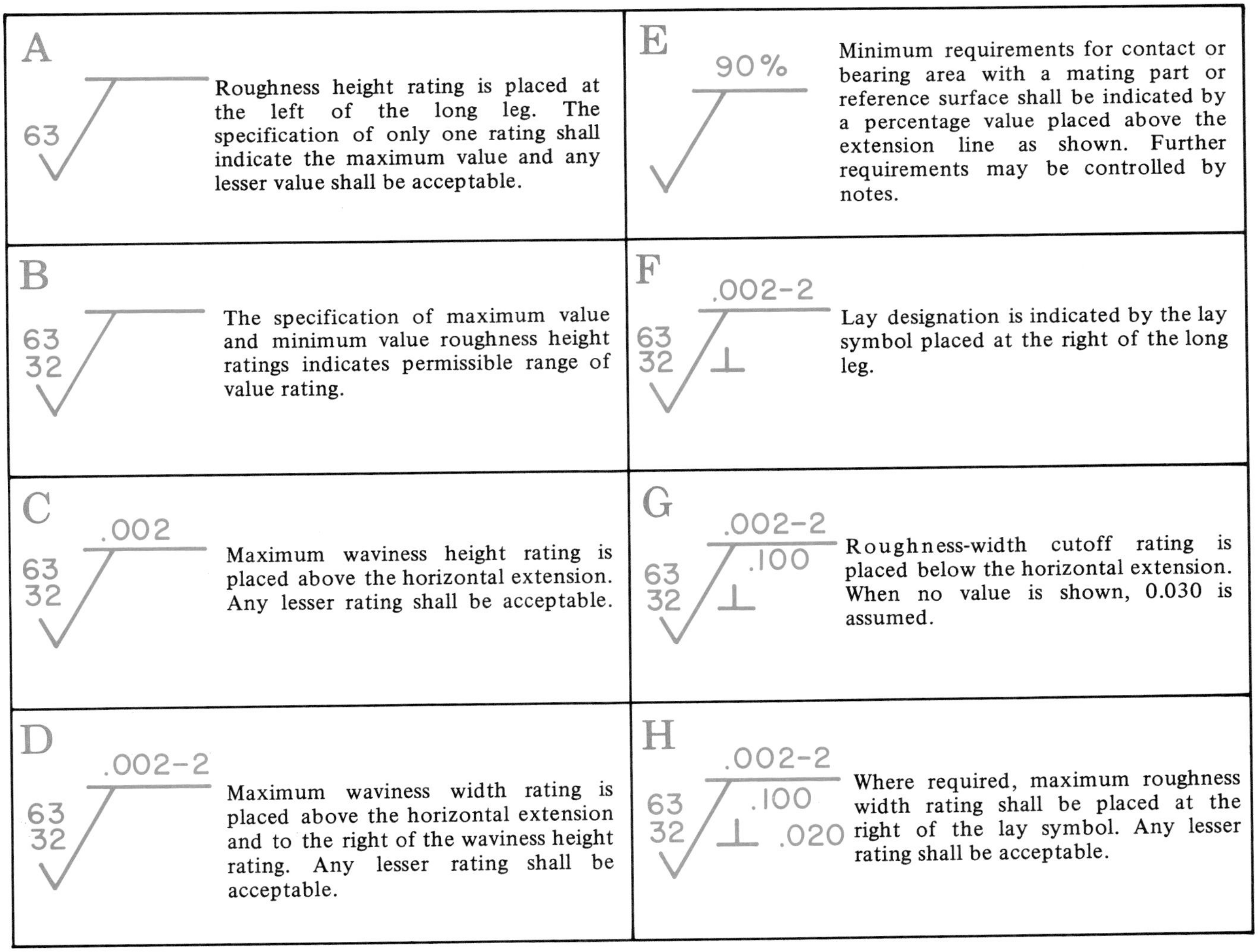

A 63 ∕	Roughness height rating is placed at the left of the long leg. The specification of only one rating shall indicate the maximum value and any lesser value shall be acceptable.
B 63 32 ∕	The specification of maximum value and minimum value roughness height ratings indicates permissible range of value rating.
C .002 63 32 ∕	Maximum waviness height rating is placed above the horizontal extension. Any lesser rating shall be acceptable.
D .002-2 63 32 ∕	Maximum waviness width rating is placed above the horizontal extension and to the right of the waviness height rating. Any lesser rating shall be acceptable.
E 90% ∕	Minimum requirements for contact or bearing area with a mating part or reference surface shall be indicated by a percentage value placed above the extension line as shown. Further requirements may be controlled by notes.
F .002-2 63 32 ∕ ⊥	Lay designation is indicated by the lay symbol placed at the right of the long leg.
G .002-2 .100 63 32 ∕ ⊥	Roughness-width cutoff rating is placed below the horizontal extension. When no value is shown, 0.030 is assumed.
H .002-2 .100 63 32 ∕ ⊥ .020	Where required, maximum roughness width rating shall be placed at the right of the lay symbol. Any lesser rating shall be acceptable.

‖	Lay parallel to the line representing the surface to which the symbol is applied.	 DIRECTION OF TOOL MARKS
⊥	Lay perpendicular to the line representing the surface to which the symbol is applied.	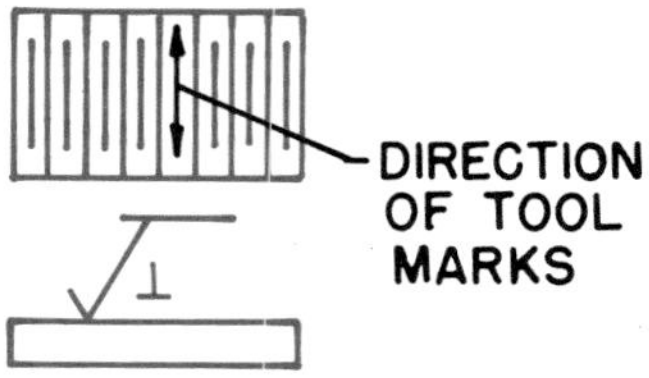 DIRECTION OF TOOL MARKS
X	Lay angular in both directions to line representing the surface to which symbol is applied.	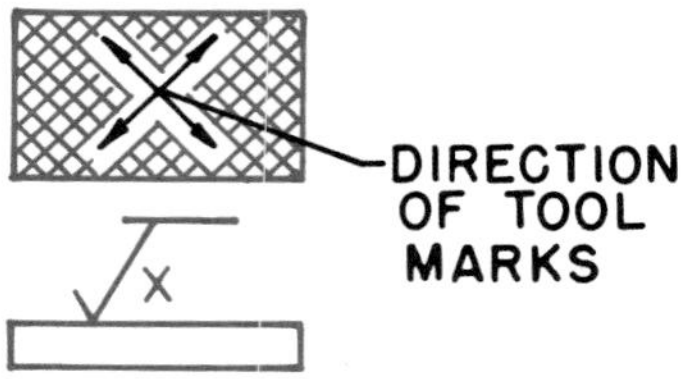 DIRECTION OF TOOL MARKS
M	Lay multidirectional.	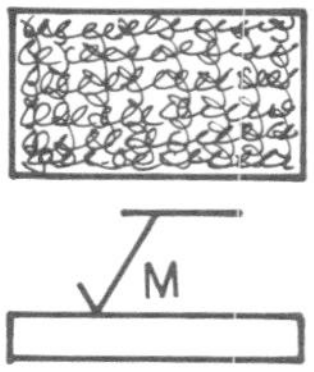
C	Lay approximately circular relative to the center of the surface to which the symbol is applied.	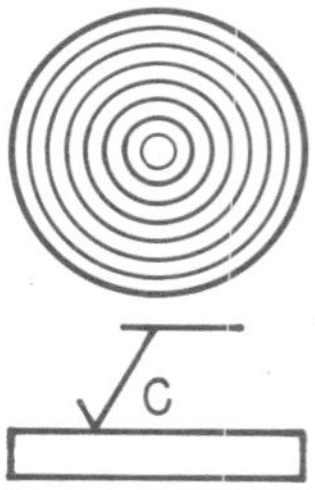
R	Lay approximately radial relative to the center of the surface to which the symbol is applied.	

Fig. 6–79 Lay symbols. (*ANSI.*)

Review

1. What are the two main elements necessary for a complete graphic description of an object?

2. What is another name for size description?

3. What unit of measure is used on metric drawings?

4. Name the method in which both English and metric units are given on the same drawing.

5. When placing dimensions on a drawing, there are two systems in common use. One is *aligned*. Name the other system.

6. There are two basic kinds of dimensions: (1) size dimensions and (2) _____ dimensions.

7. Draw the symbols for degrees, minutes, and seconds.

8. A complete circle is dimensioned as a diameter. An arc is dimensioned as a _____.

9. Name the abbreviations for the following: diameter, inch, center line, millimeter, radius, tolerance, outside diameter, countersink, material, and maximum.

10. The total amount by which a given dimension may vary is called _____.

11. There are two kinds of tolerance: unilateral and _____.

12. Specifying the maximum allowable variation in form and position is called _____.

13. Points, lines, or planes assumed to be exact for purposes of computation are called _____.

14. The surface of all material has some irregularities. This is called _____.

15. What is the finish mark used for on a drawing?

Problems for Chapter 6

Figures 6–80 through 6–92 offer a total of 54 dimensioning problems. Additional problems may be chosen from Chapters 2 through 5, or by selecting problems from later chapters.

The problems in Figs. 6–80 through 6–83 are to be done as follows:
1. Take dimensions from the printed scale at the bottom of the page, using dividers.
2. Draw the complete views.
3. Add all necessary extension and dimension lines for size and location dimensions.
4. Add arrowheads.
5. Fill in dimensions and add notes.

Figures 6–84 through 6–92 are more advanced, and will require more thought and time. For these problems use the following procedure:
1. Determine the necessary views, and prepare a freehand sketch.
2. Dimension the sketch.
3. Decide on a scale and draw views mechanically.
4. Add all necessary dimensions and notes.

Practice in decimal-inch, metric, or dual dimensioning may be obtained by converting the fractional-inch dimensions for any of the problems. It is recommended that students gain some experience in all methods.

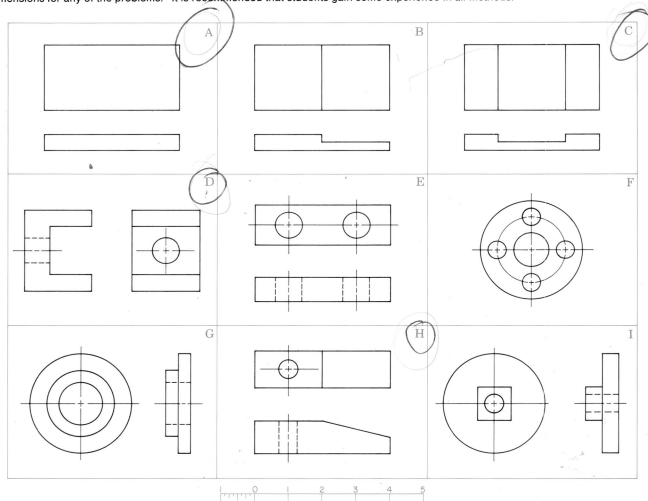

Fig. 6–80 Problems for dimensioning practice. Take dimensions from the printed scale, using dividers. Draw the views as shown and add all necessary size and location dimensions. Allow additional space for dimensions where necessary.

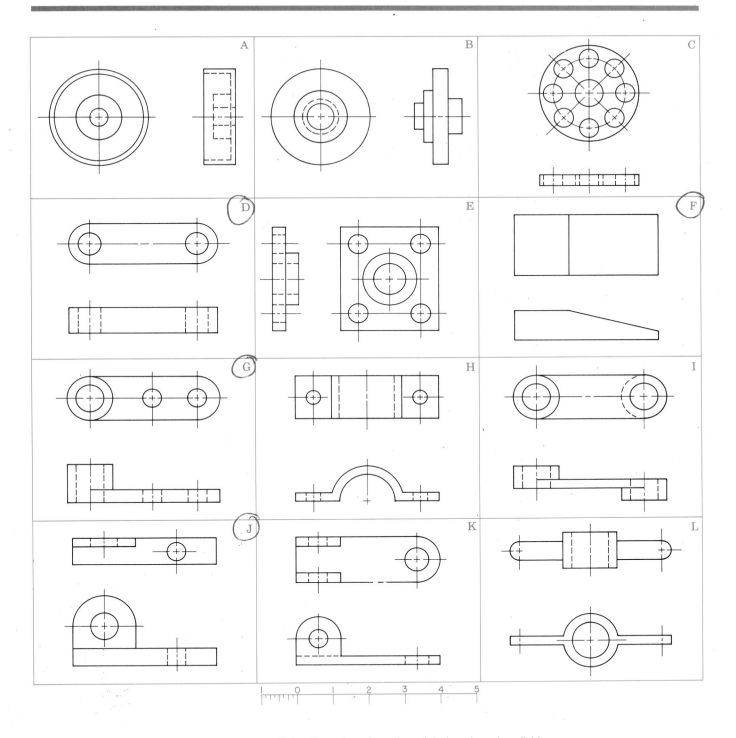

Fig. 6–81 Problems for dimensioning practice. Take dimensions from the printed scale, using dividers. Draw the views as shown and add all necessary size and location dimensions. Allow additional space for dimensions where necessary.

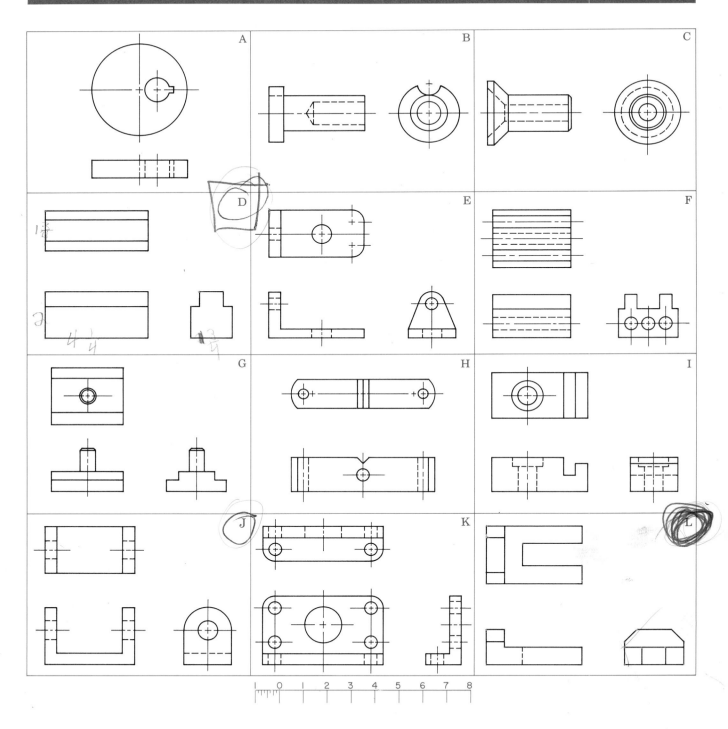

Fig. 6–82 Problems for dimensioning practice. Take dimensions from the printed scale, using dividers. Draw the views as shown and add all necessary size and location dimensions. Allow additional space for dimensions where necessary.

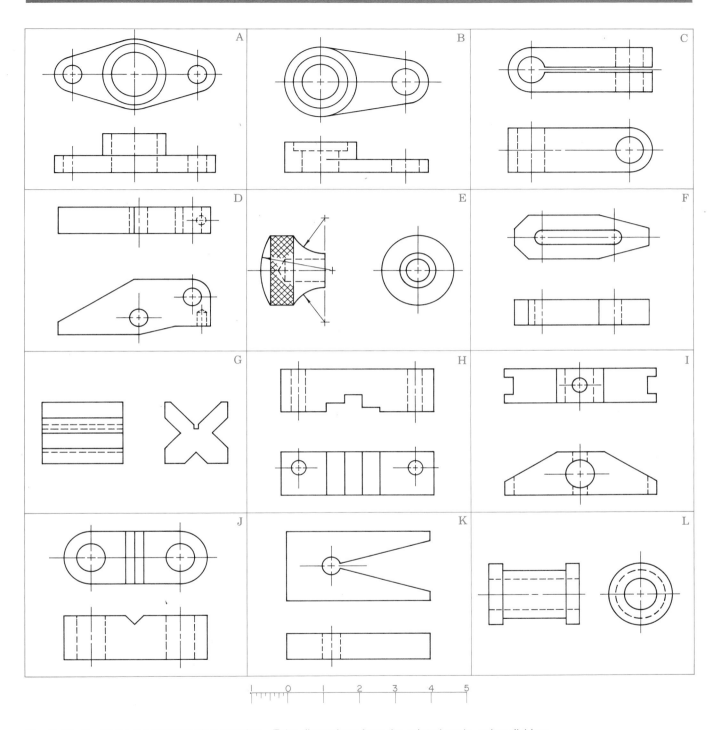

Fig. 6-83 Problems for dimensioning practice. Take dimensions from the printed scale, using dividers. Draw the views as shown and add all necessary size and location dimensions. Allow additional space for dimensions where necessary.

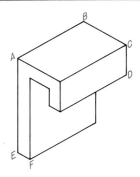

Fig. 6–84 Cut-off stop. Draw all necessary views and dimension. Scale: double size or as assigned. $AB = 40$ mm, $BC = 26$ mm, $CD = 17$ mm, $AE = 53$ mm, $EF = 7.50$ mm.

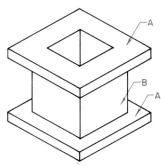

Fig. 6–85 Square guide. Draw all necessary views and dimension. Scale: full size or as assigned. $A = 5$ mm thick $\times$ 44 mm square, $B = 30$ mm square $\times$ 30 mm high, Hole = 20 mm square.

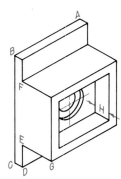

Fig. 6–86 Locator. Draw all necessary views and dimension. Scale: full size or as assigned. $AB = 40$ mm, $BC = 60$ mm, $CD = 5$ mm, $DE = 12$ mm, $EF = 36$ mm, $EG = 18$ mm, $H = 8$ mm, Hole = 10 mm DIA through, 18 mm Cbore, 2 mm deep.

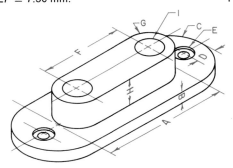

Fig. 6–87 Double-shaft support. Draw all necessary views and dimension. Scale: full size or as assigned. $A = 67$ mm, $B = 7$ mm, $C = 21$ mm R, $D = 10$ mm, $E = 5$ mm DIA through 10 mm Cbore, 2 mm deep, 2 holes, $F = 43$ mm, $G = 12$ mm R, $H = 14$ mm, $I = 12$ mm.

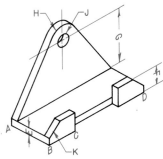

Fig. 6–88 Hanger. Draw all necessary views and dimension. Scale: $\frac{3}{4}$ size or as assigned. $AB = 1\frac{1}{4}$, $BC = \frac{13}{16}$, $BD = 3$, $E = \frac{3}{16}$, $F = \frac{1}{2}$, $G = 1\frac{9}{16}$, $H = \frac{7}{16}$ R, $J = \frac{3}{8}$ DIA, $K = 45°$.

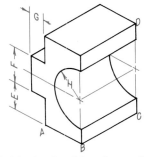

Fig. 6–89 Cradle slide. Draw all necessary views and dimension. Scale: full size or as assigned. $AB = 2\frac{3}{8}$, $BC = 3\frac{9}{16}$, $CD = 5\frac{1}{8}$, $E = 1\frac{1}{2}$, $F = 2\frac{1}{8}$, $G = \frac{7}{8}$, $H = 1\frac{5}{8}$ R.

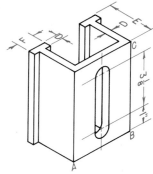

Fig. 6–90 Adjustable stop. Draw necessary views and dimension. Scale: $\frac{3}{4}$ size or as assigned. $AB = 3\frac{5}{8}$, $BC = 5\frac{1}{8}$, $D = \frac{5}{16}$, $E = 2\frac{5}{8}$, $F = 1$, slot = $\frac{7}{8}$ wide.

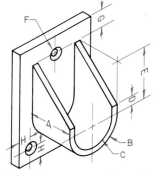

Fig. 6–91 Pipe support. Draw all necessary views and dimension. Scale: $\frac{1}{2}$ size or as assigned. Base plate = $\frac{1}{2}$ thick $\times$ $4\frac{1}{2}$ wide $\times$ $6\frac{1}{2}$ long, $A = 2\frac{3}{8}$, $B = 1\frac{1}{2}$ R, $C = 1\frac{1}{8}$ R, $D = \frac{1}{2}$, $E = 3$, $F = \frac{3}{8}$ DIA hole through, CSK to $\frac{3}{4}$ DIA, 3 holes, $G = 1$, $H = \frac{3}{4}$.

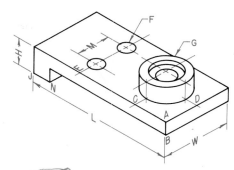

Fig. 6–92 Stop plate. Draw all necessary views and dimension. Scale: full size or as assigned. Overall sizes: $L = 4\frac{1}{4}$, $W = 2$, $H = \frac{3}{4}$. $AB = \frac{3}{8}$, $AC = 1$, $AE = 2\frac{3}{4}$, $AD = 1$, $JN = \frac{1}{2}$, $M = 1$, $F = \frac{7}{16}$ DIA, 2 holes, $G =$ Boss: $1\frac{1}{4}$ DIA $\times$ $\frac{1}{2}$ high, $\frac{1}{2}$ DIA through, $\frac{7}{8}$ Cbore = $\frac{1}{8}$ deep.

Auxiliary Views and Revolutions

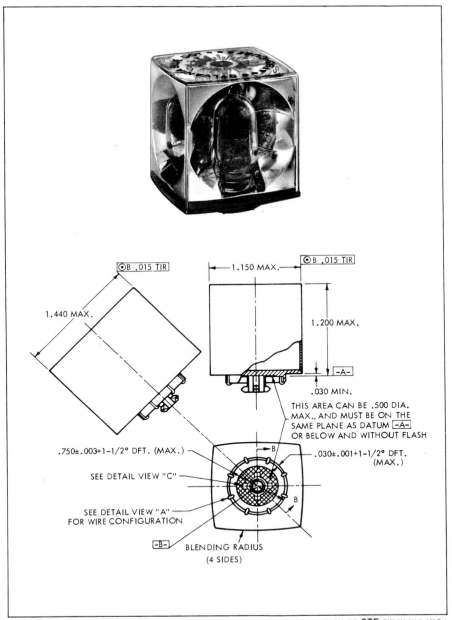

Chapter 7

⊙ B .015 TIR

1.440 MAX.

⊙ B .015 TIR

1.150 MAX.

1.200 MAX.

-A-

.030 MIN.

THIS AREA CAN BE .500 DIA. MAX., AND MUST BE ON THE SAME PLANE AS DATUM -A- OR BELOW AND WITHOUT FLASH

.750±.003+1-1/2° DFT. (MAX.)

SEE DETAIL VIEW "C"

SEE DETAIL VIEW "A" FOR WIRE CONFIGURATION

.030±.001+1-1/2° DFT. (MAX.)

B

-B-   BLENDING RADIUS (4 SIDES)

(COURTESY OF **GTE** SYLVANIA INC.)

Introduction to Auxiliary Views.
In Chapter 5, Multiview Projection, the three regular planes with the object in a normal position was explained. The three regular planes are the top, or horizontal plane; the front, or vertical plane; and the side, or profile plane. The regular planes can offer solutions to many graphic problems, but auxiliary planes of projection are necessary for describing solutions for inclined, or slanted, surfaces. In this chapter, problems with inclined surfaces are drawn on auxiliary planes that are parallel to the inclined surfaces (Fig. 7–1).

Auxiliary Views Are "Helper Views." When an object has inclined (slanting) surfaces, the regular views do not show the true shape of such surfaces (Fig. 7–2). The true size and shape of the inclined surface will not be projected on the top view, front view, or side view. However, a view on a plane parallel to the inclined surface will show the true shape of the incline surface, as at B. The views at B give a better description than the views at A.

An auxiliary view is a projection on an auxiliary plane that is parallel to an inclined (slanting) surface. It is a view looking directly at the inclined surface in a direction perpendicular to it.

The anchor pictured in Fig. 7–3 in Space A has three circular features that will appear and project as ellipses in regular views. In Space B the

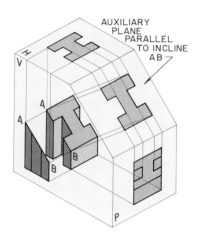

Fig. 7-1 Pictorial study of a primary auxiliary view.

projected views are more difficult to draw and to understand. The anchor in space C is completely described by two views, one of which is an auxiliary view.

Auxiliary projections are important for describing the geometric shapes which are features of inclined surfaces. They are also necessary for dimensioning such features.

The Relationship of Auxiliary Views to Regular Views. In Fig. 7-4 the regular views are illustrated at A. The simple, inclined wedge block has

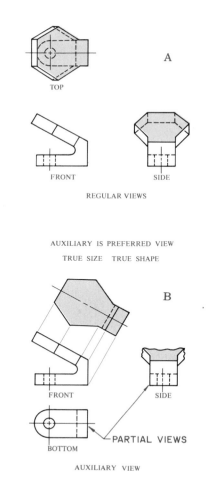

Fig. 7-2 Compare the regular views at A with the auxiliary view at B.

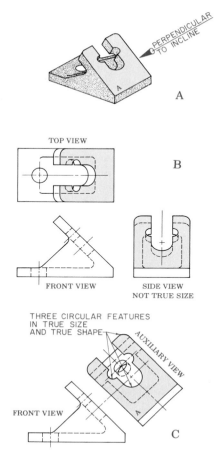

Fig. 7-3 The pictorial view at A and the three-view drawing at B are difficult to draw.

Fig. 7-4 Basic relationship of the auxiliary view to three-view drawing.

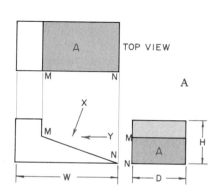

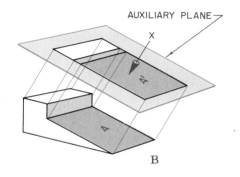

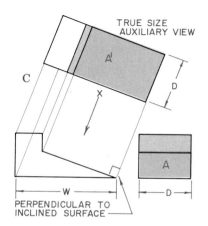

a slanted surface indicated by the line *MN* in the front view. The side view is obtained by looking in direction of arrow *Y*. The surface *A* does not show in true shape in the side view or in the top view. The line *MN* is obviously shorter in the side and top views. If we locate a plane parallel to face *A*, as at *B*, and look in the direction of arrow *X* perpendicular to face *A*, we can obtain an auxiliary plane which will show the true size and shape of surface *A* at A^1. Such an auxiliary view made on an auxiliary plane and revolved into the plane of the paper is shown at C. The true size and shape of any inclined surface may be shown in a way similar to Fig. 7–4.

The auxiliary plane in relation to the regular planes. In Fig. 7–5, Space A, the object is shown with the vertical, horizontal, and profile planes. Auxiliary views are generally projected from one of the principal planes. Therefore, it is convenient to consider the planes to be *hinged,* as illustrated. At B the planes have been revolved to show the front, top, and side views of the wedge block. At C the object is shown behind the vertical, auxiliary, and profile planes. Note that the hinge line *XY* at C and D is parallel to the inclined surface. The top and side views at D could be

omitted since the object is completely described by the front and auxiliary views.

Kinds of Auxiliary Views. The primary auxiliary views are identified with the three principal planes and the three principal dimensions. Auxiliary views are classified according to the position of the planes upon which they are drawn. There are three primary auxiliary views (Fig. 7–6): at A, the front auxiliary view, which is hinged on the front view and has the depth dimension as its primary reference; at B, the top auxiliary view, which is hinged to the top view and has the height as its primary reference; at C, the right-side auxiliary view, which is hinged to the side view and has the width as its primary reference. The three principal planes always show the auxiliary plane as an inclined line (*MN*). This line is considered to be the edge view of the plane.

Constructing an Auxiliary View. The following steps may be used for any primary auxiliary view (Fig. 7–7).

A. Examine the views that are given for an inclined surface.

B. Identify the line that is considered the edge view of the inclined plane.

C. Draw a light construction line (line of sight) perpendicular to the inclined surface in the front view.

D. Consider the auxiliary plane as being hinged to the vertical plane from which it is developed.

E. Draw the projection lines from all points labeled on the front view perpendicular to the inclined line (parallel to the line of sight).

F. Draw a reference line parallel to the edge view of the inclined surface at a convenient distance from the inclined surface.

G. Transfer the depth dimension, which is the primary reference, to the reference line as shown.

H. Project the labeled points and connect them in sequence to form the auxiliary view. The points used to identify the shape are for solving difficult problems (instructional purposes) and are not normally left on the final drawing.

The pictorial of an object needing an auxiliary view is shown in Fig. 7–8 at A. For a symmetrical object a center plane is used as a reference plane, as at B (center plane construction). The edge view of the center plane appears as a center line (Fig. 7–8B, line *XY*). Number the points on the inclined surface. Draw the

Fig. 7–5 Basic relationship of the auxiliary plane to the regular planes.

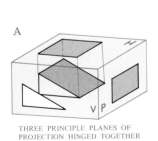

THREE PRINCIPLE PLANES OF PROJECTION HINGED TOGETHER

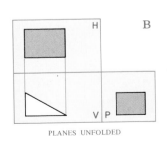

PLANES UNFOLDED

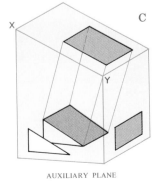

AUXILIARY PLANE

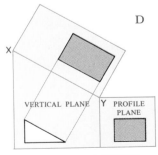

AUXILIARY PLANE REVOLVED

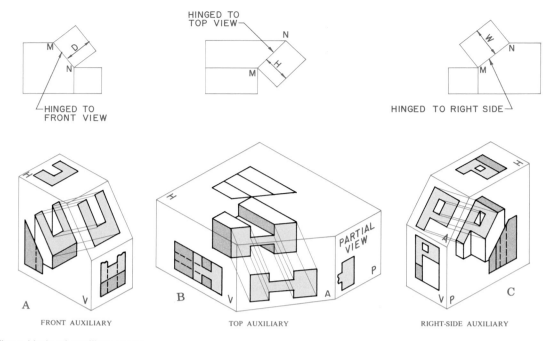

Fig. 7-6 Three kinds of auxiliary views.

FRONT AUXILIARY   TOP AUXILIARY   RIGHT-SIDE AUXILIARY

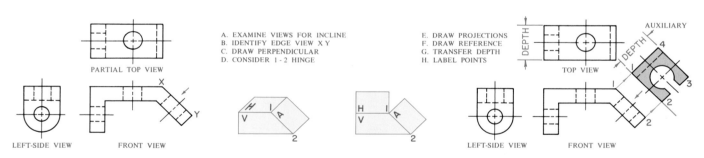

A. EXAMINE VIEWS FOR INCLINE
B. IDENTIFY EDGE VIEW X Y
C. DRAW PERPENDICULAR
D. CONSIDER 1-2 HINGE
E. DRAW PROJECTIONS
F. DRAW REFERENCE
G. TRANSFER DEPTH
H. LABEL POINTS

Fig. 7-7 Steps in constructing an auxiliary view.

Fig. 7-8 To draw an auxiliary view using the center plane reference.

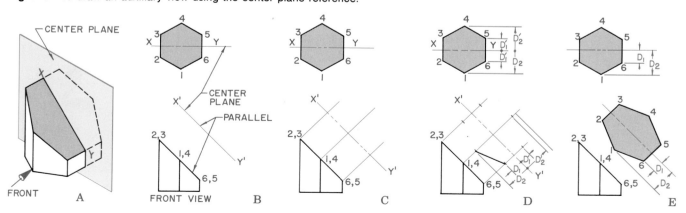

line X^1Y^1 parallel to the edge view of the inclined surface and at a convenient distance from it. Draw projecting construction lines perpendicular to the inclined surface from each point in the front view and extend them beyond X^1Y^1, as shown at C. For each of the lines just drawn at C, locate the depth measurement in the top view which is related to the points 2-3, 1-4, and 6-5 (Fig. 7-8D). Distances D_1 and D_2 are measured from the center line XY in the top view and are transferred to both sides of the center line X^1Y^1 in the front auxiliary.

Since the figure is symmetrical, the inclined surface is completed by joining the numbered points as at E. Only the inclined surface has been drawn, but the entire object could be projected from the center reference plane.

An auxiliary view with vertical reference-plane construction is drawn in Fig. 7-9. The nonsymmetrical object, such as shown in Fig. 7-9, is placed on reference planes. The plane is placed for the convenience of taking reference measurements. The reference plane can be placed in front of or in back of the object. In this case

the reference plane is placed in back of the object. The construction is similar to Fig. 7-8, except that all distances D_1, D_2, and D_3 are laid off in front of the reference plane. The entire object has been projected to the auxiliary plane. This is a front auxiliary with the depth as the primary reference dimension.

A top auxiliary view with horizontal reference-plane construction is drawn

in Fig. 7-10. The molding shown in Fig. 7-10 at A has a mitered corner (45° cut). The reference plane shown will have the height as its principal reference dimension. The reference plane XY will be placed under the molding. Locate points 1 to 6 in the right side and top views. Draw the projecting lines perpendicular to X^1Y^1 from each point in the top view. Take distances up from XY in the side view

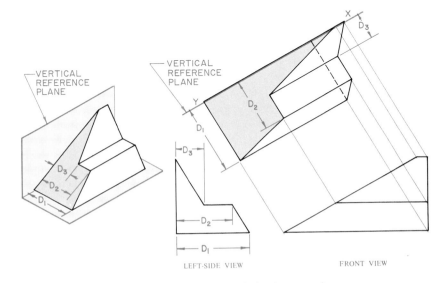

Fig. 7-9 To draw a front auxiliary using a vertical reference plane.

Fig. 7-10 To draw a top auxiliary view with a horizontal reference plane.

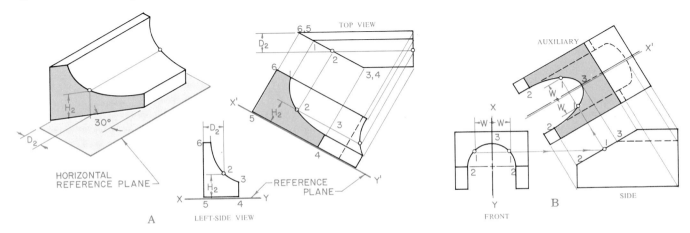

and lay them off up from X^1Y^1 in the auxiliary view, as typically shown H_2 for point 2, H_3 for point 3, etc. Locate additional points on the curve as necessary to define it accurately on the top auxiliary view.

Curves on Auxiliary Views. Auxiliary views with curved lines are obtained by locating a number of points on the curves. This is illustrated in a simple curve form in Fig. 7–10 at B. In Fig. 7–11 the auxiliary view of the inclined surface of a cylinder is represented with the cylinder in a horizontal position. In this case the vertical center line XY (representing a center reference plane) shows in the side view of the cylinder. It is the line (XY) from which measurements D_1, D_2, and so forth are made and transferred to the auxiliary reference line X^1Y^1.

Locate a convenient number of points on the circular form of the side view. The accuracy of the plotted curve will increase with the number of points located in the side view. Project these points to the edge view of the inclined surface in the front view. Draw the center line X^1Y^1 parallel to the edge view at a convenient distance from it. From the points on the inclined surface (edge view), extend projection lines perpendicular to X^1Y^1 and across the center line. Take the distance on each side of XY in the side view D_1, D_2, and D_3 and set them off from X^1Y^1, as illustrated, to form the ellipse. This is a front auxiliary with the depth as the primary reference dimension.

Partial Auxiliary Views. Figure 7–12 at A shows partial auxiliary views. Complex curves can be omitted and descriptions completed through the proper use of break lines and center lines, as illustrated. Half views can be drawn if symmetry is simple and clearly understood.

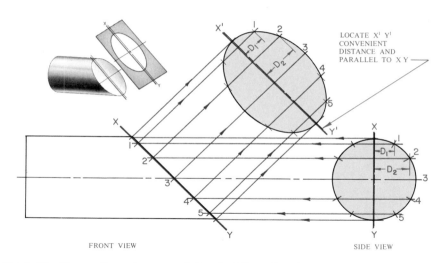

Fig. 7–11 The auxiliary-view curve (ellipse) of the cut surface of a cylinder.

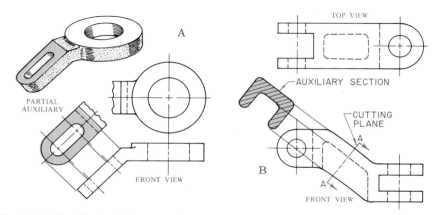

Fig. 7–12 Practical solutions—partial auxiliary views and auxiliary sections.

Auxiliary Sections. Another practical use of auxiliary views is the auxiliary section (Fig. 7–12 at B), where the inclined surface is obtained by a plane which cuts through the object. (See Sectional Views, Chapter 9.) The area in cross section (cross hatched) may be found by using the cutting-plane line AA.

Secondary Auxiliary Views. When a projection is taken from a primary auxiliary view, the result is called a *secondary auxiliary* (Fig. 7–13). Primary auxiliaries could be used to describe the true size and shape of in-

clined surfaces, but oblique surfaces (surfaces which are inclined to all three of the regular planes) must be solved with secondary auxiliary (successive auxiliary) projection.

Surface 1-2-3-4 in Fig. 7–13 is inclined to the three regular planes. At A an auxiliary has been drawn on a plane perpendicular to the inclined surface. Note at A that points 1, 2, 3, and 4 appear as a line or edge view of the plane. At B an auxiliary view has been drawn from the view A on a plane parallel to the surface 1-2-3-4. The true shape of the surface is found with a successive auxiliary.

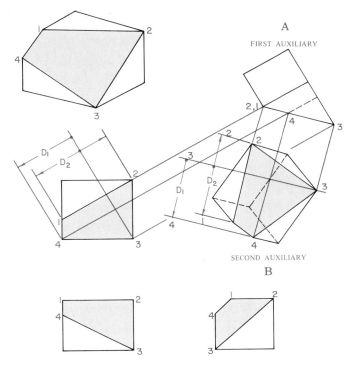

OCTAHEDRON

Fig. 7–13 Secondary auxiliary views assist in finding the true shape of a surface.

Another example is shown in Fig. 7–14. The octahedron (eight triangles composing a regular solid) is shown in three views. A first auxiliary is shown at A on a plane perpendicular to the triangle surface 0-1-2. Note that line element 1-2 in the top view appears as a point projection in the auxiliary at A and that the triangle now appears as line 0-1, 2. A secondary auxiliary view is drawn at B on a plane parallel to the edge view of plane 0-1-2. The true shape of triangle 0-1-2 is shown in the successive auxiliary at B.

Introduction to Revolutions. The basic reason for revolving an object behind a regular plane is closely related to the reason for taking an auxiliary view. Revolutions can help to obtain the true size and shape of inclined surfaces which do not show true size and shape in their regular

positions. In the revolution process the regular planes of multiview projection are maintained, and one view of the object is revolved (Fig. 7–15). The auxiliary-view process called for a new reference plane and a change in the direction of viewing the plane. The revolution process should assist in the understanding of auxiliary views, and they both will be examined for spatial problems in Chapter 8, Basic Descriptive Geometry.

The Axis of Revolution. A convenient method for thinking about an object in a revolved position is to imagine that a shaft or axis has been passed through the object. The

Fig. 7–14 True shape by secondary auxiliary projection.

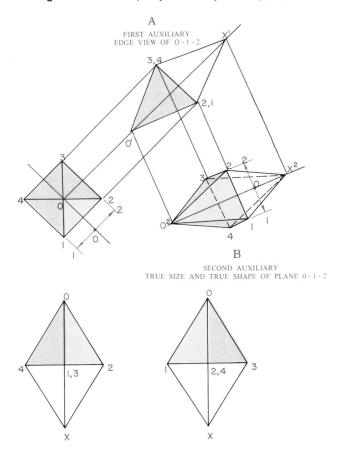

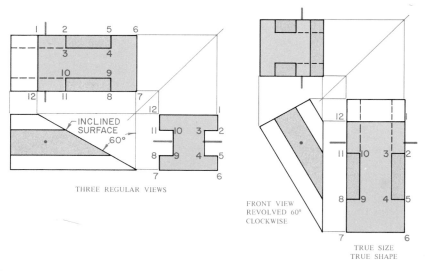

THREE REGULAR VIEWS

FRONT VIEW
REVOLVED 60°
CLOCKWISE

INCLINED
SURFACE
60°

TRUE SIZE
TRUE SHAPE

Fig. 7–15 The regular planes remain in order in revolution. The object is revolved.

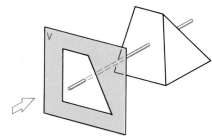

AXIS PERPENDICULAR TO HORIZONTAL

FRONT

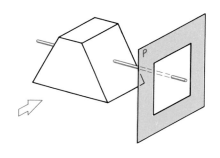

AXIS PERPENDICULAR TO VERTICAL

AXIS PERPENDICULAR TO PROFILE

Fig. 7–16 Three positions for the axis of revolution.

imaginary axis passes through the object and is perpendicular to one of the principal planes. In Fig. 7–16, the three principal planes are illustrated with an axis passing through each normal view before revolution.

An object may be revolved to the right (clockwise) or to the left (counterclockwise) about an axis perpendicular to either the vertical or the horizontal planes of projection. An object may be revolved forward (counterclockwise) or backward (clockwise) about an axis perpendicular to the profile plane.

Single Revolution. As stated, an axis of revolution may be taken perpendicular to the vertical, horizontal, or profile plane. In Fig. 7–17 at A, the usual front and top views are shown in Space 1. The views after the object has been revolved 45° clockwise about an axis perpendicular to the vertical plane are shown in Space 2. Notice that the front view is the same in size and shape as in Space 1, except that it has a new position. The new top view is obtained by projecting up from the new front view and across from the old top view in Space 1. Note

Fig. 7–17 Single revolution about each of the three axes.

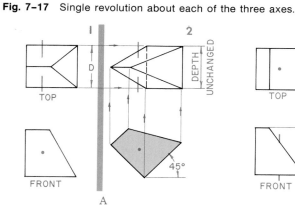

A

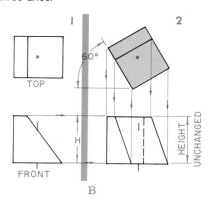

B

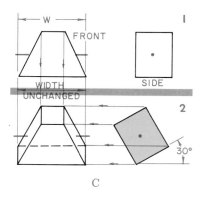

C

that the depth remains the same in both the front and top views.

In Fig. 7-17 at B, the usual top and front views are shown in Space 1. In Space 2 the top view is shown after it has been revolved to the right (clockwise) through 60° about an axis perpendicular to the horizontal plane. It remains unchanged in size and shape. The new front view is obtained by projecting down from the new top view and across from the old front view of Space 1. Note that the height remains unchanged.

In Fig. 7-17 at C, the usual front and side views are shown in Space 1. In Space 2 the side view is shown after the object has been revolved forward (counterclockwise) through 30° about an axis perpendicular to the profile plane. The new front view is obtained by projecting across from the new side view and down from the old front view in Space 1. Note that the width remains the same in Spaces 1 and 2. Revolution may be clockwise, as at Fig. 7-17B, or counterclockwise, as at A and C.

The Rules of Revolution. *One:* The view that is perpendicular to the axis of revolution is unchanged except in position. (This is true because the axis is perpendicular to the plane on which it is projected.) *Two:* Distances parallel to the axis of revolution are unchanged. (This is true because they are parallel to the plane or planes on which they are projected.) Figure 7-18 illustrates the two parts of the rule of revolution.

Revolution About an Axis Perpendicular to the Vertical Plane. Figure 7-19 shows the method of drawing a primary revolution. The imaginary axis AX is a horizontal axis which appears as a circle in the front view and as a line in the top and side views. The truncated right octagonal prism is represented by two views in the usual position at A. The object is revolved about the axis into a new position as at B. It will be observed that the front view revolved about a horizontal axis at B is the same as A, except that its position has been changed. The new side view shows the true size of the hexagonal shape and is obtained by projecting across from the new front view and by the usual transfer to depth from the top view.

Revolution About an Axis Perpendicular to the Horizontal Plane. Figure 7-20 shows the method of drawing views of an object that are revolved about an imaginary vertical axis AX. The three views are given at A. The top view is to be revolved 30° clockwise about the vertical axis AX. First, draw the top view in its new position, as at B. Since the axis is vertical, the height is not changed; so project points in the vertical plane labeled at A to similar points at B. The side view is obtained from the front and top views in the usual manner. The object shown in Fig. 7-21 is an example of one revolved counterclockwise through 45°.

Practical Revolved Views. When working drawings have inclined surfaces, the object may be described by drawing one of the views, or part of a view, in a revolved position. In Fig. 7-22 at A the top view shows the angle of the V-shaped part. The front view shows the V-shaped part revolved in the true shape position. In Fig. 7-22 at B the inclined angles of the part are shown in the front view. The inclined surfaces are revolved in the front projection and are transferred to the top view, where they appear in true size and shape.

Successive Revolutions. After an object has been revolved once about an axis perpendicular to a plane, it

Fig. 7-18 The rule of revolution.

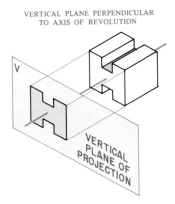

VERTICAL PLANE PERPENDICULAR
TO AXIS OF REVOLUTION

V

VERTICAL PLANE OF PROJECTION

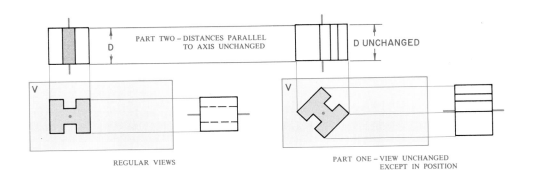

PART TWO – DISTANCES PARALLEL
TO AXIS UNCHANGED

D

V

REGULAR VIEWS

D UNCHANGED

V

PART ONE – VIEW UNCHANGED
EXCEPT IN POSITION

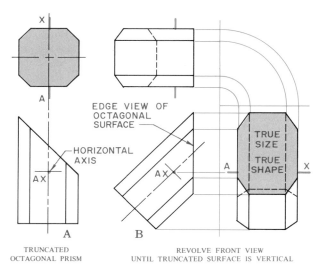

Fig. 7–19 Revolution about an axis perpendicular to the vertical.

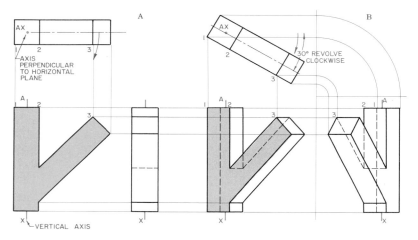

Fig. 7–20 Revolution about an axis perpendicular to the horizontal plane (clockwise).

Fig. 7–21 Revolution about an axis perpendicular to the horizontal plane (counterclockwise).

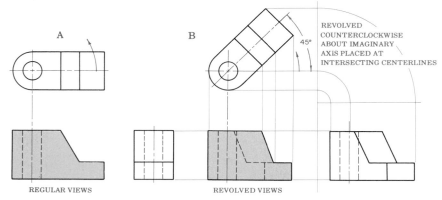

may be revolved about an axis perpendicular to another plane, as shown in Fig. 7–23 at B and C. At B the object has been revolved 30° clockwise about an axis perpendicular to the horizontal plane. At C the front view from B has been revolved 45° clockwise about an axis perpendicular to the vertical plane. At D the side view taken from C has been revolved about an axis perpendicular to the profile plane until line 3-4 appears as a horizontal line. Three revolutions occurred, one in each of the three principal planes of projection.

Auxiliary Views and Revolved Views. The true size of an inclined surface can be obtained by an auxiliary view (Fig. 7–24 at A) or by revolving the object until the inclined surface is parallel to one of the principal planes. At B and C the revolved view is similar to the auxiliary at A.

The auxiliary view causes the observer to shift into a new visual position (change direction of sight). Conversely, the revolution drawing causes

Fig. 7–22 Practical use of revolutions.

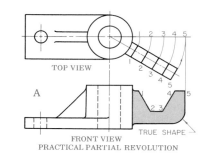

FRONT VIEW
PRACTICAL PARTIAL REVOLUTION

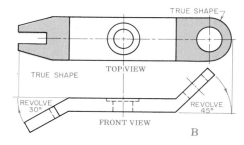

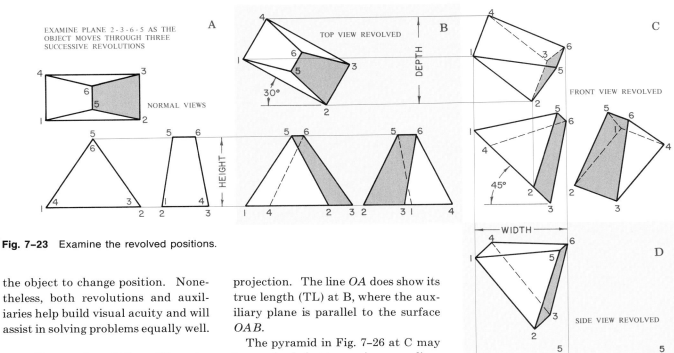

Fig. 7-23 Examine the revolved positions.

the object to change position. Nonetheless, both revolutions and auxiliaries help build visual acuity and will assist in solving problems equally well.

True Shape of an Oblique Plane by Successive Revolutions. A surface will show in its true shape when it is parallel to a plane. In Fig. 7-25 the surface 1-2-3-4 of the object pictured at D is an oblique plane because it is inclined to all three of the normal planes. At A the object is drawn in its normal position. At B the object has been revolved about an axis perpendicular to the horizontal plane until surface 1-2-3-4 is perpendicular to the vertical plane. (It appears as an edge view in the front view.) At C the object has been revolved about an axis perpendicular to the front plane until surface 1-2-3-4 is parallel to the profile plane, where it shows its true shape.

True Length of a Line. Since an auxiliary view of an inclined surface shows its true size and shape, it may be used to find the true length of a line. In Fig. 7-26 at A, the line OA does not show its true length in the top, front, or side view because it is inclined to the principal planes of

projection. The line OA does show its true length (TL) at B, where the auxiliary plane is parallel to the surface OAB.

The pyramid in Fig. 7-26 at C may be revolved about an axis perpendicular to the vertical plane until surface OAB is parallel to the profile plane. The true size of OAB will show in the side view, as will the TL of OA. A shorter method is to revolve only the surface OAB, as shown at D, to determine the TL of OA.

If the top view of the pyramid is revolved as in Fig. 7-26 at E so that the top view of line OA is horizontal (parallel to vertical plane), the front view of OA will then show its true length because it is parallel to the

vertical plane and perpendicular to the profile plane.

Instead of revolving the top of the pyramid, just line OA of the top may be revolved until it is parallel to a principal plane of projection. In Fig.

Fig. 7-24 True size shown by an auxiliary view and a revolved view.

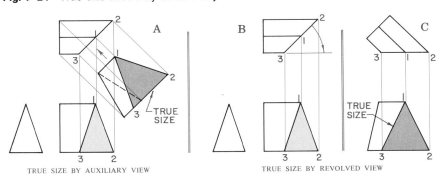

TRUE SIZE BY AUXILIARY VIEW

TRUE SIZE BY REVOLVED VIEW

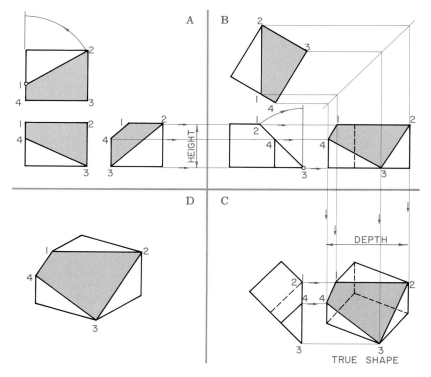

Fig. 7-25 True size of an oblique plane by revolution.

Fig. 7-26 Typical true-length problems examined.

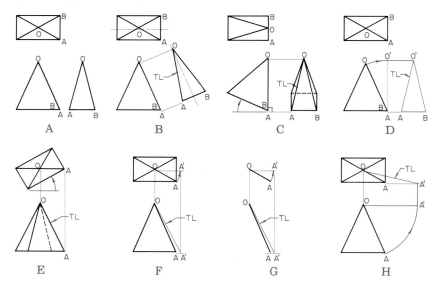

7-26 at F and G, the line has been revolved until its top view is horizontal at OA^1. The point A^1 then can be projected to the front view, where OA^1 will be in true length.

Revolving lines can occur in any view so as to make them parallel to any one of the three principal planes. Projecting the line on the plane to which it is parallel will show its true length. In Fig. 7-26 at H, the line has been revolved parallel to the horizontal plane. The true length is shown in the top view.

Industrial Applications. The following illustrations are examples of the draftsman's use of revolution. Figure 7-27 illustrates a tractor with lift positions. Figure 7-28 shows a product designer's use of a revolved position. In Fig. 7-29 surveying equipment must revolve to deal with varying topography. Figure 7-30 shows the double revolution capability of a transit.

Fig. 7-27 The profile of a tractor shows several positions of the loader. The plan view shows additional revolutions of the tractor.

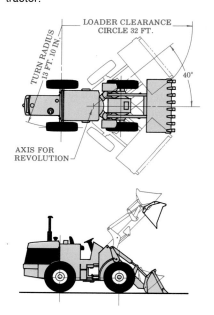

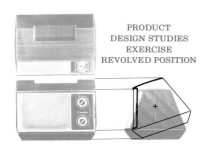

Fig. 7–28 Note that the television profile has been revolved to show the screen in true shape. (*Motorola.*)

Fig. 7–29 Civil engineering requires a knowledge of revolutions. Engineers use an automatic level to examine topography.

Fig. 7–30 Note the horizontal and vertical planes of revolution. A transit must be able to revolve in both a horizontal and a vertical plane. (*Eugene Dietzgen Co.*)

Review

1. Why are auxiliary views necessary?

2. How is the auxiliary plane placed in relationship to the inclined surface?

3. What are the two most important reasons for drawing auxiliary projections?

4. Name the three primary auxiliary projections.

5. List the steps in the construction of an auxiliary view.

6. What is meant by *reference-plane construction?*

7. Can curved lines be plotted on auxiliary views?

8. Are partial auxiliary views an accepted drawing practice according to industry standards?

9. What auxiliary projection is used for finding the true size of an oblique surface?

10. What is the basic reason for revolving the view of an object?

11. What is meant by the *axis of revolution?*

12. In your own words, describe the first rule of revolution.

13. Name the three basic single revolutions.

14. What is the primary use of successive revolutions?

15. Can both auxiliary views and revolved views be used to determine the true lengths of inclined and oblique lines?

Problems for Chapter 7

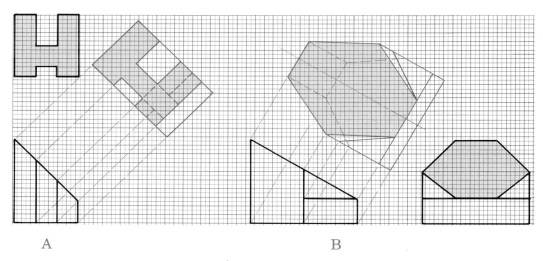

A   B

Fig. 7–31 Problem 1: Draw the front, top, and side view of each figure, A and B. Complete the front auxiliary projection. Problem 2: Change the angle of the inclined surface at A to 30° and at B to 45°. Scale: each square = ¼".

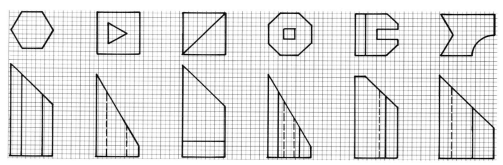

Fig. 7–32 In each of the six figures, draw the front and top views, and then complete the front auxiliary view. Scale: each square = ¼".

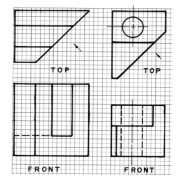

Fig. 7–33 In each of the two figures, draw the front and top views, and then complete the top auxiliary view. Scale: each square = ¼".

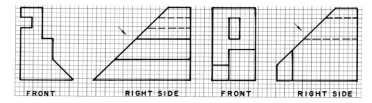

Fig. 7–34 In each of the two figures, draw the front and side views and complete the side auxiliary view. Scale: each square = ¼".

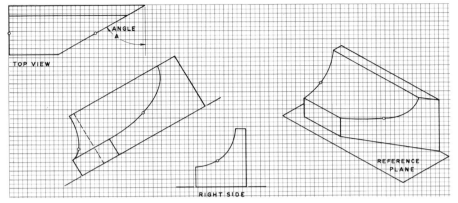

TOP VIEW

ANGLE A

RIGHT SIDE

REFERENCE PLANE

Fig. 7–35 Develop the three views, changing angle A to 45°. Complete the top auxiliary projection. Scale: each square $= \frac{1}{4}''$.

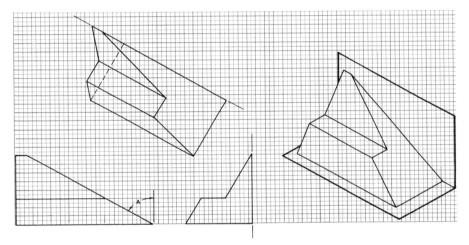

A

Fig. 7–36 Determine the necessary views, letting angle $A = 60°$. Complete the front auxiliary view. Scale: each square $= \frac{1}{4}''$.

Fig. 7–38 Draw the three normal views, using 45° for angle 0-2-1 in the front view and 0-1-2 in the side view. Find the true size of plane 0-1-2. Let the first auxiliary plane be perpendicular to line 1-2 in the top view. The second auxiliary plane is perpendicular to the edge view of plane 0-1-2. Scale: each square $= \frac{1}{4}''$.

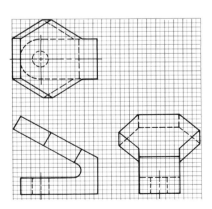

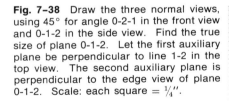

Fig. 7–37 Determine the necessary views, and develop a partial front auxiliary view. Scale: each square $= \frac{1}{4}''$.

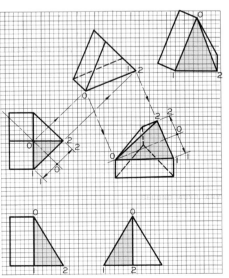

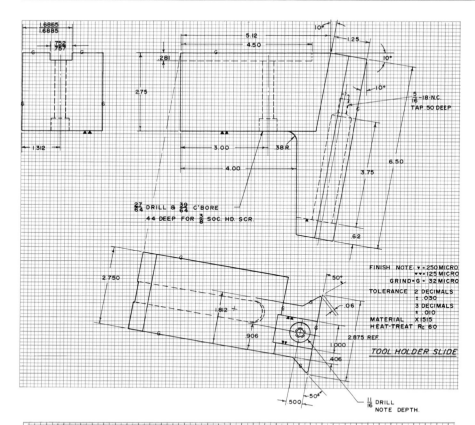

Fig. 7–39 Make a working drawing of the tool holder slide with the following revisions: change the 10° angle to 15°, change the 5.12 dimension to 4.12, change any other dimension affected by the length change to keep the counterbored hole in its present position.

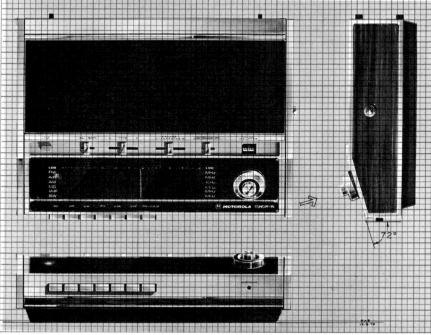

Fig. 7–40 Make a full-size three-view drawing of the table radio, using a scale of one square = $\frac{1}{4}$″. Allow space to draw a right-side auxiliary view of the dial face. (*Motorola, Inc.*)

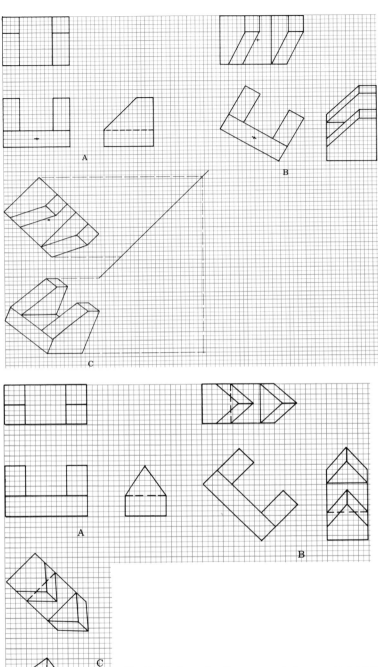

Fig. 7-41 Draw the three normal views as at A, using the scale of one square = $\frac{1}{8}$". At B, the front view is revolved 45° clockwise, showing a new top and right-side view. At C, the top view from B is revolved 45° clockwise and a new front view is shown. Draw the two views as shown at C and complete the side view. Using B for guidance, revolve the front view 30° or 60° clockwise as directed and complete the top and side views. Using C for guidance, revolve your top view 30° or 60° clockwise as directed and complete the front and side views. Using the views at C in which you have revolved the top view 30°, 45°, or 60°, revolve the side view 15° clockwise and develop new front and top views.

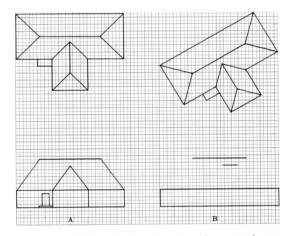

Fig. 7-43 Draw the two views as shown at A, using the scale of one square = $\frac{1}{8}$". Using B as guidance, revolve the top view 30° counterclockwise. Complete the drawing with a front and right-side view.

Fig. 7-42 Draw the three normal views as at A, using the scale of one square = $\frac{1}{8}$". Using B as guidance, revolve the front view 45° clockwise and complete the top and side views. Using C for guidance, revolve the top view from B 30° counterclockwise and complete the front and side views. Revolve the side view from C 15° counterclockwise and complete the front and top views.

Basic Descriptive Geometry | Chapter **8**

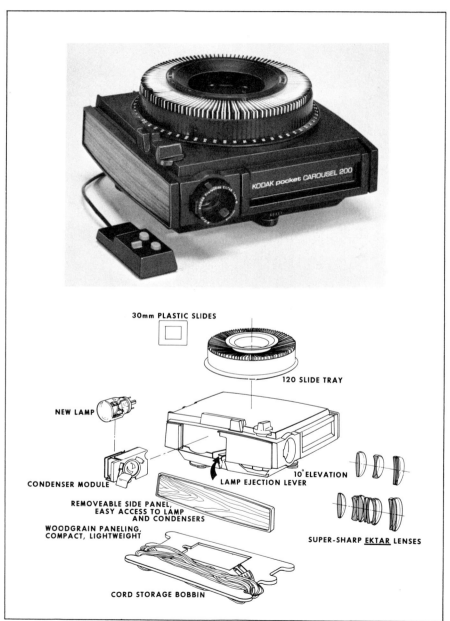

30mm PLASTIC SLIDES

120 SLIDE TRAY

NEW LAMP

CONDENSER MODULE

REMOVEABLE SIDE PANEL,
EASY ACCESS TO LAMP
AND CONDENSERS

WOODGRAIN PANELING,
COMPACT, LIGHTWEIGHT

10° ELEVATION

LAMP EJECTION LEVER

SUPER-SHARP EKTAR LENSES

CORD STORAGE BOBBIN

(EASTMAN KODAK COMPANY.)

Graphics and Mathematics. The designer working along with an engineering team can solve problems graphically with geometric elements. Structures that occupy space have three-dimensional forms made up of a combination of geometric elements (Fig. 8–1). The graphical solutions of three-dimensional forms require an understanding of the space relations which points, lines, and planes share in forming any given shape. Problems which many times require mathematical solutions can often be solved graphically with the accuracy that will allow manufacturing and construction. Basic descriptive geometry is one of the designer's methods of thinking and solving problems. In the eighteenth century a French mathematician, Gaspard Monge, developed the principles of graphically solving spatial problems related to military structures. Descriptive geometry was introduced to the U.S. Military Academy at West Point by Claude Crozet in 1816. The Mongean method of presentation has changed, but the basic principles are still taught in engineering schools throughout the world. The visual studies required in descriptive geometry develop reasoning powers that assist in graphically solving problems.

In this chapter a graphic technique is used to analyze all geometric elements. The visual examination of the geometric elements will assist in describing structures of every possible shape. The basic shape of most structures designed by man is rectangular. This is common because of the relatively simple planning needed to form this type of structure. Figure 8-2 shows the basic geometric elements, and some of the common and unusual geometric features of engineering designs.

Points. A point can be considered physically real and can be located by a small dot or a small cross. It is normally identified by two or more projections. In Fig. 8-3 at A, the small cross for point number 1 is shown in the front, top, and right-side views. At B in Fig. 8-3, the regular reference planes are shown in a pictorial view with the relationship of point 1 projected to all three planes. The reference planes are illustrated again in Fig. 8-4. Notice that the unfolding of the three planes forms a flat two-dimensional surface with the fold lines remaining. The folding lines are labeled as shown to indicate that V represents the vertical view. H represents the horizontal, or top, view and P represents the profile or right-side view. Points may also represent the intersection of two lines or the corners on an object.

Fig. 8-1 Geometric space-frame structure. Franklin Park Mall, Toledo, Ohio. (*Unistrut Corp.*)

Fig. 8-2 Basic geometric elements and shapes.

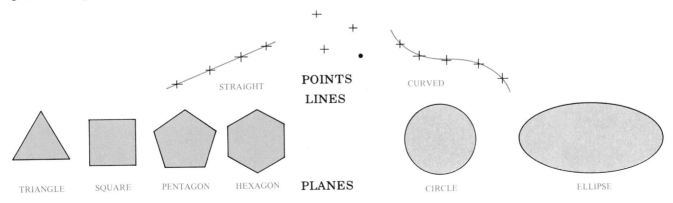

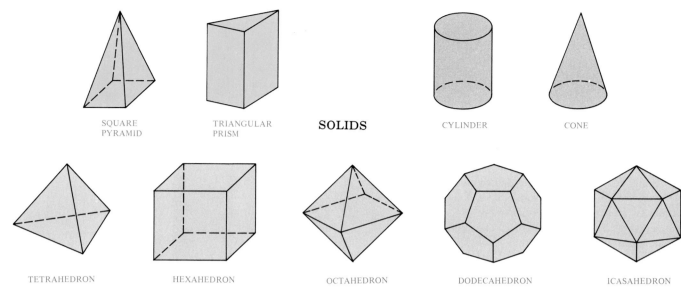

SOLIDS

SQUARE PYRAMID   TRIANGULAR PRISM   CYLINDER   CONE

TETRAHEDRON   HEXAHEDRON   OCTAHEDRON   DODECAHEDRON   ICASAHEDRON

FIVE BASIC SOLIDS

Fig. 8–2 (continued). Basic geometric elements and shapes.

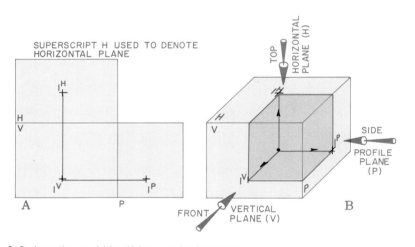

Fig. 8–3 Locating and identifying a point in space.

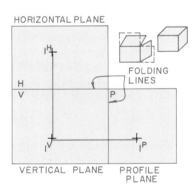

Fig. 8–4 Points identified on unfolded reference planes.

Fixed Points. Figure 8–5 shows a group of points. Points are related to each other by the distance and direction measured on the coordinated reference planes. The vertical height relation may be seen in the front and side views. The relative width dimensions may be seen in the front and top views. The relative depth dimensions may be seen in the top and side views. Note that the three basic dimensions are indicated by H, W, and D (Fig. 8–5).

Lines. If a point moves in a path from a fixed starting position, a trace of its path describes a line. A line can be defined as having location, direction, and length. A straight line can be described as moving in only one direction. Circular lines and straight lines can be drawn easily, but irregular curves must be plotted very carefully. A straight line may be deter-

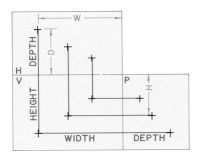

Fig. 8–5 Points are related to coordinated reference planes.

mined by two points or by one point and a fixed direction.

The Basic Lines. The coordinating reference planes allow straight lines to be grouped into three classes.

1. *Normal Lines.* A line which is perpendicular to a reference plane will project as a point on that plane (Fig. 8–6 at A, B, and C). Therefore any line which is parallel to any two of the three reference planes, as shown in Fig. 8–6 at D, E, and F, is considered true length (TL) as noted.

2. *Inclined Lines.* A line that appears inclined in one plane as shown in Fig. 8–7 at A, B, and C but is parallel to any one of the other two principal reference planes will appear foreshortened in two planes. The inclined line will be true length (TL) as noted.

3. *Oblique Lines.* A line *AB* that appears inclined in all three reference planes is an oblique line that makes an angle with all principal planes. (It is not perpendicular or parallel to any of the three planes, Fig. 8–8.) The true length is not shown in any of these views, and the angles of direction cannot be measured on the principal reference planes.

Auxiliary Reference Plane—True Length of an Oblique Line. The true length of a normal line and the inclined line have line projections

A LINE HAS LOCATION, DIRECTION, AND LENGTH

NORMAL LINES - PERPENDICULAR TO A PRINCIPAL REFERENCE PLANE

NORMAL LINES - PERPENDICULAR AND PARALLEL WILL ALWAYS BE TRUE LENGTH (T. L.) AS SHOWN.

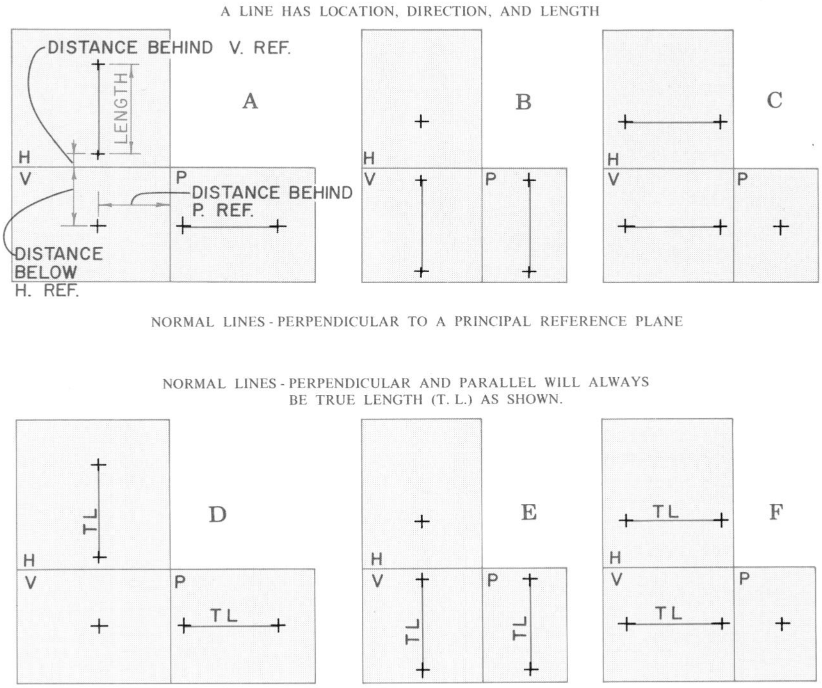

Fig. 8–6 Normal lines in true length are parallel to two reference planes.

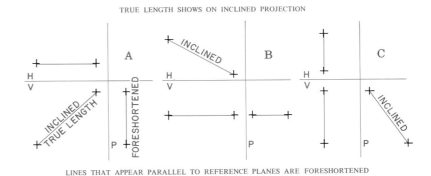

TRUE LENGTH SHOWS ON INCLINED PROJECTION

LINES THAT APPEAR PARALLEL TO REFERENCE PLANES ARE FORESHORTENED

Fig. 8–7 Inclined lines will be parallel to one reference plane.

which are parallel to a principal plane of projection. It is therefore concluded that a line parallel to a plane of projection shows true length in that projection. Since an oblique line is not parallel to any of the three principal reference planes, an auxiliary reference plane can be placed parallel to any one of the oblique lines, as shown

in Fig. 8–9 at A, B, and C. The auxiliary projection and regular plane of projection will have the same relationship as any two principal planes of projection. First, they must always be perpendicular to each other (see Fig. 8–9 at D), and secondly, they must be measured in relation to the previous plane on which they are re-

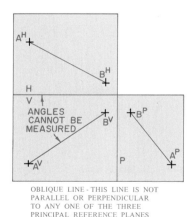

OBLIQUE LINE - THIS LINE IS NOT
PARALLEL OR PERPENDICULAR
TO ANY ONE OF THE THREE
PRINCIPAL REFERENCE PLANES

Fig. 8–8 Oblique lines appear inclined in all projections.

lated. The true length (TL) is obtained as noted.

Line Terminology. Lines drawn on paper can seem to have little meaning or practical value. They are, however, related to real things, and are continually in use. The following terms are related to mining, geology, engineering, and navigation.

Slope. A line that makes an angle with the horizontal plane has a slope measured in degrees. In Fig. 8–10 at A, slope is illustrated in the front view when the line is in true length, or, as at B, slope is found for an oblique line in an auxiliary projection on a plane perpendicular to a horizontal reference when the line is in true length.

Bearing. The angle a line makes in the top view with a north and south line is considered its bearing. The north and south line is generally a vertical line and north is assumed to be at the top. Therefore, to the right is an easterly direction and to the left it would be a westerly direction. The measurement should be made in the horizontal projection and dimensioned in degrees as shown in Fig. 8–11 at A and B.

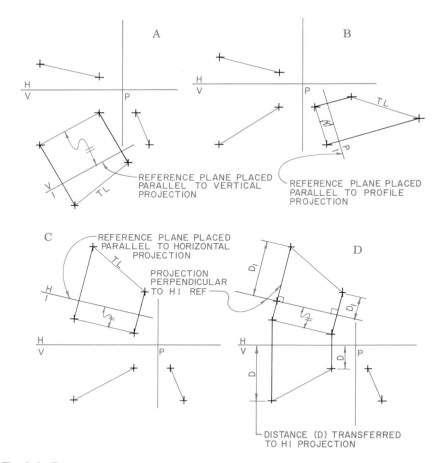

Fig. 8–9 True length of an oblique line by auxiliary projection.

Azimuth. A measurement that defines the direction of a line off due north is the azimuth. The reading is always measured off the north-south line in the horizontal plane, and clockwise dimensioning is always used as shown in Fig. 8–11 at C.

Grade. The inclined measurement common to preparing grade variations is determined on a percentage base. Figure 8–12 illustrates the scale for highway construction of a +12% grade. The grade rises 12 feet in 100 feet of horizontal construction.

Lines in Space. If two lines located in space intersect, there will be a common point on both lines. Figure 8–13

illustrates the alignment necessary for checking out the point of intersecting elements. In Fig. 8–14 the apparent points of intersection in the H and V projection are not aligned, and the intersection is incomplete. How would the intersection look completed? (Note that the cross symbol for locating points will no longer be used in graphic examples.)

Figure 8–15 shows the relationship of parallel lines in a three-view study. All line projections are parallel if they appear parallel in all three reference planes. Note that the lines in Fig. 8–16 are apparently parallel in the front and top views but are not parallel in the side view.

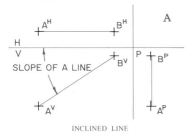

THE SLOPE OF LINE A B IS
A (+) PLUS SLOPE UPWARDS

INCLINED LINE

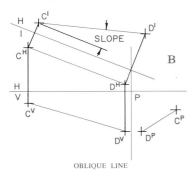

THE SLOPE OF LINE C D IS
A (-) MINUS SLOPE DOWNWARDS

OBLIQUE LINE

Fig. 8-10 Slope of a line in the vertical projection.

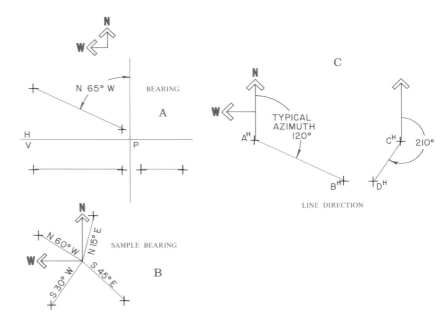

BEARING

SAMPLE BEARING

LINE DIRECTION

Fig. 8-11 Bearing in the horizontal projection, bearing readings, and azimuth related to due north.

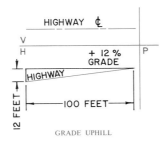

GRADE UPHILL

Fig. 8-12 Grade is measured in the vertical projection.

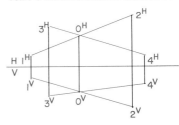

POINT 0 INDICATES ALIGNED INTERSECTION

Fig. 8-13 Intersecting points.

Fig. 8-14 Lines in space examined for intersection.

Fig. 8-15 Lines are parallel when all three reference projections are parallel.

Fig. 8-16 Lines in space examined for parallel relationship.

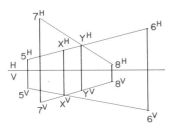

APPARENT POINTS OF INTERSECTION
ARE NOT ALIGNED IN TWO PROJECTIONS

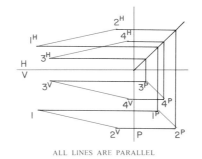

ALL LINES ARE PARALLEL

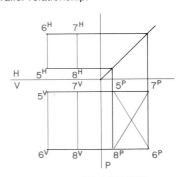

LINES MUST APPEAR PARALLEL
IN ALL THREE VIEWS TO BE
PARALLEL

Perpendicular lines are examined in Fig. 8–17 at A and B. When examining construction of perpendicular relationships, it will be necessary to find the true length of one line in order to know if the angle between them is actually a right angle. Note that the TL at A indicates that one line is parallel to a principal plane of projection. The oblique lines at B are examined in an auxiliary projection to determine true length and the right angle.

Planes. When a line moves from a fixed position, a trace of its path describes a plane. Planes for practical studies are considered to be without thickness and can be extended without limit. A plane may be represented or determined by intersecting lines, two parallel lines, a line and a point, three points, or a triangle.

Basic Planes. The three basic planes are identified by their geometric relationship to the three principal reference planes. Plane one is perpendicular to two reference planes and parallel to the third reference plane. Plane two is perpendicular to one reference plane and inclined to the other two reference planes. Plane three is not perpendicular or parallel to any one of the three basic reference planes.

Basic plane one. Sometimes this type is referred to as a *normal plane* and can be examined in Fig. 8–18 at A, B, and C. The three examples exhibit similar properties. Two of the principal reference planes in each example show the edge-wise view (a line) of the plane. At A the plane is parallel to the vertical reference plane and perpendicular to the horizontal and profile planes. At B the plane is parallel to the horizontal reference plane and perpendicular to the verti-

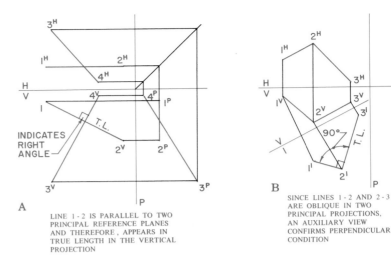

A LINE 1-2 IS PARALLEL TO TWO PRINCIPAL REFERENCE PLANES AND THEREFORE, APPEARS IN TRUE LENGTH IN THE VERTICAL PROJECTION

B SINCE LINES 1-2 AND 2-3 ARE OBLIQUE IN TWO PRINCIPAL PROJECTIONS, AN AUXILIARY VIEW CONFIRMS PERPENDICULAR CONDITION

Fig. 8–17 Perpendicular lines.

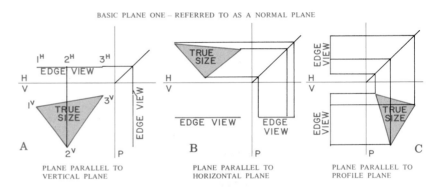

BASIC PLANE ONE – REFERRED TO AS A NORMAL PLANE

A PLANE PARALLEL TO VERTICAL PLANE

B PLANE PARALLEL TO HORIZONTAL PLANE

C PLANE PARALLEL TO PROFILE PLANE

Fig. 8–18 Normal planes, three examples.

cal and profile reference planes. Finally, at C the plane is parallel to the profile reference and perpendicular to the vertical and horizontal reference planes.

Basic plane two. This type is referred to as the *inclined plane* and can be examined in Fig. 8–19 at A, B, and C. Once again the three examples exhibit similar properties. In one of the principal reference planes, the plane shows as a line (edge view), and so it is perpendicular to that plane. The other two reference planes show the plane as a foreshortened surface

or not in true size. At A the inclined plane is perpendicular to the vertical reference plane and inclined to the horizontal and profile planes, where the planes are not true size. At B the inclined plane is perpendicular to the horizontal reference plane, where it appears as a line. The other two reference planes show the plane foreshortened and therefore not in true size. At C the inclined plane is perpendicular to the profile reference plane where it appears as a line. The other two reference planes show the plane foreshortened and therefore not in true size.

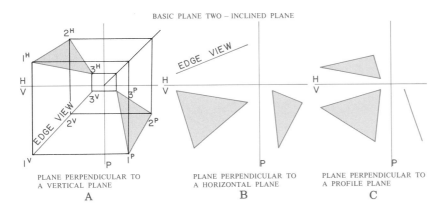

BASIC PLANE TWO — INCLINED PLANE

EDGE VIEW

PLANE PERPENDICULAR TO
A VERTICAL PLANE
A

PLANE PERPENDICULAR TO
A HORIZONTAL PLANE
B

PLANE PERPENDICULAR TO
A PROFILE PLANE
C

Fig. 8–19 Inclined planes, three examples.

Basic plane three. This third type is referred to as an *oblique plane* and can be examined in Fig. 8–20 at A. The oblique plane is not perpendicular to any of the three principal reference planes. Therefore it cannot be parallel to any one of the three planes and consequently appears as a foreshortened plane in each of the three regular views. Figure 8–20 at B describes the same oblique plane from A in a pictorial position.

A Point on a Line. In Fig. 8–21 at A, the line *AB* in the vertical plane contains a point *X*. To place the point on the line in the other two reference planes, it is necessary to project construction lines perpendicular to the folding lines, as shown at B. The construction lines are projected across to $A^H B^H$ and $A^P B^P$ to locate point *X* in the horizontal and profile projections. Straight lines may be extended to new points on

either end as required for the solution of problems, as shown in Fig. 8–22. A point may appear to be on a line in one view, although another view may indicate it is actually in front, on top, or in back of the line, as shown in Fig. 8–23 at A, B, and C.

A Line in a Plane. A line lies in a plane if it intersects two lines of a plane or if it intersects one line and is parallel to another line of that plane. Figure 8–24 at A, B, and C shows how lines may be added to planes. At A the line *RS* is obviously a part of plane *ABC,* inasmuch as *R* was on line *AB* and *S* was located on *BC.* Line *RS* is identified as an oblique line by observing that the line is not parallel to a principal plane of projection and is obviously not perpendicular to the reference planes.

At B a horizontal line *MN* (referred to as a *level line*) is constructed in the vertical projection of plane *ABC.* The projection of *MN* to the respec-

Fig. 8–21 A point on a line.

Fig. 8–20 Oblique plane in three-view projection and pictorial.

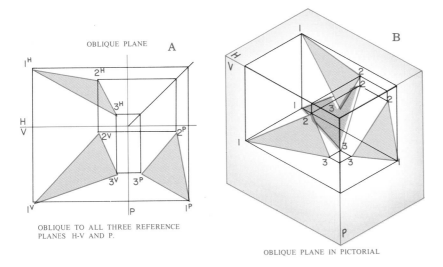

OBLIQUE PLANE
A

OBLIQUE TO ALL THREE REFERENCE
PLANES H-V AND P.

OBLIQUE PLANE IN PICTORIAL

B

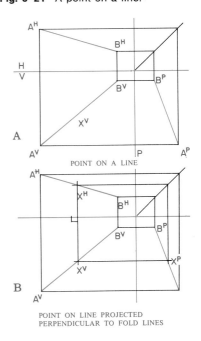

POINT ON A LINE

A

POINT ON LINE PROJECTED
PERPENDICULAR TO FOLD LINES

B

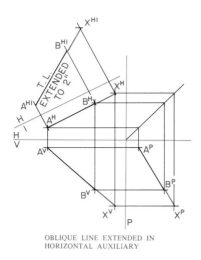

OBLIQUE LINE EXTENDED IN
HORIZONTAL AUXILIARY

Fig. 8–22 Straight lines may be extended.

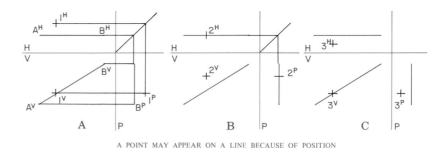

A POINT MAY APPEAR ON A LINE BECAUSE OF POSITION

Fig. 8–23 Point-line relationship.

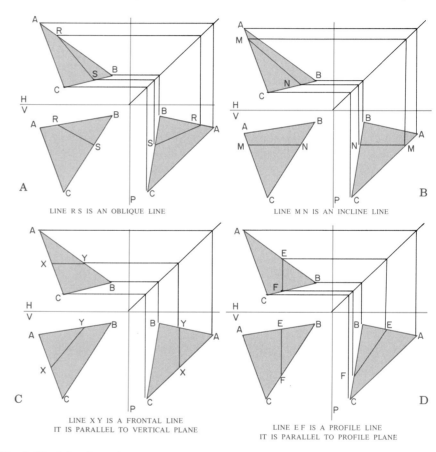

LINE R S IS AN OBLIQUE LINE

LINE M N IS AN INCLINE LINE

LINE X Y IS A FRONTAL LINE
IT IS PARALLEL TO VERTICAL PLANE

LINE E F IS A PROFILE LINE
IT IS PARALLEL TO PROFILE PLANE

Fig. 8–24 A line in a plane.

tive lines in the horizontal projection indicates that this is an inclined line, and the top view will show the true length (TL).

At C a line XY is constructed parallel to the HV reference line in the horizontal reference plane. The line that is projected into the vertical plane appears as an inclined line and will be in true length (TL). (This line is referred to as a *frontal line,* inasmuch as it is parallel to the vertical plane.)

At D a vertical line EF is constructed within plane ABC, and it is parallel to the profile reference plane. The line EF projected to the profile reference appears in true length TL and is called a *profile line.*

Locating a Point in a Plane. A point may be located in a plane by adding a line containing the point to the plane. Figure 8–25 at A shows that a point O appears within the plane ABC. At Fig. 8–25 at B the line AX has been assumed as a projection containing point O. The line XY at Fig. 8–25 at C has been projected to ABC in the horizontal reference

plane, and point O is located on the line by construction of a vertical projection to the horizontal reference plane.

A Point View of a Line. If a line is perpendicular to a reference plane, it

will be projected as a point on that plane. In Fig. 8–26 at A, the line AB is parallel to two principal reference planes and therefore appears as a point in the third or horizontal reference plane. Note the three conditions illustrated.

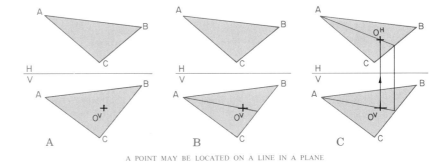

Fig. 8-25 Locating a point in a plane.

A POINT MAY BE LOCATED ON A LINE IN A PLANE

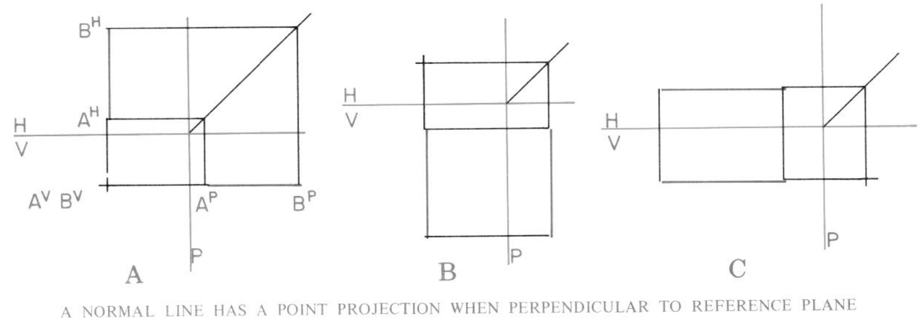

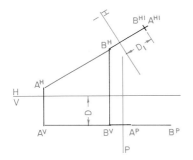

A NORMAL LINE HAS A POINT PROJECTION WHEN PERPENDICULAR TO REFERENCE PLANE

Fig. 8-26 A point view of a line.

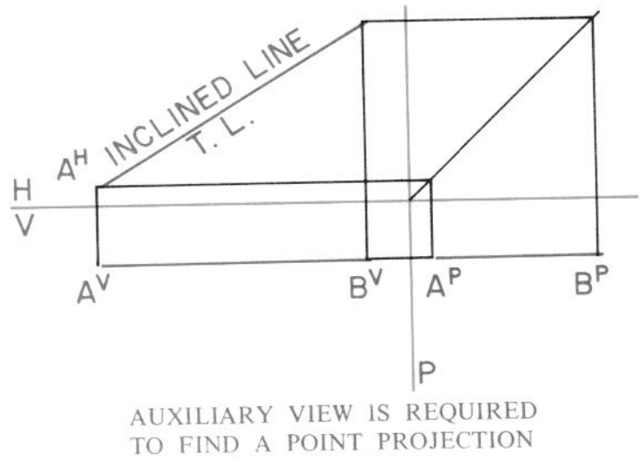

AUXILIARY VIEW IS REQUIRED
TO FIND A POINT PROJECTION

Fig. 8-27 Point projection obtained by auxiliary projection.

Fig. 8-28 Transferring a line to the auxiliary projection.

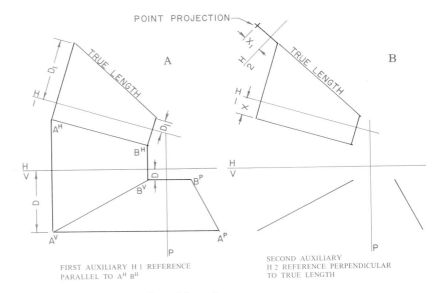

FIRST AUXILIARY H 1 REFERENCE
PARALLEL TO $A^H B^H$

SECOND AUXILIARY
H 2 REFERENCE PERPENDICULAR
TO TRUE LENGTH

Fig. 8-29 Point projection of an oblique line.

When a line is parallel to only one principal reference plane and inclined to the other two, as in Fig. 8-27, the point projection is obtained by auxiliary projection. A reference plane is placed perpendicular to the inclined line at a convenient distance and labeled H/1, as shown in Fig. 8-28. The distance X is transferred as shown for a vertical auxiliary projection or a horizontal auxiliary projection.

If a line appears inclined in all three reference planes (it is an oblique line), the point projection may be obtained by using two auxiliary projections. Set up the first auxiliary reference plane parallel to the oblique inclined line selected and find the true length, as shown in Fig. 8-29 at A. The second auxiliary reference plane will be placed perpendicular to the true-length line just obtained in the first auxiliary. In Fig. 8-29 at B the point projection is located by transferring the distance D as shown.

Parallel Lines. The point projection of two parallel lines is one method of showing the shortest distance between them. In Fig. 8-30 the parallel lines *MN* and *RS* are considered oblique, and two auxiliary projections are required to find the point projections. The first auxiliary reference plane V/1 is parallel to *MN* and *RS,* and the second auxiliary reference plane V/2 is perpendicular to the true-length lines found in the first auxiliary.

In Fig. 8-31 a second method of showing the shortest distance between two parallel lines is determined by considering the lines *AB* and *CD* as a part of a plane. Connect the points *A*, *B*, *C*, and *D* to form a plane. Draw a horizontal line in the top view *DX*. Project the point *X* into the vertical view. Draw the first reference plane V/1 perpendicular to *AX* in the vertical view. The edge view of the plane *ABCD* is determined by transferring distances 1, 2, 3, and 4, as shown. In the second auxiliary V/2, the true lengths of *AB* and *CD* show the true length, and the plane formed is in true size. The true distance between the

lines is measured perpendicular from *AB* to *CD*, as shown.

Point-Line Relations. To find the shortest distance from a point to a line, the line must be found as a point projection. Point *A* and oblique line *CD* in Fig. 8-32 are projected into the first auxiliary projection H/1, and the reference plane H/2 is parallel to line *CD*. In H/1 the true length of *CD* is labeled (TL). The second auxiliary reference plane H/2 is placed perpendicular to line *AB*, and the distance between points in this projection is true length, as shown.

Shortest Distance Between Skew Lines. Skew lines *AB* and *CD* in Fig. 8-33 are nonparallel and nonintersecting lines. They are basically oblique and appear inclined in all principal views. The shortest distance will be a perpendicular distance, as shown in Fig. 8-33 at B. The first

step is to find the true length of one of the lines in the first auxiliary by placing a V/1 reference line parallel to line *CD*. The second auxiliary reference 1/2 is placed perpendicular to the true length of line *CD*, and the point projection of line *CD* is determined as shown. Construct a perpendicular line from the point projection of *CD* to line *AB*. The perpendicular line intersects *AB* extended at point *X*. The intersecting projection may be transferred back to the first auxiliary as shown on *AB* extended.

The True Size of an Inclined Plane. In Fig. 8-34 the plane *ABC* shows as an edge view in the top view. The auxiliary reference plane H/1 is placed conveniently parallel to the edge view. Perpendicular projectors are erected, and the distances *XY* and *Z* transferred as shown to find the true size of the plane in the first auxiliary projection.

The True Size of an Oblique Plane. The plane *ABC* in Fig. 8-35 at A will show an edge view in the first auxiliary if projected upon a plane perpendicular to any line in the figure. In the top view a line *BX* is drawn parallel to the reference plane. The front view of *BX* is now projected into a point projection in the first auxiliary with reference line V/1 placed per-

Fig. 8-31 Distance between lines forming a plane.

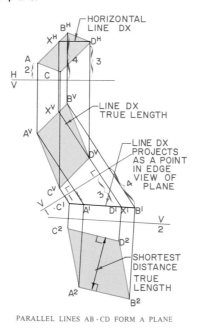

PARALLEL LINES AB-CD FORM A PLANE

Fig. 8-30 Distance between parallel lines.

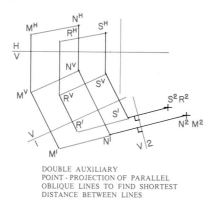

DOUBLE AUXILIARY
POINT-PROJECTION OF PARALLEL
OBLIQUE LINES TO FIND SHORTEST
DISTANCE BETWEEN LINES

Fig. 8-32 Distance from a point to a line.

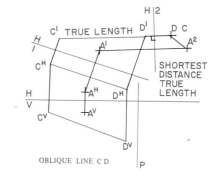

OBLIQUE LINE CD

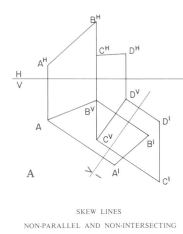

SKEW LINES
NON-PARALLEL AND NON-INTERSECTING

Fig. 8–33 Distance between skew lines.

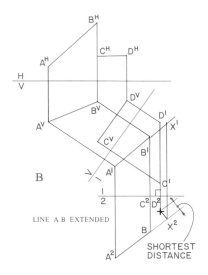

LINE A B EXTENDED

Fig. 8–34 True size of an inclined plane.

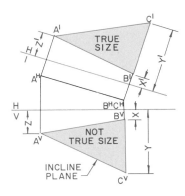

Fig. 8–34 True size of an inclined plane.

pendicular to line *BX*. The point projection will be in the edge-wise view of plane *ABC*, as shown. In the secondary auxiliary in Fig. 8–35 at B, the reference line V/2 has been placed parallel to the edge-wise view. The projection of plane *ABC* in the secondary auxiliary will show the true size.

True Angle Between Lines. The angle between two lines will appear in its true value when the lines show in true length. In Fig. 8–36A the two lines show as an inclined plane, inasmuch as the vertical view shows that lines *AB* and *AC* coincide (lie in a single line). The V/1 auxiliary reference is placed parallel to the vertical view. The two lines are perpendicular to the vertical reference plane. The first auxiliary will show the true angle between the lines, and each line is in true length.

The oblique condition of lines *MN* and *NS* does not show in an edge-wise view, and so Fig. 8–36 at B shows how the problem may be solved by using two auxiliary planes. The first reference plane is perpendicular to the plane formed by the lines, and the second reference plane is parallel to

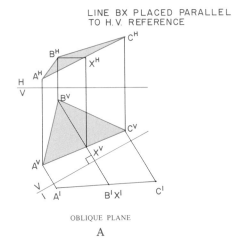

OBLIQUE PLANE

A

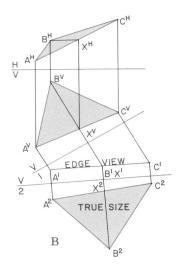

B

Fig. 8–35 True size of an oblique plane.

Fig. 8–36 True angle between lines.

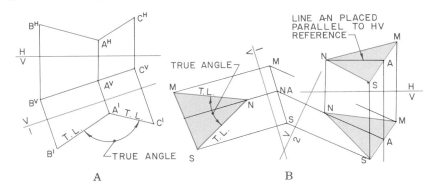

A   B

the first auxiliary plane (parallel to the edge view of the lines *MN* and *NS*).

Piercing Points. A line must intersect a plane if it is not in the plane or parallel to the plane. The point common to the plane and line when intersection occurs is called a *piercing point*. The line can be thought of as piercing the plane.

Edge-view System. The edge view of a plane contains all the points in the plane. Therefore a line crossing the edge view will show the point where a line pierces the plane. If a line lies in a plane or is parallel to the plane, it cannot intersect the plane. In Fig. 8–37 the straight line is neither in the plane nor parallel to the plane; it will intersect the plane at a point common to both the plane and line. The edge view of plane *ABC* is shown in the vertical plane, with the line *RS* piercing the plane at *P*. The line *RS* in the horizontal plane will pierce the plane at point *P* when it is projected in the usual manner. When the visibility of line *MN* is examined closely in the vertical projection, it will be noted that element *A* of the triangle is lower than point *R* of the piercing line; therefore, the dashed portion of line *RS* is invisible.

The piercing point of a line *MN* intersecting an oblique plane *ABC* can be solved with the edge-view system, as shown in Fig. 8–38. The first auxiliary will determine the edge view of plane *ABC*, and the piercing point *P* of line *MN* can be carried back to the vertical and horizontal with the projection techniques shown by arrows.

Cutting-Plane System. When a line *RS* intersects an oblique plane *ABC*, a cutting plane which contains the line of intersection will determine

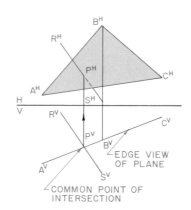

Fig. 8–37 Piercing-point–edge-view system.

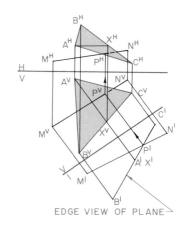

Fig. 8–38 Piercing point—a line with an oblique plane.

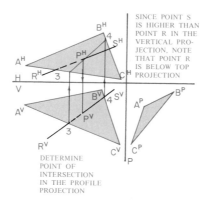

Fig. 8–39 Piercing-point–cutting-plane system.

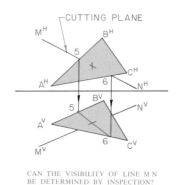

Fig. 8–40 Piercing point developed.

additional working points, as shown in Fig. 8–39. The cutting plane appears to intersect two elements of the triangle in the front view at points 3 and 4. The points 3 and 4 are projected to the top view on lines *AC* and *BC*, and when connected across the plane, the piercing point *P* is determined. The point *P* is projected to the front view, and the proper visibility is determined and hidden lines are added. Figure 8–40 illustrates the cutting plane in the horizontal projection intersecting lines *AB* and *AC* at points 5 and 6, respectively. Can the visibility be determined by inspection?

Angle Between Intersecting Planes. A dihedral angle is formed when planes intersect. Two planes that intersect have a straight line common to the two planes. The dihedral angle formed can only be measured perpendicular to the line of intersection. When a point view of the line of intersection is found, the planes will be shown as edges, and the angle between them will be in true size. In Fig. 8–41 at A, the planes *ABC* and *ACD* have a common line of intersection *AC*. The first auxiliary projection H/1 in Fig. 8–41B is taken in the horizontal plane of line

AC. The H/1 auxiliary allows *AC* to be drawn in true length (TL). In the second auxiliary, the point projection of the true-length line *AC* also shows the two given planes as edges. The true angle is measured in the second auxiliary, as shown in Fig. 8–41 at B.

Angle Between a Line and a Plane. The view which shows a plane in the edge-view along with a true-length line will also show the true angle between the two.

Plane method. In Fig. 8–42, oblique line *XY* intersects the oblique plane *ABC*. The first step is to determine the edge view of the oblique plane ABC. The first auxiliary is determined by placing H/1 perpendicular to a true-length line in the horizontal projection. After projecting the edgeview in H/1, the reference plane H/2 is placed parallel to the plane *ABC*. In the second auxiliary the true size of plane *ABC* is plotted. In the third auxiliary 2/3, the reference line is placed parallel to line *XY*. The new edge view of the plane and the true-length line form the true angle between a line and a plane.

Line method. In Fig. 8–43, the oblique plane *ABC* and oblique line *XY* intersect as shown. The first auxiliary V/1 is placed parallel to the line *XY*, and the true length of *XY* is determined. In the second auxiliary a reference V/2 is placed perpendicular to the true length of line *XY*, and the point projection of *XY* is found. Finally, in the third auxiliary the 2/3 reference line is set perpendicular to the true length of B^2O^2, thus showing plane *ABC* as an edge view in the third auxiliary. The intersection of the line and plane will show the true size of the angle in the third auxiliary.

Revolution. Problems in the preceding chapter discussed the rule of revolution. Basic problems can be solved

many times by changing the position of the object so it can reveal required information. Descriptive geometry problems can be solved by revolution.

True size of an oblique plane by revolution. In Fig. 8–44 the plane *ABC* is obviously inclined to all principal planes. A line *AX* is placed in the plane *ABC* parallel to the horizontal reference line. Line *AX* projected to the top view appears as an inclined line in true length. The first

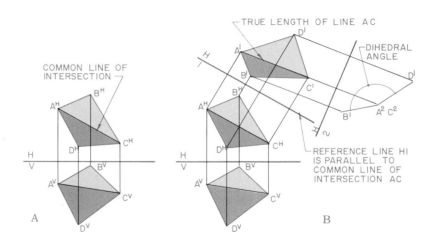

Fig. 8–41 Angle between intersecting planes.

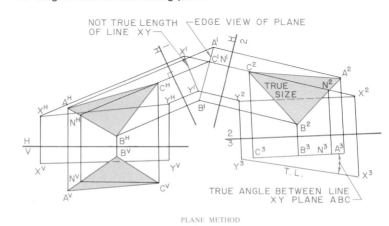

PLANE METHOD

Fig. 8–42 Angle between a line and a plane (plane method).

Fig. 8–43 Angle between a line and a plane (line method).

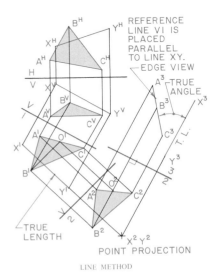

POINT PROJECTION

LINE METHOD

auxiliary projected in the horizontal will have an H/1 reference perpendicular to line AX. The point view of AX is found in the first auxiliary, and line ABC appears as an edge view, with X within the line. At point AX the edge view EV is revolved so as to appear parallel to reference line H/1 (dashed line). The projection of points B and C to the horizontal projection will now allow the true size of plane ABC to be drawn in the top view as shown.

Almost all problems studied in descriptive geometry can be worked out by using the auxiliary planes. The basic approach for solving problems (Fig. 8-45) revolves about knowing how to

1. find the true length of a line
2. find the point projection of a line
3. find the edge view of a plane, and
4. find the true size of a plane figure.

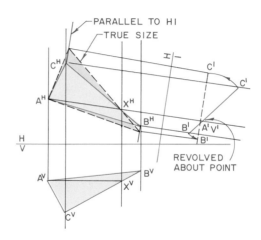

Fig. 8-44 True size of an oblique plane by revolution.

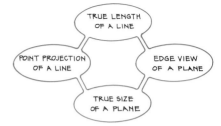

Fig. 8-45 Four basic approaches for solving problems.

Review

1. Any given shape may be formed by which three basic geometric elements?

2. What is the basic shape of most structures designed by man, and why is this form common?

3. Do you believe that basic descriptive geometry is one of the designer's methods of thinking and of solving problems? Why?

4. How is a point located for graphical analysis?

5. What are the three characteristics of a line?

6. Name the three basic lines used in the study of descriptive geometry.

7. What is the relationship of the normal line to the three principal planes of projection?

8. What is the relationship of the inclined line to the three principal planes of projection?

9. In how many of the principal planes of projection will an inclined line appear in true length?

10. What is the relationship of the oblique line to the three principal planes of projection?

11. How many auxiliary projections are required to find the point projection of an oblique line?

12. How can you determine if two lines located in space actually intersect?

13. What are the major characteristics that describe a plane located in space?

14. Name the three basic planes.

15. Describe the difference between an inclined plane and an oblique plane.

16. Describe how a line may be added to a plane in space.

17. How can you determine the shortest distance between two parallel inclined lines located in space?

18. How is the auxiliary reference plane placed in determining the true size of an inclined line?

19. What conditions must exist before you can determine the true angle between intersecting lines?

Problems for Chapter 8

Figures 8–46 through 8–71 Using the $\frac{1}{8}''$ grid, determine the relationship of lines and points to the HVP planes and lay out the following problems. Letter all points $\frac{1}{8}''$ high using the proper superscript.

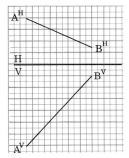

Fig. 8–46 Determine the true length of line AB.

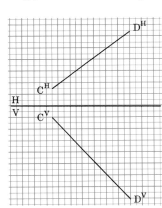

Fig. 8–47 Determine the true length and slope of line CD.

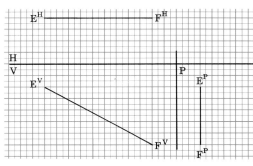

Fig. 8–48 Line *EF* is the center line of a pipeline. Scale: 1″ = 40′-0″. Locate line *X* 20′ below *E* on all three projections. Determine the grade of line *EF*. Determine the true distance from *E* to *X*.

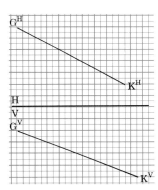

Fig. 8–49 Determine the point projection of line *GK*.

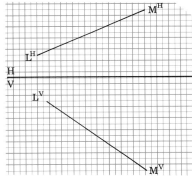

Fig. 8–50 Determine the angle *LM* makes with the vertical plane. What is the bearing?

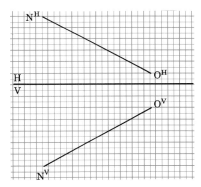

Fig. 8–51 Determine the slope of line *NO*. Extend *NO* to measure $2\frac{1}{4}$ long. Draw all three views.

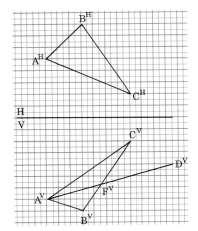

Fig. 8–52 Locate point *D* in the plan (horizontal) projection.

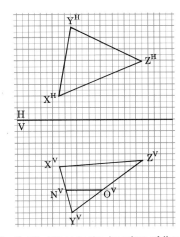

Fig. 8–53 What is the bearing of line *NO* located on plane *XYZ*?

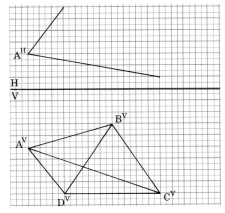

Fig. 8–54 Complete the plan view of plane *ABCD* and develop a side view.

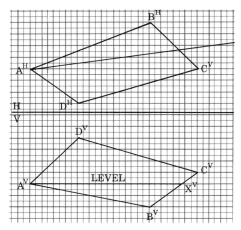

Fig. 8-55 Find the edge view of plane *ABCD* and determine the angle that it makes with the horizontal plane.

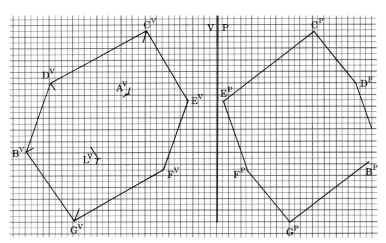

Fig. 8-56 Determine the visibility of the edges relating to points *A* and *L* of the parallelepiped.

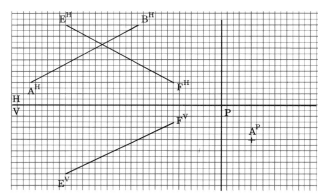

Fig. 8-57 Complete the three views showing the intersection of lines *AB* and *EF*.

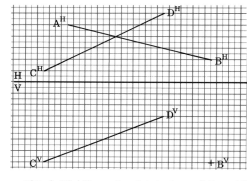

Fig. 8-58 Draw the front view of line *AB* which intersects line *CD*. What is the distance from *C* to *A*?

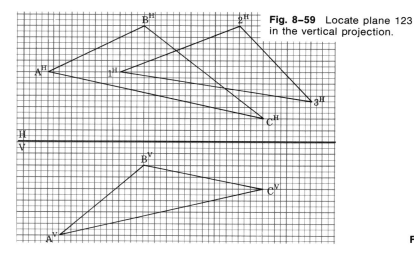

Fig. 8-59 Locate plane 123 in the vertical projection.

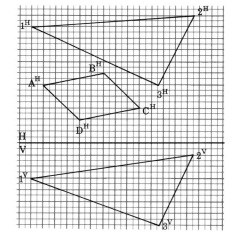

Fig. 8-60 Construct plane *ABCD* parallel to plane 123.

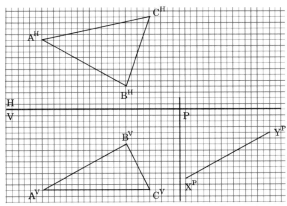

Fig. 8–61 Determine the true size of oblique plane *ABC*. Draw line *XY* parallel to plane *ABC* in the plan view.

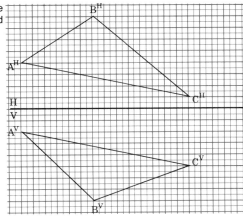

Fig. 8–62 Determine the true size of plane *ABC* and label its slope.

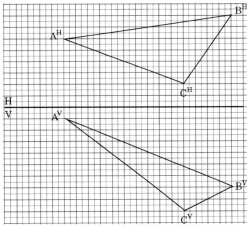

Fig. 8–63 Draw the true size of plane *ABC* and dimension the three angles.

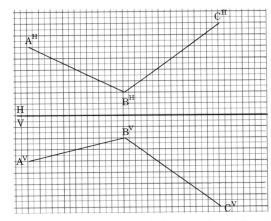

Fig. 8–64 Find the true angle between lines *AB* and *BC*.

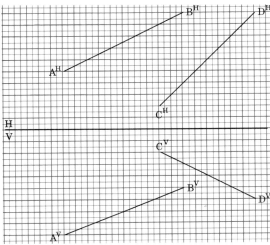

Fig. 8–65 Determine and locate the shortest distance between skew lines *AB* and *CD*.

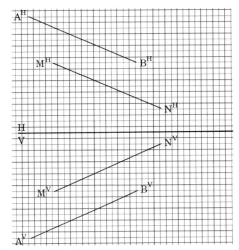

Fig. 8–66 Determine the shortest distance between parallel lines *AB* and *MN*.

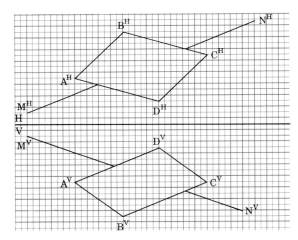

Fig. 8-67 Determine if line *MN* pierces line *ABCD*.

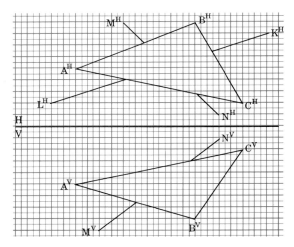

Fig. 8-68 Determine if line *MN* pierces plane *ABC*. Locate line *KL* so that it pierces the center of plane *ABC*.

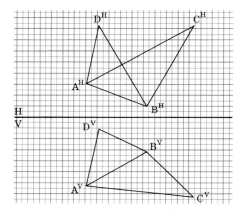

Fig. 8-69 Determine the visibility and angle between planes *ABC* and *ABD*.

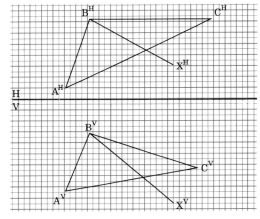

Fig. 8-70 Determine the angle between line *BX* and plane *ABC*.

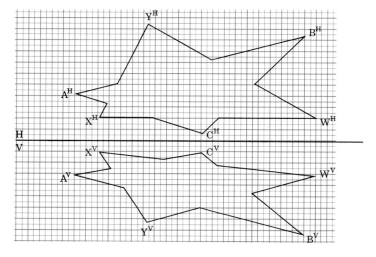

Fig. 8-71 Determine the visibility and intersection of planes *ABC* and *WXY*.

Sectional Views and Conventions | Chapter 9

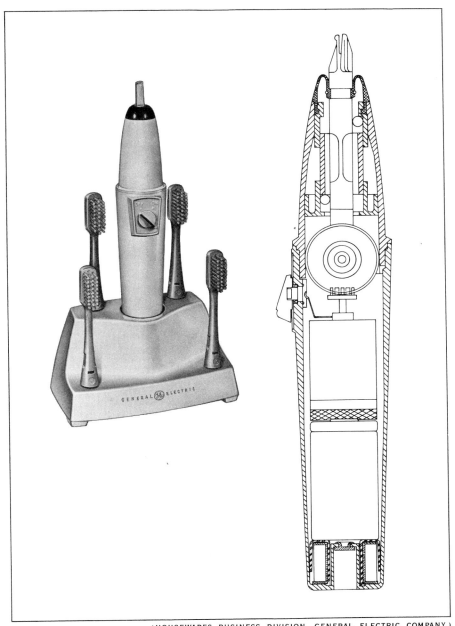

(HOUSEWARES BUSINESS DIVISION, GENERAL ELECTRIC COMPANY.)

Sectional Views. Interior or other parts of an object which cannot be seen may be shown by using *hidden lines* made with short dashes (Chapter 5). This method is satisfactory if the hidden detail is rather simple. However, objects that have complicated interior shapes cannot easily be shown by using dashed lines. An attempt will very often result in confusion, as shown in Fig. 9–1. In such cases special views called *sections,* or *sectional views,* are drawn. Sectional views show the inside of an object by imagining that part of it has been cut away to expose interior (hidden) details (Fig. 9–2).

A sectional view is obtained by supposing an imaginary *cutting plane* cuts through the object, and everything in front of the plane is removed to show the cut surface and the interior details (Fig. 9–2). The location of the cutting plane is shown on the normal view by using a special line called a *cutting-plane line* (Fig. 9–3). The cut surface in the sectional view is shown by *section lining* (also called *crosshatching*) composed of uniformly spaced thin lines.

Section Lining and Symbols. The general-purpose symbol (cast iron) is used for most purposes (Fig. 9–4). This is especially true on drawings of separate pieces. American National Standard sectioning symbols representing different materials may also be used (Fig. 9–4). These are especially important on assembly drawings involving a variety of materials.

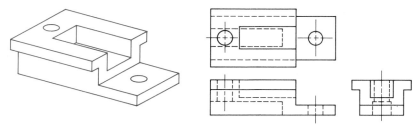

Fig. 9-1 Pictorial view of object and the three normal views.

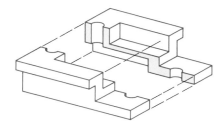

Fig. 9-2 Object cut to show inside details. The front of the object has been removed.

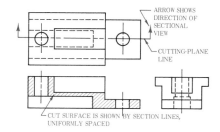

ARROW SHOWS DIRECTION OF SECTIONAL VIEW

CUTTING-PLANE LINE

CUT SURFACE IS SHOWN BY SECTION LINES, UNIFORMLY SPACED

Fig. 9-3 Three views with cutting-plane line and section lining.

However, exact specifications for each material must be given in a note, materials list, or otherwise.

Spacing of Section Lines. Section lines may be close together or far apart, depending upon the area (Fig. 9-5). American National Standards suggests approximately $\frac{1}{32}$ in. to $\frac{1}{8}$ in. or more, uniformly spaced, and generally at 45°. Wider spacing is preferred when it can be used to show the sectioned area clearly. A neater drawing will result and time will be saved if section lines are *not* spaced too close together. For most purposes the distance between lines can be about $\frac{3}{32}$ in., spaced by eye. For small areas use closer spacing ($\frac{1}{16}$ in. or less), and for large areas use wider spacing (up to $\frac{1}{8}$ in. or more). When the sectioned area is very small, as for thin plates, sheets, and structural shapes, blacked-in (solid black) sections may be used, as in Fig. 9-6. Note the white space between the parts.

A timesaving method for indicating a large sectioned area is to use outline sectioning, as shown in Fig. 9-7. This method is often used on design drawings with the section lines drawn freehand and widely spaced. Still other methods are to gray the sectioned area (Fig. 9-8), to outline with a pencil (Fig. 9-9), or to rub pencil dust over the area. A fixative may be applied to prevent smudging.

Fig. 9-4 American National Standard symbols for section lining.

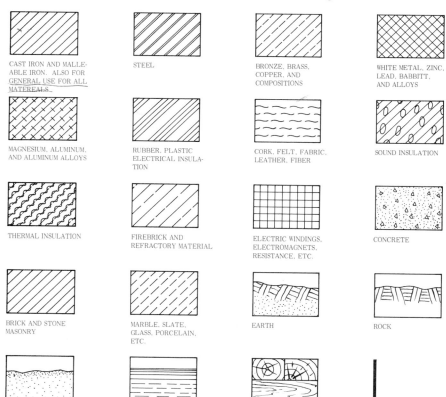

CAST IRON AND MALLEABLE IRON. ALSO FOR GENERAL USE FOR ALL MATERIALS.

STEEL

BRONZE, BRASS, COPPER, AND COMPOSITIONS

WHITE METAL, ZINC, LEAD, BABBITT, AND ALLOYS

MAGNESIUM, ALUMINUM, AND ALUMINUM ALLOYS

RUBBER, PLASTIC ELECTRICAL INSULATION

CORK, FELT, FABRIC, LEATHER, FIBER

SOUND INSULATION

THERMAL INSULATION

FIREBRICK AND REFRACTORY MATERIAL

ELECTRIC WINDINGS, ELECTROMAGNETS, RESISTANCE, ETC.

CONCRETE

BRICK AND STONE MASONRY

MARBLE, SLATE, GLASS, PORCELAIN, ETC.

EARTH

ROCK

SAND

WATER AND OTHER LIQUIDS

WOOD ACROSS GRAIN WOOD WITH GRAIN

THIN PARTS

Fig. 9-5 Section lines are spaced by eye. Their distance apart varies according to the size of the space to be sectioned.

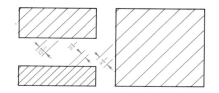

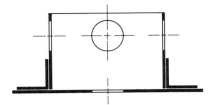

Fig. 9-6 Thin section.

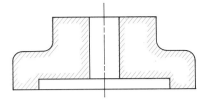

Fig. 9-7 Outline sectioning.

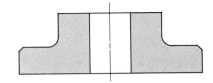

Fig. 9-8 Cut surface may be grayed.

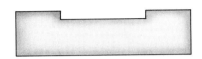

Fig. 9-9 Cut surface may have grayed outline.

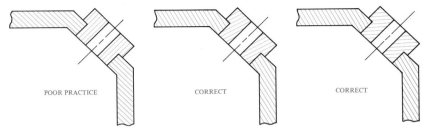

POOR PRACTICE   CORRECT   CORRECT

Fig. 9-10 Do not draw section lines parallel or perpendicular to a main line of the view.

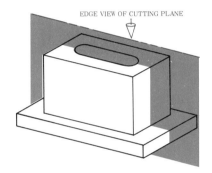

EDGE VIEW OF CUTTING PLANE

Fig.9-11 The cutting-plane line represents the edge view of the cutting plane.

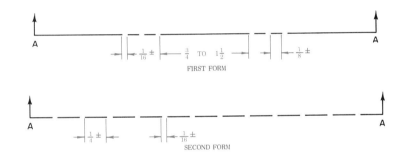

Fig. 9-12 American National Standard cutting-plane lines.

On any section avoid section lines parallel or perpendicular to an important visible line (Fig. 9-10). If necessary, section lines may be drawn using any suitable angle and spacing to identify the parts.

The Cutting-plane Line. The cutting-plane line represents the edge view of the cutting plane (Fig. 9-11). Two forms are approved for American National Standard (Fig. 9-12). The first form is generally preferred. The second form shows up well on complicated drawings. Short lines at right angles, with arrowheads to show the direction for looking at the sec-

tion, are drawn at the ends of the cutting-plane line. Bold capital letters are placed at the corners as shown if needed for reference to the section.

The cutting-plane line need not be used when it is clear that the section is taken on the main center line of an object or other clearly seen location (Fig. 9-13).

Sections Through Assembled Pieces. When two or more pieces are shown together in a sectional view, section lines are drawn in different directions for each piece (Figs. 9-14 and 9-15). However, each separate

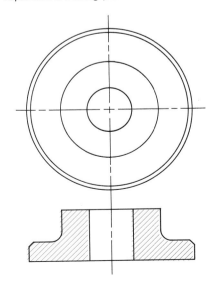

Fig. 9-13 A center line may be used to represent a cutting-plane line.

piece has the section lines in the same direction wherever any part of it appears as a cut surface, as shown in Fig. 9–15.

Full Sections. A full sectional view is obtained when the object is cut completely apart, as in Fig. 9–16. Such views are generally referred to simply as *sections*. The two most common types of full sections are *vertical* and *profile* (Figs. 9–17 and 9–18).

Offset Sections. The cutting plane is usually taken straight through the object, but it may be offset at one or more places in order to show some detail or to miss some part (Fig. 9–19). Here the plane is not continuous but is offset to pass through the bolt hole. The hole is shown more clearly than if the section were con-

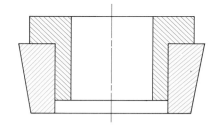

Fig. 9–14 Two pieces. Section lines in two directions.

tinued through the center. Offsets are shown in the edge view of a cutting plane but *not* in the sectioned view. If reference letters are necessary, they are placed at the ends of the cutting-plane line, opposite the arrowheads.

Half Sections. A half section is obtained when two cutting planes at right angles are used to cut out and

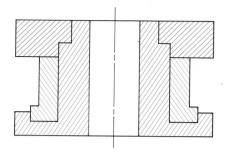

Fig. 9–15 Three pieces. Section lines in three directions.

remove one quarter of the piece, as in Fig. 9–20. The half section at E shows one-half the view as a section and the other half as an exterior view. Such views can be used to advantage with symmetrical pieces to show both the interior and the exterior in one view. A *center line* is used where the exterior and the sectional half views meet because the piece is not actually cut. Also, the complete top view is drawn

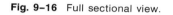

Fig. 9–16 Full sectional view.

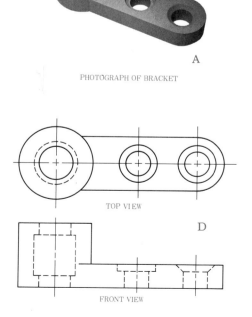

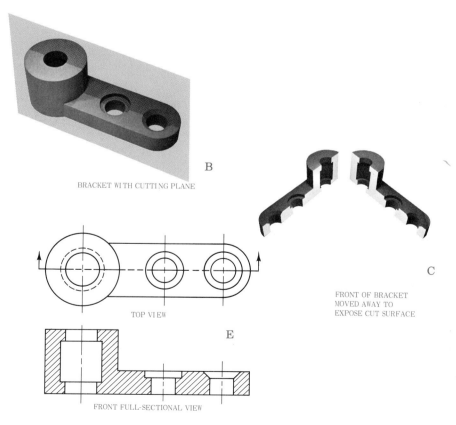

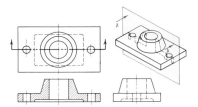

Fig. 9–17 Vertical section.

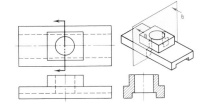

Fig. 9–18 Profile section.

Fig. 9–19 An offset section.

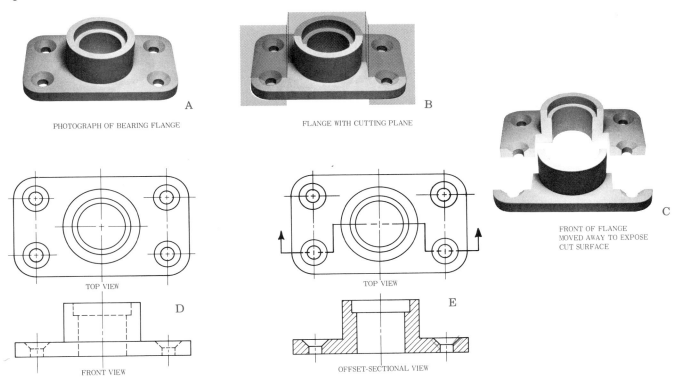

PHOTOGRAPH OF BEARING FLANGE

A

FLANGE WITH CUTTING PLANE

B

FRONT OF FLANGE
MOVED AWAY TO EXPOSE
CUT SURFACE

C

TOP VIEW

D

FRONT VIEW

TOP VIEW

E

OFFSET-SECTIONAL VIEW

because no part is actually removed. If dimensions are to be shown, it will be necessary to draw some, or all, hidden lines in the exterior half. If sight direction is required, only one arrow is used, as shown at E. The cutting-plane line could be omitted here, since the location of the section is evident.

Hidden and Visible Lines on Sectional Views. Hidden lines should not be drawn on sectional views ex-

cept when needed for dimensioning or for clearly describing the shape. In Fig. 9–21 the hub is clearly described at A with no hidden lines. Compare it with the view at B.

On sectional assembly drawings (where several parts are shown assembled, or put together), most of the hidden lines should be omitted to keep the drawing from becoming complicated and hard to read (Fig. 9–22). Sometimes a half section or part section can be used to advantage to avoid

the use of hidden lines on an assembly drawing.

Under usual conditions all visible lines on or beyond the plane of the section should be drawn on a sectional view (Fig. 9–23 at A). Observe the numbered lines which match the lines on the picture at B. A drawing without these lines, as at C, would have no value and should *never* be used. Without the visible lines beyond the plane of the section in Fig. 9–24, the view would have no meaning.

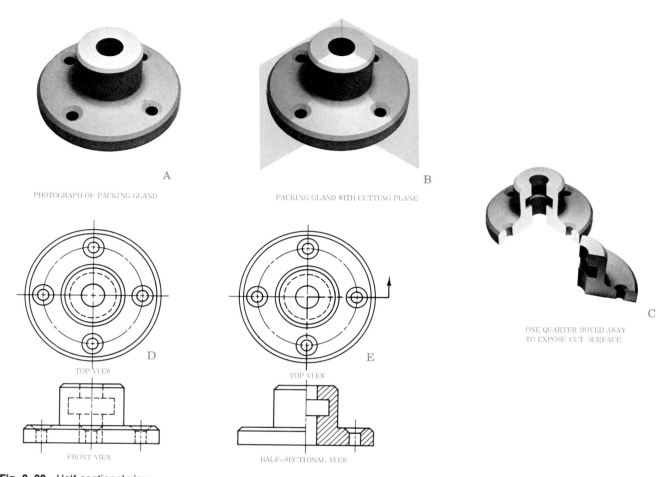

PHOTOGRAPH OF PACKING GLAND

A

PACKING GLAND WITH CUTTING PLANE

B

ONE QUARTER MOVED AWAY
TO EXPOSE CUT SURFACE

C

TOP VIEW

D

TOP VIEW

E

FRONT VIEW

HALF–SECTIONAL VIEW

Fig. 9-20 Half sectional view.

Fig. 9-21 Omit hidden lines when not needed for clearness or dimensioning.

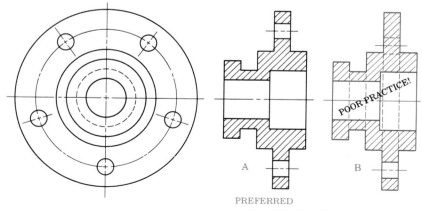

POOR PRACTICE!

A

B

PREFERRED

HIDDEN LINES NOT SHOWN

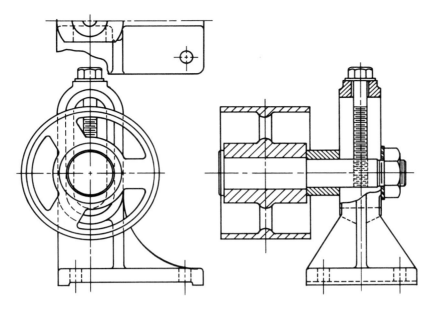

Fig. 9-22 Omit hidden lines to keep the drawing from becoming confusing.

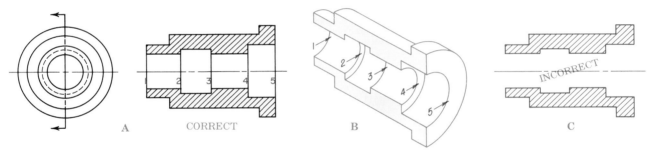

Fig 9-23 Show all visible lines beyond the sectioned surface.

Fig. 9-24 Correct and incorrect uses of visible lines beyond the plane of the section.

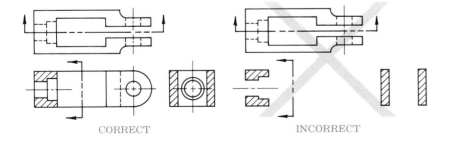

CORRECT   INCORRECT

Broken-out Sections. Broken-out sections may be used when it is desirable to show some interior detail without drawing a full or half section (Fig. 9-25). In such cases a cutting plane is assumed to be passed through the desired detail and "broken out" in front of the plane. Note that the broken-out section is limited by a "break line."

Break lines are drawn freehand and are the same thickness as visible lines. Figure 9-26 shows some additional examples of broken-out sections.

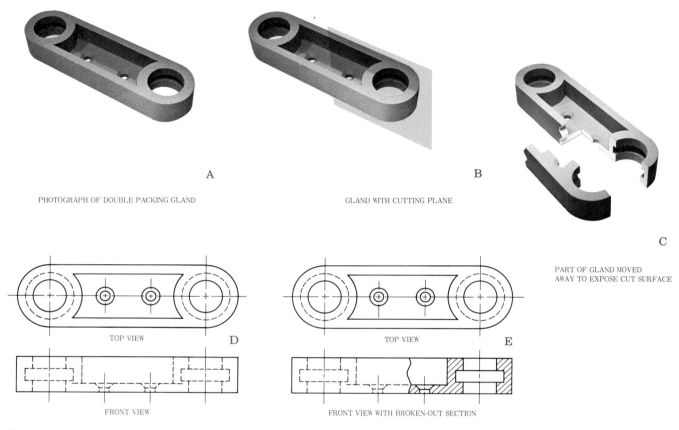

A

PHOTOGRAPH OF DOUBLE PACKING GLAND

B

GLAND WITH CUTTING PLANE

C

PART OF GLAND MOVED
AWAY TO EXPOSE CUT SURFACE

TOP VIEW   D

FRONT VIEW

TOP VIEW   E

FRONT VIEW WITH BROKEN-OUT SECTION

Fig. 9–25 Broken-out section.

Revolved Sections. When the cutting plane passes through a portion of the object (Fig. 9–27) and the cut surface is revolved 90°, the result is a *revolved section* (also called a *rotated section*). This is shown in Fig. 9–28.

The method shown in Fig. 9–29 is useful for long, slender parts when the cross-sectional shape is uniform. In such cases the length of the view may be shortened and given by a dimension. This allows larger revolved sectional views to be drawn.

Removed Sections. In some cases it is necessary to take a sectional view from its normal position on the view and move it to another position on the drawing sheet. When this is done, the

Fig. 9–26 Additional examples of broken-out sections.

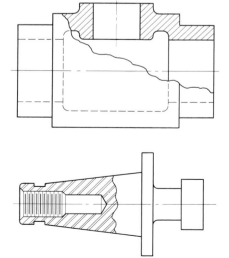

Fig. 9–27 Cutting plane in position for revolved section.

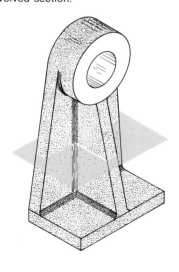

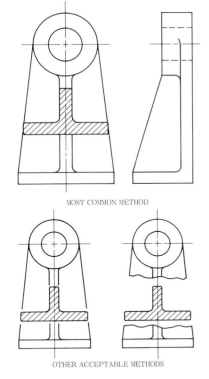

MOST COMMON METHOD

OTHER ACCEPTABLE METHODS

Fig. 9–28 Revolved section.

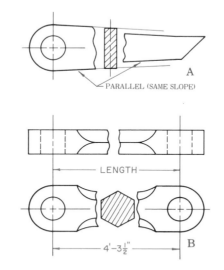

PARALLEL (SAME SLOPE)

A

LENGTH

4'–3½"

B

Fig. 9–29 Revolved sections in long parts.

result is a *removed sectional view* (also called a *removed section*). However, the removed section will be more easily understood if it is kept in the same relative position as it normally would be if it were placed on the view. In other words, the removed section should not be rotated in any direction. Fig. 9–30 shows right and wrong ways to position revolved sections. Bold letters should be used to identify the removed section and the location of the cutting plane on the regular view. This is also shown in Fig. 9–30.

A removed section may be a sliced section (the same as a revolved section), or it may contain some additional detail beyond the cutting plane. It may be drawn to an enlarged scale if necessary to show details clearly and to provide more space for dimensions.

Sometimes a removed exterior view (nonsectioned) may be desirable. It may be to the same or an enlarged scale and may be complete or used to show certain features.

Auxiliary Sections. When a cutting plane is passed at an angle, as in Fig. 9–31 at A, the resulting sectional view, parallel to the cutting plane, is an auxiliary section. The method of

Fig. 9–30 Correct and incorrect positions of revolved sections.

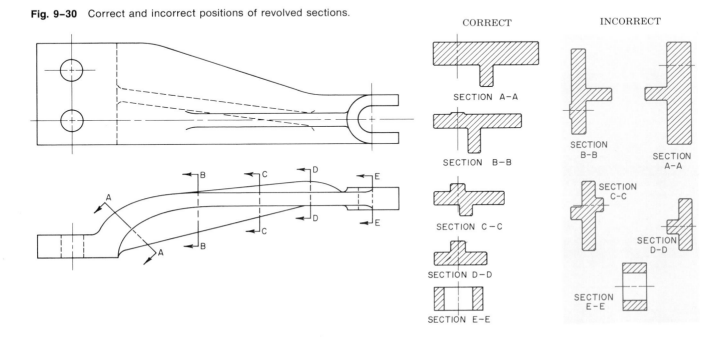

CORRECT

INCORRECT

SECTION A–A

SECTION B–B

SECTION C–C

SECTION D–D

SECTION E–E

SECTION B–B

SECTION A–A

SECTION C–C

SECTION D–D

SECTION E–E

SECTIONAL VIEWS AND CONVENTIONS 181

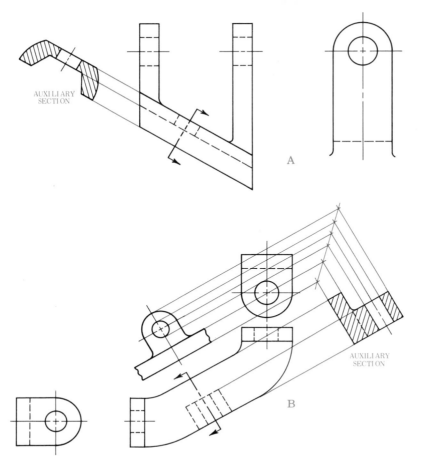

Fig. 9–31 Auxiliary section.

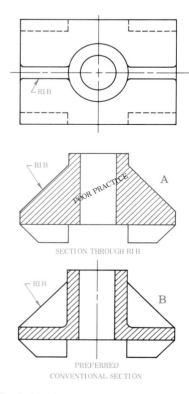

Fig. 9–32 Ribs in section.

drawing is the same as for any auxiliary view (Chapter 7).

The usual practice on working drawings is to show only the cut section; however, any part, or all beyond the auxiliary cutting plane, may be shown if needed. In Fig. 9–31 at B, one hidden line is shown on the auxiliary section, and there are three incomplete views.

Ribs and Webs in Section. Ribs and webs are thin flat parts of an object used to brace or strengthen another part of the object. When a cutting plane passes through a rib or web parallel to the flat side, the sectioning is omitted, as in Fig. 9–32 at B. Think

of the plane as being just in front of the rib. A true section, as at A, would give the idea of a very heavy, solid piece, which would not be a true description of the part.

If a cutting plane passes perpendicular to the flat side of a rib, web, or other thin feature, sectioning is shown. Figure 9–33 is an example.

Alternate Section Lining. Alternate section lining may be used when a rib (or similar flat feature) does not show clearly in a sectional view. The sectional view at A in Fig. 9–34 shows the eccentric as it would be drawn without a rib. It also shows the usual way it would be drawn by not section-

ing a rib. This view could mean there is no rib or there is a rib. The top and bottom of the rib is even with the surfaces it joins. However, at B, alternate section lines are used with hidden lines to show the extent of a rib.

Alternate section lining is a method of indicating a rib by leaving out every other section line on the rib. Alternate section lines are useful to show a rib in one-view drawings of parts or in assembly drawings.

Other Parts Usually Not Sectioned. Section lining is omitted from spokes and gear teeth when the cutting plane passes through them.

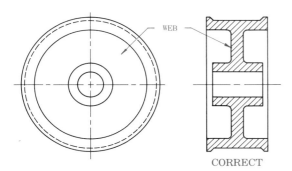

Fig. 9-33 Web in section.

Examples of these features are shown in Fig. 9-35. When a cutting plane passes lengthwise (through the axis) of shafts, bolts, pins, rivets, or similar elements, sectioning is not used (Fig. 9-36). Such elements are left in full because there is no interior detail and because sectioning might give a wrong idea of the part. The exterior view is easier to read and requires less time to draw. However, when such parts are cut across the axis, they should be sectioned (Fig. 9-37). The sectional assembly (Fig. 9-38) shows and names a number of features which are not sectioned.

Phantom (Hidden) Sections. The phantom, or hidden, section (Fig. 9-39) is used when it is desired to show on one view the interior and the exterior of a part that is not completely symmetrical. Note that the circular boss is on only one side of the piece. A half section could not be used in this case. A partial phantom section is sometimes useful for showing a detail in section on an exterior view instead of a broken-out section.

Rotated Features in Section. Sections or elevations of symmetrical pieces would sometimes be hard to read (as well as hard to draw) if drawn in true projection. In Fig. 9-40 ribs and lugs have been rotated parallel to the vertical plane. The ribs show true

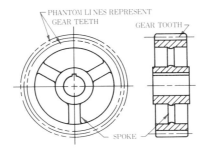

Fig. 9-35 Spokes and gear teeth not sectioned.

Fig. 9-34 Alternate, or wide, section lining.

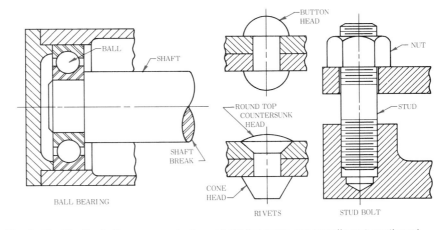

Fig. 9-36 Shafts, bolts, screws, rivets, and similar parts are usually not sectioned.

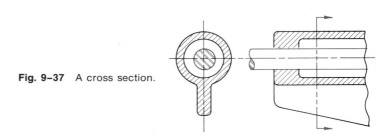

Fig. 9-37 A cross section.

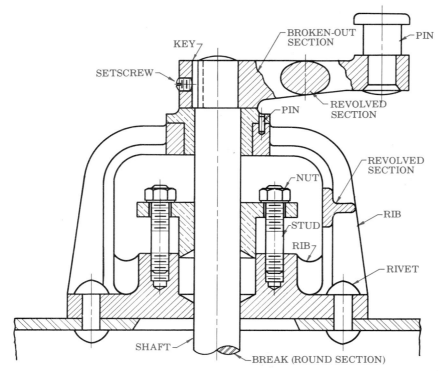

Fig. 9-38 Examples of what should not be sectioned.

Labels on Fig. 9-38: KEY, SETSCREW, BROKEN-OUT SECTION, PIN, REVOLVED SECTION, PIN, REVOLVED SECTION, NUT, RIB, STUD, RIB, RIVET, SHAFT, BREAK (ROUND SECTION)

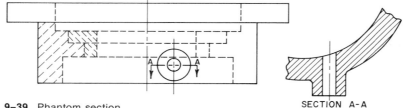

Fig. 9-39 Phantom section.

SECTION A-A

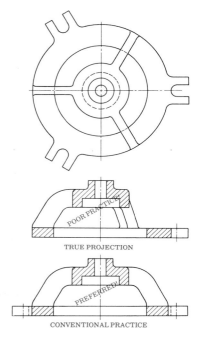

POOR PRACTICE

TRUE PROJECTION

PREFERRED

CONVENTIONAL PRACTICE

Fig. 9-40 Some feature should be rotated to show true shape.

shape. Only the parts which extend all the way around the vertical axis are sectioned. In Fig. 9-41 the lugs are rotated but not sectioned.

When a section passes through spokes, the spokes are not sectioned (Fig. 9-42 at A). Compare this with the drawing for a solid web (Fig. 9-42 at B) and notice the sectioning which indicates that the web is solid rather than spoked.

When a section or elevation of a part with holes on a circle is drawn

(Fig. 9-43), the holes are rotated until parallel with the vertical plane. The holes then show the true distance from the center, regardless of where they would project.

The principle of rotated features has many uses when it is desirable to show true conditions or true distances even though the views do not project. This may be illustrated by the bent lever (Fig. 9-44), where a part is rotated or stretched out in one of the views.

Conventional Breaks and Symbols. Conventional breaks and symbols are used to make the object easier to draw and easier to understand. Figure 9-45 shows the methods used to show long, uniformly shaped parts and for *breaking out,* or shortening, the drawing of parts. This allows the views to be drawn to a larger scale. Since the cross-sectional shape of the part is shown by the break, end views are usually not necessary. The length is given by a dimension. The symbols for conventional breaks are usually drawn freehand. However, on larger drawings it is sometimes necessary to use instruments to give a neat appearance. Figure 9–46 shows how the break is drawn for cylinders and pipes.

Intersections in Section. An intersection is the result of two parts joining at a given point (Fig. 9–47). A true projection of the intersection is difficult to draw and takes too much time. Also, it is of little or no value to the blueprint reader. Approximated and preferred sections are shown in Fig. 9–48.

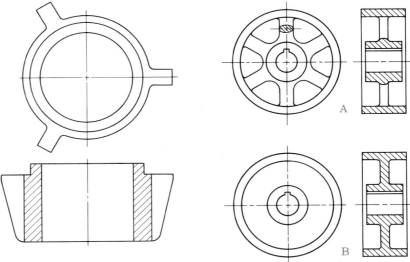

Fig. 9–41   Do not section lugs.

Fig. 9–42   Section through spokes.

Fig. 9–43   Good and poor practices for showing holes.

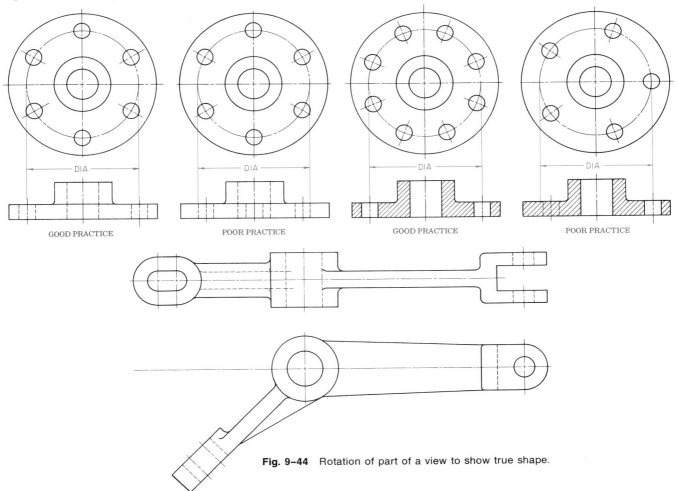

GOOD PRACTICE   POOR PRACTICE   GOOD PRACTICE   POOR PRACTICE

Fig. 9–44   Rotation of part of a view to show true shape.

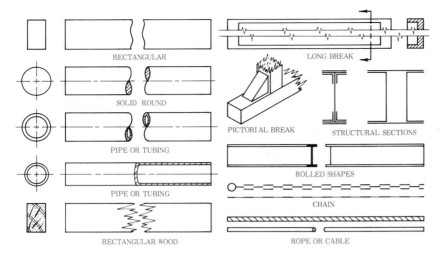

RECTANGULAR

SOLID ROUND

PIPE OR TUBING

PIPE OR TUBING

RECTANGULAR WOOD

LONG BREAK

PICTORIAL BREAK

STRUCTURAL SECTIONS

ROLLED SHAPES

CHAIN

ROPE OR CABLE

Fig. 9–45 Conventional breaks and symbols.

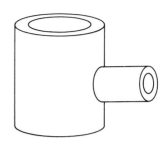

Fig. 9–47 Intersecting parts.

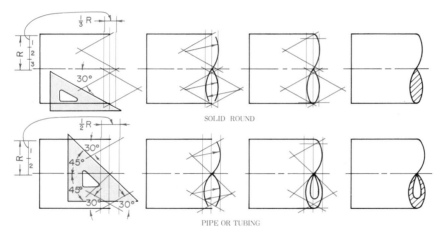

SOLID ROUND

PIPE OR TUBING

Fig. 9–46 Drawing the break symbols for cylinders and pipes.

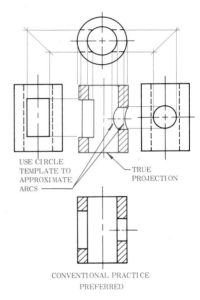

USE CIRCLE TEMPLATE TO APPROXIMATE ARCS

TRUE PROJECTION

CONVENTIONAL PRACTICE PREFERRED

Fig. 9–48 Approximated and preferred sections.

Review

1. Sectional views are used to show _____ details.

2. The location of the cutting plane is shown on a normal view by using a line called a _____.

3. The cut surface in a sectional view is shown by _____.

4. Name four types of sectional views.

5. Thin, flat parts of an object used to brace or strengthen another part of the object are called _____ or _____.

6. When a rib (or similar flat feature) does not show clearly in a sectional view, _____ section lining may be used.

7. A sectional view is obtained by supposing an imaginary _____ passes through the object to expose interior details.

8. The section lining symbol for cast iron can also be called the _____ symbol.

9. When the object is cut completely apart, a _____ sectional view is obtained.

10. When two cutting planes at right angles are used to cut out and remove one quarter of the piece, a _____ sectional view results.

Problems for Chapter 9

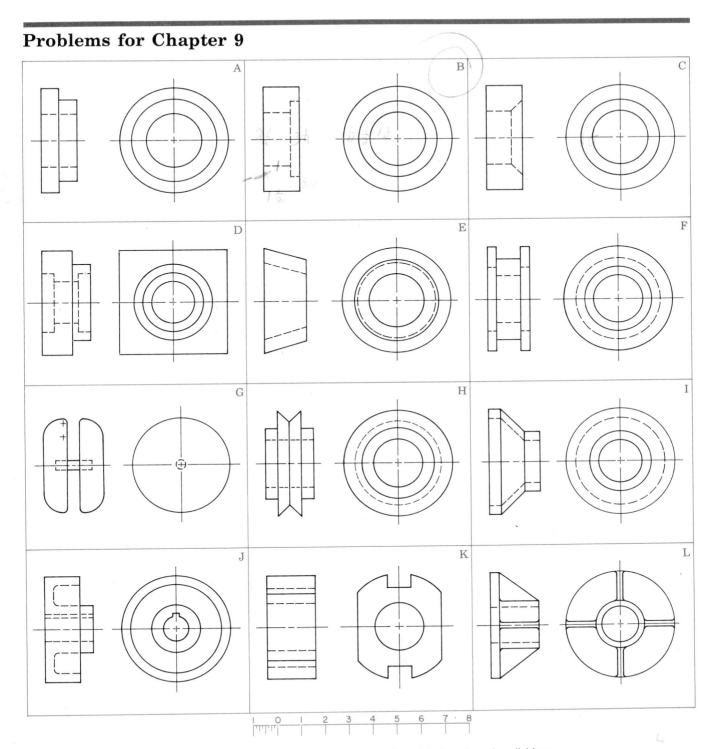

Fig. 9–49 Problems for practice in sectioning. Take dimensions from the printed scale, using dividers. Draw both views. Make full- or half-sectional view as assigned. Add dimensions if required. Estimate the size of radii.

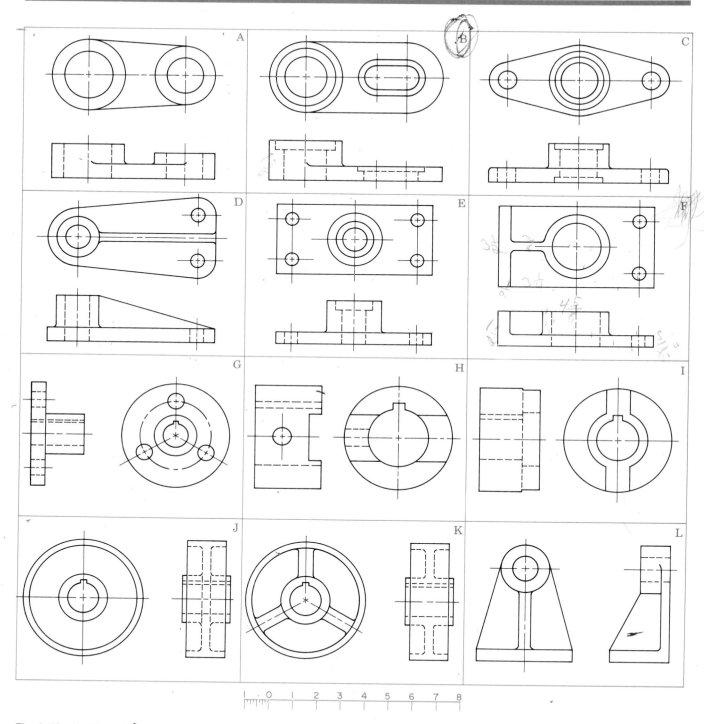

Fig. 9-50 Problems for practice in sectioning. Take dimensions from the printed scale, using dividers. Draw both views. Make full- or half-sectional view as assigned. Add dimensions if required. Estimate the size of radii.

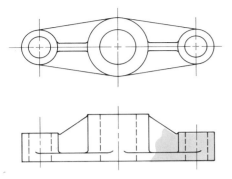

Fig. 9-51 Rod guide. Scale: full size or as assigned. Make top view and broken-out section as indicated by the red screen.

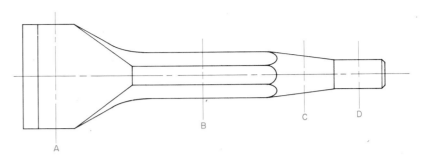

Fig. 9-52 Chisel. Scale: full size or as assigned. Make revolved or removed sections on red center lines. A is a $\frac{1}{2}'' \times 4\frac{1}{2}''$ rectangle, B is a 2'' (across flats) octagon, C is a circular cross section, D is a circular cross section.

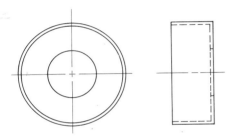

Fig. 9-53 Grease cap. Scale: full size or as assigned. Make front and right full- or half-sectional views as assigned.

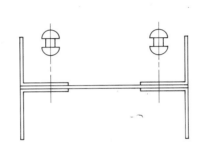

Fig. 9-54 Structural joint. Scale: $\frac{3}{4}$ size or as assigned. Make full-sectional view of joint with rivets moved into their proper position on the center lines.

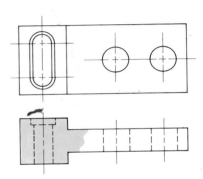

Fig. 9-55 Adjusting plate. Scale: full size or as assigned. Draw front and top views. Make broken-out section as indicated by the red screen.

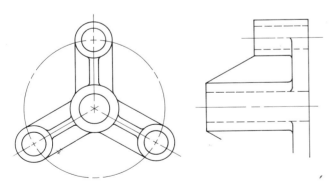

Fig. 9-56 Rotator. Scale: full size or as assigned. Complete the right-side view and make full- or half-sectional view.

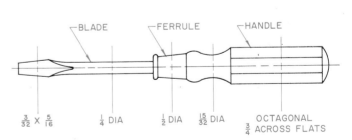

Fig. 9-57 Screwdriver. Scale: full size or as assigned. Draw the view as shown. Add removed sections on the red center lines. Use section lining to indicate materials. Blade is steel, ferrule is sheet steel, handle is wood.

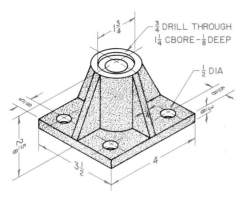

Fig. 9-58 Base plate. Scale: full size or as assigned. Material: cast iron.

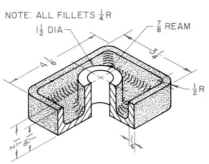

Fig. 9-59 Shaft base. Scale: ³⁄₄ size or as assigned. Material: cast iron.

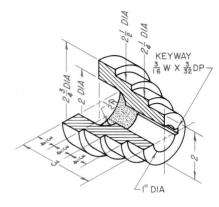

Fig. 9-60 Step pulley. Scale: full size or as assigned. Material: cast iron.

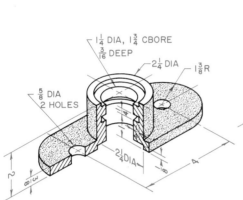

Fig. 9-61 Lever bracket. Scale: full size or as assigned. Material: cast iron.

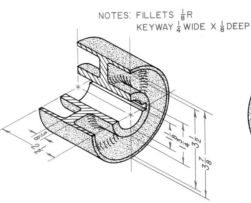

Fig. 9-62 Idler pulley. Scale: full size or as assigned. Material: cast iron.

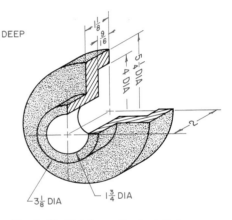

Fig. 9-63 Retainer. Scale: full size or as assigned. Material: cast aluminum.

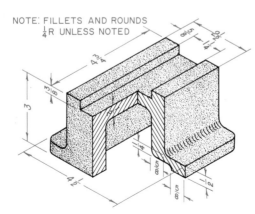

Fig. 9-64 Rest. Scale: ³⁄₄ size or as assigned. Material: cast aluminum.

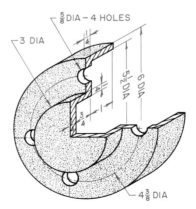

Fig. 9-65 End cap. Scale: full size or as assigned. Material: cast iron.

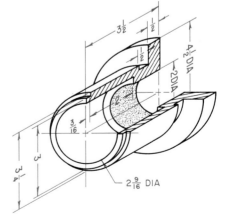

Fig. 9-66 Flange. Scale: full size or as assigned. Material: cast aluminum.

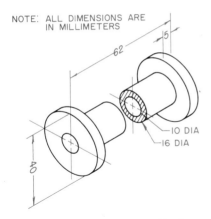

NOTE: ALL DIMENSIONS ARE
IN MILLIMETERS

Fig. 9-67 Spool. Scale: double size or as
assigned. Draw necessary views and add
revolved or removed section on center
portion. Material: wood.

Fig. 9-68 Link. Scale: full size or as assigned. Draw necessary views and add
revolved or removed section as assigned.
Material: cast iron. Add dimensions if required.

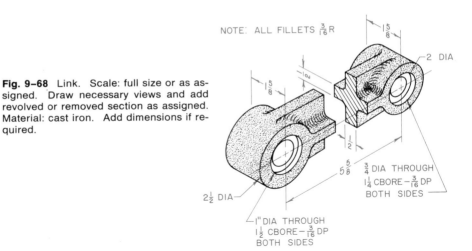

NOTE: ALL FILLETS $\frac{3}{16}$ R

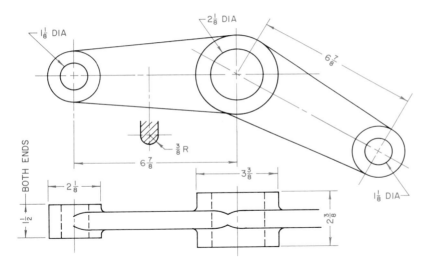

Fig. 9-69 Offset arm. Scale: $\frac{3}{4}$ size or as assigned. Complete the front view and add a revolved or removed section on the top view. Material: cast iron. Add dimensions if required.

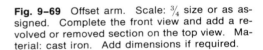

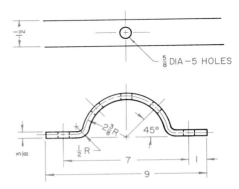

Fig. 9-70 Slide bar. Scale: full size or as assigned. Draw front view and revolved section.
Material: cast aluminum. Add dimensions if required.

Fig. 9-71 Clamp. Scale: full size or as
assigned. Complete the top view by
drawing it as though it were stretched out
in one plane. Material: steel. Add dimensions if required.

Fasteners | Chapter **10**

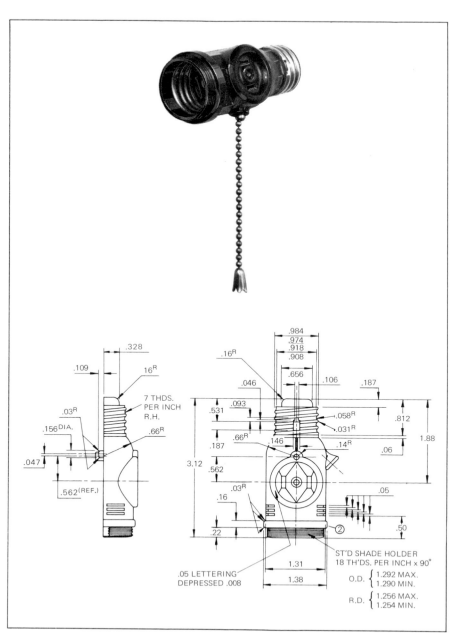

(WIRING DEVICE DEPARTMENT, GENERAL ELECTRIC COMPANY.)

The Function of Fasteners. The term *fasteners* is a family name that includes all methods of holding parts together. Fasteners include such things as screws, bolts and nuts, rivets, welding, brazing, soldering, adhesives, collars, clutches, keys, and many other special types. Each of these has its own special function to perform and may be used for (1) permanent fastening and (2) removable fastening, or adjustable fastening.

Screws and Screw Threads. The use of the principle of a screw thread is so old that its origin is unknown. Archimedes (287–212 B.C.) is credited with first applying the principle of the screw to practical use in a device for raising water. The same principle is used today in screw conveyors for moving flour and sugar in commercial bakeries, for raising wheat to grain elevators, for moving coal in stokers, and for many other purposes.

Screws and other fasteners are used in so many ways and have become so important that it is essential for the engineer, draftsman, and technician to become familiar with their different forms (Fig. 10-1) In addition, it is important to be able to draw and specify each type correctly.

The True Shape of a Screw Thread. The shape of all screw threads is based on a *helix*, or *helical curve*. Technically, a helix is a curve generated by a point moving uniformly around a cylinder and uniformly parallel to the axis of the cylinder. In simpler terms, if a wire is wrapped around a cylinder in uniformly spaced coils, it forms helical curves. A coil spring, for example, is made up of a series of helical curves.

Another means of understanding the shape of screw threads is by wrapping a right triangle, cut from paper, around a cylinder as shown in Fig. 10-2. The hypotenuse of a right triangle will form one turn of a helix if wrapped around a cylinder if the base of the triangle is equal to the circumference of the cylinder (Fig. 10-2 at A). The altitude will be the pitch of the helix. A right triangle and the projections of the corresponding helix are shown in Fig. 10-2 at B.

The method of drawing the projections of a helix is shown in Fig. 10-3 in Spaces A and B. Draw two projections of a cylinder (Space A). Divide the circumference into a number of equal parts and the pitch into the same number of equal parts. From each point in the circumference draw lines parallel to the axis to meet lines

perpendicular to the axis drawn through the corresponding divisions of the pitch (Space B). A smooth curve drawn through the points thus found will give the projection of the helix.

The application of the helix is shown in Space C, which is the actual projection of a square thread. Such drawings are seldom made, since they require too much time and are no

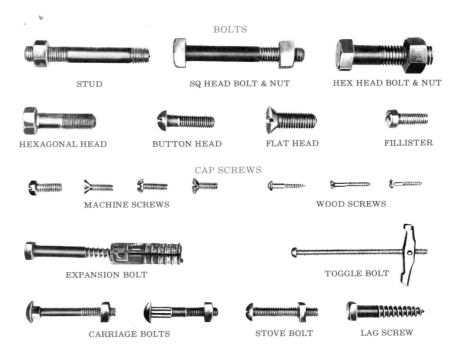

Fig. 10-1 Examples of some threaded fasteners.

Fig. 10-2 Picture of a helix at A and a projection of a helix at B.

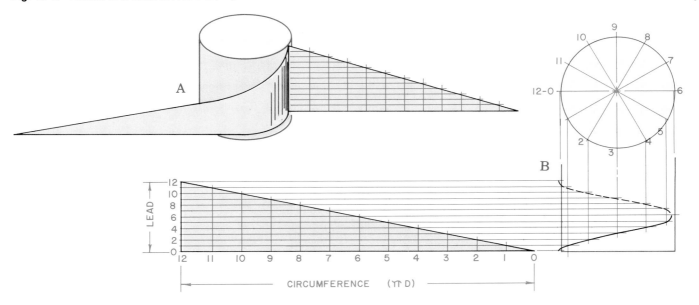

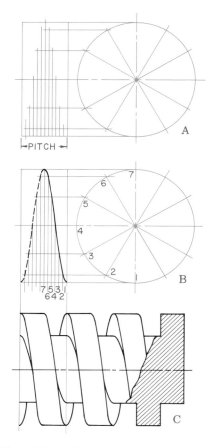

Fig. 10-3 Helix and square thread.

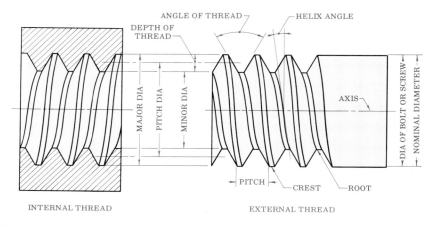

Fig. 10-4 Screw-thread terms.

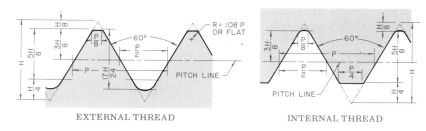

Fig. 10-5 Unified screw-thread terms.

better, practically, than the conventional representations commonly used.

Screw-Thread Standards. The first screws were made to suit a particular purpose and without any thought of how anyone else might make one of the same diameter. With industrial development, it became necessary to provide for uniformity in the form and number of threads per inch on screws of any given diameter.

Screw-thread standards in the United States were developed from a system presented by William Sellers in a report to the Franklin Institute in Philadelphia in 1864. Screw-thread standards in England came from a paper presented to the Institution of Civil Engineers in 1841 by Sir Joseph Whitworth. These two standards were not interchangeable.

Work on standards has continued with the growth of industry and the need for uniformity and interchangeable manufacture. In 1948 the Unified Thread Standards were agreed upon by standardization committees of Canada, the United Kingdom, and the United States. The Unified Standards are now the basic American National Standards and are described in the *American National Standard Unified Screw Threads for Screws, Bolts, Nuts and Other Threaded Parts* (ANSI B1.1–1960) and in Handbook H-28, *Federal Screw Thread Specifications*.

Since 1948, additional progress has been made toward the unification of thread systems. In 1968 the International Organization for Standardization (ISO) adopted the unified standard as ISO inch screw-thread system to go along with an ISO metric screw-thread system standard. Both systems have a number of characteristics in common.

These similarities are very important as all countries of the world move toward one common system of measurement, the metric system. As an aid in helping to move toward total unification, the American National Standards Institute has published the *American National Standard Unified Screw Threads—Metric Translation* (ANSI B1.1a–1968).

Screw-Thread Terms. Figure 10–4 shows several important screw-thread terms. The Unified and American (National) screw-thread profile shown in Fig. 10–5 is the form used for gen-

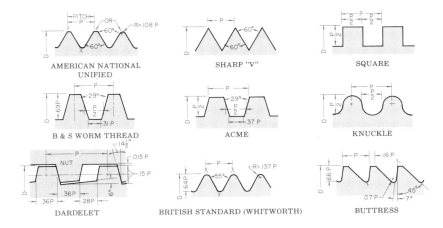

Fig. 10–6 Some of the various screw-thread profiles.

eral fastening purposes. Other forms of threads are used to meet various requirements, and some of these are illustrated in Fig. 10–6. The sharp V is seldom used. The square thread and similar forms (worm thread and acme thread) are designed to transmit motion or power and to hold the forces in line with the axis. The knuckle thread is familiar on electric-light sockets, and so forth, and as a "cast" thread. The Dardelet thread is a self-locking thread designed by a French military officer. The former British Standard (Whitworth) has rounded crests and roots and a 55° angle. The former United States Standard had flat crests and roots and a 60° angle. The buttress thread takes pressure in one direction

only: against the surface at an angle of 7° with a perpendicular to the axis.

Definition of a Screw Thread. "A screw thread is a ridge of uniform section in the form of a helix on the external or internal surface of a cylinder, or in the form of a conical spiral on the external or internal surface of a cone or frustum of a cone."

The pitch of a thread P is the distance from a point on the thread form to the corresponding point on the next form, measured parallel to the axis (Fig. 10–7). The lead L is the distance the threaded part would move parallel to the axis during one complete rotation in relation to a fixed mating part (the distance a screw would enter a threaded hole in one turn).

Single and Multiple Threads. Most screws have single threads (Fig. 10–7 at A). Unless otherwise specified, a single thread is understood. A single thread has a single ridge in the form of a helix. Therefore, one complete turn (revolution) will advance the threaded part into the nut a distance equal to the pitch. This is called *lead*.

A double thread (Fig. 10–7 at B) has two ridges side by side in the form of helixes, and the lead is twice the pitch. A triple thread (Fig. 10–7 at C) has three ridges side by side, and the lead is three times the pitch.

Multiple threads are used where quick assembly of two parts is required. Fountain pens and toothpaste tube caps are examples.

Right- and Left-Hand Threads. Threads are always right hand unless marked otherwise. A right-hand thread is one which turns in a clockwise direction to enter a threaded part when viewed from the outside end (Fig. 10–8 at A). A left-hand thread is one which turns in a counterclockwise direction when viewed from the outside end (Fig. 10–8 at B). Left-hand threads are always indicated by the initials LH. A turnbuckle (Fig. 10–9) is an example of right- and left-hand threads. Threads on bicycle pedals is another example. The left-hand pedal has left-hand threads; the right-hand pedal, right-hand threads.

Fig. 10–7 Single, double, and triple threads.

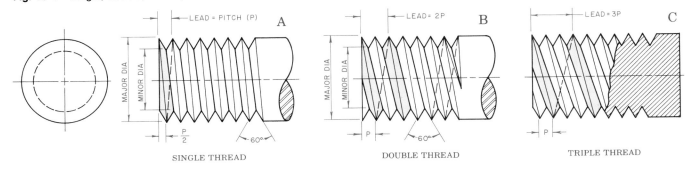

Representations of Screw-Threads on Drawings. The same symbols are used for coarse or fine threads and for right-hand or left-hand threads, with notes to give the necessary information.

Three types of representation are provided by the ANSI:

1. Detailed representation approximates the actual appearance of threads (Fig. 10–10). The pitch is not often drawn to scale but is approximated for convenience in drawing. The helixes are drawn as straight lines. The threads are drawn as sharp V's. In general, this representation would not be used for diameters of less than 1 in. It is a somewhat realistic (or pictorial) representation but is not usual practice on working drawings, except where it might be needed for clearness.

2. The schematic representation omits drawing the V's (Fig. 10–11). The crest and root lines are spaced by eye (estimating the pitch) to look good for the given diameter. The crest and root lines may be perpendicular to the axis or slanted to approximate the helix angle (Fig. 10–4). American National Standards shows thin crest lines and thick root lines, but all lines are often drawn the same to save time, especially on regular pencil working drawings.

3. The simplified representation is similar to the schematic representation. Crest and root lines are drawn on dotted lines except where either of them would show as a visible solid line (Fig. 10–12 at A, C, F, G, H, I, and J). The simplified representation saves a large amount of the draftsman's time by omitting unnecessary detail.

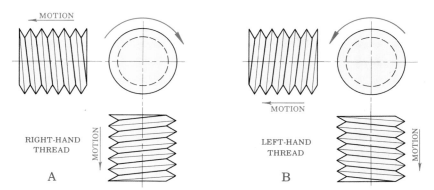
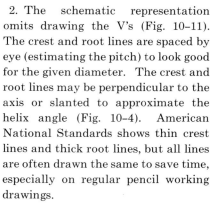

Fig. 10–8 A right-hand screw thread at A and left-hand screw thread at B.

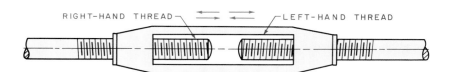

Fig. 10–9 A turnbuckle uses right- and left-hand screw threads.

Fig. 10–10 Detailed representations of screw threads.

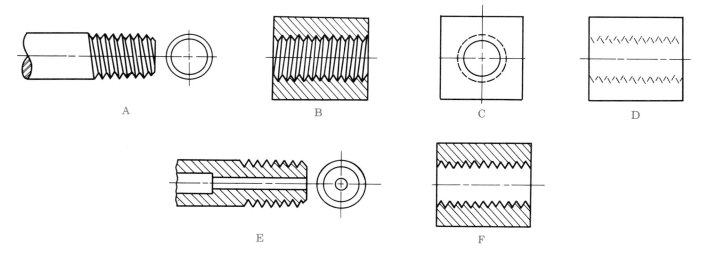

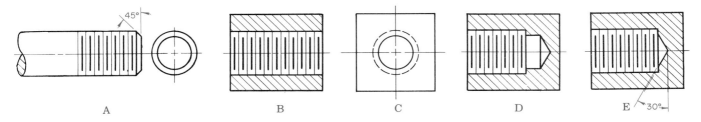

Fig. 10–11 Schematic representations of screw threads.

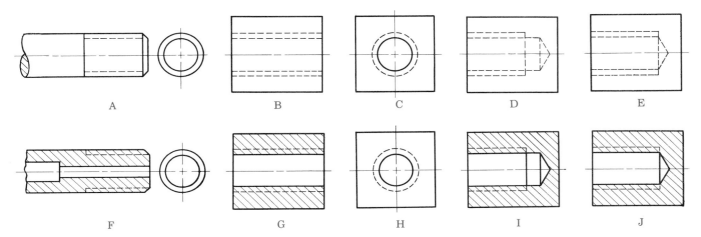

Fig. 10–12 Simplified representations of screw threads.

To Draw the Detailed Representation of Screw Threads. The detailed representation uses the sharp-V profile. Straight lines are used to represent the helixes of the crest and root lines. The order of drawing the V-form thread is shown in Fig. 10–13. The pitch is seldom drawn to scale; generally it is approximated and drawn to look good. In Fig. 10–13 at A, lay off the pitch P and the half pitch $P/2$, as shown. Adjust the triangle to the slope and draw the crest lines (if a drafting machine is used, set the ruling arm to the slope of the crest line). At B draw one side of the V for the threads with the 30°–60° triangle (if a drafting machine is used, set the ruling arm for the 30° angle). Then reverse the triangle, or ruling arm, and complete the V's. At C set the triangle, or the ruling arm of the

Fig. 10–13 To draw the detailed representation of screw threads.

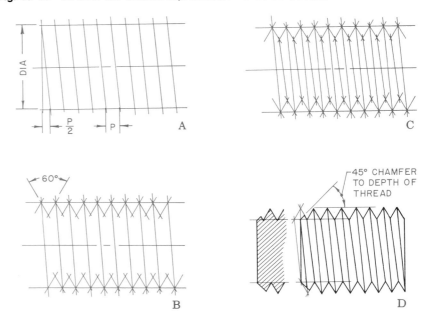

drafting machine, to the slope of the root lines and draw as shown. Notice that the root lines are not parallel to the crest lines because the root diameter is less than the major diameter. At D draw 45° chamfer lines as shown by the construction in red.

The sectional view of a threaded hole with a right-hand thread is shown in Fig. 10–14. Notice that the slope of the thread lines is opposite to that of a right-hand external thread, since they must match the far side of the screw.

The realistic, or V-form, thread representation (Fig. 10–15) is sometimes useful to provide clarity where two or more threaded pieces are shown in section.

To Draw the Schematic Representation of Screw Threads. At A in Fig. 10–16, the outside diameter of the

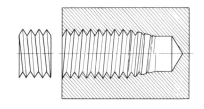

Fig. 10–14 Internal threads in section (threaded hole).

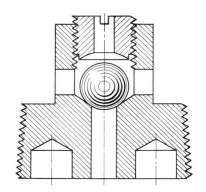

Fig. 10–15 Threads in section on assembled pieces.

screw thread is laid off. At B a method of construction is shown for the thread depth and the chamfer. At C the thin crest lines are drawn perpendicular to the axis. At D the thick root lines are drawn parallel to the crest lines.

The crest and root lines may be drawn at a slope (Fig. 10–17 at A) where single threads are suggested by a slope of half the pitch. On pencil drawings the crest and root lines are generally drawn the same width (Fig. 10–17 at B). On most drawings schematic representations are drawn to look good, with no attempt to lay off the pitch to scale.

To Draw the Simplified Representation of Screw Threads. At A in Fig. 10–18, the outside diameter of the screw is laid off. At B a method of construction is shown for the screw-

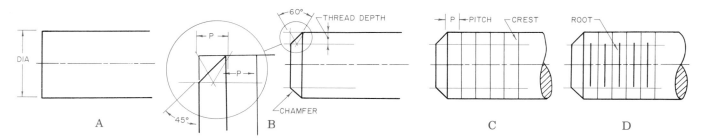

Fig. 10–16 To draw the schematic representation of screw threads.

Fig. 10–17 Slope-line representation at A and uniform-width lines at B.

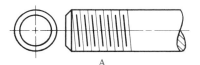

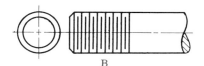

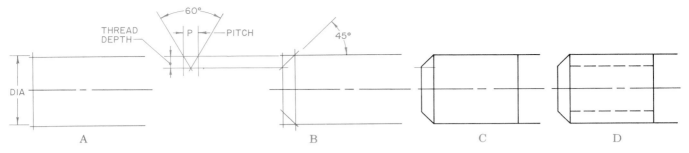

Fig. 10–18 To draw the simplified representation of screw threads.

Fig. 10–19 To draw square threads.

Fig. 10–20 To draw acme threads.

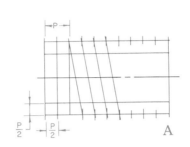

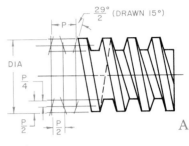

thread depth and the chamfer. At C the chamfer has been drawn and the length of the thread is indicated. At D dash lines have been drawn to complete the simplified representation of the threads.

To Draw Square Screw Threads. The depth of the square thread is one-half the pitch. In Fig. 10–19 at A, lay off the diameter; the pitch *P* one-half pitch spaces; and the depth of thread lines. At B draw the crest lines. At C draw the root lines as shown. At D the internal square thread is drawn in section.

To Draw Acme Screw Threads. The depth of the acme thread is one-half the pitch (Fig. 10–20). The stages in drawing acme threads are shown at A. The pitch diameter is midway between the outside diameter and the root diameter and locates the pitch line. On the pitch line, lay off one-half pitch spaces and draw the thread profile. Draw the crest lines; then, draw the root lines to complete the view. The construction shown at B is enlarged.

A sectional view of an internal acme thread is shown at C. Other representations sometimes used for internal threads by hidden lines and in sections are shown at D.

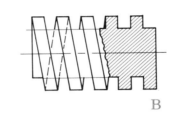

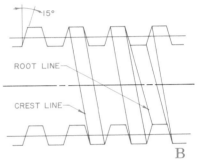

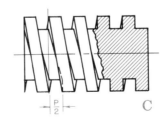

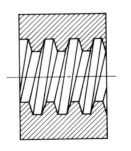

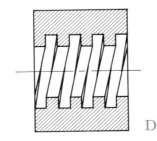

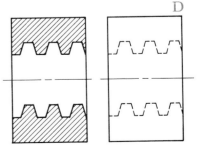

Thread Series for Unified and American National Standard Screw Threads. The number of threads per inch for a given diameter varies according to the purpose for which the screw is used. Several *series* of threads are provided in ANSI B1.1. Letter symbols are used to designate the different series.

Coarse-thread Series (UNC or NC). The pitch is relatively large for a specific diameter and is used for general engineering purposes.

Fine-thread Series (UNF or NF). The pitch is smaller (greater number of threads per inch) for a specific diameter than for the coarse-thread series. It is used where a finer thread is required, as in the automotive, aircraft, and similar industries.

Extra-fine-thread Series (UNEF or NEF). The pitch is smaller than for the fine-thread series. It is used where the depth of thread must be kept very small, as on aircraft equipment or thin-walled tubes.

In addition, various constant-pitch-thread series, UN, with 4, 6, 8, 12, 16, 20, 28, and 32 threads per inch offer a variety of combinations for those purposes where the coarse, fine, and extra-fine series are not appropriate. However, it is recommended that when any of the constant-pitch series are selected, preference should be given to the 8-, 12-, or 16-thread series. The most common use of the constant-pitch threads is on parts that require repeated assembly and disassembly where reconditioning of threads becomes necessary.

Eight-thread Series (8UN or 8N). This series uses 8 threads per inch for all diameters.

Twelve-thread Series (12UN or 12N). This series uses 12 threads per inch for all diameters.

Sixteen-thread Series (16UN or 16N). This series uses 16 threads per inch for all diameters.

Special Threads (UNS, UN, or NS). These are nonstandard, or special, combinations of diameter and pitch.

American National Standard thread series have been largely replaced by the Unified standard. The letter symbols for the American National series are listed for reference and identification.

Coarse-thread Series (NC). This series is used for bulk production of screws, bolts, nuts, and general use.

Fine-thread Series (NF). This series has smaller pitch than NC for the same diameter and is used where the coarse series is not suitable.

Extra-fine-thread Series (NEF). This series is used where a smaller pitch than NF is desired, as on thin-walled tubes.

The symbol NS is used to specify special threads.

Eight-thread Series (8N) is a uniform-pitch series for large diameters.

Twelve-thread Series (12N) is a uniform-pitch series where a medium-fine pitch is required for large diameters.

Sixteen-thread Series (16N) is a uniform-pitch series where still smaller pitch is required.

Notice that the Unified system has eight constant-pitch series, while the American National system has only three. However, as mentioned earlier, the 8-, 12-, and 16-pitch series are most commonly used.

Classes of Fits for Unified and American National Screw Threads. The amounts of tolerance and allowance specified to meet requirements for screw-thread fits are provided by *screw-thread classes.* Specific requirements can be met by the selection of series and class. In brief, the classes for Unified threads are: *Classes 1A, 2A, and 3A* for ex-

ternal threads only; *Classes 1B, 2B, and 3B* for internal threads only.

Classes 1A and 1B replace American National Class 1 for new designs. These classes have a large allowance (loose fit) and are intended for purposes where quick and easy assembly is required.

Classes 2A and 2B are the thread standards most used for general purposes, such as the production of bolts, screws, nuts, and similar threaded items.

Classes 3A and 3B provide for more accurate work, where closer tolerances are desired than in Classes 2A and 2B.

Classes 2 and 3 are American National Standard. Description and tabular information are given in Appendix 1 of ANSI B1.1, from which the following is quoted: "Gaging practice for minimum Class 3 external threads is the same as for Class 3A threads; that for minimum Class 2 external threads is the same as for Class 2A; and that for maximum Classes 2 and 3 internal threads is the same as for Classes 2B or 3B."

Screw-thread Specifications. All screw threads are specified by diameter (nominal, or major, diameter), number of threads per inch, length of thread, initial letters of the series, and class of fit. Threads are understood to be single right hand. If threads are to be left hand, the letters LH should follow the class symbol. For double or triple threads, the word "double" or "triple" should be given.

Some examples using fractional sizes follow:

$1\frac{1}{4}$—7UNC—1A ($1\frac{1}{4}$ diameter, 7 threads per inch, Unified threads, coarse threads, series 1, external).

$\frac{1}{2}$—13UNC—2A ($\frac{1}{2}$ diameter, 13 threads per inch, Unified threads, coarse threads, series 2, external).

$\frac{7}{8}$—14UNF—2B ($\frac{7}{8}$ diameter, 14 threads per inch, Unified threads, fine threads, series 2, internal).

$1\frac{5}{8}$—18UNEF—3B—LH ($1\frac{5}{8}$ diameter, 18 threads per inch, Unified threads, extra-fine threads, series 3, internal, left-hand).

Tapped (threaded) holes are specified by a note giving the diameter of the tap drill ($\frac{27}{64}''$); depth of hole ($1\frac{3}{8}''$); thread information ($\frac{1}{2}$ diameter, American National threads, Class 2); and length of thread ($1''$), as:

$\frac{27}{64}$ DRILL $\times 1\frac{3}{8}$ DEEP
$\frac{1}{2}$—13NC—2 $\times$ 1 DEEP

Metric Translation. On drawings where the metric system is used, the thread information is specified as follows:

Nominal size (expressed in decimal inches)
Number of threads per inch
Thread series symbol
Thread class symbol
Supplemental dimensions (limiting dimensions in this case) expressed in metric units
Equivalent inch dimensions enclosed in parentheses (optional)
Example:
0.375—16 UNC—2A
PD 8.460–8.349
(0.3331–0.3287 in.) Optional

On inch drawings, the thread information is specified as follows:

Nominal size (expressed in decimal inches)
Number of threads per inch
Thread series symbol
Thread class symbol
Supplemental dimensions expressed in decimal inches (optional)
Equivalent metric dimensions enclosed in parentheses (optional)
Example:
0.375-16 UNC-2A
PD 0.331–0.3287
(8.460–8.349 MM) Optional

When a conversion from decimal-inch units to millimeters is required, the following rules should be applied:

1. Multiply the decimal figure by 25.4 (1 in. = 25.4 mm).

2. Round the value obtained to three decimal places.

Example:
Convert 0.4675 in. to millimeters.
$0.4675 \times 25.4 = 11.87450$, or 11.874, millimeters

When a conversion from millimeters to decimal-inch units is required, the following rules should be applied:

1. Divide millimeters by 25.4.

2. Round the value obtained to four decimal places.

Example:
Convert 5.5240 millimeters to inches.

$$\frac{5.5240}{25.4} = 0.2175 \text{ in.}$$

Threaded Fastenings. Fasteners are made in many forms for different uses. Enough information is included in the following paragraphs to enable the student to identify and draw the threaded fasteners in most common use on machines, engineering and other projects, and constructions. These include: square- and hexagonal-head bolts, square and hexagonal nuts, studs, machine screws, cap screws and setscrews, etc. Tables of dimensions for drawing purposes for bolts, nuts, and some of the other generally used threaded fasteners are given in the Appendix.

Certain bolt and nut dimensions have been designated as Unified Standard for use in the United States, Great Britain, and Canada. For complete information, the latest American National Standard should be consulted.

American Standard Square and Hexagon Bolts and Nuts. These are of such importance that the student should learn the principal terms used (Fig. 10–21) and be able to draw the

Fig. 10–21 Bolt and nut terms.

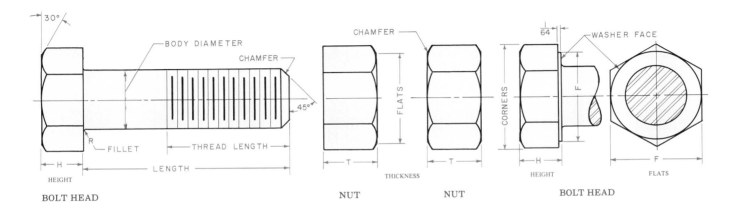

BOLT HEAD   NUT   NUT   BOLT HEAD

necessary views from the instructions which follow. In general, bolts and nuts may be regular or heavy and square or hexagon. Regular bolts and nuts are used for the general run of work. Heavy bolts and nuts are somewhat larger than regular and are used where a larger bearing surface is required or where a relatively larger hole is required in the part being held. Regular forms include square bolts, hexagon bolts, semifinished hexagon bolts, square nuts, hexagon nuts, and semifinished hexagon nuts. Heavy forms include hexagon bolts

and nuts, semifinished hexagon bolts and nuts, finished hexagon bolts, and square nuts.

Regular bolts and nuts are not finished on any surface. Semifinished bolts and nuts are processed to have a flat bearing surface. "Finished bolts and nuts" refers to the quality of manufacture and the closeness of tolerance and does not mean that the surfaces are completely machined. Semifinished boltheads and nuts (Fig. 10–22) have a washer-faced bearing surface or have chamfered corners with a diameter equal to the distance

across the flats. The thickness of the washer face is approximately ¹⁄₆₄ in.

Regular Boltheads and Nuts. For drawing purposes the dimensions may be obtained from the proportions given in Figs. 10–23 and 10–24 or from the Appendix. The chamfer angle may be drawn at 30° for either the hexagon or the square forms. (The standard indicates 25° for the square form.) Radii for the arcs may be found by trial. Note that one-half the distance across corners, *ab*, may be found by the construction shown.

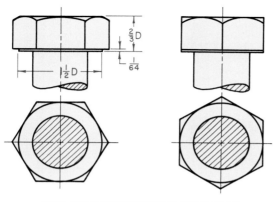

HEX. BOLT HEAD WITH WASHER FACE

Fig. 10–22 Semifinished boltheads and nuts.

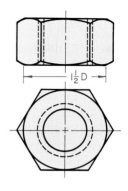

HEX. NUT WITH
WASHER FACE

HEX. NUT WITH
CHAMFER FACE

Fig. 10–23 Regular hexagonal bolthead and nut.

Fig. 10–24 Regular square bolthead and nut.

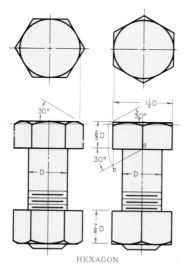

HEXAGON

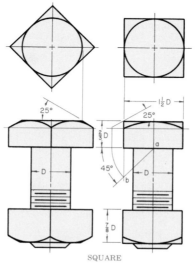

SQUARE

Heavy Hexagonal Boltheads and Nuts. For drawing purposes the dimensions may be obtained from the proportions given in Fig. 10–25 or from the Appendix. Also see Appendix V-1962 of ANSI B18.2-1960 and B18.2.1-1965.

A new standard covers hexagonal structural bolts, which are made of high-strength steel. These are used for structural-steel joints. Dimensions for drawing are given in the Appendix.

To Draw a Regular Square Bolthead Across Flats. (Square boltheads are not made in heavy sizes.) For regular sizes the width across

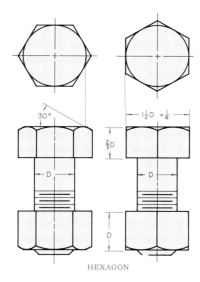

Fig. 10-25 Heavy hexagonal bolthead and nut.

HEXAGON

flats = $W = 1\frac{1}{2}D$, where D is the major diameter of the bolt (Fig. 10-26). The height of the head = $H = \frac{2}{3}D$. In Space A draw the center line and start the top view by drawing the chamfer circle with a diameter equal to the distance across the flats. Draw a square about this circle. In Space B draw a horizontal line representing the bearing surface of the head and lay off the height of the head, H. Draw the top line of the head and project down from Space A to obtain the vertical edges. Draw ox in the top view (Space A), revolve to y, and project down to Space B. From a in Space B draw the 30° chamfer line ab and project across to c and d. In Space C the chamfer arc is drawn through c, e, and d by radius R. Radius R can be found by trial, or $R = W$ may be used. Complete the view as in Space D.

To Draw a Regular Square Bolthead Across Corners. (Square boltheads are not made in heavy sizes.) For regular sizes, $W = 1\frac{1}{2}D$ and $H = \frac{2}{3}D$. In Space A of Fig. 10-27

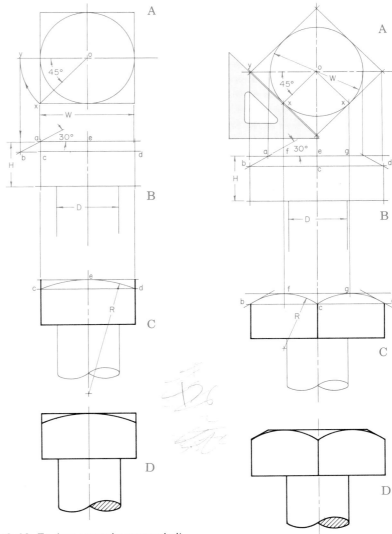

Fig. 10-26 To draw a regular square bolthead across flats.

Fig. 10-27 To draw a regular square bolthead across corners.

draw the center line and start the top view by drawing the chamfer circle with a diameter equal to the distance across the flats. About this circle, draw a square with the 45° triangle, as shown. Draw a horizontal line in Space B representing the bearing surface of the head and lay off the height of the head, H. Draw the top line of the head and project the diameter of the chamfer circle from the top view and draw 30° chamfer lines as shown (Space B). Project point x down to

obtain points f and g in Spaces B and C and draw line bcd. Then draw the chamfer arcs through bfc and cgd by radius R. Radius R can be found easily by trial, or $R = yo = \frac{1}{2}$ distance across corners may be used. Now, complete the view as in Space D. You can use the same method to draw a nut except that $T = \frac{7}{8}D$ for regular nuts, and $T = D$ for heavy nuts.

Boltheads and nuts are usually drawn across corners on all views of design drawings, regardless of projection. This is done to show the largest space required in order to permit turning space, or clearance. It also prevents hexagon heads and nuts being confused with square heads and nuts.

To Draw a Hexagon Bolthead Across Corners. Start the top view as in Space A (Fig. 10–28) by drawing the chamfer circle with a diameter equal to W, the distance across the flats. For a head across corners, draw a hexagon as indicated by the lines 1, 2, 3, 4, 5, 6. For the front view (Space B) draw a horizontal line representing the bearing or undersurface of the head. Lay off the height of the head and the top surface. Then project from the top view and draw the chamfer line.

Draw line *abcd* (Space C) to locate the chamfer intersections. Radius R_1 can be found by trial so that the arc will pass through points *b* and *c* and be tangent to the top line. Complete the front view as in Space D by drawing arcs with radii R_2 (tangent to top line and through points *a* and *b* at the left, and *c* and *d* at the right, by trial).

To draw a hexagon bolthead across flats, proceed as illustrated in Fig. 10–29. You can draw hexagon nuts with the same construction, but note the difference between the height of the head and thickness of the nut.

Boltheads and Nuts. Boltheads and nuts have dimensions so well standardized that they are seldom dimensioned on drawings. For a standard bolt the necessary information is given in a note, as in Fig. 10–30 at A, which specifies 1-in. diameter, 8 threads per inch, Unified coarse-thread series, Class 2A fit, 2¾ in. long, regular hex-head bolt. A bolt may

hold a part in place by passing through another part and using a nut, as at B, or by passing through a part and screwing into a threaded hole, as at C.

A stud or stud bolt (Fig. 10–30 at D) has threads on both ends and is used where a bolt is not suitable and for parts which must be removed often. The length of thread from each

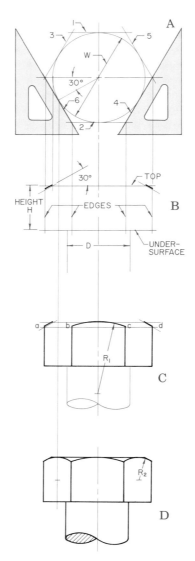

Fig. 10–28 To draw a regular hexagonal bolthead across corners.

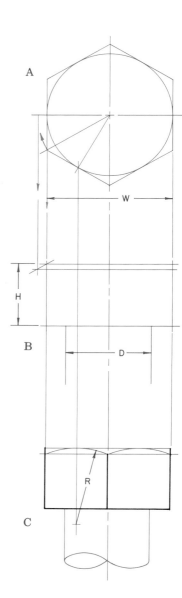

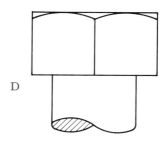

Fig. 10–29 To draw a regular hexagonal bolthead across flats.

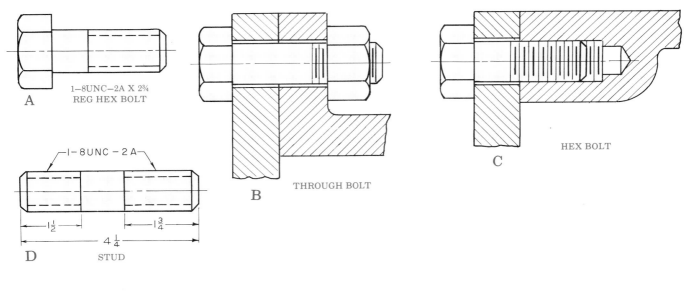

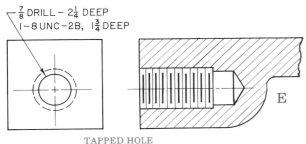

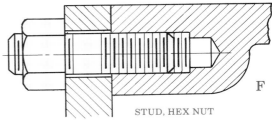

A — 1–8UNC–2A X 2¾ REG HEX BOLT

B — THROUGH BOLT

C — HEX BOLT

D — I – 8 UNC – 2 A — STUD — 1½ — 1¾ — 4¼

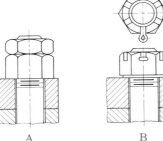

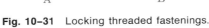

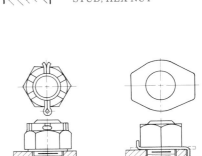

E — ⅞ DRILL – 2¼ DEEP / I–8 UNC–2B, 1¾ DEEP — TAPPED HOLE

F — STUD, HEX NUT

Fig. 10–30 Notes for bolts, studs, and threaded holes.

end is given by dimensions as shown. A tapped (threaded) hole is dimensioned as at E. One end of a stud is screwed permanently into place as at F, and a nut is screwed onto the projecting end. Under certain conditions a stud may be passed through two parts with a nut on each end.

Lock Nuts. Lock nuts and various devices are used to prevent nuts, or bolts and screws, from working loose. Many special devices are available.

Some forms of lock nuts are shown in Fig. 10–31. There are many special forms designed to provide positive locking.

A self-locking fastener, the Jay-Lock (Fig. 10–32), consists of an epoxy

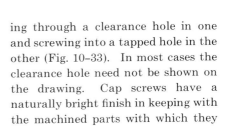

A   B   C   D

Fig. 10–31 Locking threaded fastenings.

chemical locking agent which combines with a hardening agent. When bolts or screws are engaged, it provides a strong, vibrationproof band.

Cap Screws. Cap screws are used for fastening two pieces together by pass-

ing through a clearance hole in one and screwing into a tapped hole in the other (Fig. 10–33). In most cases the clearance hole need not be shown on the drawing. Cap screws have a naturally bright finish in keeping with the machined parts with which they are used. Coarse, fine, or 8 threads may be on cap screws with Class 3A on the socket-head type and Class 2A on the others. Dimensions for drawing various sizes of American National Standard cap screws are listed in the Appendix.

Machine Screws. Machine screws are used where small diameters are required (Fig. 10–34). Sizes below ¼ in. in diameter are specified by number. They may screw into a tapped hole or extend through a clearance hole and into a nut (square nuts are used). The finish is bright. The ends are flat, as illustrated. Machine screws up to 2 in. long are threaded full length. Coarse or fine threads and Class 2 may be used on machine screws.

Setscrews. Setscrews are used for holding two parts in a desired position relative to each other by screwing through a threaded hole in one piece and bearing against the other (Fig. 10–35). There are two general types: square head and headless. The square-head setscrews are a source of accidents when used on rotating parts and violate safety codes. Headless setscrews may have either a slot or a socket. Any of the points may be used on any setscrew.

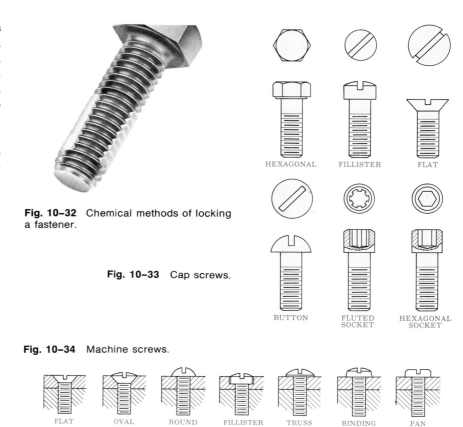

Fig. 10–32 Chemical methods of locking a fastener.

Fig. 10–33 Cap screws.

Fig. 10–34 Machine screws.

Wood Screws. Wood screws are made of steel, brass, or aluminum and are finished in various ways (Fig. 10–36). Steel screws may be bright (natural finish), blued, galvanized, or copper plated; both steel and brass screws are sometimes nickel plated.

Round-head screws are set with the head above the wood; flat-head screws are set flush, or countersunk. Wood screws may be drawn as illustrated. They are specified by number, length, style of head, and finish. Length of flat-head screws is measured overall;

Fig. 10–35 Setscrews.

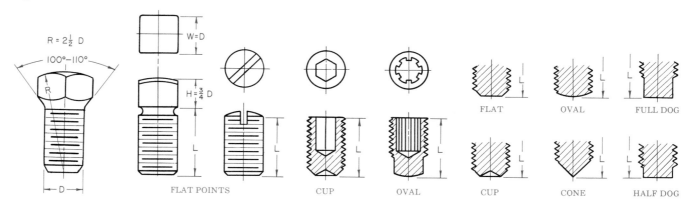

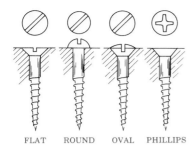

Fig. 10-36 Wood screws.

round-head screws, from under head to point; oval-head screws, from largest diameter of countersink to point. For sizes and dimensions see the Appendix.

Some Miscellaneous Threaded Fastenings. These are shown in Fig. 10-37. The names indicate the purposes for which they are used. Screw hooks and screw eyes are specified by diameter and overall length.

A lag screw, or lag bolt, is used for fastening machinery to wood supports and for heavy wood constructions when a regular bolt cannot be used. It is similar to a regular bolt but has wood-screw threads. Lag bolts are specified by the diameter and the length from under the head to the point. The proportions of the head are the same as the proportions found on regular boltheads.

Materials for Threaded Fasteners. Threaded fasteners are usually made from steel, brass, bronze, aluminum, cast iron, wood, and nylon. Nylon screws and bolts are made in various bright colors, such as red, blue-green, yellow, white, etc.

Keys. Keys are used to secure pulleys, gears, cranks, and similar parts to a shaft (Fig. 10-38). The form of the key is selected to suit the duty that it must perform. This ranges from the saddle key for light duty to

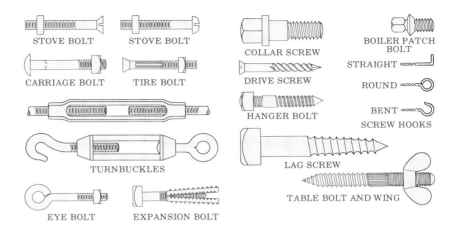

Fig. 10-37 Miscellaneous threaded fastenings.

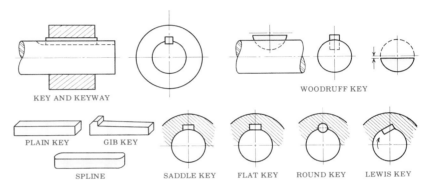

Fig. 10-38 Keys.

special forms, such as two square keys, for heavy duty. The common sunk key may have a breadth of about one-fourth the shaft diameter and a thickness of from five-eighths the breadth to the full breadth. The Woodruff key is much used in machine-tool work. It is made in standard sizes and is specified by number (see Appendix). Special forms of pins have been developed to take the place of keys for some purposes. These pins require only a drilled hole instead of the machining which is necessary in order to make keys.

Rivets. Sheet-metal plates, structural-steel shapes, boilers, tanks, and many other classes of work are put together with rivets as permanent fastenings. Rivets are rods of metal with a preformed head on one end. The rivet, heated red hot, is placed through the parts to be joined and held in place while a head is formed on the projecting end. The rivet is said to be "driven."

Large rivets (Fig. 10–39) have nominal diameters ranging in size from $\frac{1}{2}$ to $1\frac{3}{4}$ in. Small rivets (Fig. 10–40) range from $\frac{1}{16}$ or $\frac{3}{32}$ to $\frac{7}{16}$ in. in diameter.

There are many forms of "blind" rivets for use where one side of the plates cannot be reached or where the space is too small to use a regular rivet. One type is the du Pont explosive rivet (Fig. 10–41), which has a small explosive charge in the cavity when inserted. After the charge is exploded, a head is formed. This makes blind riveting possible, since the head can be formed inside closed or inaccessible places.

Sometimes it is desirable to have clear surfaces on plates which are fastened together. This requires flush riveting (Fig. 10–42) on one or both sides; it occurs on airplanes, automobiles, spacecraft, etc.

Riveted joints (Fig. 10–43), used for joining plates, may have lap joints or butt joints and may have single or multiple riveting. (See Appendix for American National Standard rivet dimensions.)

For some purposes, as in tanks, steel buildings, etc., high-strength structural bolts are coming into use. Welding is also in wide use.

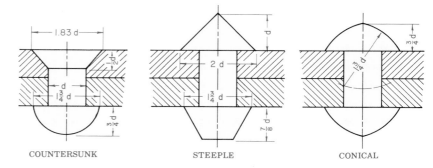

COUNTERSUNK   STEEPLE   CONICAL

Fig. 10–39   Large rivets.

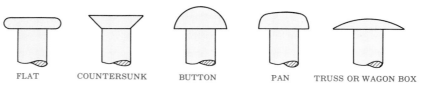

FLAT   COUNTERSUNK   BUTTON   PAN   TRUSS OR WAGON BOX

Fig. 10–40   Small rivets.

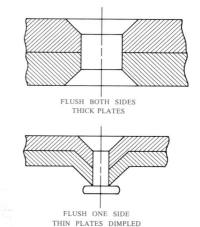

FLUSH BOTH SIDES
THICK PLATES

FLUSH ONE SIDE
THIN PLATES DIMPLED

Fig. 10–42   Flush rivets.

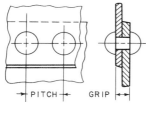

Fig. 10–41   Explosive rivet. (*Explosives Department, E. I. du Pont de Nemours & Company.*)

Fig. 10–43   Riveted joints.

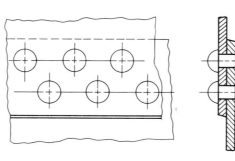

SINGLE-RIVETED LAP JOINT

PITCH   GRIP

DOUBLE-RIVETED LAP JOINT
STAGGERED RIVETING

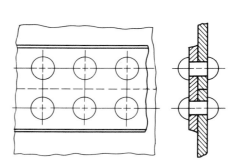

SINGLE-RIVETED BUTT JOINT

Review

1. Name five types of fasteners.

2. The first known application of the principle of the screw was for use in _____.

3. The true shape of a screw thread is based on a _____ curve.

4. One complete turn will advance a threaded part into the nut a distance equal to the _____ on a single thread.

5. What type of thread is used for quick assembly of parts?

6. Left-hand threads are indicated by the initials _____.

7. Name the three types of thread representation.

8. What system of measurement uses millimeters?

9. The three most common thread series are coarse, _____ and _____.

10. Name the classes of fits for Unified screw threads.

Problems for Chapter 10

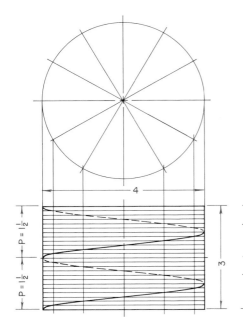

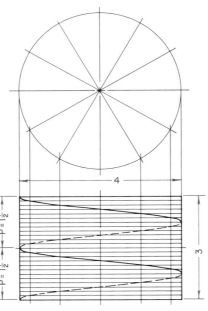

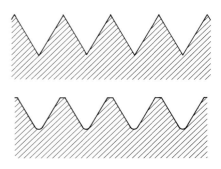

Fig. 10–46 Draw the profiles of the sharp "V" and the American National Unified thread. Letter the name of each under it. Pitch = 1".

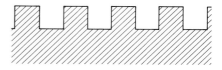

Fig. 10–44 Draw two complete turns of a right-hand helix as shown above. Use dimensions indicated and work full size. Number all points to avoid errors.

Fig. 10–45 Draw two complete turns of a left-hand helix as shown above. Use dimensions indicated and work full size. Number all points to avoid errors.

Fig. 10–47 Draw the profile of the square thread. Letter the name under it. Pitch = 1".

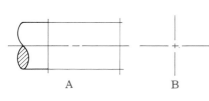
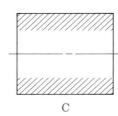
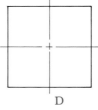
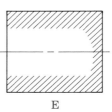
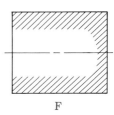

Fig. 10–48 Schematic representation of screw threads. Take dimensions from the printed scale at the bottom of the page, using dividers. Draw the views as shown and complete each as follows: A = schematic representation showing 1"-8UNC-2A threads; B = end view of A; C = schematic representation of section through 1"-8UNC-2B (internal) threads; D = right-side view of C; E = schematic representation of section through ⅞ drill × 1½ deep, 1"-8UNC-2B × 1⅛ deep; F = schematic representation of section through ⅞ drill × 1½ deep, 1"-8UNC-2B × 1½ deep.

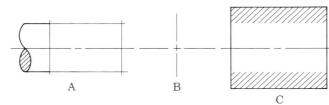
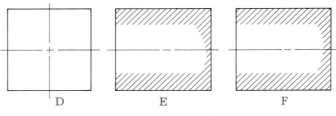

Fig. 10–49 Simplified representation of screw threads. Take dimensions from the printed scale at the bottom of the page, using dividers. Draw the views as shown and complete each as follows: A = simplified representation showing 1"-8UNC-2A threads; B = end view of A; C = simplified representation of section through 1"-8UNC-2B (internal) threads; D = right-side view of C; E = simplified representation of section through ⅞ drill × 1½ deep, 1"-8UNC-2B × 1⅛ deep; F = simplified representation of section through ⅞ drill × 1½ deep, 1"-8UNC-2B × 1½ deep.

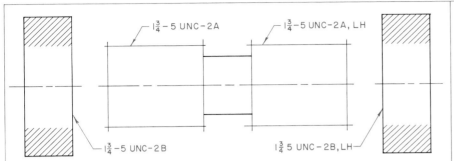

Fig. 10–50 Detailed representation of screw threads. Take dimensions from the printed scale at the bottom of the page, using dividers. Draw the views as shown and complete each according to the specifications noted on each. Use detailed thread representation.

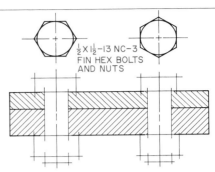

Fig. 10–51 Regular hexagonal bolt and nut. Draw the views and complete the bolts and nuts in the sectional view. See Appendix for bolt and nut detail sizes.

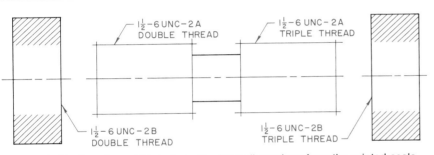

Fig. 10–52 Double and triple threads. Take dimensions from the printed scale at the bottom of the page, using dividers. Draw the views as shown and complete each according to the specifications noted on each. Use detailed thread representation.

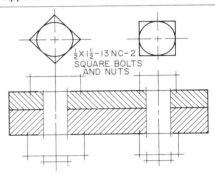

Fig. 10–53 Regular square bolt and nut. Draw the views and complete the bolts and nuts in the sectional view. See bolt and nut detail sizes.

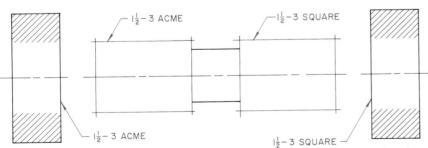

Fig. 10–54 Acme and square threads. Take dimensions from the printed scale at the bottom of the page, using dividers. Draw the views as shown and complete each according to the specifications noted on each. Use detailed thread representation.

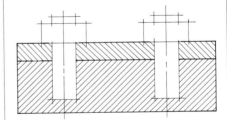

Fig. 10–55 Studs. Draw the view as shown and complete the $\frac{1}{2}'' = 1\frac{3}{4}''$ studs and regular semifinished hexagonal nuts. Check the Appendix for specific nut sizes. Other dimensions may be taken from the printed scale at the bottom of the page. Use schematic thread representation.

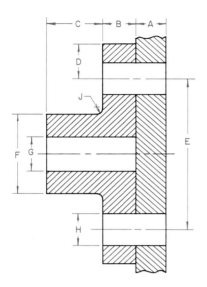

Fig.	Bolt DIA	A	B	C	D	E	F	G	H	J
10-56	1/4	1/4	5/16	5/8	1/2	2	3/4	3/8	9/32	3/32
10-57	5/16	5/16	3/8	5/8	5/8	2	3/4	3/8	11/32	3/32
10-58	3/8	3/8	7/16	5/8	5/8	2 1/4	3/4	3/8	13/32	1/8
10-59	7/16	7/16	1/2	3/4	5/8	2 1/2	1	1/2	15/32	1/8
10-60	1/2	1/2	9/16	7/8	5/8	2 3/4	1 1/4	5/8	9/16	1/8
10-61	9/16	9/16	5/8	1	3/4	3	1 1/2	3/4	5/8	1/8
10-62	5/8	5/8	11/16	1 1/8	3/4	3 1/4	1 3/4	3/4	11/16	1/8
10-63	3/4	3/4	13/16	1 1/4	7/8	3 1/2	2	7/8	13/16	3/16
10-64	7/8	7/8	1	1 3/8	1	3 3/4	2 1/4	1	15/16	1/4
10-65	1	1	1 1/8	1 1/2	1 1/8	4	2 1/2	1 1/8	1 1/8	1/4
10-66	1 1/8	1 1/8	1 1/4	1 5/8	1 1/4	4 1/4	2 3/4	1 1/4	1 1/4	1/4

Figs. 10-56 to 10-66 Take all dimensions from the table for the problem assigned and draw the flange and head plate as shown. On the red center lines, draw American National Standard bolts and nuts (hex or square) as assigned. Place bolt head at the left and show bolt head across flats; nut across corners.

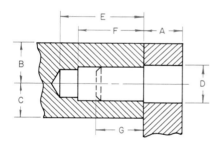

Fig.	Stud DIA	Nut	A	B	C	D	E	F	G
10-67	3/4	Hex	13/16	7/8	3/4	13/16	1 3/4	1 3/8	1
10-68	7/8	Sq	15/16	1 1/4	7/8	15/16	2	1 9/16	1 1/8
10-69	1	Hex	1 1/8	1 1/4	1	1 1/8	2 1/4	1 3/4	1 1/4
10-70	1 1/8	Sq	1 1/4	1 1/2	1 1/8	1 1/4	2 3/4	2 1/8	1 1/2

Figs. 10-67 to 10-70 On the center line shown, draw a stud with hexagonal or square nut, across flats or corners, as directed by the instructor. Take dimensions from the table.

Fig.	Bolt DIA	Head Style	A	B	C
10-71	3/8	Button	1/2	1 1/2	2 1/2
10-72	1/2	Button	3/4	1 3/4	3
10-73	5/8	Button	1	2	3 1/2
10-74	3/8	Flat	1/2	1 1/2	2 1/2
10-75	1/2	Flat	3/4	1 3/4	3
10-76	5/8	Flat	1	2	3 1/2
10-77	3/8	Fillister	1/2	1 1/2	2 1/2
10-78	1/2	Fillister	3/4	1 3/4	3
10-79	5/8	Fillister	1	2	3 1/2

Figs. 10-71 to 10-79 Take dimensions from the table for the problem assigned and draw the figure shown at the left. Refer to the Appendix for sizes and draw the assigned style of head and size of cap screw. Also, draw a top view of the screw head.

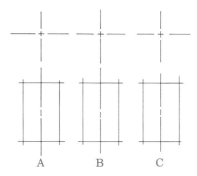

Fig. 10–80 Setscrews. Draw three setscrews: A = $\frac{3}{4}$ DIA × $1\frac{1}{4}$ long, square head, flat point. B = $\frac{3}{4}$ DIA × $1\frac{1}{4}$ long, slotted head, oval point. C = $\frac{3}{4}$ DIA × $1\frac{1}{4}$ long, socket head, cup point. Use schematic thread representation. Draw top view of each. Do not section.

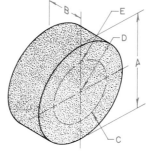

Fig. 10–81 Spacer. Draw two views of the spacer. Use schematic representation to show the threaded holes. A = 4″ DIA, B = $1\frac{1}{2}$″, C = $2\frac{1}{2}$″ DIA, D = 1″-8UNC-2B (through), E = $\frac{3}{8}$-16UNC-3B (through). Add notes and all necessary dimensions.

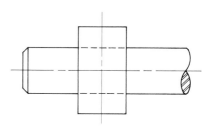

Fig. 10–82 Draw the 1″ DIA × $3\frac{1}{2}$″ long shaft and $1\frac{7}{8}$″ DIA × 1″ collar as shown. Draw the collar in full section and add a No. 4 × 2″ American National Standard taper pin on the red center line. Estimate sizes not given. Materials: shaft—steel; collar—cast iron.

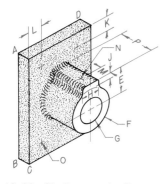

Fig. 10–83 Shaft support. Draw necessary views. At *N*, draw a $\frac{5}{16}$″ setscrew (square head, flat point). At *O* (four locations), draw $\frac{3}{8}$″ coarse threads (simplified representation). All fillets and rounds = $\frac{1}{8}$ R. AB = $4\frac{1}{2}$, BC = $\frac{1}{2}$, AD = 3, E = $1\frac{1}{2}$, F = $\frac{7}{8}$ R, G = 1″ DIA, H = $\frac{3}{4}$, J = $\frac{1}{4}$, K = $\frac{5}{8}$, L = $\frac{5}{8}$, M = $\frac{1}{2}$, P = $1\frac{1}{2}$.

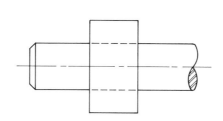

Fig. 10–84 Draw the 1″ DIA × $3\frac{1}{2}$″ long shaft and 2″ DIA × 1″ collar as shown. Draw the collar in half section and add an ANSI No. 404 Woodruff key at the top of the shaft. Estimate sizes not given. Materials: shaft—steel; collar—cast aluminum.

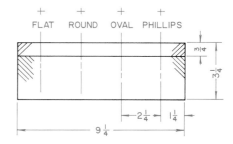

Fig. 10–85 Wood screws. Draw the view shown above. Add $2\frac{1}{2}$″ #12 wood screws on the red center lines. Show the four head types as indicated and draw a top view of each on the center mark above the view.

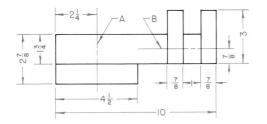

Fig. 10–86 Draw the view shown. On center line at A, draw $\frac{1}{2}$″ × $2\frac{1}{2}$″ fillister-head cap screw (head at top). At B, draw a $\frac{3}{8}$″ × 4″ flat-head cap screw. Show entire view in section. Material: steel. Use schematic thread representation.

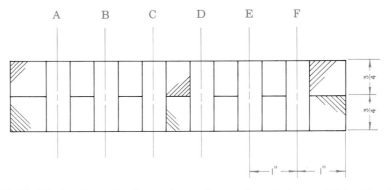

Fig. 10–87 Rivets. Draw the figure shown above. Overall sizes = $1\frac{1}{2}$″ × 7″. Refer to Chapter 10 and the Appendix and draw the six types of heads (top and bottom) for $\frac{1}{2}$″ DIA rivets. Do not dimension. A = button head; B = high button head; C = cone head; D = flat-top countersunk head; E = round-top countersunk head; F = pan head.

Working Drawings | Chapter 11

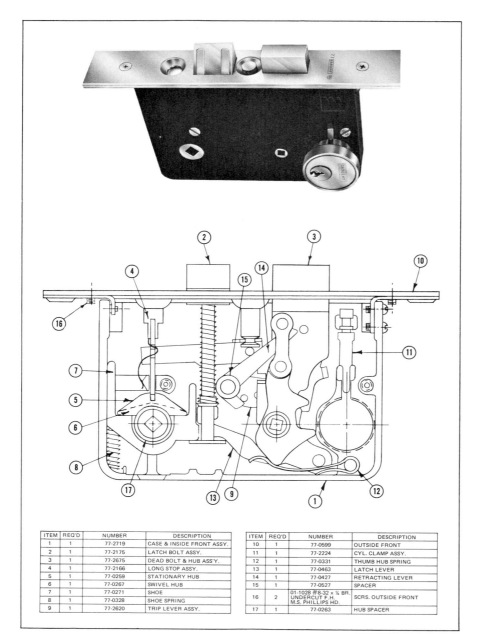

ITEM	REQ'D	NUMBER	DESCRIPTION
1	1	77-2719	CASE & INSIDE FRONT ASSY.
2	1	77-2175	LATCH BOLT ASSY.
3	1	77-2675	DEAD BOLT & HUB ASS'Y.
4	1	77-2166	LONG STOP ASSY.
5	1	77-0259	STATIONARY HUB
6	1	77-0267	SWIVEL HUB
7	1	77-0271	SHOE
8	1	77-0328	SHOE SPRING
9	1	77-2620	TRIP LEVER ASSY.

ITEM	REQ'D	NUMBER	DESCRIPTION
10	1	77-0599	OUTSIDE FRONT
11	1	77-2224	CYL. CLAMP ASSY.
12	1	77-0331	THUMB HUB SPRING
13	1	77-0463	LATCH LEVER
14	1	77-0427	RETRACTING LEVER
15	1	77-0527	SPACER
16	2	01-1028 #8-32 x ¼ BR. UNDERCUT F.H. M.S. PHILLIPS HD.	SCRS. OUTSIDE FRONT
17	1	77-0263	HUB SPACER

(SARGENT & COMPANY.)

Definitions. A working drawing is one that gives all the information necessary for making a single part or a complete machine or structure. The working drawing completely describes shape and size and gives specifications for the kinds of material to be used, the methods of finish, and the accuracy required. A pictorial drawing and a working drawing of a simple machine part are shown in Fig. 11–1.

The term *working drawing* is a family name that includes *detail drawings, assembly drawings,* and *assembly working drawings.* Each of these terms will be thoroughly described throughout this chapter.

Working Drawings. Working drawings are usually multiview drawings with complete dimensions and notes added. Care must be taken in preparing working drawings so that all necessary information is included. Nothing must be left to guess.

A good working drawing must conform with the style and practices followed by the office or industry where it is made. Most industries follow the standards recommended by the American National Standards Institute (ANSI). By doing so, plans can be easily read and understood from one industry to another. However, in some cases, shortcuts are used to shorten drafting time, as described in Chapter 16.

Regardless of the individual styles and practices used, certain characteristics must be followed by all. Good

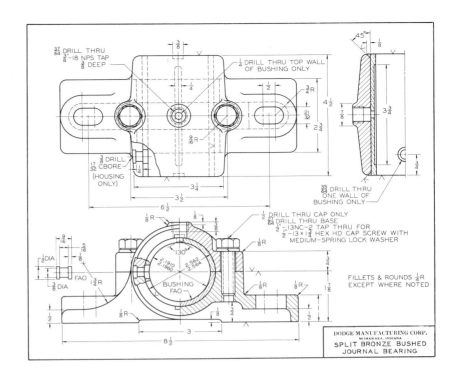

Fig. 11–1 Pictorial drawing and a working drawing of a split bronze-bushed journal bearing. (*Dodge Manufacturing Corporation.*)

contrast through proper line technique is extremely important. Figures that are easy to read, uniform lettering, and the use of standard terms and abbreviations are essential. When completed, a working drawing must be thoroughly checked by the draftsman for errors and possible improvements before it is submitted to the supervisor or checker for approval.

Making Drawings for Industrial Use. In the language of drawing, an object is described by telling its shape and its size. All drawings, whether for steam or gas engines, machines, buildings, airplanes, automobiles, missiles, or satellites, are made on the same principles.

Sometimes an unfavorable comparison is made between a student's drawing and a "real drawing." The finished appearance of a real drawing, as made by a draftsman or engineer,

is due to a thorough knowledge of engineering drafting and its use in industry. The correct order of going about the work and some of the procedures that draftsmen usually follow are described in this chapter. The student must become thoroughly familiar with this practice if his drawings are to have the style and good form that are necessary in industry today.

A large percentage of the drawings done in industry are pencil drawings. Good technique is essential. In pencil work, neatness, few erasures, and dense, sharp lines are essential.

While most industrial drawing is done in pencil, there is a growing need for high-quality ink drawings. Modern techniques in microreproduction, storage, and retrieval make it necessary for draftsmen to prepare drawings of a higher quality with thicker, sharper lines and larger, neater lettering.

Detail Drawings. The drawing of a single piece that gives all the information necessary for making it is called a *detail drawing,* as for the simple part in Fig. 11–2. A detail working drawing must be a complete and accurate description of the piece, with carefully selected views and well-located dimensions (Fig. 11–3).

When a large number of machines are to be manufactured, it is usual to make a detail drawing for each part on a separate sheet, especially when some of the parts may be used on other machines. When several parts are used on a single machine, it is common practice in some industries to detail a number of parts on one drawing. Sometimes separate detail drawings are made for the use of different workmen, such as the patternmaker, machinist, or welder. Such drawings have only the dimensions and information needed by the workmen for whom the drawing is made. Figure

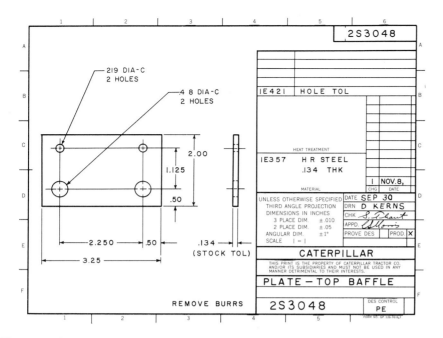

Fig. 11-2 A working drawing of a simple detail. (*Caterpillar Tractor Company.*)

11-4 shows a forging as formed and after it has been machined. The working drawing of the superspacer latch pinion made by the Hartford Special Machinery Company is shown in Fig. 11-5. Notice how the parts to be removed after all work is done are shown; also, notice the detailed list of machine operations. Figure 11-6 shows a combination drawing, or two-part detail drawing, for an oil-pump drive gear. The right-hand half gives the dimensions for the forging, and the left-hand half gives the machining dimensions and information. A detail drawing may contain calculated data, as shown for the air cleaner strap in Fig. 11-7.

Standard detail drawings are often made for parts which have the same shape but not the same dimensions. When such parts are used often, the views are drawn with blank spaces for

Fig. 11-3 A one-view working drawing. The view and the extra section provide a complete description.

Fig. 11-4 Index-plunger operating handle—forging and finished part. (*The Hartford Special Machinery Company.*)

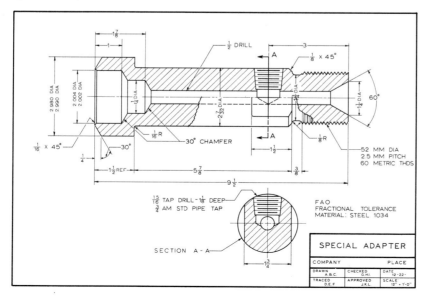

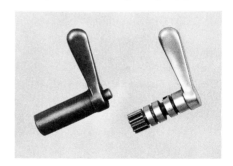

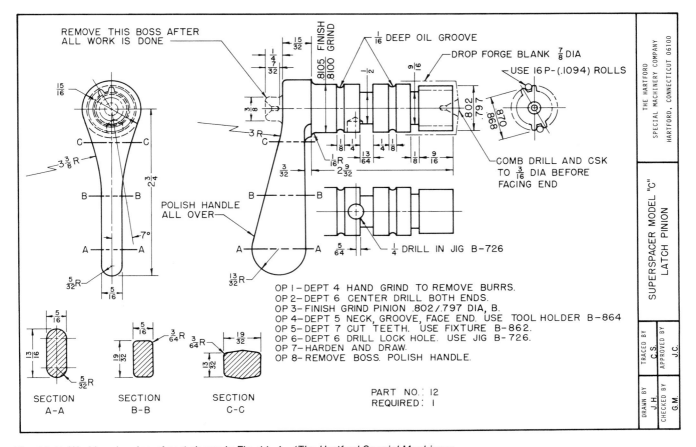

Fig. 11-5 Working drawing of part shown in Fig. 11-4. (*The Hartford Special Machinery Company.*)

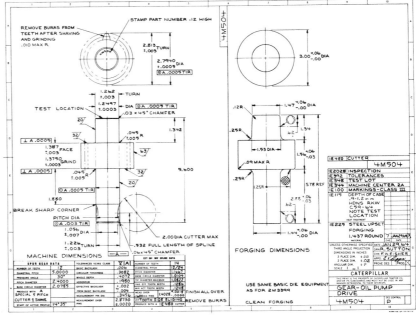

Fig. 11-6 A two-part detail drawing showing separate information for forging and machining. (*Caterpillar Tractor Company.*)

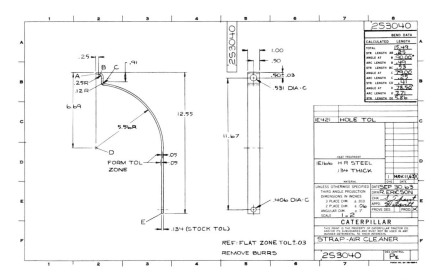

Fig. 11-7 A detail drawing which includes calculated data. (*Caterpillar Tractor Company.*)

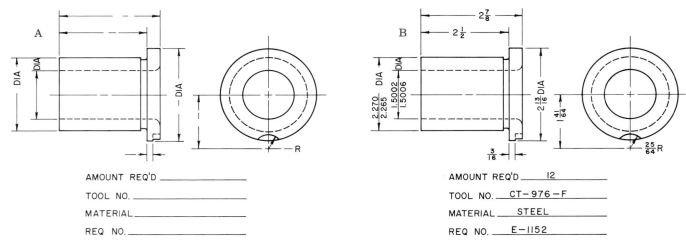

AMOUNT REQ'D _____

TOOL NO. _____

MATERIAL _____

REQ NO. _____

AMOUNT REQ'D ___12___

TOOL NO. __CT-976-F__

MATERIAL ___STEEL___

REQ NO. __E-1152__

Fig. 11-8 Detail drawing of a standard part with dimensions blank at A and filled in at B.

dimensions and notes (Fig. 11-8 at A) to be filled in with the required information (as at B). Of course, the views will not be to scale, except perhaps for one size. A similar kind of drawing is sometimes used with letters used for dimensions. A table is placed on the drawing to give the dimensions for different sizes of the part. Either all or part of the dimensions may be used on a tabulated (also called *tabular*) drawing (Fig. 11-9).

Fig. 11-9 Tabulated (tabular) drawing.

BUSHING					
PART NO.	A	B	C	D	E
CB 1	1.500	2.000	1.000	0.500	0.375
CB 2	1.625	2.125	1.125	0.625	0.437
CB 3	1.750	2.250	1.250	0.750	0.500
CB 4	1.875	2.375	1.375	0.875	0.562
CB 5	2.000	2.500	1.500	1.000	0.625

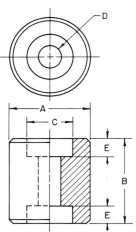

Assembly Drawings. A drawing of a completely assembled construction is called an *assembly drawing*. Such drawings vary greatly in respect to completeness of detail and dimensioning. Their particular value is in showing the way in which the parts go together, to show the appearance of the construction as a whole, and to give dimensions necessary for installation, space necessary, foundation, electrical or hydraulic connections, and so forth. When complete information is given, assembly drawings may be used for working drawings. This is possible when there is little or no complex detail; Fig. 11–10 shows such a drawing. Furniture and other wood construction can often be shown in assembly working drawings by adding necessary enlarged details or partial views (Fig. 11–11).

Assembly drawings of machines are generally made to small scale. They have selected dimensions to tell overall distances, important center-to-center distances, and local dimensions. All, or almost all, hidden lines may be left out; and if drawn to a very small scale, unnecessary detail may be omitted (Fig. 11–12 is an example). Either exterior or sectional views may be used. When the general appearance is the main purpose of the drawing, only one or two views need be used. Because of the size of some assembled constructions, it may be necessary to draw different views of the assembly on separate sheets. The same scale should be used on all sheets.

A special assembly drawing (Fig. 11–13) is made for reference to identify parts to be used for assembly. Note the tabular list in the upper right-hand corner. Note also the selection of dimensions given on the drawing.

Many other kinds of assembly drawings are made for special purposes: part assemblies for a group of parts, drawings for use in assembling or erecting a machine, drawings to give directions for maintenance and

Fig. 11–10 An assembly working drawing for a belt tightener.

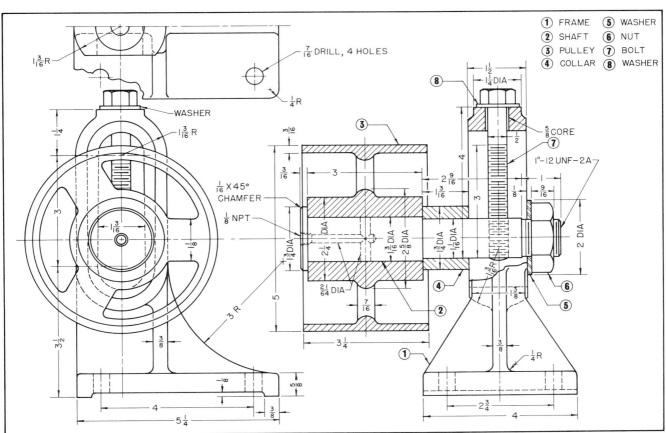

PICTORIAL RENDERING
OF STEREO CABINET

operations, and so forth. A most important kind of assembly drawing is the *design layout,* from which the detail drawings are made.

Choice of Views. A drawing can be used more easily if the draftsman makes the proper selection of views. For the complete description of an object, at least two views are generally required. Although a drawing is not a picture it is always advisable to select the views that require the least effort to read. Each view must have a part in the description; otherwise, it is not needed and should not be drawn. In some cases one view is all that is necessary, provided a note is added or the shape and size are standard or evident. Complex pieces may require more than three views, some of which may be partial views, aux-

Fig. 11–11 An assembly working drawing with enlarged details and partial views.

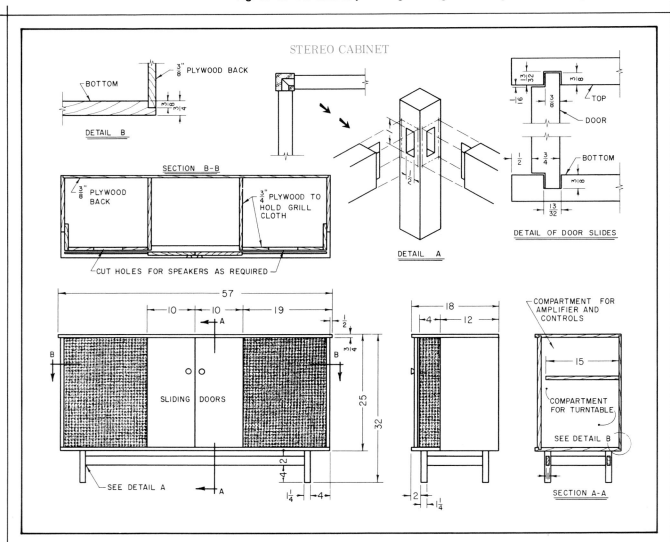

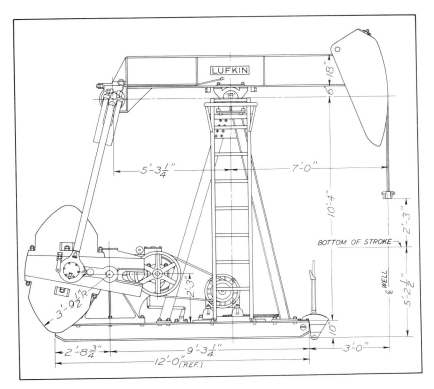

Fig. 11–12 An outline assembly drawing. *(Lufkin Foundry and Machine Company.)*

Fig. 11–13 A reference assembly drawing. *(Link-Belt Company.)*

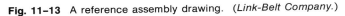

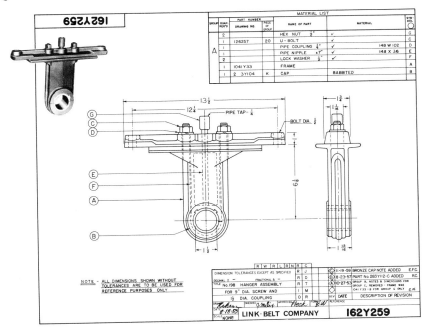

iliary views, and sectional views. The reason for making the drawing must always be kept in mind when a question arises. The final test of the value of a drawing is its clearness and exactness in giving the complete information necessary for the intended purpose.

Choice of Scale. The choice of scale for a detail drawing is governed by three factors:

1. The size necessary for showing all details clearly.

2. The size necessary for carrying all dimensions without crowding.

3. The size of paper used.

In most cases it is desirable to make detail drawings to full size. Other scales commonly used are half, quarter, and eighth. Such scales as $2'' = 1'$, $4'' = 1'$, and $9'' = 1'$ are to be avoided. If a part is very small, it is sometimes drawn to an enlarged scale, perhaps twice full size or more.

When a number of details are drawn on one sheet, they should, if possible, be made to the same scale. If different scales are used, they should be noted near each drawing. A detail, or part detail, drawn to a larger scale may often be used to advantage on drawings. This will save the making of separate detail drawings. General assembly drawings can be made to the scale that will show the desired amount of detail and work up well on the size of sheet used. Sheet-metal pattern drawings for practical use are always made full size, although practice models may be constructed from small-scale layouts.

Complete assemblies generally use a small scale, often fixed by the standard size of sheet selected by the company for assemblies. Part assemblies use a scale selected to suit the purpose, such as: to show how the parts are put together, to identify the

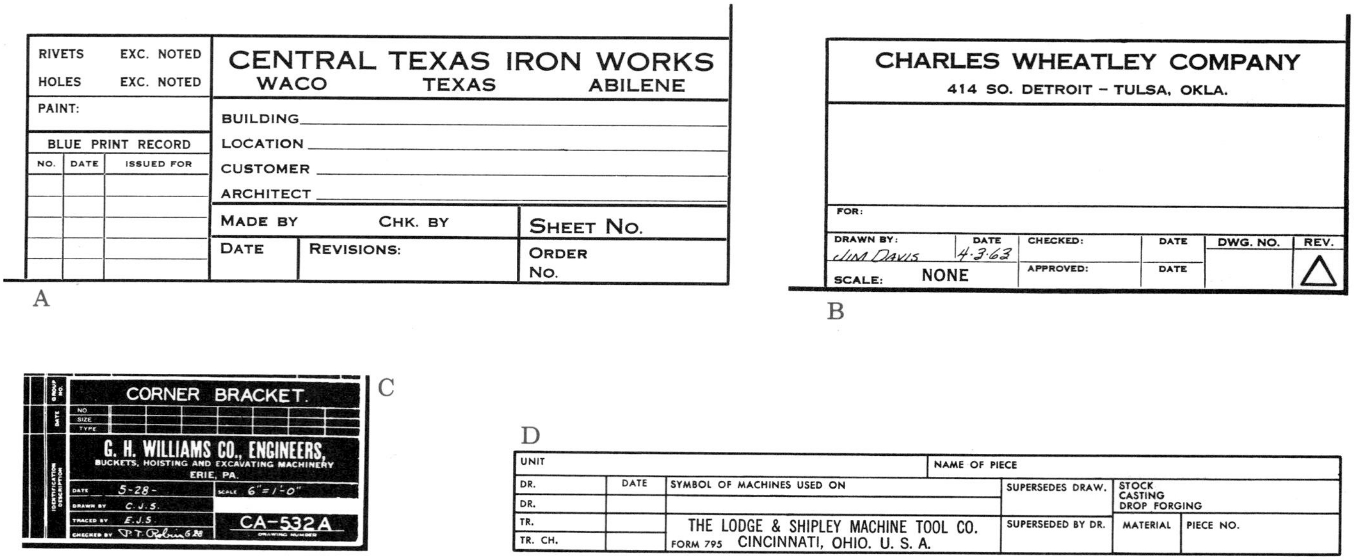

Fig. 11–14 Titles. Boxed titles at A, B, and C. A strip title at D.

parts, to explain an operation, or to give other information.

Titles. Every sketch and drawing must have some kind of title; the form, completeness, and location vary. On working drawings the title is usually boxed in the lower right-hand corner (Fig. 11–14 at A, B, and C) or included in the record strip extending across the bottom or end of the sheet (Fig. 11–14 at D) as far as needed.

The title gives as much as is necessary of the following information:

1. The name of the construction, machine, or project.

2. The name of the part or parts shown, or simple details.

3. Manufacturer, company, or firm name, and address.

4. Date, usually date of completion.

5. Scale or scales.

6. Heat treatment, working tolerances, etc.

7. Numbers of the drawing of the shop order or customer's order, according to the system used.

8. Drafting-room record: names or initials, with dates, of draftsman, tracer and checker, and approval of chief draftsman, engineer, etc.

9. A revision block or space for recording changes, when required, should be placed above or at the left of the title block.

A form for a basic layout of a title block (Fig. 11–15) is extracted from American National Standards Drafting Practices, *Section 1, Size and Format* (ANSI Y14.1–1957) with the permission of the publisher, The American Society of Mechanical Engineering, 345 East 47th Street, New York, NY. The arrangement, size, and content may vary.

In large drafting rooms the title is generally printed on the paper, cloth, or film, leaving spaces to be filled in. Separate printed adhesive titles are used by many firms.

List of Material or Parts List. It is necessary, or desirable, to include on most drawings a list of parts, the materials of which they are made,

identification numbers, or other information. Such drawings may include assembly drawings of various kinds or detail drawings, where a number of parts are shown on the same sheet.

The names of parts, material, number required, part numbers, and so forth, may be given in notes near the views of each part. The preferred method is to place part numbers near the views with a leader and then collect all the necessary information in tabulated lists called a *list of material, bill of material,* or *parts list* (Fig. 11–16). This list is sometimes placed above the title, but American National Standards recommends placing it in the upper right corner of the sheet. Sometimes it may be lettered or typewritten on a separate sheet titled "Parts List for Drawing No. 00" in order to identify it. The American National Standards (ANSI Y14.1) presents the form shown in Fig. 11–17. The column widths may be varied to suit the data to be entered therein.

Grouping and Placing Parts. When a number of details are used for only

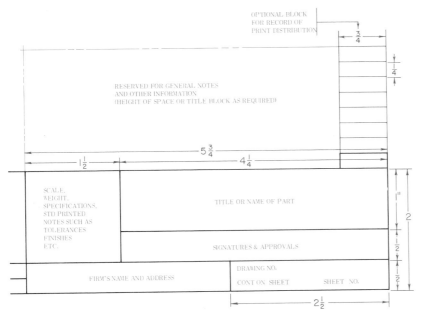

Fig. 11–15 Basic layout for a title block. (*From American National Standards Institute ANSI Y14.1–1957.*)

one machine, they are often grouped on a single sheet or set of sheets. A convenient arrangement is to group the forging details together, the casting details together, the brass details together, and so on for other materials. In general, it is well to show parts in the position that they will occupy in the assembled machine, with related parts near each other. Long pieces, however, such as shafts and bolts, are drawn with their long dimensions parallel to the long dimension of the sheet (Fig. 11–18).

Notes and Specifications. Information that cannot be represented graphically must be given in the form of lettered notes and symbols. Trade information that is generally understood by those on the job is often given in this way. Such notes include the following items: number required, material, kind of finish, kind of fit, method of machining, kinds of screw threads, kinds of bolts and nuts, sizes of wire, and thickness of sheet metal.

BILL OF MATERIAL FOR IDLER PULLEY			
NAME	REQ.	MAT'L	NOTES
IDLER PULLEY	1	C.I.	
IDLER PULLEY FRAME	1	C.I.	
IDLER PULLEY BUSHING	1	BRO.	
IDLER PULLEY SHAFT	1	C.R.S.	
$\frac{5}{8}$ SAE HEX NUT	1		$\frac{3}{8}$ HIGH PURCHASED
WOODRUFF KEY 405	1		PURCHASED
$\frac{1}{8}$ OILER	1		PURCHASED

Fig. 11–16 A bill of material.

Fig. 11–17 Recommended form for a list of materials. (*From American National Standards Institute ANSI Y14.1–1957.*)

LIST OF MATERIAL				
GROUP NO. AND QUANTITY	PART NO.	NAME	DRAWING NO. OR DESCRIPTION	.25

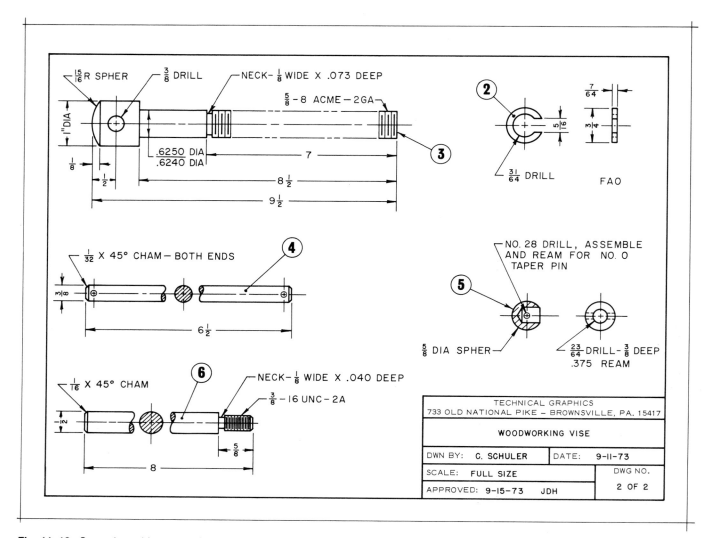

Fig. 11–18 Several machine parts drawn on one sheet.

The materials in general use are wood, plastic, cast iron, wrought iron, steel, brass, aluminum, and various alloys. All parts to go together must be of the proper size so that they will fit. Some pieces may be left rough, partly finished, or completely finished. The wood used for making furniture is first shaped with woodworking tools and machines. Many metals, such as cast iron, brass, aluminum, and so forth, are given the required form by molding, casting, and machining. First, a wooden pattern of the shape and size required is made and placed in sand to make an impression, or mold, into which the molten metal is poured. Wrought iron and steel are made into shapes by rolling or forging in the rolling mill or blacksmith shop. Some kinds of steel may be cast.

There are many interesting ways of forming metals for special purposes and many special alloys that cannot be described in a drawing book, but the student will learn much by observing the shapes of parts of machinery and the materials of which they are made.

After a part is cast or forged, it must be machined on all surfaces that are to fit other surfaces. Round surfaces are generally formed on a lathe. Flat surfaces are finished or smoothed on a planer, milling machine, or shaper. Drill presses, boring mills, or lathes are used for making holes. Extra metal is allowed for surfaces that are to be finished. To specify such surfaces, the V-symbol is placed on the lines which represent the edges of surfaces to be finished. If the entire piece is to be finished, a note such as "Finish All Over" (or "FAO") may be used and all other marks omitted.

Specifications as to methods of machining, finish, and other treatment are given in the form of notes, such as "Spot face," "Grind," "Polish," "Knurl," "Core," "Drill," "Ream," "Countersink," "Harden," "Caseharden," "Blue," and "Temper." It is often necessary to add notes in regard to assembling, order of doing work, or other special directions on a drawing.

Checking a Drawing. After a drawing has been completed, it must be very carefully examined before it is used. This is called *checking the drawing.* It is very important work and should be done by someone who has not worked on the drawing in order to better spot errors.

Thorough checking requires a definite order of procedure and consideration of the following items:

1. See that the views completely describe the shape of each piece.

2. See that there are no unnecessary views.

3. See that the scale is sufficiently large to show all detail clearly.

4. See that all views are to scale and that correct dimensions are given.

5. See that there will be no interferences during assembly or operation and that necessary clearances are provided.

6. See that sufficient dimensions are given to define the sizes of all parts completely and that no unnecessary or duplicate dimensions are given.

7. See that all necessary location or positioning dimensions are given with necessary precision.

8. See that necessary tolerances, limits and fits, and other precision information is given.

9. See that the kind of material and the number required of each part are specified.

10. See that the kind of finish is specified, that all finished surfaces are marked, and that a finished surface is not called for where one is not needed.

11. See that standard parts and stock items, such as bolts, screws, pins, keys or other fastenings, handles, catches, etc., are used where suitable.

12. See that all necessary explanatory notes are given and that they are properly placed.

Each draftsman is expected to inspect his own work for errors or omissions before the drawings are turned over to the checker.

Review

1. What is the name given to a drawing that supplies all information necessary for making a single part or a complete machine?

2. Are most industrial drawings done in pencil or in ink?

3. A drawing of a single part that gives all information necessary for making it is called a(n) _____ .

4. A drawing of a completely assembled construction is called a(n) _____ .

5. Name three factors that help determine the scale to be used on a drawing.

6. Name two types of title blocks.

7. What scale is usually used for sheet-metal pattern drawings?

8. What is another name for a *parts list?*

9. If an entire part is to be finished, what general note may be used?

10. What symbol may be used to identify specific surfaces that are to be finished?

Problems for Chapter 11

Fig. 11–19 Arbor. Assignment 1: Make a working drawing of each part shown. Scale—optional. Dimension. The flanges are die-cast aluminum. Shaft is cold rolled steel. Assignment 2: Make an assembly drawing of the arbor with a 6″ DIA × 1″ grinding wheel between the flanges. Show sectional views where necessary. Draw all fasteners. Estimate all sizes and details not given.

Fig. 11–20 Trammel. Assignment 1: Make a working drawing of each part shown. Scale—optional. Dimension. Indicate "2 REQD" for the point, body, and knurled screw. The point is to be heat treated after machining. Assignment 2: Make an assembly drawing of the trammel. Estimate all sizes and details not given.

Fig. 11–21 Power expansion bit. Assignment 1: Make a working drawing of each part shown. Scale—optional. Dimension. Cutter—tool steel. Body—cast iron. Use sectional views where necessary. Assignment 2: Make an assembly drawing of the power expansion bit. Dimension if required. Estimate all sizes and details not given.

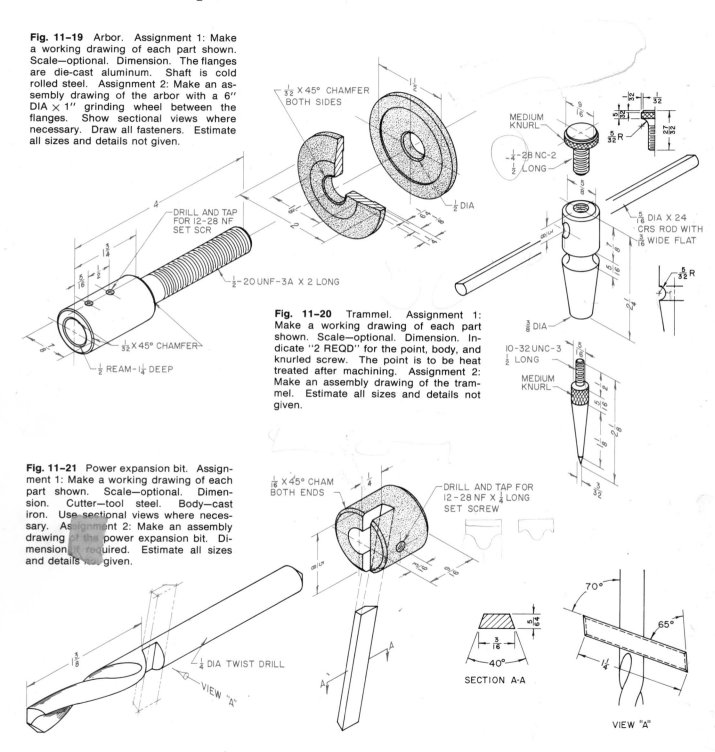

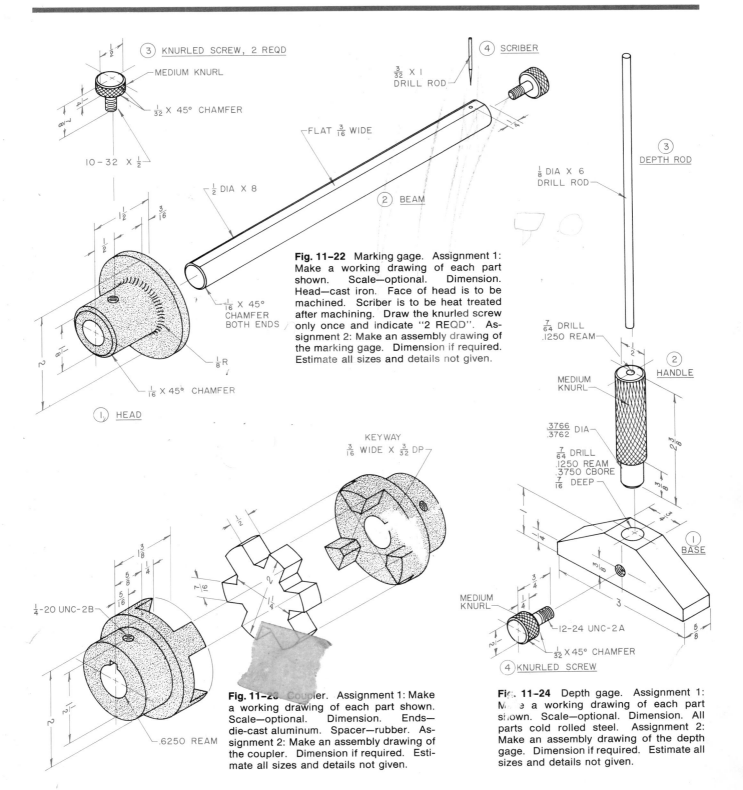

③ KNURLED SCREW, 2 REQD

MEDIUM KNURL

$\frac{1}{2}$

$\frac{1}{4}$

$\frac{7}{8}$

$\frac{1}{32}$ X 45° CHAMFER

10 - 32 X $\frac{1}{2}$

$\frac{3}{32}$ X 1 DRILL ROD

④ SCRIBER

FLAT $\frac{3}{16}$ WIDE

$\frac{1}{2}$ DIA X 8

② BEAM

$\frac{1}{4}$

$\frac{1}{16}$ X 45° CHAMFER BOTH ENDS

$\frac{1}{2}$ $\frac{3}{16}$

$1\frac{1}{2}$

$\frac{1}{2}$

$\frac{1}{8}$

2

$\frac{1}{8}$ R

$\frac{1}{16}$ X 45° CHAMFER

① HEAD

Fig. 11-22 Marking gage. Assignment 1: Make a working drawing of each part shown. Scale—optional. Dimension. Head—cast iron. Face of head is to be machined. Scriber is to be heat treated after machining. Draw the knurled screw only once and indicate "2 REQD". Assignment 2: Make an assembly drawing of the marking gage. Dimension if required. Estimate all sizes and details not given.

$\frac{1}{8}$ DIA X 6 DRILL ROD

③ DEPTH ROD

$\frac{7}{64}$ DRILL .1250 REAM

$\frac{1}{2}$

② HANDLE

MEDIUM KNURL

$\frac{.3766}{.3762}$ DIA

$\frac{7}{64}$ DRILL .1250 REAM .3750 CBORE $\frac{7}{16}$ DEEP

$2\frac{3}{8}$

$\frac{3}{8}$

$\frac{1}{8}$

$\frac{3}{8}$

$\frac{3}{4}$

① BASE

3

$\frac{5}{8}$

KEYWAY $\frac{3}{16}$ WIDE X $\frac{3}{32}$ DP

$\frac{1}{2}$

$\frac{3}{8}$

$\frac{1}{4}$

$\frac{5}{8}$

$\frac{5}{16}$

$\frac{7}{16}$

2

$1\frac{1}{4}$

$\frac{1}{4}$-20 UNC-2B

.6250 REAM

$1\frac{1}{2}$

2

MEDIUM KNURL

$\frac{3}{4}$

$\frac{1}{4}$

$\frac{1}{2}$

12-24 UNC-2A

$\frac{1}{32}$ X 45° CHAMFER

④ KNURLED SCREW

Fig. 11-23 Coupler. Assignment 1: Make a working drawing of each part shown. Scale—optional. Dimension. Ends—die-cast aluminum. Spacer—rubber. Assignment 2: Make an assembly drawing of the coupler. Dimension if required. Estimate all sizes and details not given.

Fig. 11-24 Depth gage. Assignment 1: Make a working drawing of each part shown. Scale—optional. Dimension. All parts cold rolled steel. Assignment 2: Make an assembly drawing of the depth gage. Dimension if required. Estimate all sizes and details not given.

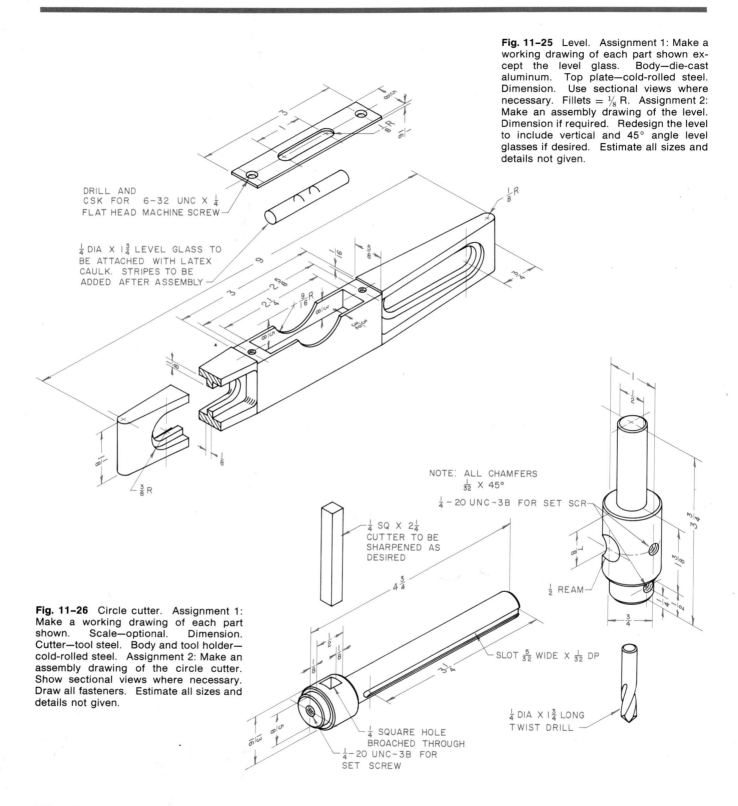

Fig. 11-25 Level. Assignment 1: Make a working drawing of each part shown except the level glass. Body—die-cast aluminum. Top plate—cold-rolled steel. Dimension. Use sectional views where necessary. Fillets = ⅛ R. Assignment 2: Make an assembly drawing of the level. Dimension if required. Redesign the level to include vertical and 45° angle level glasses if desired. Estimate all sizes and details not given.

DRILL AND CSK FOR 6-32 UNC X ¼ FLAT HEAD MACHINE SCREW

¼ DIA X 1¾ LEVEL GLASS TO BE ATTACHED WITH LATEX CAULK. STRIPES TO BE ADDED AFTER ASSEMBLY

NOTE: ALL CHAMFERS 1/32 X 45°

¼ - 20 UNC - 3B FOR SET SCR

½ REAM

¼ SQ X 2¼ CUTTER TO BE SHARPENED AS DESIRED

SLOT 5/32 WIDE X 1/32 DP

Fig. 11-26 Circle cutter. Assignment 1: Make a working drawing of each part shown. Scale—optional. Dimension. Cutter—tool steel. Body and tool holder—cold-rolled steel. Assignment 2: Make an assembly drawing of the circle cutter. Show sectional views where necessary. Draw all fasteners. Estimate all sizes and details not given.

¼ SQUARE HOLE BROACHED THROUGH

¼ - 20 UNC - 3B FOR SET SCREW

¼ DIA X 1¾ LONG TWIST DRILL

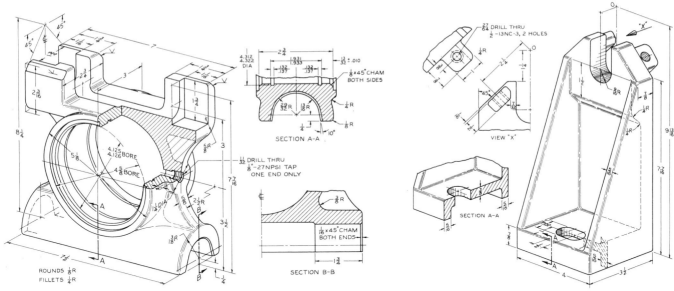

Fig. 11-27 Housing. Assignment: Make a working drawing of the housing. Scale—optional. Dimension. Use partial and sectional views where necessary. Material—cast iron. Estimate all sizes and details not given.

Fig. 11-28 End base. Assignment: Make a working drawing of the end base. Scale—optional. Dimension. Use partial and sectional views where necessary. Material—cast iron. Estimate all sizes and details not given.

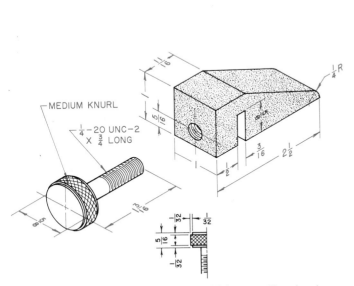

Fig. 11-29 Stop clamp. Assignment 1: Make a working drawing of the stop clamp. Scale—optional. Dimension. Body—die-cast aluminum. Knurled screw—cold-rolled steel. Assignment 2: Make an assembly drawing of the stop clamp. Dimension if required. Estimate all sizes not given.

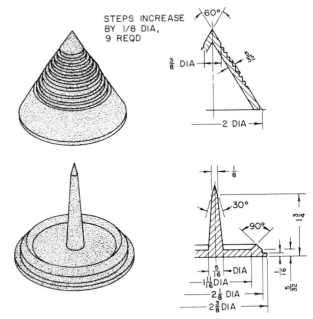

Fig. 11-30 Blade balance. Assignment 1: Make a working drawing of each part shown. Scale—optional. Dimension. Both parts are die-cast aluminum. No machining is required. Assignment 2: Make an assembly drawing of the blade balance. Estimate all sizes and details not given. Show the top half in half or full section.

Pictorial Drawing | Chapter 12

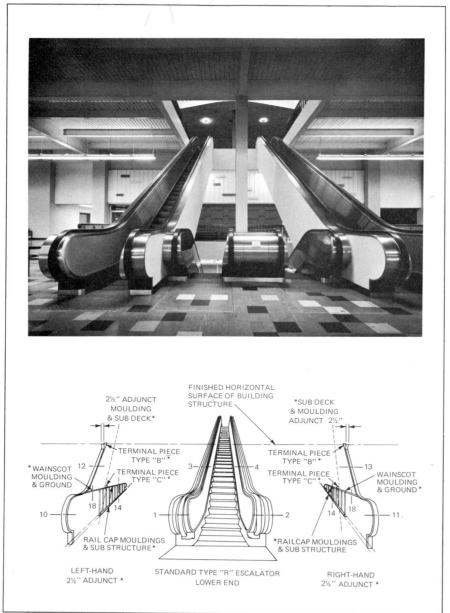

2½" ADJUNCT
MOULDING
& SUB DECK*

FINISHED HORIZONTAL
SURFACE OF BUILDING
STRUCTURE

*SUB DECK
& MOULDING
ADJUNCT 2½"

TERMINAL PIECE
TYPE "B"*

*WAINSCOT
MOULDING
& GROUND

12

3

TERMINAL PIECE
TYPE "C"*

4

TERMINAL PIECE
TYPE "B"*

13

WAINSCOT
MOULDING
& GROUND*

10

18   14

1

TERMINAL PIECE
TYPE "C"*

2

14   18

11

RAIL CAP MOULDINGS
& SUB STRUCTURE*

*RAILCAP MOULDINGS
& SUB STRUCTURE

LEFT-HAND
2½" ADJUNCT *

STANDARD TYPE "R" ESCALATOR
LOWER END

RIGHT-HAND
2½" ADJUNCT *

(OTIS ELEVATOR COMPANY.)

Uses of Pictorial Drawing. Pictorial drawing is an essential part of the graphic language in engineering, architecture, science, electronics, technical illustration, and in every profession. It is used in all kinds of technical literature as well as in catalogs and assembly, service, and operating manuals. Architects use pictorial drawings to show what the finished building will look like (Fig. 12–1). Advertising agencies use pictorial drawings to show the consumer a new product.

Pictorial drawings are often used to show exploded views on production and assembly drawings to illustrate parts lists (Fig. 12–2), to explain the operation of machines, apparatus, and equipment, and for many other commercial and technical purposes. In addition, most people use some form of pictorial sketches to help in communicating ideas that are difficult to describe in words.

Types of Pictorial Views. Figure 12–3 shows examples of several kinds of pictorial drawings. Persons familiar with the subject often make perspective-view sketches. These views show the object as it would actually appear to the eye. An easier view to sketch, although the result is not so pleasing in appearance as a well-made perspective, is one of the pictorial methods of projection, such as isometric or oblique. These methods all show three faces in one view. Their advantage is that the principal lines

Fig. 12-1 Pictorial drawings show objects as they appear. (*Inland Steel Urban Development Corp.*)

Fig. 12-2 An exploded assembly drawing may be used to illustrate a parts list. (*Sargent & Company*)

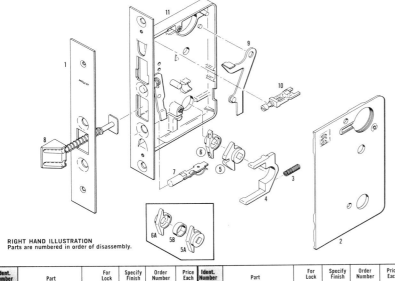

RIGHT HAND ILLUSTRATION
Parts are numbered in order of disassembly.

Ident. Number	Part	For Lock	Specify Finish	Order Number	Price Each	Ident. Number	Part	For Lock	Specify Finish	Order Number	Price Each
1	Outside Front	7714 7715	x	77-0106	$2.20	6	Swivel Hub	7714 7746		77-0267	$0.60
		7746	x	77-0111	2.20	7	Long Stop Assembly	7714 7746	x	77-2165 77-2166	1.00 1.00
2	Cap	7715 7714 7746		77-0274 77-2222	1.90 1.90	8	Latch Bolt Assembly	All	x	77-2175	4.00

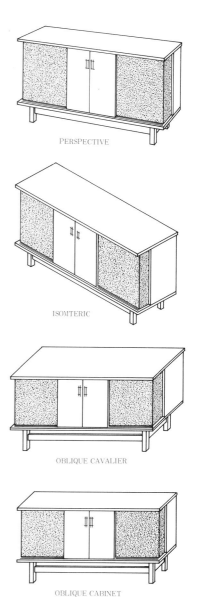

PERSPECTIVE

ISOMTERIC

OBLIQUE CAVALIER

OBLIQUE CABINET

Fig. 12-3 Stereo cabinet drawn in various types of pictorial views.

can be measured directly. Although similar in effect, these three methods should not be confused.

Cabinet drawing is a form of oblique drawing in which distances on the receding axes are reduced one-half. In other words, it is drawn only half as deep as a normal oblique view.

Axonometric Projection. Isometric projection is one form of axonometric projection. *Axonometric* is the family name for *isometric, dimetric,* and *trimetric.* The theory behind each type is the same; the angle of projection is different. Figure 12-4 shows the basic differences in the three types of axonometric projections. In isometric projection, the three axes form equal angles of 120° to the plane of projection, and only one scale is needed for measurements along each of the three axes. This is the easiest type of axonometric drawing to make. In dimetric projection, only two of the angles are equal, and two special foreshortened scales are needed to make measurements. Trimetric projection requires three foreshortened scales, and all three angles are different.

Because of the complexity of constructing dimetric and trimetric drawings, they are used less often than other types of pictorial drawing. As a result, this chapter will deal only with isometric, oblique, and perspective types of pictorial drawing.

Isometric Projection and Isometric Drawing. There is a difference between isometric projection and isometric drawing, although many people use the terms to mean the same thing. The differences will be explained in this section. While it is not very likely that you will be required to make pictorial drawings using isometric projection, it is probably a good idea for you to understand the theory behind it.

An isometric projection may be obtained by a method called *revolution* (revolved views). Figure 12-5A shows three normal views (multiview drawing) of a cube. In Fig. 12-5B, each of the three views has been revolved 45°. Notice that the front and side views now show as two equal rectangles. On the side view a diagonal is drawn from point *O* to point *B*. This is called the *body diagonal.* It is the longest straight line that can be drawn in a cube.

In Fig. 12-5C, the cube is revolved upward until the body diagonal is horizontal, as shown. The front view is an isometric projection. Notice that the cube has been revolved 35°16′ in order to raise the body diagonal to a horizontal position. When this happens, the lower edges of the front form an angle of 30° to the horizontal.

Since the cube is revolved and the pictorial view is made by projection, the lines become foreshortened. The actual difference is 0.8165 to 1 in. In other words, 1 in. on the cube in Fig. 12-5A is equal to 0.8165 in. on the isometric projection in Fig. 12-5C.

In Fig. 12-5D, an isometric drawing and an isometric projection of the same cube are shown. In isometric drawing, all edges are drawn their true length instead of the shortened length. This variation in size does not affect the pictorial value of the view for shape description, but it does simplify the drawing of the view because all measurements are made with a regular scale. This makes it possible to draw a pictorial view at once without projecting from other views or using a special scale.

Isometric drawing. A multiview drawing of a filler block is shown in Fig. 12-6 at A. The isometric drawing is built upon a skeleton of three lines representing the edges of a cube, as shown at B. These three lines form three equal angles, each of 120° (3 × 120° = 360°) and are called *isometric axes.* Axis (line) *OA* is drawn vertically. Axes *OB* and *OC* are drawn with the 30°–60° triangle. The intersection of these lines represents the upper front corner *O* of the block, as shown at C. Measuring the width *W*, the depth *D*, and the height *H*, of the block on the three axes and drawing lines through the points parallel to the axes will give the isometric drawing of the block. To locate the rectangular hole shown at D, lay off 1 in. along *OC* to *c*, and from *c* lay off 2 in. to c^1. Through *c* and c^1 draw lines parallel to *OB*. In a like manner, locate *b* and b^1 on axis *OB* and draw lines parallel to *OC*. Draw a vertical line from corner 3. NOTE: *The dimensions, letters, and numerals are for instructional purposes and normally should not be placed on your drawing.* Brighten all necessary lines to complete the drawing (Fig. 12-6E).

The purpose of a pictorial drawing, in general, is to show how something looks. Hidden edges (lines) are not "part of the picture"; therefore, they

Fig. 12–4 The three types of axonometric projections.

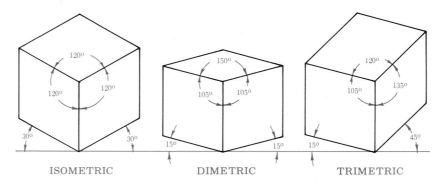

ISOMETRIC   DIMETRIC   TRIMETRIC

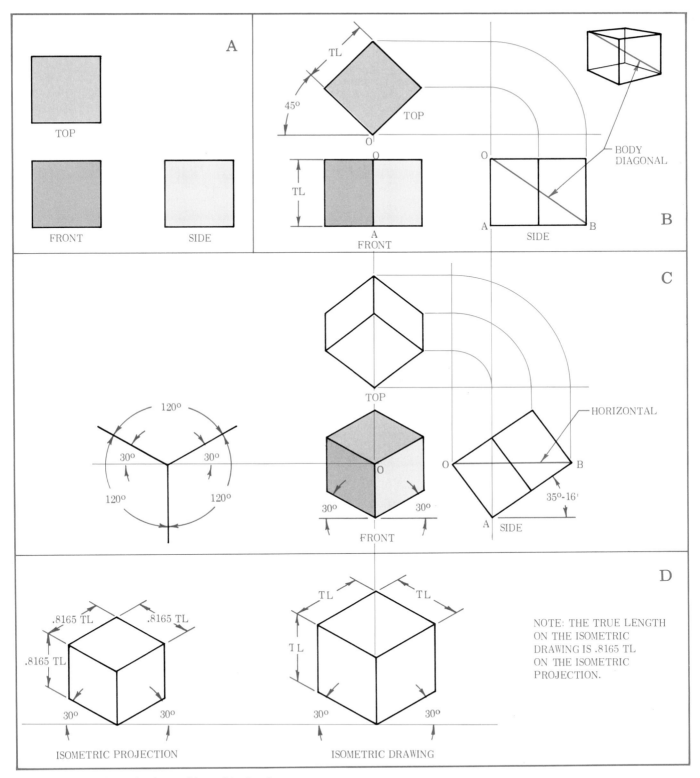

Fig. 12–5 Isometric projection and isometric drawing.

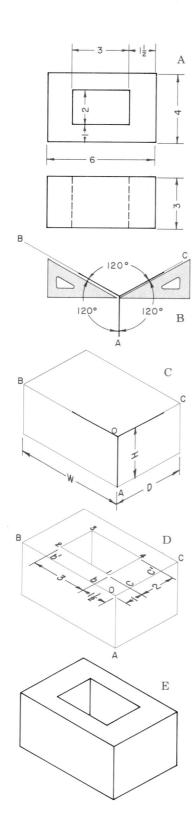

Fig. 12-6 Steps in making an isometric drawing.

are omitted, except in some particular case when a certain feature might be indicated for explanation.

Position for the isometric axes. The axes may be arranged in different ways, provided their relative positions are not changed (120° with each other). Several positions shown and identified in Fig. 12-7 will be applied later in this chapter.

The arrangement of the axes in Fig. 12-6 is the first position and is based upon the three edges of the cube which meet at the upper front corner. It is often more convenient to place the axes in the second position, which starts at the lower corner (Fig. 12-8).

Any line of an object parallel to one of the edges of a cube is drawn parallel to an isometric axis and is called an *isometric line.* An important rule of isometric drawing is: *Measurements can be made only on isometric lines.*

Nonisometric lines. Lines which are not parallel to any of the isometric

axes are called *nonisometric lines* (Fig. 12-9). Such lines will not show in their true length and cannot be measured; they must be drawn by locating their two ends first and then connecting the points. Angles between lines on isometric drawings do not show in their true size and cannot be measured in degrees.

Figure 12-10 shows how nonisometric lines are located and drawn. A multiview drawing of a packing block is shown at A. NOTE: *The colored lines and letters are for instructional purposes only and should not normally appear on the finished drawing.* Use the following procedure for making an isometric drawing of the block.

A. Block in the overall sizes of the packing block as shown at B.

B. Use dividers or a scale to transfer *AG* and *HB* from the multiview drawing to the isometric layout to locate points *G* and *H*. The nonisometric lines are then drawn by connecting point *D* with *G* and *C* with *H*. This is shown in Fig. 12-10C.

Fig. 12-7 Positions for isometric axes.

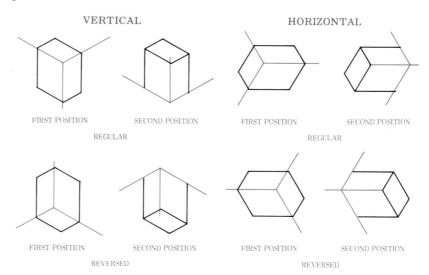

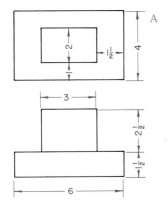

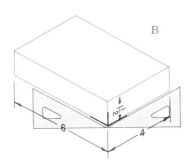

Fig. 12-9 Nonisometric lines.

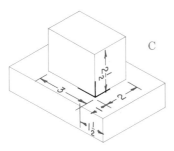

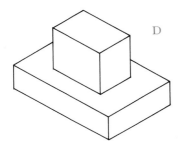

Fig. 12-8 Isometric drawing with the axes in the second position. See also Fig. 12-6.

C. Complete the layout by drawing *GJ* and *HI* and connecting points *E* and *J* (the second nonisometric line), as shown in Fig. 12-10D.

D. Darken all necessary lines, and erase construction lines to complete the drawing (Fig. 12-10E).

Angles in isometric. Angles cannot be measured with a protractor in isometric drawing. To draw an angle given in degrees, such as the 40° angle shown in Fig. 12-11A, use the following procedure.

A. Make *AO* and *OB* any convenient length. Draw *AB* perpendicular to *AO* at any convenient location.

B. Use dividers or compass to lay off *AO* along the base of the cube (Fig. 12-11B). Draw vertical line *AB*.

C. Draw a line through points *O* and *B* to complete the isometric angle. If you check the isometric angle with a protractor, you will find that it does not measure 40°.

Follow the same procedure to construct the angle on the top of the isometric cube. This method can be used to lay out any angle on any isometric plane.

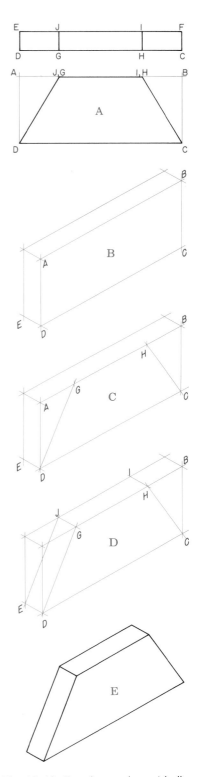

Fig. 12-10 Drawing nonisometric lines.

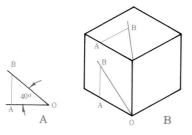

Fig. 12–11 Constructing angles in isometric drawing.

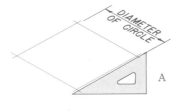

Draw an isometric square with the sides equal to the diameter of the circle.

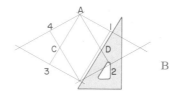

Use a 30°–60° triangle to locate points A, B, C, D, and 1, 2, 3, 4.

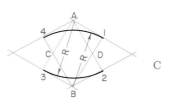

With A and B as centers and a radius equal to A2, draw arcs as shown.

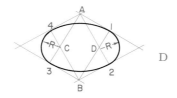

With C and D as centers and a radius equal to C4, draw arcs to complete the isometric circle (ellipse).

Fig. 12–13 Steps in drawing an isometric circle.

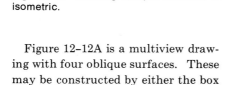

Fig. 12–12 Drawing oblique surfaces in isometric.

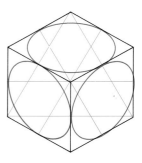

Fig. 12–14 Isometric circles on a cube.

Figure 12–12A is a multiview drawing with four oblique surfaces. These may be constructed by either the box or skeleton method, as shown at B and C.

Isometric circles. In isometric drawing, circles appear as ellipses. Since it is time consuming to draw a true ellipse, a four-centered approximation can be used. The procedure for constructing such an ellipse is shown and explained in Fig. 12–13. Figure 12–14 shows isometric circles on the three surfaces of a cube.

Figure 12–15 shows the construction for an isometric drawing of a cylin-der. Notice that the radii for the arcs at the bottom are identical to those at the top.

The construction for quarter rounds is the same as for one-quarter of a circle. This is illustrated in Fig. 12–16. Notice that the radius is measured along the tangent lines from the corner in each case and that the actual perpendiculars are then drawn to locate the centers for the isometric arcs. It will be observed that r_1 and r_2 are found in the same way as the short and long radii of a complete isometric circle.

When an arc is more or less than a quarter circle, it is sometimes possible to draw all or part of a complete isometric circle and use as much of it as is needed.

Figure 12–17 shows the construction for an outside corner arc and an inside corner arc. Note the tangent points, *T*, and centers *1* and *1¹* and *2* and *2¹*.

Irregular curves in isometric. Irregular curves in isometric cannot be drawn using the four-center method. Points must first be plotted and connected using a French curve, as shown in Fig. 12–18.

Isometric templates. Isometric templates are made in a variety of forms. They are convenient and time saving when many isometric drawings have to be made. Many of them have openings for drawing ellipses and 30°, 60°, and 90° guiding edges. Simple

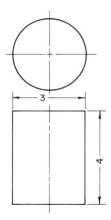

Multiview drawing of a cylinder.

Fig. 12-15 Steps in drawing an isometric cylinder.

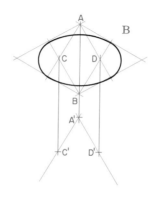

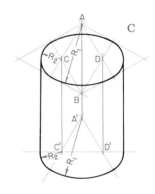

Draw an ellipse as described in Fig. 12–13. Drop centers A, C, and D a distance equal to the height of the cylinder (4 in. in this case). Draw lines A^1C^1 and A^1D^1.

A line through C^1D^1 will locate the points of tangency. Draw the arcs using the same radii as in the ellipse at the top. Draw the vertical lines to complete the cylinder.

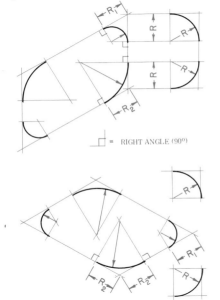

☐ = RIGHT ANGLE (90°)

Fig. 12-16 Drawing quarter rounds in isometric.

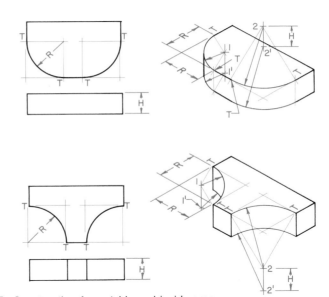

Fig. 12-17 Construction for outside and inside arcs.

homemade guides (Fig. 12–19) are convenient for straight-line work in isometric. Ellipse templates are very convenient for drawing true ellipses (Fig. 12–20). If available, they give a better appearance and save the extra time required to draw approximate ellipses. Refer to Chapter 3 for information on the use and care of templates.

Making an isometric drawing. The following procedure refers to the isometric drawing of the guide shown in Fig. 12–21. A multiview drawing of the guide is shown in Fig. 12–21A. Study the size, shape, and relationship of views before you proceed.

A. Draw the axes AB, AC, and AD in second position (Fig. 12–21B). Turn

back to Fig. 12-7 if you do not recall the positions of the isometric axes.

Measure from A the length 3 in. on AB.

Measure from A the width 2 in. from AC.

Measure from A the thickness $5/8$ in. on AD.

Through these points draw isometric lines, blocking-in the base.

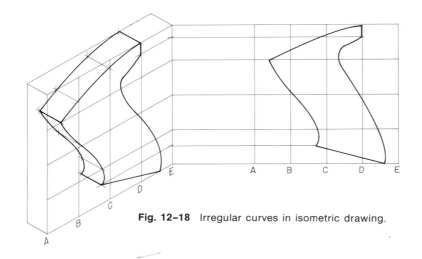

Fig. 12–18 Irregular curves in isometric drawing.

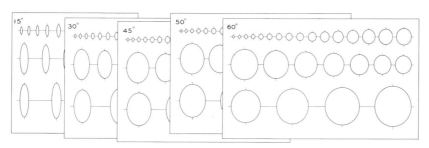

Fig. 12–20 Ellipse templates.

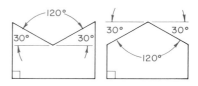

Fig. 12–19 Simple isometric templates.

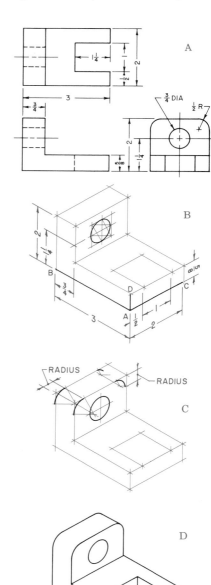

Fig. 12–21 Steps in making an isometric drawing.

B. Block-in the upright part, making two measurements of only 2 in. and ¾ in.

C. Locate the center of the hole and draw center lines as shown. Block-in a ¾-in. isometric square and draw the hole as an approximate ellipse. At the upper corners measure the ½-in. radius on each line (Fig. 12–21C). Draw real perpendiculars to find the centers of the quarter circles. Refer to Fig. 12–16 for information on drawing isometric quarter rounds.

D. Darken all necessary lines, and erase all construction lines to complete the isometric drawing of the guide, as shown in Fig. 12–21D.

Isometric sections. Isometric drawings are generally made as outside views, but sometimes a sectional view is needed. The section is taken on an isometric plane. That is, on a plane parallel to one of the faces of the cube. Figure 12–22 shows isometric full sections taken on a different plane for each of three objects. Note the construction lines indicating the part that has been cut away. Isometric half sections are illustrated in Fig. 12–23. The construction lines of A are for the object shown in B. The construction lines of C are for the object shown in D. Notice the outlines of the cut surfaces in A and C. The cut method is to draw the complete outside view and the isometric cutting plane. The part of the view that has been cut away is then removed. A second method is to draw

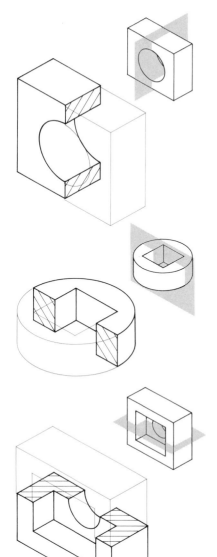

Fig. 12-22 Examples of isometric full sections.

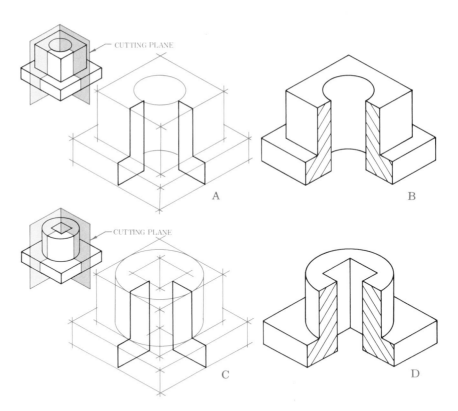

Fig. 12-23 Examples of isometric half sections.

the section on the isometric cutting plane and then work from it to complete the view.

Reversed axes. Sometimes it is desirable to represent a part as viewed from below. This is done by reversing the axes, as in Fig. 12-24. The multiview drawing is shown at A. Fig. 12-24B shows the axes drawn in re-versed position. Dimensions are taken from the multiview drawing to complete the layout. Lines are darkened to complete the reversed-axis isometric drawing shown in Fig. 12-24C.

Long axis horizontal. Long pieces may be drawn with one isometric axis in the horizontal position as shown in Fig. 12-25. For the object shown at A, the axes are shown by heavy black lines at B. Measurements are taken from the multiview drawing at A and the various isometric lines are developed. Remember! Circles are first drawn as isometric squares and completed by using the four-center method or with an ellipse template. Except for the position of the axes, this type of isometric view is drawn in the same way as any other isometric view.

Dimensioning isometric drawings. There are two general methods of placing dimensions on isometric drawings. The older method is to place dimensions in the isometric planes, or extensions of them, and to adjust letters, numerals, and arrowheads to isometric shapes, as shown at A in Fig. 12-26. The unidirectional system is a simpler method. Numerals and lettering are read from the bottom of the sheet (Fig. 12-26B). However, since isometric drawings are usually not used as working drawings, they are seldom dimensioned.

Oblique Projection and Oblique Drawing. When the projectors make an angle other than 90° with the picture plane, an oblique projection results. There is usually no distinction made between oblique projection and

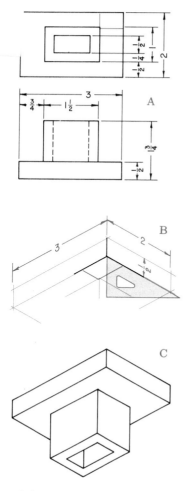

Fig. 12-24 Steps in making an isometric drawing with reversed axes.

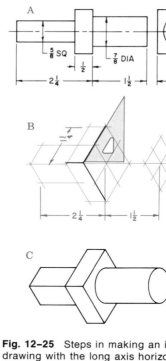

Fig. 12-25 Steps in making an isometric drawing with the long axis horizontal.

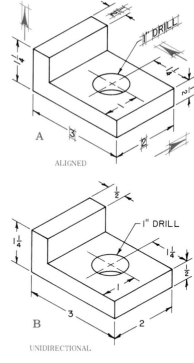

ALIGNED

UNIDIRECTIONAL

Fig. 12-26 Two methods of dimensioning isometric views.

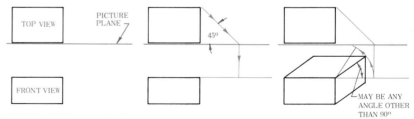

Fig. 12-27 Oblique projection.

oblique drawing as there is with isometric. If the projectors are drawn full length, the oblique projection is called *cavalier*. If the projectors are drawn one-half size, the projection is called *cabinet*. Many draftsmen use three-quarter size, which is sometimes called *normal*, or *general*, *oblique*. However, the family name *oblique drawing* is most commonly used for all three types. Figure 12-27 shows how an oblique projection is developed. While it is not very likely that

you will be required to make pictorial drawings using oblique projection as shown in the illustration, it is probably a good idea for you to understand the theory behind it.

Oblique drawing. Oblique drawings are constructed in the same manner as isometric drawings, that is, on three axes. However, two of the axes always make right angles with each other (Fig. 12-28).

The same methods and rules that were used in isometric drawing apply

to oblique drawing, but compared with isometric, oblique drawing has the distinct advantage of showing one face without distortion. Thus, objects with irregular outlines can be drawn by this method much more easily and effectively than in isometric. Some draftsmen prefer it for practically all pictorial work. The first rule in oblique drawing is:

Place the object so that the irregular outline or contour faces the front (Fig. 12-29A).

The second rule is:

Place the object so that the longest dimension is parallel to the picture plane (Fig. 12–29B).

Positions for the oblique axes. In Fig. 12–30, two of the axes, *AO* and *OB*, are drawn at right angles. The oblique axis, *OC*, may be drawn at any angle to the right, left, up, or down, as illustrated. It is usually desirable to place the object in the most natural position from which it would normally be viewed.

Angles and inclined surfaces on oblique drawings. Angles which are parallel to the picture plane show in their true size. Other angles can be laid off by locating the ends of the inclined line of the angle.

A plate with the corners cut off at angles is shown in Fig. 12–31 at A. An oblique drawing with the angles parallel to the picture plane is shown at B. At C the angles are parallel to the profile plane, and at D they are parallel to the horizontal plane. In each case the angle is laid off by measurements parallel to oblique axes, as shown by the construction lines.

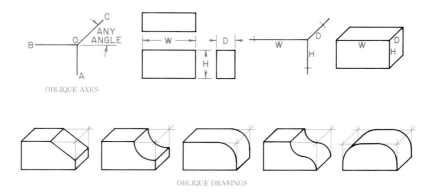

Fig. 12–28 The oblique axes and oblique drawings.

Fig. 12–29 Two general rules for oblique drawings.

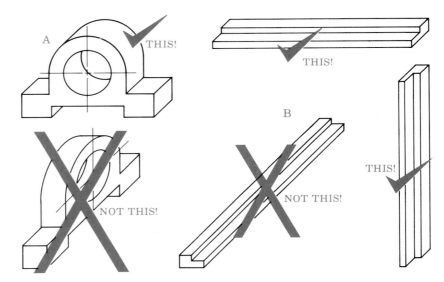

Fig. 12–30 Positions for oblique axes.

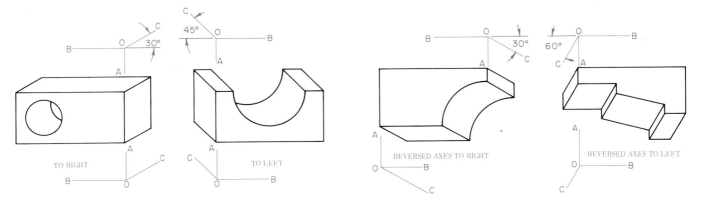

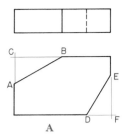
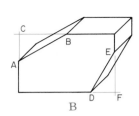
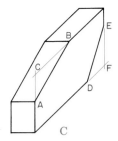
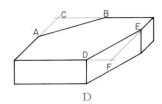

Fig. 12–31 Angles on oblique drawings.

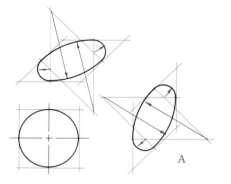
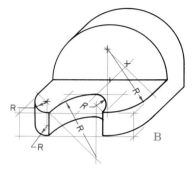
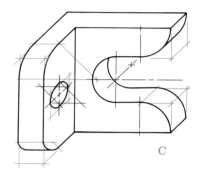

Fig. 12–32 Circles parallel to the picture plane are true circles; on other planes, ellipses.

Oblique circles. On the front face, circles and curves show in their true shape (Fig. 12–32). On other faces the four-center method may be used to draw ellipses. In Fig. 12–32A, a circle is shown as it would be drawn on a front plane, a side plane, and a top plane.

In Fig. 12–32B, the oblique drawing shown has some arcs in a horizontal plane. In Fig. 12–32C, the oblique drawing shown has some arcs in a profile plane.

Circles not parallel to the picture plane when drawn by the approximate method are not pleasing but are satisfactory for some purposes. Ellipse templates, when available, give much better results. If a template is used, the oblique circle should first be blocked-in as an oblique square in order to locate the proper position of the circle. Blocking-in the circle first

also helps the draftsman select the proper size and shape of the ellipse. If a template is not available, the ellipse can be plotted as shown in Fig. 12–33.

To make an oblique drawing. The procedure for making an oblique drawing is shown in Fig. 12–34. Notice that all but two small circles can be shown in their true shape.

Oblique sections. Oblique drawings are generally made as outside views, but sometimes a sectional view is necessary. The section is taken on a plane parallel to one of the faces of an oblique cube. Figure 12–35 shows an oblique full section and an oblique half section. Note the construction lines which indicate the part that has been cut away.

Cabinet drawings. A cabinet drawing is an oblique drawing in which the distances parallel to the

oblique (receding) axis are drawn one-half size. Figure 12–36 shows a bookcase drawn in cavalier, normal oblique, and cabinet drawing. Cabinet drawings get their name from the fact that they are extensively used in the furniture industry.

Fig. 12–33 To plot oblique circles.

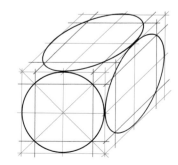

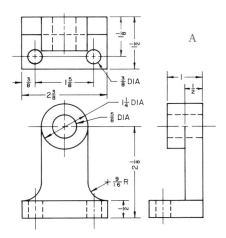

A

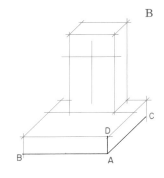

B

Multiview drawing of the object to be drawn in oblique.

Draw the axes *AB*, *AC*, and *AD* for the base in second position and on them measure the length, width, and thickness of the base. Draw the base. On it, block-in the upright, omitting the projecting boss as shown.

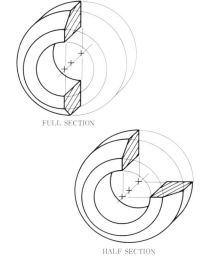

FULL SECTION

HALF SECTION

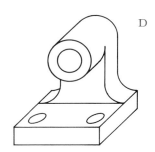

C

D

Block-in the boss and find the centers of all circles and arcs. Draw the circles and arcs.

Darken all necessary lines, and erase construction lines to complete the drawing.

Fig. 12–35 Oblique full and half sections.

Fig. 12–34 Steps in making an oblique drawing.

Fig. 12–36 Three types of oblique drawings.

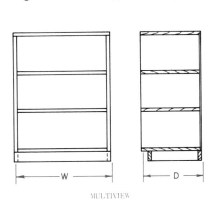

MULTIVIEW

CAVALIER

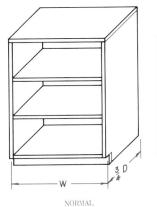

NORMAL

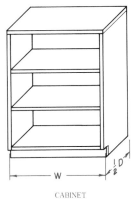

CABINET

erspective Drawing. Perspective drawing (Fig. 12-37) is a three-dimensional representation of an object as it appears to the eye from a particular point. As a result, perspective drawings more closely resemble photographs than do other types of pictorial drawing. The main difference between perspective drawing and other types of pictorial drawing is that in perspective, parallel edges on the receding planes are not drawn parallel as they are in isometric and oblique drawing.

Definitions of terms. The projectors for isometric and oblique views are perpendicular to the plane of projection and are parallel to each other. The projectors for a perspective view converge at a single point, the eye of the observer, and are called *visual rays.* They may be thought of as lines drawn from points on the object to the eye, as indicated in Fig. 12-38, where other perspective terms are illustrated. The *picture plane* (PP) is the plane upon which the view (picture) is projected and drawn. The *station point* (SP) is the position of the observer when looking at the object. The horizontal plane passes through the observer's eye and intersects the picture plane to form the *horizon line* (HL). The ground plane upon which the observer stands intersects the picture plane to form the *ground line* (GL). The *center of vision* (CV) is the point in which the *line of sight* (LS) (visual ray from the eye perpendicular to the picture plane) pierces the picture plane. The *vanishing point* (VP) is a point at which the receding axes converge. The eye level may be at any elevation on, above, or below the ground. If viewed from a high elevation, an *aerial,* or *bird's-eye, view* is obtained. If on the ground, a *ground,* or *worm's-eye, view* is obtained. In Fig.

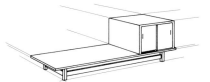

Fig. 12-37 Perspective drawing of a music center.

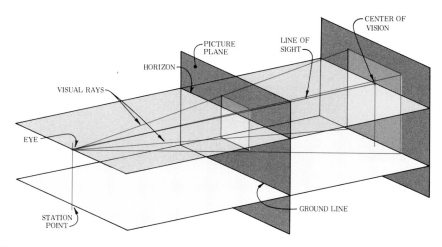

Fig. 12-38 Some perspective terms.

12-38, the perspective of a card has been drawn on the picture plane. Notice that the center of vision is on the card (not above or below). This type of view is called a *normal,* or *man's-eye, view.* Figure 12-39 shows the three types of perspective views.

Factors that affect appearance. Two factors affect the appearance of an object as it is viewed in perspective: the *distance* from the object to the viewer and the *position* (angle) of the object in relationship to the viewer.

How distance affects the view. You know that the size of an object appears to change as your distance from it changes. The farther away from the object you become, the smaller the object appears. The closer you get, the larger it appears to become. A graphic explanation of the effect of distance is shown in Fig. 12-40. If an object is placed against

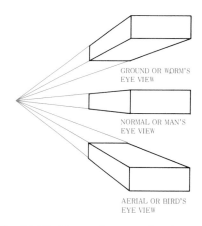

Fig. 12-39 Types of perspective views.

a scale at a normal reading distance from the viewer, it appears to be the size indicated by the scale. However, if the object is moved back from the scale, doubling the distance from the viewer, the object appears only half as large. Notice that each time the dis-

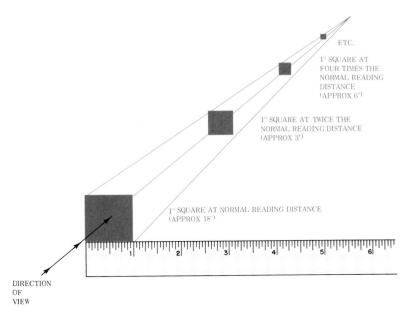

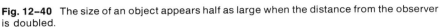

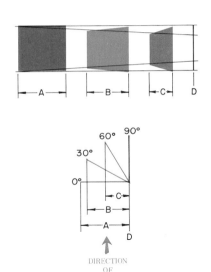

DIRECTION
OF
VIEW

Fig. 12–40 The size of an object appears half as large when the distance from the observer is doubled.

Fig. 12–41 The position of the object in relationship to the observer affects its appearance.

Fig. 12–42 The lines of the porch rail, deck, roof, and building's side appear to converge at a distance. (*Designed by Howard A. Friedman & Associates.*)

tance is doubled, the apparent size is reduced by one-half.

How position affects the view. If the object is parallel to the observer (0°), the top and bottom edges are parallel (Fig. 12–41). As the object is rotated, the top and bottom edges appear to converge. The difference in distance from the observer to the right- and left-hand (near and far) edges of the object causes the foreshortening (see Fig. 12–40). Notice also that the view becomes narrower as the object is rotated away from the observer. Once again, the foreshortening is caused by the difference in distance between the near and far edges of the object.

One-point perspective. One-point perspective means *one vanishing point.* The effect of one-point perspective can be seen in Fig. 12–42. The procedure for making a one-point (also called *parallel*) perspective drawing of the object in Fig. 12–43 is shown

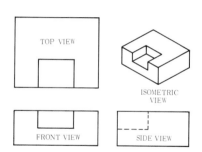

Fig. 12-43 Multiview and isometric drawings of an object to be drawn in single-point perspective.

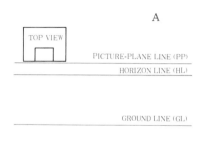

Decide on the scale to be used and draw the top view near the top of the drawing sheet. A more interesting view is obtained if the top view is drawn slightly to the right or to the left of center. Draw an edge (top) view of the picture plane (PP) through the front edge of the top view. Draw the horizon line (HL). The location will depend upon whether you want the object to be viewed from above, on, or below eye level. Draw the ground line. Its location in relation to the horizon line will determine *how far* above or below eye level the object will be viewed.

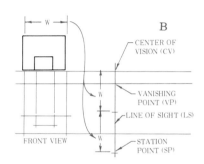

Locate the station point (SP). (a) Draw a vertical line (line of sight) from the picture plane toward the bottom of the sheet. Draw the line slightly to the right or to the left of the top view. (b) Set your dividers at a distance equal to the width (*W*) of the top view. (c) Begin at the center of vision on the picture plane and step off two to three times the width (*W*) of the top view, along the line of sight, to locate the station point (SP). Project downward from the top view to establish the width of the front view on the ground line. Complete the front view.

in Fig. 12-44. Figures 12-45 and 12-46 show the same object drawn in different positions. Notice that one face of the object is generally placed on the picture plane (thus the name parallel perspective) and will remain in true size and shape. True-scale measurements may be made on this surface.

Two-point perspective. Two-point perspective means *two vanishing points*. It is also called *angular perspective* since none of the faces are drawn parallel to the picture plane. The effect of two-point perspective can be seen in Fig. 12-47.

The procedure for making a two-point, or angular, perspective drawing of the object shown in Fig. 12-48 is shown in Fig. 12-49. Figures 12-50 and 12-51 show the same object drawn in different positions.

Inclined surfaces. Inclined surfaces in perspective are constructed by locating the ends of inclined lines and connecting them as shown in Fig. 12-52.

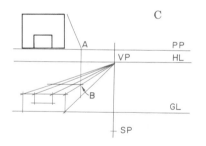

The vanishing point (VP) is the intersection of the line of sight (LS) and the horizon line (HL). Project lines from points on the front view to the vanishing point. Establish depth dimensions in the following way: (a) Project a line from the back corner of the top view to the station point. (b) At point A on the PP, drop a vertical line to the perspective view to establish the back vertical edge. (c) Draw a horizontal line through point B to establish the back top edge.

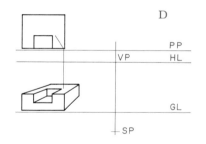

Proceed as in the previous step to lay out the slot detail. Darken all necessary lines, and erase construction lines as desired to complete the drawing.

Fig. 12-44 Procedure for making a single-point, or parallel, perspective drawing (bird's-eye view).

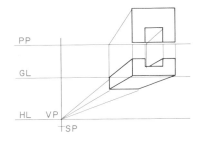

Fig. 12-45 Single-point perspective, worm's-eye view.

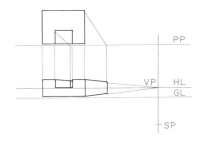

Fig. 12-46 Single-point perspective, normal, or man's-eye, view.

Fig. 12-47 When a building is viewed at an angle, two sides can be seen. The top and ground lines of each side appear to converge toward points. This is the effect in two-point, or angular, perspective. (*Bruning Division, Addressograph Multigraph Corp.*)

Fig. 12-48 Multiview and isometric drawings of object to be drawn in two-point perspective.

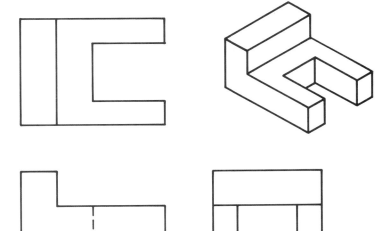

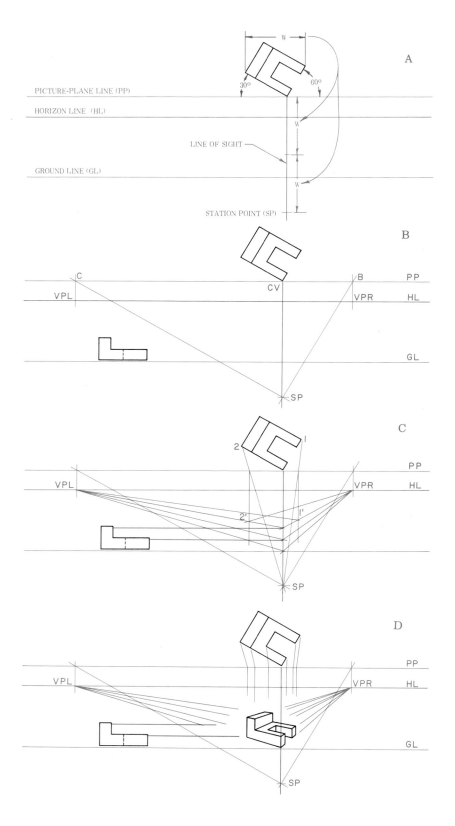

Draw an edge view of the picture plane (PP). Allow enough space at the top of the sheet for the top view. Draw the top view with one corner touching the PP. In this case, the front and side of the top view form angles of 30° and 60°, respectively. Other angles may be used, but 30° and 60° seem to give the best appearance on the finished perspective drawing. The side with the most detail is usually placed along the smaller angle for a better view. Draw the horizon line (HL) and the ground line (GL). (Follow the procedure given in Fig. 12–44.) Draw a vertical line (line of sight) from the center of vision (CV) toward the bottom of the sheet to locate the station point. (Follow the procedure given in Fig. 12–44.)

Draw line *SP-B* parallel to the end of the top view and line *SP-C* parallel to the front of the top view. (Use a 30°–60° triangle.) Drop vertical lines from the picture plane (PP) to the horizon line (HL) to locate vanishing point left (VPL) and vanishing point right (VPR). Draw the front or side view of the object on the ground line as shown.

Begin to block-in the perspective view by projecting vertical dimensions from the front view to the line of sight (also called measuring line) and then to the vanishing points. Finish blocking in the view as follows: (a) Project lines from points 1 and 2 on the top view to the station point. (b) Where these lines cross the picture plane (PP), drop vertical lines to the perspective view to establish the length and width dimensions. (c) Project point 1^1 to VPL and 2^1 to VPR.

Add detail by following the procedure described in the previous two main steps. Darken all necessary lines and erase construction lines as desired.

Fig. 12–49 Procedure for making a two-point perspective drawing (bird's-eye view).

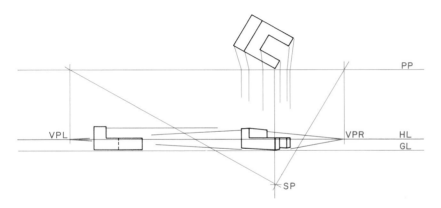

Fig. 12–50 Two-point perspective, normal, or man's-eye, view.

Fig. 12–51 Two-point perspective, worm's-eye view.

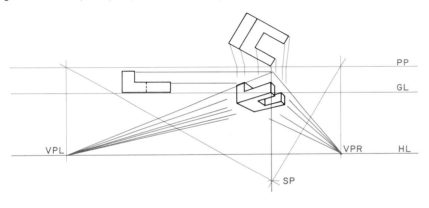

Fig. 12–52 Two-point perspective with an inclined surface.

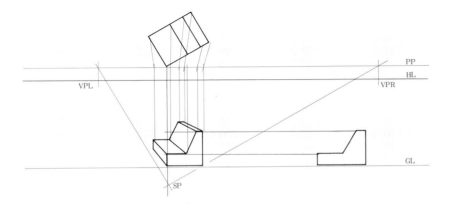

Circles and arcs in perspective.
Figure 12–53 shows a perspective drawing of an object with a cylindrical surface. Notice that points are located on the front and top views and projected to the perspective view. The intersections of these lines form a path through which the perspective arc is drawn with a French curve or an ellipse template.

Perspective drawing shortcuts.
Perspective drawing is time consuming because of the amount of layout work needed before the actual perspective view can be started. Also, a large drawing surface is often necessary in order to locate distant points. The disadvantages can be kept to a minimum by using various shortcuts described in this section.

Perspective grids. One of the simpler shortcuts is the perspective grid shown in Fig. 12–54. While there are many advantages in using grids, there is one major disadvantage: grids are limited to one type of view based on one set of points and one view location. However, in many industrial drafting rooms, this may be satisfactory if a variety of views is not necessary.

Perspective grids are available commercially or can be prepared by the draftsman. It is practical for the draftsman to prepare his own grid only if he has a number of perspective drawings to make in a special style.

Perspective drawing boards. The Klok perspective board (Fig. 12–55) is another timesaving device for making accurate perspective drawings. A special T-square with the top edge of the blade centered on the head is provided for drawing converging lines. Perspective scales on the board eliminate the need for extensive projection or calculation.

A homemade perspective drawing board (Fig. 12–56) can be made by using a regular drawing board, some

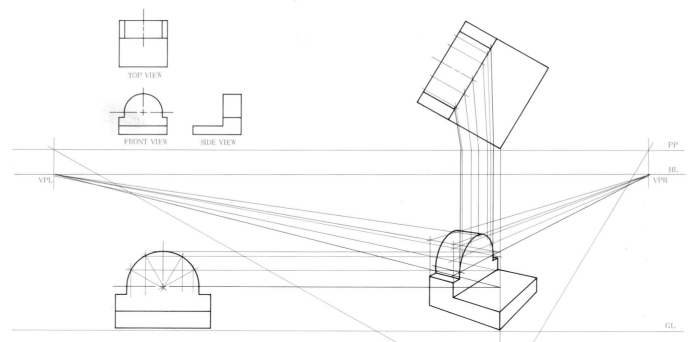

Fig. 12–53 Two-point perspective with a cylindrical surface.

TOP VIEW

FRONT VIEW   SIDE VIEW

PP

HL

VPL   VPR

GL

SP

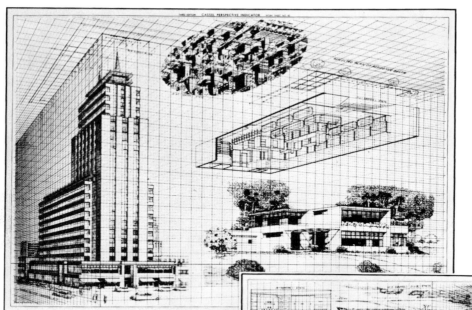

Fig. 12–54 Perspective grid. (*Bruning Division, Addressograph Multigraph Corp.*)

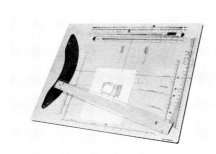

Fig. 12-55 Klok perspective drawing board. (*Modulux Division, United States Gypsum Co.*)

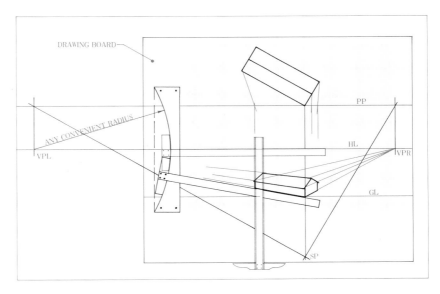

Fig. 12-56 A homemade perspective drawing board.

thin pieces of wood, cardboard, or hardboard, and by making or modifying a T-square. Like the perspective grid, this device is practical only if many drawings of about the same size and style are to be drawn, *e.g.*, perspective drawings of houses.

The procedure for making a perspective board is as follows:

A. Follow the procedure in Fig. 12-48 for locating the vanishing points. Use the top view of an object similar in size to the ones that will be drawn on the finished board. The layout should be done on a large sheet of paper in order to locate distant vanishing points.

B. Place a drawing board within the layout, as shown in Fig. 12-56. Heavy cardboard or thin wood may be used for guides. Fasten the guide material in place and strike the arcs, using the vanishing points as centers and any convenient radius.

C. Cut the arcs with a sharp knife to form the guides. Draw the PP, HL, GL, LS, and SP on the board. The board may now be removed from the layout sheet.

D. Construct the T-square as shown. Notice that the top edge of the blade falls on the center line of the heads. Thin material must be used for the head as well as for the blade.

To use the perspective board:

A. Place a sheet of tracing paper in position between the guides (Fig. 12-56). Tracing paper will allow the lines drawn on the board to be visible without redrawing them.

B. Draw the top view of the desired object in its proper position.

C. Proceed as in Fig. 12-48. Lines that project toward VPL will be drawn with the T-square head against the left guide. For those that project toward VPR, the right-hand guide is used. Vertical lines are drawn with the T-square head against the bottom edge of the board.

Review

1. Axonometric projection is the family name for _____ , _____ , and _____ .

2. Is an isometric projection larger or smaller than an isometric drawing?

3. Name the three most common types of pictorial drawing.

4. Since it is time consuming to draw a true ellipse, what method is normally used?

5. What is a line called that is not parallel to any of the isometric axes?

6. Describe the method used for drawing an irregular curve in pictorial drawings.

7. What is the difference between cavalier and cabinet type oblique drawings?

8. What are the two most common types of sectional views used in pictorial drawing?

9. Which type of pictorial drawing most closely resembles a photograph?

Problems for Chapter 12

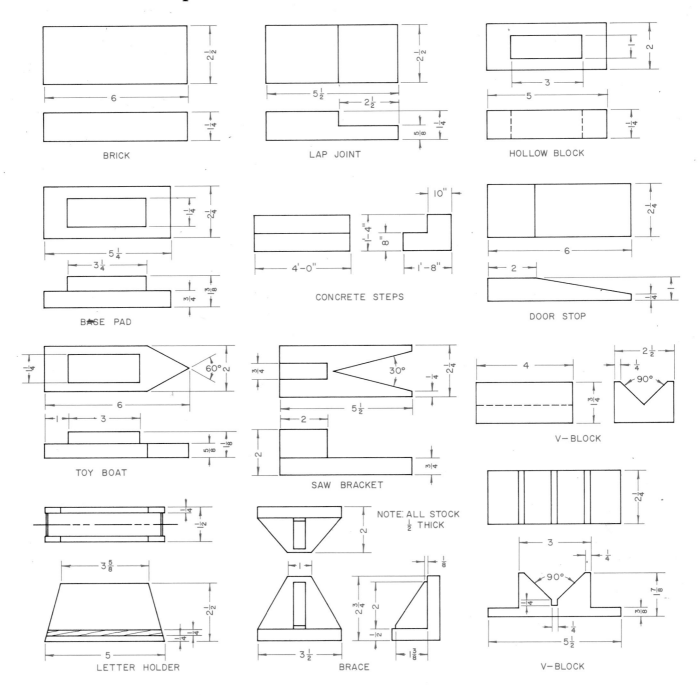

BRICK

LAP JOINT

HOLLOW BLOCK

BASE PAD

CONCRETE STEPS

DOOR STOP

TOY BOAT

SAW BRACKET

NOTE: ALL STOCK ½ THICK

V-BLOCK

LETTER HOLDER

BRACE

V-BLOCK

Fig. 12–57 Isometric drawing problems. Scale: Optional. Assignment 1: Make an isometric drawing of the object assigned. Assignment 2: Make an isometric half- or full-sectional view as assigned. Do not dimension unless instructed to do so. NOTE: These problems may also be used for oblique and perspective drawing practice.

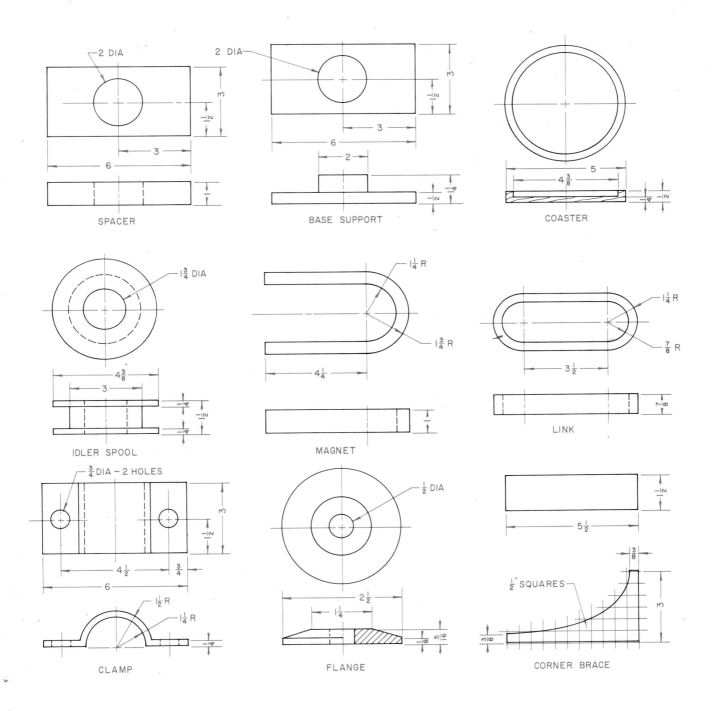

Fig. 12-58 Isometric drawing problems. Scale: Optional. Assignment 1: Make an isometric drawing of the object assigned. Assignment 2: Make an isometric half- or full-sectional view as assigned. Do not dimension unless instructed to do so. NOTE: These problems may also be used for oblique and perspective drawing practice.

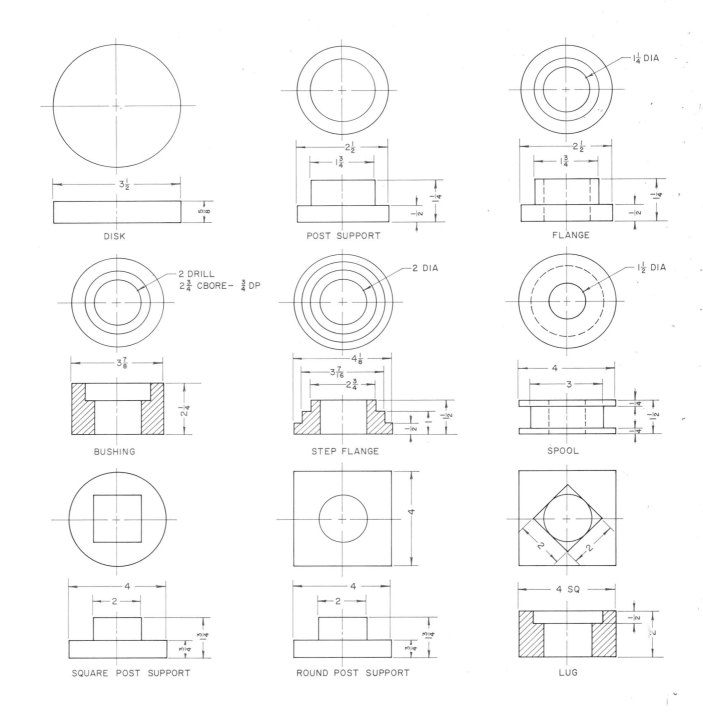

DISK

POST SUPPORT

FLANGE

2 DRILL
2 3/4 CBORE- 3/4 DP

BUSHING

2 DIA

STEP FLANGE

1 1/2 DIA

SPOOL

SQUARE POST SUPPORT

ROUND POST SUPPORT

LUG

Fig. 12–59 Oblique drawing problems. Scale: Optional. Assignment 1: Make an oblique drawing of the object assigned. Assignment 2: Make an oblique half- or full-sectional view as assigned. Do not dimension unless instructed to do so. NOTE: These problems may also be used for isometric and perspective drawing practice.

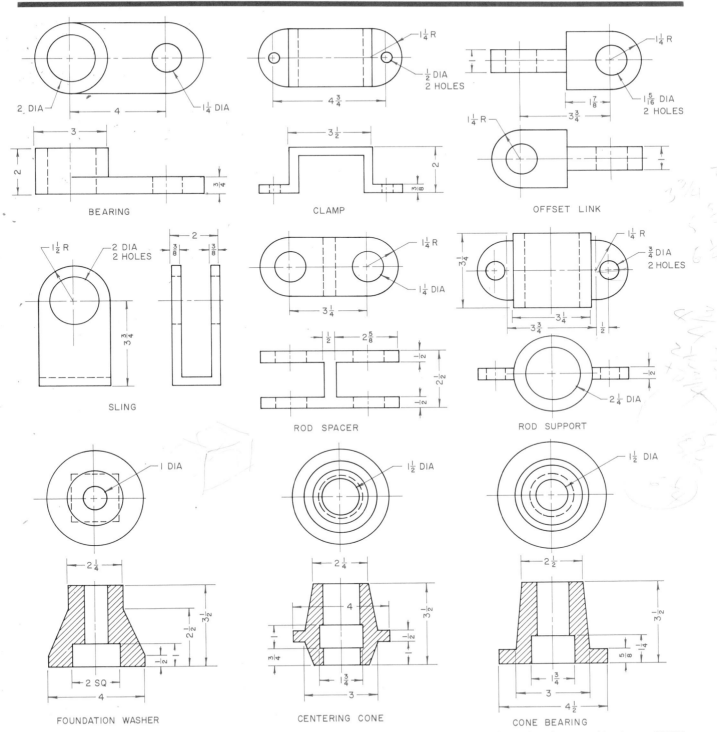

Fig. 12–60 Oblique drawing problems. Scale: Optional. Assignment 1: Make an oblique drawing of the object assigned. Assignment 2: Make an oblique half- or full-sectional view as assigned. Do not dimension unless instructed to do so. NOTE: These problems may also be used for isometric and perspective drawing practice.

BEARING

CLAMP

OFFSET LINK

SLING

ROD SPACER

ROD SUPPORT

FOUNDATION WASHER

CENTERING CONE

CONE BEARING

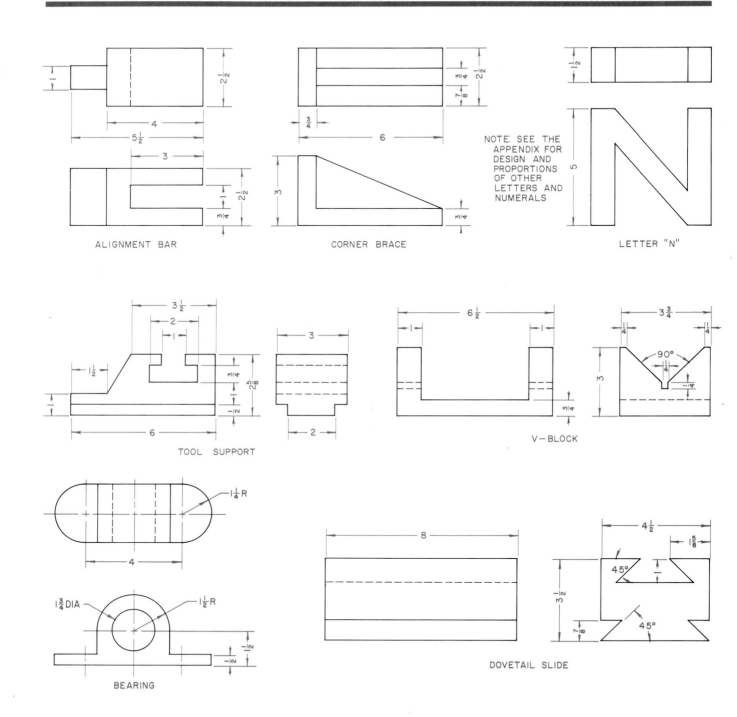

ALIGNMENT BAR

CORNER BRACE

NOTE: SEE THE APPENDIX FOR DESIGN AND PROPORTIONS OF OTHER LETTERS AND NUMERALS

LETTER "N"

TOOL SUPPORT

V — BLOCK

BEARING

DOVETAIL SLIDE

Fig. 12–61 Perspective drawing problems. Make a one-point-perspective or a two-point-perspective drawing of the object assigned. Use light, thin lines for the construction lines and do not erase them. Brighten the finished perspective drawing lines. Use any suitable scale. The locations of the PP, HL, GL, etc., are optional.

Inking | Chapter **13**

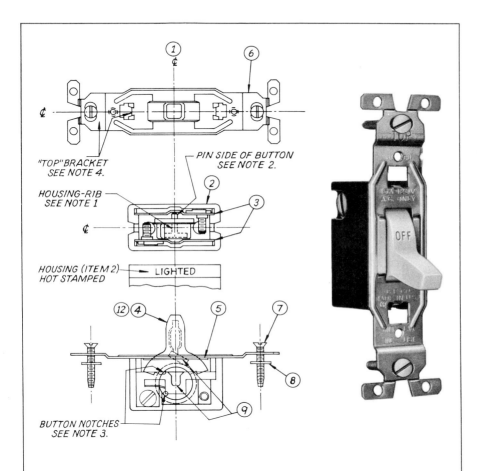

"TOP" BRACKET
SEE NOTE 4.

PIN SIDE OF BUTTON
SEE NOTE 2.

HOUSING-RIB
SEE NOTE 1

HOUSING (ITEM 2)
HOT STAMPED ← LIGHTED

BUTTON NOTCHES
SEE NOTE 3.

▲ ASSEMBLY SEQUENCE
PARTS MUST BE ASSEMBLED IN THE
FOLLOWING RELATION TO EACH OTHER.
NOTE:
1. HOUSING: RIB TOWARD OPERATOR AT ASM
2. BUTTON: PIN SIDE ORIENTED AS NOTED.
3. TRIGGER: "OFF" SIDE TO LEFT AND IN
PROPER BUTTON NOTCH AS SHOWN.
4. BRACKET: "TOP" SIDE TO LEFT.
FOLD DOWN TABS AT ASSEMBLY.
TABS MUST BE CONTAINED WITHIN
NOTCH.

▲ LUBRICATION
A. BUTTON-ON FLAT SHELL, PIN
ENDS & LOWER NOTCH.
B. FLAT SPRING-ON SIDES THAT
CONTACT TRIGGER.
C. HOUSING-INSULATION SADDLE.

(WIRING DEVICE DEPARTMENT, GENERAL ELECTRIC COMPANY.)

Nature and Purpose of Inked Drawings. Most technical drawings are made with pencil on a good quality tracing paper (vellum) or on drafting film. However, the use of ink in preparing high-quality reproducible tracings is extensive and continues to increase. As the demand for ink-quality drawings increases, the development of more advanced inking techniques and equipment increases. For example, the development of the tube-type technical pen with all its accessories and attachments has increased the draftsman's efficiency in inking while improving the quality of the finished drawing.

Microfilming and other microreproduction (also called micrographics) methods have made it necessary to modernize inking methods and improve the quality of ink drawings (see Chapters 14 and 15). In the process of photographically reducing and enlarging technical drawings, some definition (detail and sharpness) is lost. As a result, a higher quality original will always produce a higher quality reproduction. If the original drawing is of poor quality, the reproduction may not be usable.

Drawing Ink. Inks used for technical drawings must have certain special characteristics. For example, they must dry fast and yet be free-flowing. They must have excellent adhesion qualities and must not chip, crack, peel, or smear after drying (Fig. 13-1). Also, technical drawing ink (also called India ink) must be completely opaque in order to produce good uniform line tone.

Waterproof and nonwaterproof inks are available for use on high-quality paper, cloth, polyester film, and illustration board. The waterproof inks are best suited for mechanical drawings, and nonwaterproof inks are ideal for fine-line pictorial illustrations. Both types of ink can produce opaque lines which result in superior reproductions when photographed or copied by diazo machines, as described in Chapter 14.

Acid-base inks are used on high-gloss acetate film. Special pens must be used with acid-base ink, and the drawing lines are etched into the surface. While etching is usually considered permanent, some manufacturers have developed special techniques for making corrections and revisions. Other special inks are available in color, such as the liquid inks for ball-point pens used on incremental plotters controlled by computers or tapes. They are generally waterproof and are available for all media (Fig. 13-2).

Basic Inking Instruments. The ruling pen (Fig. 13-3) has blades which can be adjusted to draw different line widths. They can be used to draw both straight and curved lines. Ruling pens are sometimes referred to as *all-purpose pens.* The special pens illustrated in Fig. 13-4 are convenient for drawing double lines, contour lines, and thick (heavy) lines. They may be filled from a cartridge tube (Fig. 13-5), a squeeze bottle, or a dropper cap (Fig. 13-6).

A few simple rules should help the draftsman produce a high-quality ink drawing with a minimum of difficulty. Be sure to study them carefully before attempting to use the ruling pen.

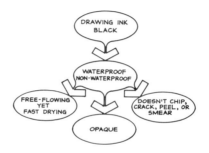

Fig. 13-1 Ink quality for technical drawings.

1. Do not hold the pen over the drawing when filling.

2. Do not overfill the pen; about $\frac{1}{4}$ to $\frac{3}{8}$ in. is generally adequate.

3. Never dip the pen into an open bottle or allow ink to get on the outside of blades.

4. Keep blades clean by frequent wiping with a soft towel, tissue, or pen cleaner.

5. While inking a line, keep both nibs in contact with the drawing surface.

6. Keep the blades parallel to the direction of the pen stroke.

7. Do not press the pen nibs hard against the straight edge. Excessive tension will cause line widths to vary.

8. Do not press the nibs too hard against the drawing surface. The blades may damage the drawing surface and quickly dull the pen.

Typical faulty lines can result from many causes. Figure 13-7 illustrates some that can be avoided.

Technical fountain pens have revolutionized the inking of technical drawings (Fig. 13-8). A variety of point sizes that produce various line widths is available (Fig. 13-9). These points can assure constant line widths because of the quality of design and manufacture which provides a steady, nonclogging flow of ink.

The technical pen may have a refillable cartridge for storing ink (Fig.

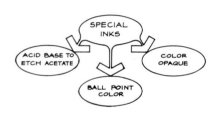

Fig. 13-2 Special inks for new methods of presentation.

Fig. 13-3 The all-purpose ruling pen.

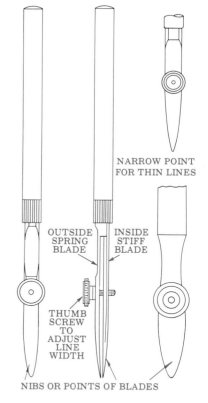

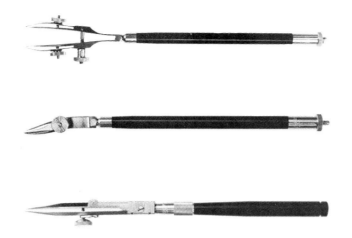

Fig. 13–4 Special ruling pens. (*Teledyne Post.*)

Fig. 13–5 Filling the pen from a cartridge tube.

Fig. 13–6 Filling the pen from a dropper cap. (*Higgins Ink Co.*)

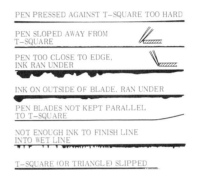

Fig. 13–7 Causes of faulty lines from a ruling pen.

PEN PRESSED AGAINST T–SQUARE TOO HARD

PEN SLOPED AWAY FROM T–SQUARE

PEN TOO CLOSE TO EDGE, INK RAN UNDER

INK ON OUTSIDE OF BLADE, RAN UNDER

PEN BLADES NOT KEPT PARALLEL TO T–SQUARE

NOT ENOUGH INK TO FINISH LINE INTO WET LINE

T–SQUARE (OR TRIANGLE) SLIPPED

Fig. 13–8 Technical fountain pens. (*J. S. Staedtler, Inc.*)

Fig. 13–9 The range of lines available from technical pen points. (*J. S. Staedtler, Inc.*)

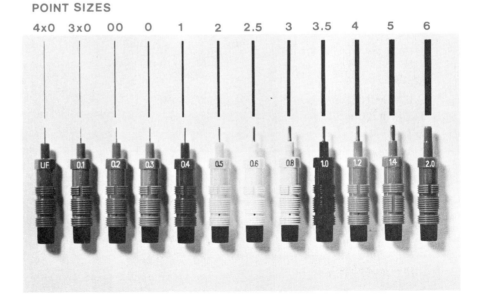

POINT SIZES

4x0 3x0 00 0 1 2 2.5 3 3.5 4 5 6

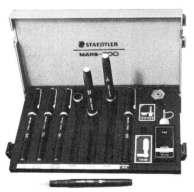

13–10) or may be provided with a piston-action filling system which can easily be refilled from a standard drawing-ink bottle (Fig. 13–11).

The performance of technical pens is determined by the quality of the point and the drawing surface. There are generally three classes of points for use on all media:

1. Hardened stainless steel for general use on all media.

2. Self-polishing jewel points for highly abrasive surfaces, such as drafting film.

3. Wear-free tungsten carbide points designed for drafting film and the excessive wear common to programmed drafting equipment.

The points are designed with a shoulder, as shown in Fig. 13–12, to prevent smudging. The pen shoulder is barrel-shaped for use with curved or straight guiding instruments. The technical fountain pen has several distinct advantages.

1. Uniform line width is produced with ease (no line adjustments are required).

2. The barrel-shaped shoulder eliminates smearing because point and shoulder are offset (Fig. 13–13).

3. The cartridge is easily loaded with ink and needs refilling less frequently than a ruling pen.

4. The ink laid down by a fountain pen may dry more uniformly because of equal flow (Fig. 13–14).

5. The technical pen seldom needs cleaning because of a point which has a weighted cleaning wire inside the capillary tube (Fig. 13–15).

The technical pen may be stored in a container which maintains humidity (Fig. 13–16), or in some cases, the pen has a hermetic seal in the cap which provides an airtight storage when the

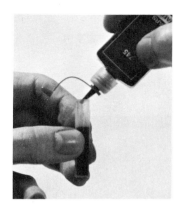

Fig. 13–10 Refillable cartridge. (*J. S. Staedtler, Inc.*)

Fig. 13–11 Piston-action filling system. (*J. S. Staedtler, Inc.*)

Fig. 13–12 The offset of the barrel shoulder moves easily along all guiding edges. (*J. S. Staedtler, Inc.*)

RULING PEN   TECHNICAL PEN

Fig. 13–14 Section view of typical ink flow from a ruling pen and a technical pen.

Fig. 13–15 Clicking sound from cleaning wire assures smooth flow of ink. (*J. S. Staedtler, Inc.*)

POINT   SAFETY SCREW

POINT HOUSING   FLOW REGULATING DEVICE

Fig. 13–13 The technical pen in use. Note that the shoulder keeps the barrel of the pen above the curve.

Fig. 13–16 Humidified container for technical pens. (*Keuffel & Esser Co.*)

cap is tightened with a sharp twist. When proper care is exercised in filling and cleaning a pen, good results are obtained.

Inking Straight Lines. The ruling pen or technical fountain pen may be used as shown in Figs. 13–17 and 13–18. The correct position for drawing lines with a ruling pen is illustrated so errors, as shown in Fig. 13–7, may be avoided (Fig. 13–17). Note the direction of the stroke and the angle of the ruling pen. The technical fountain pen is held in a vertical position for the most uniform line (Fig. 13–18).

Making Circles and Arcs. Circle templates and compasses may be used with technical fountain pens (Fig. 13–19). A ruling pen cannot be used conveniently with most templates. Many compasses have ruling-pen attachments (Fig. 13–20) that are used for inking arcs and circles. Others may have technical-pen adaptors, as shown in Fig. 13–21. The pencil leg is removed from the compass, and the ruling-pen leg or the fountain-pen adaptor is inserted. The needle point should be carefully adjusted, as shown in Fig. 13–22 so that the legs are perpendicular to the drawing surface. Always draw the circle in one stroke. The compass may be inclined slightly when using a ruling pen but should be

Fig. 13–17 Correct position of the ruling pen when drawing lines.

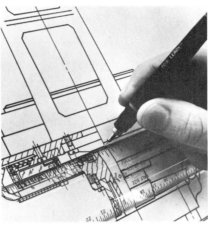
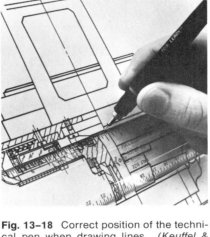

Fig. 13–18 Correct position of the technical pen when drawing lines. (*Keuffel & Esser Co.*)

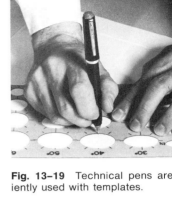

Fig. 13–19 Technical pens are conveniently used with templates.

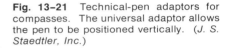
Fig. 13–21 Technical-pen adaptors for compasses. The universal adaptor allows the pen to be positioned vertically. (*J. S. Staedtler, Inc.*)

Fig. 13–20 Compass and ruling-pen attachment. (*Teledyne Post.*)

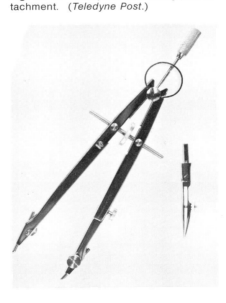

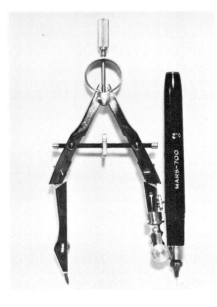

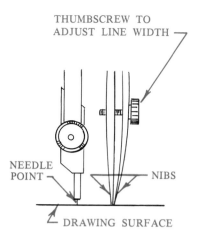

Fig. 13-22 Needle point of compass adjusted so that the ruling nibs and point are perpendicular to the drawing surface.

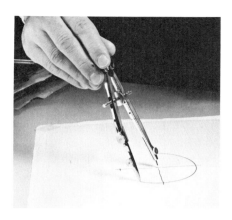

Fig. 13-23 Inking with a compass. (*Teledyne Post.*)

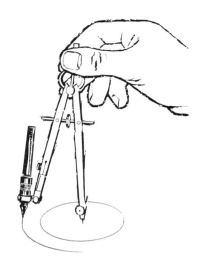

held vertically when using a fountain-pen adaptor (Fig. 13-23). The lengthening bar, or beam compass, may be used for large circles, as shown in Fig. 13-24.

Inking Curved Lines. Irregular (French) curves may be used for guiding the pen when curves other than circular arcs are inked. Fixed curved templates and irregular adjustable curves (Fig. 13-25) may be used for guiding the pen. The following procedure may be helpful.

1. Locate the tangent points that change the curve's direction or join a straight line to a curve (Fig. 13-26 at A).

2. Pencil-in the curve lightly, noting the direction from a smaller to a larger radius.

3. Ink over the middle of the penciled line (Fig. 13-26 at B).

Lettering Guides and Equipment. The lettering set illustrated in Fig. 13-27 consists of three basic tools for inking: The technical fountain pen is

Fig. 13-24 Use of lengthing bar for large circles.

Fig. 13-25 Fixed and adjustable irregular curves.

selected according to the width in proportion to the letter height. Template guides are selected on a size and style basis from about $\frac{1}{16}$ to 2 in. high. A scriber has a tailpin to follow the horizontal line in the bottom of the guide and a tracer pin to follow the engraved letter. The technical pen is set in the barrel slot and simply reproduces the letter above the template (Fig. 13-28). Adjustable scribers (Fig. 13-29) are available for height and slant control. Numerous styles of templates (Fig. 13-30) are also

available for refined inked work which is suitable for reproduction.

Erasing Techniques. The ink especially formulated for polyester drafting film is waterproof but can be removed easily from the film. A slightly moistened plastic eraser will remove or lift the ink without any pressure in rubbing. The polyester film does not absorb ink, and therefore it allows ink to dry on its highly finished surface (Fig. 13-31). Inks used on other media, such as tracing vellum or illus-

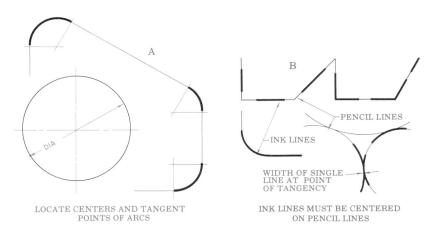

LOCATE CENTERS AND TANGENT
POINTS OF ARCS

INK LINES MUST BE CENTERED
ON PENCIL LINES

Fig. 13-26 Ink curves to points of tangency first (A). Ink lines must be centered over pencil lines (B).

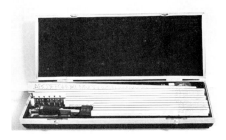

Fig. 13-27 One of the draftsman's most useful tools is a lettering set. (*Keuffel & Esser Co.*)

Fig. 13-28 The three basic parts are the pen, the template, and the scriber. (*Keuffel & Esser Co.*)

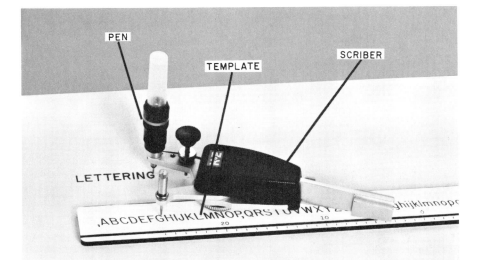

PEN

TEMPLATE

SCRIBER

LETTERING

tration board, can be removed with conventional ink erasers but care must be exercised. Light pressure with strokes in the direction of the line will remove ink caked on the surface. Excessive pressures will disturb the drawing surface and revision will be difficult. The tooth surface illustrated in Fig. 13-32 retains the ink; thus, the best erasing techniques may be different for each and so should be studied or experimented with for each specific medium.

The Alphabet of Lines. Line symbols used in making drawings may be considered as a graphic alphabet. The American National Drafting Standards Manual recommends three widths of lines: thick, medium, and thin. The symbols for ink lines are shown in Figure 13-33. Each line is used for a definite purpose and must not be used for anything else. Detail drawings should have fairly wide (thick) outlines with thin center and dimension lines so that the drawing will have contrast and be easy to read. Hidden lines have a medium width. If all the lines are of the same width, the drawing will have a flat appearance and will be hard to read.

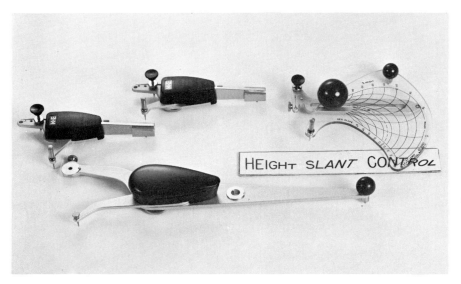

Fig. 13-29 Adjustable scribers provide lettering flexibility. (*Keuffel & Esser Co.*)

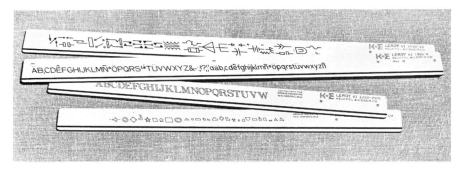

Fig. 13-30 Templates are available in various styles of lettering. (*Keuffel & Esser Co.*)

Fig. 13-31 The results of good (left) and poor (right) ink adhesion.

Fig. 13-32 Section showing ink in contact with drawing surface.

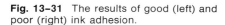

INK MUST COME IN DIRECT CONTACT WITH SURFACE

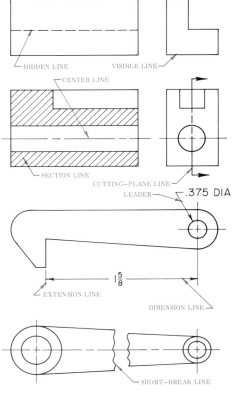

.375 DIA

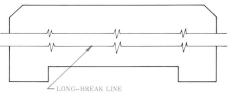

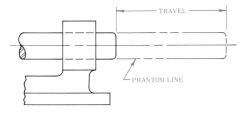

Fig. 13-33 Alphabet of lines.

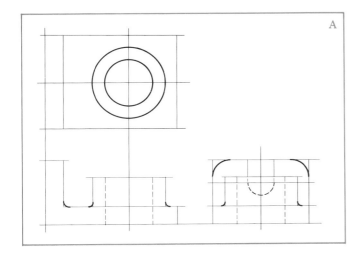

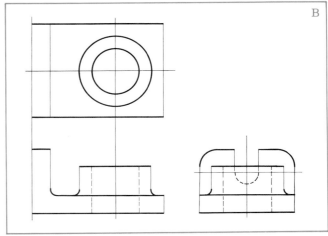

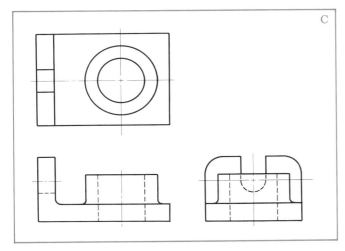

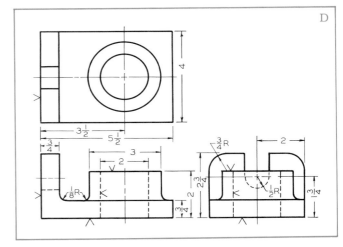

Fig. 13–34 Order of inking or tracing a drawing.

Order of Inking. Good inking is the result of two things: careful practice and a definite order of working. Smooth joints and tangents, sharp corners, and neat fillets not only improve the appearance of a drawing but make it easier to read.

The general order of inking or tracing is shown in Fig. 13–34. The arcs are inked first, as in Space A, and should center over the pencil lines. Horizontal lines should be inked as in Space B and the drawing completed as in Space C. Then the dimension lines, arrowheads, finish marks, and so forth, are added and the dimensions filled in as in Space D. The order of inking is:

1. Ink main center lines.
2. Ink small circles and arcs.
3. Ink large circles and arcs.
4. Ink hidden circles and arcs.
5. Ink irregular curves.
6. Ink horizontal full lines.
7. Ink vertical full lines.
8. Ink inclined full lines.
9. Ink hidden lines.
10. Ink center lines.
11. Ink extension and dimension lines.
12. Ink arrowheads and figures.
13. Ink section lines.
14. Letter notes and titles.
15. Ink border lines.
16. Check drawing carefully.

Ultrasonic Cleaner. The technical fountain pen and the ruling pen may need periodic cleaning. The ultrasonic cleaner brings greater efficiency to ink drawing because there is very little down time. The cleaning action normally takes 20 to 30 seconds, and pens are ready to perform without excessive care. The pen point is dipped into the liquid cleaner and dried ink is freed when 60,000 to 80,000 cycles per second are acting on the blocked point (Fig. 13–35).

Thorough cleaning of technical pens may require disassembly. The cap and ink cartridge may be removed and the point allowed to soak in a special cleaning fluid. After removing the dried ink that normally accumu-

la es, flush the parts in a stream of col l water. The best inking results are achieved after all parts have been dried so that the ink will not become diluted and result in gray line work. Careful storage and proper handling of inking equipment will insure good results and very little repair time.

Fig. 13–35 An ultrasonic cleaner. (*Keuffel & Esser Co.*)

Review

1. Name the two types of ink pens used in drafting.

2. Describe the best position in which to hold a technical pen.

3. Name two methods for inking circles and arcs.

4. Name the instruments used to guide the pen when drawing curved lines other than true circular areas.

5. Describe the procedure for removing India ink from drafting film.

6. How many widths of lines are recommended for use on ink drawings? Name them.

7. What type of ink is used on acetate film?

8. Waterproof inks are used for mechanical drawings; nonwaterproof inks are used for _____.

9. Do not overfill a ruling pen; about _____ to _____ in. is generally adequate.

Problems for Chapter 13

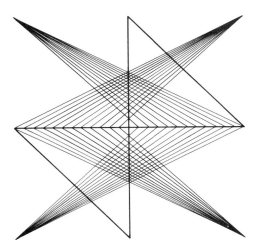

Fig. 13-36 Using light pencil construction lines, lay off a 6″ square and divide it equally into four 3″ squares. On the horizontal midline, mark off 24 $\frac{1}{4}$″ units. Proceed to ink in the lines to construct the given design.

Fig. 13-37 Construct in ink the intersecting units that have an illusion of depth. The axis to the left is 20°, and to the right 25°. The depth of a unit is 2″ to the left and 2$\frac{3}{4}$″ to the right. The height is 2″.

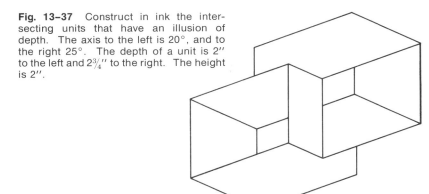

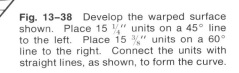

Fig. 13-38 Develop the warped surface shown. Place 15 $\frac{1}{4}$″ units on a 45° line to the left. Place 15 $\frac{3}{8}$″ units on a 60° line to the right. Connect the units with straight lines, as shown, to form the curve.

Fig. 13-39 In ink, construct the 2 logos shown as geometric forms. Prepare a straight-line logo and a curved-line logo of your own design. The lettering "Grand Opening" is based on circular forms. Design a similar sign of your own word choice in this 4-line circular technique. Design your own lettering style and prepare a sign in that style.

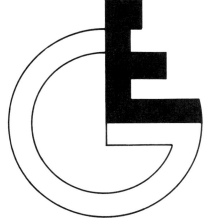

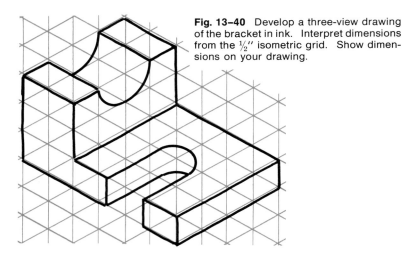

Fig. 13–40 Develop a three-view drawing of the bracket in ink. Interpret dimensions from the ½″ isometric grid. Show dimensions on your drawing.

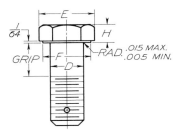

Fig. 13–41 Develop two views of the aircraft bolt. $D = 1''$, $E = 1.65625$, $F = 1.4375$, $H = \frac{1}{2}D$. Threads per inch $= 14$. Ink all lines.

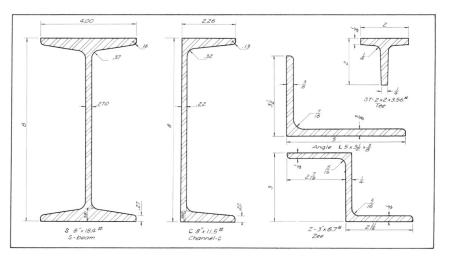

Fig. 13–42 Prepare ink drawings of the five structural shapes at full scale on a B-size sheet.

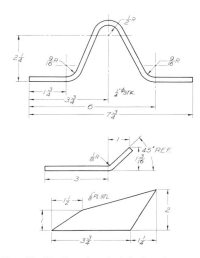

Fig. 13–43 Develop in ink the shapes of the ¼″ round barstock and the sheet stock shown. Dimension the drawings.

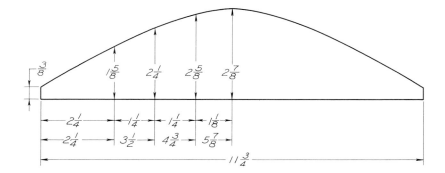

Fig. 13–44 Plot the symmetrical curve and develop it in ink with an irregular curve. Dimension the part with a Leroy set, if available.

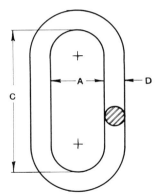

Fig. 13–45 Prepare one view of the wire rope short link with revolved section. Use dimensions selected by the instructor. Dimension your drawing. (*United States Steel.*)

Wire Rope Link

A	C	D
1	3	$3/8$
$1\frac{1}{4}$	4	$1/2$
$1\frac{1}{2}$	4	$5/8$
$1\frac{3}{4}$	5	$3/4$
2	5	$7/8$

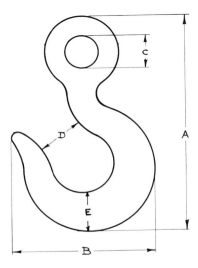

Fig. 13–46 Prepare a wire rope hook using the dimensions selected by the instructor. Determine radii necessary for smooth tangencies. Dimension your drawing. (*United States Steel.*)

Wire Rope Hook

A	B	C	D	E
$4\frac{15}{16}$	$3\frac{3}{16}$	$7/8$	$1\frac{1}{16}$	$27/32$
$5\frac{13}{32}$	$3\frac{16}{32}$	1	$1\frac{1}{8}$	$29/32$
$6\frac{1}{4}$	$4\frac{3}{32}$	$1\frac{1}{8}$	$1\frac{1}{4}$	$1\frac{1}{8}$
$6\frac{7}{8}$	$4\frac{17}{32}$	$1\frac{1}{4}$	$1\frac{3}{8}$	$1\frac{5}{16}$
$7\frac{5}{8}$	$4\frac{7}{8}$	$1\frac{3}{8}$	$1\frac{1}{2}$	$1\frac{3}{8}$
$8\frac{19}{32}$	$5\frac{3}{4}$	$1\frac{1}{2}$	$1\frac{11}{16}$	$1\frac{9}{16}$
$9\frac{1}{2}$	$6\frac{3}{8}$	$1\frac{5}{8}$	$1\frac{7}{8}$	$1\frac{11}{16}$

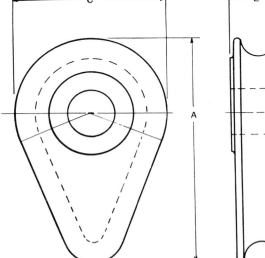 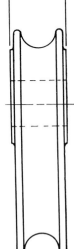

Fig. 13–47 Prepare a two-view drawing of the solid cast-steel thimble with dimensions selected by the instructor. Select suitable radii for fillets and rounds. Note the points of tangency to determine the tapered shape. Dimension your drawing. (*United States Steel.*)

Cast Steel Thimble

Wire Rope DIA	A	C	E
$1/2$	$3\frac{1}{16}$	$2\frac{1}{8}$	$7/8$
$5/8$	$5\frac{1}{16}$	$3\frac{3}{8}$	$1\frac{1}{4}$
$3/4$	$5\frac{1}{16}$	$3\frac{3}{8}$	$1\frac{1}{4}$
$7/8$	$6\frac{9}{16}$	$4\frac{1}{2}$	$1\frac{1}{2}$
1	$6\frac{9}{16}$	$4\frac{1}{2}$	$1\frac{1}{2}$

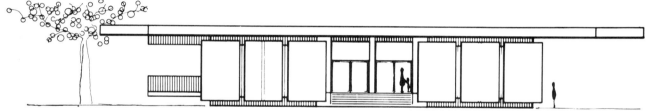

Fig. 13–48 Examine the architectural elevation. Develop proportions around the following data: stair risers 6″, and doors 3′ × 7′. The facia, wall panels, and balcony (shown left with grill railing) are to be determined by the student. Scale: $\frac{1}{8}$″ = 1′-0″. Ink all lines.

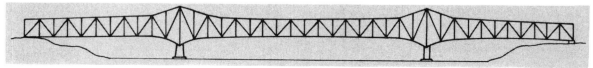

Fig. 13–49 Prepare a schematic elevation of the Warren-type truss bridge. Work out proportions through the diagonal system. Ink all lines. (*United States Steel.*)

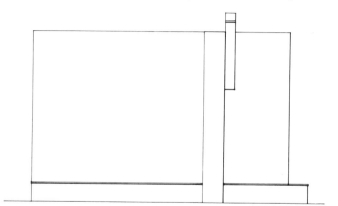

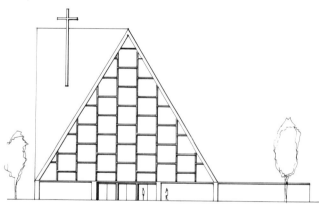

Fig. 13–50 Develop a two-view drawing of the house of God. Examine the triangular shape and determine appropriate dimensions for this design. Select a suitable scale. Propose a floor plan. Ink all lines.

Figs. 13–51 and 13–52 Examine the product illustrations for inking technique. Note that the pocket camera has white lines for interior corners and the movie camera has a textured pattern. Develop ink illustrations of either as assigned by the instructor. Develop a pictorial or multiview drawing as assigned. (*Eastman Kodak Co.*)

Drafting Media and Reproduction

Chapter 14

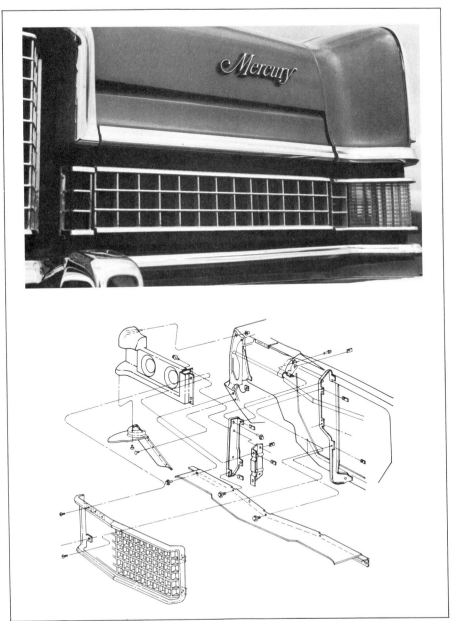

(FORD MOTOR COMPANY.)

Drafting Media. The graphic communications typically developed in the drafting room today are prepared on paper, illustration board, cloth, or film (Fig. 14-1). The basic media have been greatly improved in recent years to meet an ever increasing standard of quality demanded by industry. The application of these materials to the four levels of communication described in Chapter 1 is generally as follows:

Level one—Creative communication. Sketch paper and transparent paper are used for refining the ideas captured on the original layout. Grid paper is used for controlling proportions and developing schematic layout diagrams.

Level two—Technical communication. Sketch refinement is done on tracing paper. Charts are made on grid paper or special graph paper for pictorials or diagrams.

Level three—Market communication. Often, dramatically refined sketches which are colored for presentation are used at this level. Illustration paper or board is used for good ink and color renderings of pictorials or cutaways which explain the details for the client.

Level four—Construction or manufacturing communication. These drawings are prepared on high-grade tracing paper, vellum, cloth, or polyester film. The ageproof papers and film are considered essential for permanent records.

Fig. 14-1 Drafting media.

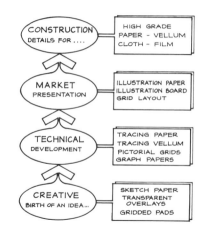

Fig. 14-2 Typical media for communications.

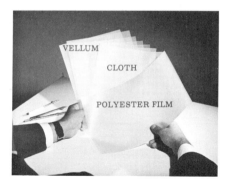

Fig. 14-3 Translucent media for good reproduction. (*Teledyne Post.*)

The engineering design team requires the characteristics unique to all the basic media for specific but various uses (Fig. 14-2).

Paper. A wide range of opaque and translucent paper is available. Opaque drawing paper is considered to have more reproduction limitations than translucent paper for designers. High-grade opaque paper may be used for black and white and colored pictorial renderings which later may be photographed for reproduction. The opaque paper may be lined for graphs, diagrams, or plotting graphical mathematical solutions.

Tracing papers that have good translucent qualities are used to develop original detail drawings. This medium must also have a good surface quality to take pencil and withstand repeated erasing along with a lot of handling by the engineering team (Fig. 14-3).

Translucent tracing papers. Natural tracing papers are made of rag and have adequate strength but limited transparency. The natural translucent paper without rag content is recommended only for sketching purposes because of limited life expectancy. Advances have been made in preparing thin tracing paper which will permit reasonable transparency and have good tooth (roughness) surface for retaining pencil and ink.

Transparentized tracing papers. These prepared papers are sometimes referred to as *vellum.* They are ideal

Fig. 14-4 Transparentized vellum is a chemically treated translucent paper. (*Bruning Division, Addressograph Multigraph Corp.*)

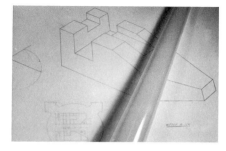

Fig. 14-5 Tracing cloth is translucent and highly durable. (*Bruning Division, Addressograph Multigraph Corp.*)

for quick reproduction (Fig. 14-4). This rag content paper has been formulated with resins so as to give the thicker paper a better transparency. It has a fast print-making quality and also gives clearer and sharper prints. Good erasability, longer life, and high strength are additional qualities of

transparentized papers. The one disadvantage to date is that they will show fold marks even though they are considered ageproof.

Cloth. Cloth, or linen as it was referred to many years ago, has qualities that surpass paper. Today this is an expensive medium that is excellent for pencil or ink. Permanence, erasability, and high reproduction qualities are all built around cotton fiber sized with starch. Moisture-resistant cloths have made this medium competitive with others so that it now offers stable characteristics necessary for the engineering design team (Fig. 14-5).

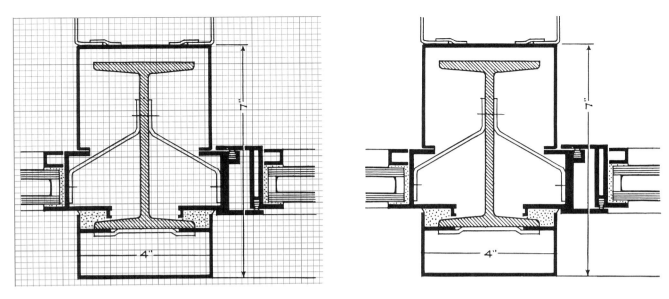

Fig. 14-6 Preprinted grid lines do not show on reproduction copies. (*Keuffel & Esser Co.*)

Film. The finest medium developed for the contemporary engineering team in recent years is polyester film. It has dimensional stability (will not shrink), a high degree of transparency, and is ageproof (will not discolor) as well as waterproof. The clear image produced by pencil or by ink is the result of a superior surface that is stable even after repeated erasing. Film is available according to thickness and matte (not glossy) surface (one side or two sides). Acetate tracing films and adhesive transparent film are some of the special films available.

The satisfactory results from the various media are partly due to the improved quality of lead, ink, and erasers available. Each medium recently developed or improved has a surface quality that is workable and reworkable.

Fade-out grid and lined stock. The higher grades of tracing paper and film are available with a light-blue grid pattern that will not show up on reproduction copies (Fig. 14-6). The general pattern is ten squares to the inch, eight squares to the inch, and four squares to the inch. Cross-section lined paper is available in rolls, sheets, and pads, generally with a choice of colored lines. Some types include orthographic, isometric, and millimeter squares.

Illustration board. The artist's drawing and illustration board used in preparing drawings for the client's approval are usually of varying surface quality. The high surface is smooth for ink renderings, and the medium and regular surfaces are preferred for pencil and color wash drawings (Fig. 14-7). All patent office drawings are required to be on a 10 × 15 in., three-ply bristol board. Official layout of patent drawings requires specific guide lines, which are in a booklet entitled *A Guide for Patent Draftsmen,* available from the U.S. Government Printing Office.

Reproduction. Many copies of completed drawings are required for various purposes. The engineering team needs sets for development, revision, and reference (Fig. 14-8). The factory

Fig. 14-7 Architectural study prepared on illustration board.

needs copies to manufacture and inspect the parts. Subcontractors, other companies which supply parts, also need copies for manufacturing and inspecting the parts they supply. The purchasing department needs copies to order standard parts not made in the shop, such as bolts, nuts, washers, gaskets, and the like. If the item is assembled in a location other than the shop, copies are required for assembly. The design team also often needs copies of their intermediate drawings during the development stage.

Fast, accurate, and economical reproduction processes have been available for many years. The selection of the proper process to satisfy a particular need is important. Several factors to be considered in making the selection include:

1. Input of the originals—size, weight, color, and opacity of medium used.

2. Output quality—degree of legibility expected, transparent copies, opaque copies, and workable copy (erasable and reproducible).

3. Size of reproduction—enlarged copy, reduced copy, same-size copy, and copy that is automatically folded.

4. Speed of reproduction—number of copies needed, single copy on demand, multiple copies, and speed of copying machine.

5. Color of reproduction—blue line and blueprint.

6. Cost of reproduction—material, operational costs, and overhead.

The variety of copying equipment available today has expanded to include blueprint, diazo, electrostatic, thermographic, and photographic processing of engineering drawings (Fig. 14-9).

Blueprint. This is an inexpensive light-sensitive contact process which can produce a copy the same size as the original. The copy is like a negative, in that a white image is produced on a blue background. This is the oldest copying process. It requires the prints to be water-washed after light exposure, treated with a potash (potassium dichromate) solution to intensify the blue, and rewashed before drying. It is a continuous, automated process which leaves only trimming to be done with hand shears.

Diazo. Sometimes the diazo process is called the *whiteprint*. The diazo

Fig. 14-8 Reproduction copies for the engineering team. (*Bruning Division, Addressograph Multigraph Corp.*)

print has an opposite (positive) look to the blueprint, that is, a dark image produced on a white background. It is made from a drawing having dark lines on a transparent medium. Diazo printing is available in a few colors, including blueline, black line, and red line. The paper, cloth, or film stock is coated with a light-sensitive emulsion which can be developed by a dry process or a semimoist process.

Diazo dry process. The print paper is passed through a light source when it is in contact with the original transparent drawing (Fig. 14-10). After the drawing and print paper have been exposed to light, the dye coating of the paper remains where light did not pass through the original drawing. The print with the remaining dye coating is passed through an ammonia vapor where the line image is developed. The original drawing is returned without passing through the developer.

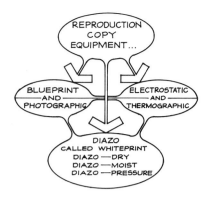

Fig. 14-9 Kinds of reproduction processes.

Diazo moist process. The print paper is coated with a light-sensitive emulsion which is developed with a liquid solution that merely dampens the paper. The contact of drawing and print paper to light is similar to the dry process, but developing takes place when the print paper is fed through fine-groove rollers where a film of solution develops the image. The original drawing does not pass through the developer (Fig. 14-11).

Diazo prints are produced on paper, cloth and film. The developing speed can be regulated to fast, medium, or slow, according to the coating applied to the print material. Regulation of the speed is necessary to produce prints with good contrast. The cloth and films are available in matte finish (free from gloss) and are reworkable on either side. These are used as alternate originals or for changing design schemes. The pressure diazo is a revolutionary new white printing process which is dry, odorless, and developed with a dry activator (Fig. 14-12).

Electrostatic reproduction. Xerography is a form of dry-copy process which uses an electrostatic force to deposit dry powder to produce black images on copy paper (Fig. 14-13). Xerographic prints are typically posi-

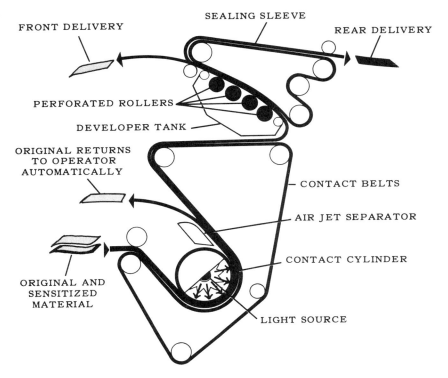

FRONT DELIVERY

SEALING SLEEVE

REAR DELIVERY

PERFORATED ROLLERS

DEVELOPER TANK

ORIGINAL RETURNS
TO OPERATOR
AUTOMATICALLY

CONTACT BELTS

AIR JET SEPARATOR

CONTACT CYLINDER

ORIGINAL AND
SENSITIZED
MATERIAL

LIGHT SOURCE

Fig. 14–10 Diazo dry-process flow diagram. (*Bruning Division, Addressograph Multigraph Corp.*)

Fig. 14–11 Diazo moist-process flow diagram. (*Bruning Division, Addressograph Multigraph Corp.*)

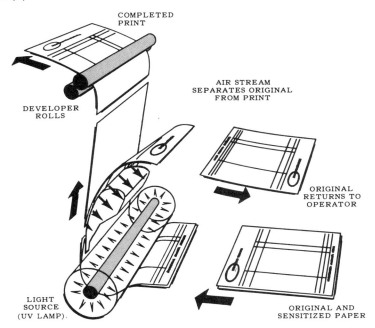

COMPLETED
PRINT

AIR STREAM
SEPARATES ORIGINAL
FROM PRINT

DEVELOPER
ROLLS

ORIGINAL
RETURNS TO
OPERATOR

LIGHT
SOURCE
(UV LAMP)

ORIGINAL AND
SENSITIZED PAPER

tive prints formed by an electrostatically charged selenium-coated plate. When light strikes the charged surface, the charge is dissipated; but where there is an opaque condition caused by lines and lettering, the charge is retained on the surface.

The exposed surface has a developer of magnetic powder which produces the image. Copy paper is then passed to the image, and electrical discharge occurs which transfers the image from the plate to the paper. The image is fused to the copy paper with heat.

Electrostatic transfer can enlarge, reduce, or produce the same-size copy. Long-distance transmission and printout of graphic originals, such as an engineering drawing, is available with translated electronic signals. These signals can be controlled by computer and transmitted by microwave radio, or telephone. The newest reproduction processes are electrostatic, and copies from translucencies to photographs are now available.

Thermographic reproduction. Thermofax is a trade name of the original manufacturer of thermographic copying machines. This kind of copying operates on the principle that a surface with a dark substance will absorb more heat than a surface with a light substance. The print paper is sensitive to heat instead of light. The original engineering drawing therefore does not have to be on transparent paper. Carbon or metallic markings capable of being heated by infrared light are essential to the process. Developers are not used. Dry processing allows copies in color or on translucent sheets.

Photographic reproduction. The major film suppliers have developed ways to assist in reproducing engineering drawings. Copy cameras (Fig. 14–14) create images on light-sensitive material (silver halide coatings on film). The images are developed per-

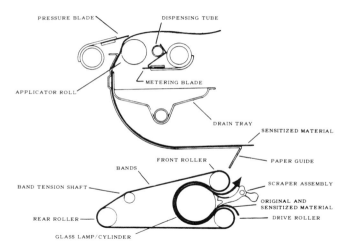

Fig. 14-12 Pressure diazo-process flow diagram. (*Bruning Division, Addressograph Multigraph Corp.*)

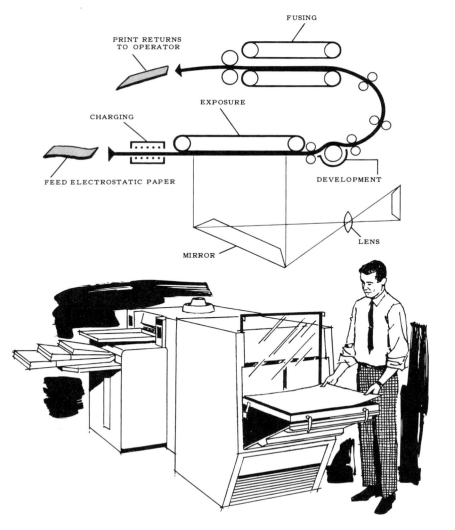

manently with solutions and fixing techniques. Reduction and enlargement of images are easily regulated. High-contrast films are used to provide uniform line widths and blackness (Fig. 14–15).

Film for reproduction produces better prints than the commonly accepted diazo process. The film is produced with a matte finish, and additions can be made with pen and ink or pencil. Intermediates, discussed in Chapter 15, are the most typical use of photographic film in reproduction.

Microfilm. This form of film provides an excellent engineering record system. The greatest advantage of microfilm is in the handling of engineering information. Film systems for records may include various microfilms (Fig. 14–16). Roll film, jacketed film, micro-opaque microfilm, and aperture cards are all commonly used for engineering records.

Fig. 14–13 Electrostatic reproduction machine and flow diagram. (*Bruning Division, Addressograph Multigraph Corp.*)

Fig. 14–14 Copy camera for photographic reproduction. (*Keuffel & Esser Co.*)

Rollfilm is available in 16 mm, 35 mm, 70 mm, and 105 mm widths. Scanning devices for study and copying are available from reproduction manufacturers.

Jacketed film is ideal for filing systems which are reviewed regularly and need frequent changes. The micro-opaque film is mounted on opaque card stock and is used with readers which can magnify its positive form.

Microfiche is a sheet of transparent film that contains many microimages arranged in rows (Fig. 14–17). Some of the sizes for storage are 3 × 5 in., 4 × 6 in., and 5 × 8 in. Copies can be printed after selecting the proper microimage on a screen reader. The reproduction may be variable in size.

Fig. 14–15 This rendering has been photographically reproduced. While being reduced in size, clarity of line has not been lost. (*E. Zavada, photo.*)

Fig. 14–16 A microfilm system includes these elements. (*Bruning Division, Addressograph Multigraph Corp.*)

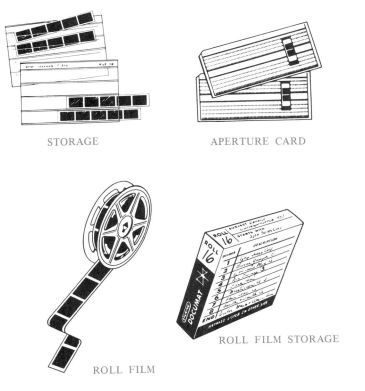

STORAGE

APERTURE CARD

RECORDS FOR FILE

ROLL FILM

ROLL FILM STORAGE

MICROFILM ROLL INDEX

Department: *National Sales Office*

Class of Records: *General Correspondence*

File Period	Subject	Roll No.	Divider
4·95·75	*Ajax Sales Corp.*	16	1
4·95·75	*Burns, Joseph L.*	16	2
4·95·74	*Donahue Mfg. Co.*	16	3
4·95·71	*Key Transportation Co.*	16	4
4·95·72	*Lark, Inc.*	16	5

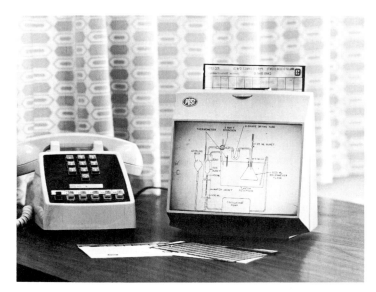

Fig. 14-17 Microfiche in a screen reader. (*Teledyne Post.*)

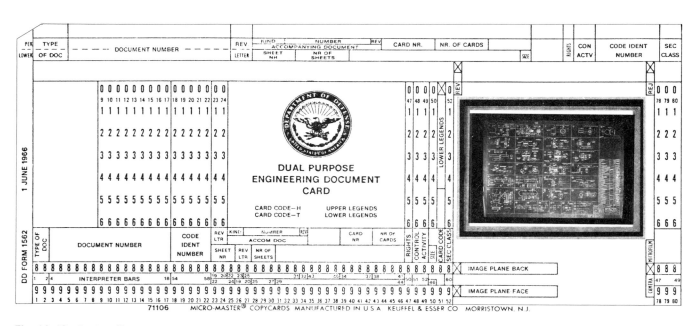

Fig. 14-18 A microfilm aperture card. (*Keuffel & Esser Co.*)

Aperture card. An aperture card is a standard-size data file card having a rectangular slot which can hold a framed microfilm (Fig. 14-18). The most common card is $3\frac{1}{4} \times 7\frac{3}{8}$ in., and data is punched in the card except in the area near the film. This data card can provide engineering information relative to the film and assist in its storage. Cards can be sorted, filed, and retrieved automatically, and reproductions of the microimage may be developed to a desired size needed for a particular use.

Drafting Materials and Systems. The draftsman as a member of the engineering team is responsible for selecting the appropriate drawing media for the level of communication relative to the assignment undertaken. Technology has improved the quality of older media and has introduced new media, which gives the draftsman a wider and better choice than he ever had. This has increased the ease of producing graphic communications. This in turn is complemented by the development of sophisticated reproduction equipment which can produce many types of copies through mechanized systems. Instant storage, retrieval, copying, and folding machines have taken the normal office routines out of the modern engineering office. New developments will continue to take place in all areas, and the draftsman will constantly be learning to use the new tools of his profession.

Review

1. Name the media most appropriate for each of the four levels of communication.

2. What are the preferred uses for opaque paper?

3. Name five good qualities of vellum and one disadvantage of its use for drafting.

4. Tracing cloth (has/does not have) some qualities which are better than paper.

5. Name some advantages of using drafting film.

6. List several factors which affect the selection of a reproduction process.

7. What is the difference between the copy made on a blueprint machine and on a diazo machine?

8. List several capabilities of electrostatic copiers which are not common to other reproduction processes.

9. Microfilm is not a good storage system for original drawings because the image is too small to see properly. True or false?

10. List reasons why a large company would consider installing a microfilm storage and retrieval system in its engineering department.

Systems for Graphic Communications

Chapter 15

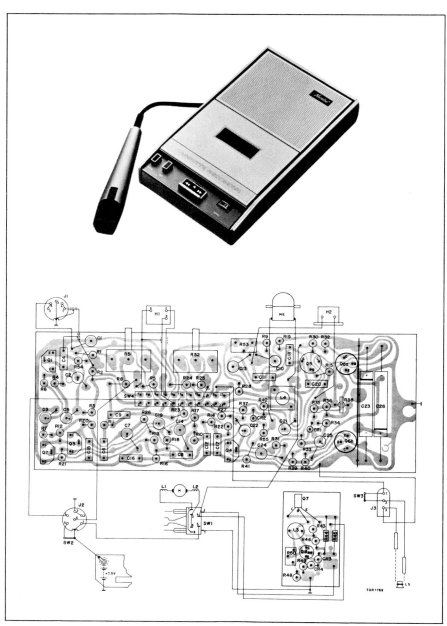

(NORTH AMERICAN PHILIPS CORPORATION.)

Organizing Graphic Communications. The drafting room is generally the processing center for all the drawings and reports that are prepared by the engineering team. The two systems employed in processing design projects are written and graphic.

System one: Written communication. The system used for controlling graphic forms of communication is generally a written form of communication. These are considered technical reports and fall into five categories:

1. *Design report,* which includes the stated needs of the client and the preliminary analysis of the project by designers.

2. *Technical report,* which is a design analysis of all factors affecting the research, design, and development of a project.

3. *Market report,* which is a formal report including recommendations issued before a decision is reached to proceed with manufacturing or construction (also called client report).

4. *Contract report,* which includes all the detailed construction documents or manufacturing procedures after an agreement to proceed has been reached.

5. *Progress report,* which is a periodic check to insure quality and a review of contract specifications. Usually it is submitted at specified time intervals or as predetermined stages of completion are reached.

These reports are essential to the success of a design team and effectively complement the graphic communications which are necessary to design, sell, start, and finish a project (Fig. 15–1).

System two: Graphic communications. The four levels of graphic communication described in Chapter 1 are used repeatedly as design teams work together on design projects (Fig. 15–2).

Level one: creative communication. This is also termed *the birth of an idea.* Its form is usually the designer's sketches which represent new ideas to be evaluated.

Level two: technical communication. Technical personnel combining their knowledge of materials and methods graphically present solutions which at times are difficult to describe verbally. Sketches, schematic diagrams, and graphs can do this job.

Level three: market communications. The refined technical study is illustrated for the client to evaluate style, form, and function. Pictorials, cutaways, and exploded views can help the client make a decision.

Level four: construction communication. Detailed drawings which are prepared for contracts, estimating, construction, or manufacturing. They must be complete and accurate. They are often complemented by written specifications.

The levels of graphic communication produced by the design team are related to the organization of the team. Major industries organize design teams to solve problems which will contribute to the advancement of our technology and help raise our standard of living. All the consumer products, as well as countless services in demand today, are available as the result of the engineering design team

Fig. 15–1 System one: Written communications.

at work. The members of this team include the scientist, engineer, technician, craftsman, designer, and draftsman (Fig. 15–3).

The Engineering Team. A large architectural design firm and an automobile manufacturer are two good examples of industries that require engineering teams. Those responsible for managing large firms must use the services of the design team to meet the needs of clients or future customers. The services of the team must be scheduled properly so that problems can be economically solved. From the preliminary study to the final working drawing, all progress is scheduled, whether it be for the construction of a skyscraper or the manufacture of a sports car (Fig. 15–4). The manager of a design project may develop a flowchart that helps him regulate the design process.

Organizing the Levels of Drafting Work. The organization chart shown in Fig. 15–5 illustrates an established procedure for processing work through the drafting room. The design process is dependent upon the principles of mechanical drawing at all levels. It is this fundamental technical skill which is used to develop any design project. The draftsman can assist the design team at all four levels of communication.

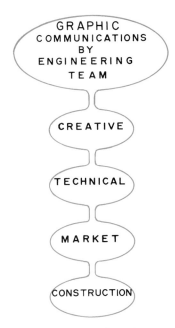

Fig. 15–2 System two: Graphic communications.

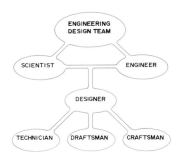

Fig. 15–3 The graphics communication team.

Level one: The layout draftsman assists the designer by working closely with the design sketches, transforming them into mechanical drawings.

Level two: The contribution of the senior draftsman is through his experience, which will aid his judgement as sketches, layouts, and diagrams are studied for design decisions and recommendations.

Level three: The draftsman-illustrator generally assists in the presen-

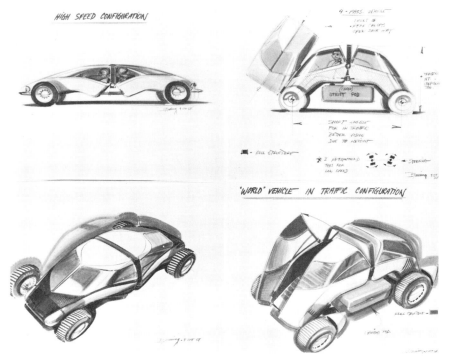

HIGH SPEED CONFIGURATION

"WORLD" VEHICLE IN TRAFFIC CONFIGURATION

Fig. 15–4 A preliminary graphic study. (*General Motors Corporation Design Staff.*)

Fig. 15–5 Design team manager and levels of the design team. (*Teledyne Post.*)

tation studies. He is a draftsman who specializes in pictorial views and renderings developed for client interpretation and approval.

Level four: The senior and junior detail draftsmen prepare the completed construction and manufacturing drawings from the layout work, under the supervision of the chief draftsman. The final details are closely checked, often by a senior detailer assigned to the job of drawing checker.

System for Drafting Procedures. As the design team develops a project which requires drafting services, a record is maintained of the progress under a contract or job number. The numbers will be required information on all reports and drawings. The numbers assigned become a part of the drawing-title-block information.

Title Blocks. The basic title block recommended for industrial drawings is shown in Fig. 15–6. The drawing-number block and the drawing title are necessary for cataloging and cross indexing. The date, scale, signatures, initials, tolerance-information, material-specification, and revision blocks are essential for interpretation and manufacturing.

Revision Block. Changes on a drawing are recorded in the revision block, which is usually located in the lower right-hand corner of the drawing sheet. Alphabetical or numerical symbols are used to identify the revised portion of the drawing. They are essential for finding the change which occured and for identifying the person who authorized the change.

Basic Sheet Sizes. Title blocks and revision blocks may be slightly different in size and location, depending upon the size of the drawing sheet used. Design teams in all industries use standard sized drawing sheets which are available in multiples of 8½″ by 11″, or 9″ by 12″. The several sizes are shown in Fig. 15–7 and are identified by the letters A, B, C, D, and E.

Checking Report. An important part of the procedures system is the checking of working drawings to ensure accuracy. The checking is usually done by an experienced detailer who reports to the chief draftsman. The checker reviews the work on a print, marking errors for correction or approval as appropriate. The job is important because all the detail work must conform to standard drafting practices.

Progress Report. Drafting departments record the progress of the team assigned to detail a project. The individual draftsman may be required to report his progress either daily or weekly. The report will record the job contract number and the number of hours spent working on individual drawings. Engineering departments are generally allotted an estimated time for a given project, and therefore, the reports are important for keeping the job within the estimated time and for the assignment of draftsmen to specific drawings.

Drafting Standards. A major industry or a large company will generally establish a system for processing construction or manufacturing drawings. Some of the large manufacturing companies publish standard drafting practices which assist their own production as well as the suppliers and users who are associated with them. Engineering teams from International Harvester, Caterpillar

Fig. 15–6 The basic title block. *(Camco.)*

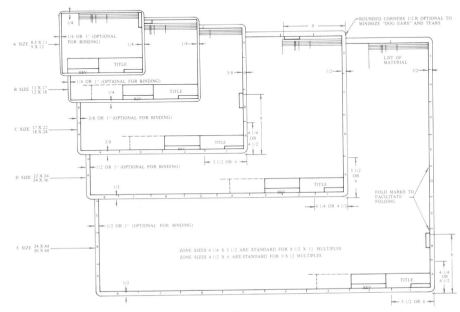

Fig. 15–7 Basic drawing sheet sizes. *(ANSI.)*

Tractor, and International Business Machines are just a few of the major companies which have standards that are published.

Early in the twentieth century, five American engineering societies formed an organization to coordinate the development of national standards for engineering practices. The American Standards Association (ASA) has served American industry for many years. In 1966 the ASA was reorganized as the United States of America Standards Institute and used the symbol USASI. Today this leading organization is called the American National Standards Institute, Incorporated (ANSI) as a result of the expansion that took place in October of 1969. There are over 4000 standards which they have approved. There are 19 major sections devoted to Standard Drafting Practices. The scope of each of the major sections was developed by national committees to assist the designer and draftsman with preferred drafting practices. The Y14 series of standards is of particular interest to designers, architects, and draftsmen. A list of the Y14 series titles appears as a chart on the following page.

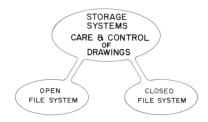

Fig. 15–8 Systems for storage.

1. Develop a closed catalog system that requires expert handling.

2. If original drawings have to be revised or changed extensively, handle them carefully and work off a reproduced copy if possible. An erasable intermediate is preferred. This is a print of the original drawing made on an erasable drawing paper. This paper is different from ordinary blueprint paper in that other prints can be made from it, while prints cannot be made from other blueprints.

3. The original drawing should be used only for making prints. The original drawing should never be used as a desk reference.

4. Flat storage of original drawings is preferred. Originals should not be folded because creases will show up as lines on blueprints.

5. Cleanliness in the drafting room is essential. Vellum and paper stock can absorb moisture, and stains can result; therefore, clean, dry hands and clean working surfaces are essential.

A good drafting-room system requires easy access to the drafting files, but regulated control is essential.

The drafting standards by ANSI are necessary because of the broad scope of industrial work performed across the country. Management must consider standards important in achieving the best communications possible. The Society for Automotive Engineers (SAE) is a well-established organization which has also compiled standards that complement the design and drafting profession. The tremendous number of drawings produced in the drafting room each day are not only standardized to meet a national practice, but they are stored, filed, or controlled according to specific industrial standards.

Care and Control of Drawings. A system is generally developed for cataloging and storing original drawings (Fig. 15–8). Large companies keep the original drawings for a specific contract grouped together under a job or contract number. By doing so, they have easy access to the drawing for making additional prints or for revision.

In some companies a *closed catalog system* is a part of the standards department. In this system the number of persons who are allowed to handle original drawings is rigidly controlled. Responsibility for the safekeeping of the drawings is given to one person. That person may have a staff of people responsible to him who handle the documents for such purposes as filing originals and making prints when necessary. Other companies favor an *open file system* within the engineering department where the original drawings are always readily available. The open file system is generally not a desirable system because easy access to the original drawings by anyone in the department can lead to mishandling. Since the original drawings are valuable documents, and the planning and drawing time for producing such documents is costly, a company could incur needless expense by having to retrace or redraw original drawings.

Some guidelines for efficient and safe storage of originals follow:

Typical Drawing Storage. Original drawings are stored in vertical files, as illustrated in Fig. 15–9, or in horizontal file drawers, as shown in Fig. 15–10. The vertical files offer a letter-file convenience in filing drawings. All important documents should be maintained flat and smooth

Fig. 15-9 Vertical file storage with a capacity of 3000 to 5000 drawings. (*Ulrick Plan File.*)

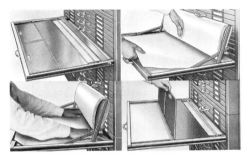

Fig. 15-10 Horizontal file storage. (*Bruning Division, Addressograph Multigraph Corp.*)

and protected from water and dust. A large volume of small drawings from size A ($8\frac{1}{2}'' \times 11''$), B ($12'' \times 18''$), and C ($18'' \times 24''$) are conveniently stored in vertical drawer files.

Folding the prints. The originals are not to be folded, but the prints filed for reference are folded with the top edge creased so that other prints will not be caught in the open fold. Figure 15-11 illustrates the ANSI system for folding B-, C-, D-, and E-size prints. Large industrial centers will use a print-folding machine, as illustrated in Fig. 15-12. When prints of current work are stored in the draft-

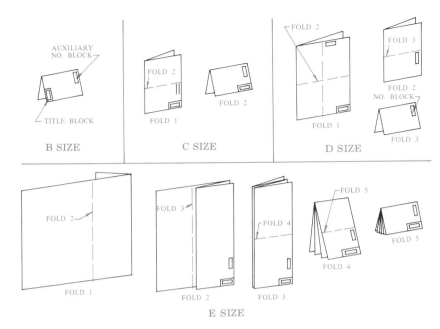

B SIZE   C SIZE   D SIZE

E SIZE

Fig. 15-11 ANSI print-folding system.

Fig. 15-12 This machine will reproduce, reduce, fold, and sort prints. (*Xerox Corp.*)

Fig. 15–13 A quick-access filing system. (*Plan Hold Corp.*)

Fig. 15–14 Microfilm reader. (*Teledyne Post.*)

ing room for quick reference, various systems are used. Figure 15–13 illustrates a compact quick-access filing system. It is portable and is sometimes called a roll filing system. The way to eliminate the storage of folded prints is to microfilm drawings. Drawings must be of a high quality to be suitable for microfilming. The many reference details and supply catalogs for parts and equipment are also available on microfilm. Major industries use reader printers like the one shown in Fig. 15–14 for design reference information.

Microfilming and data control. The typical storage of drawings is space consuming and requires great care and expense in handling. Microfilming was introduced to the drafting room for convenient storage and reproduction. The National Microfilm Association (NMA) has been established to promote standards in the field of micrographics in the United States. They have coordinated their efforts for graphic standards with the American National Standards Institute (ANSI).

Microfilming Process. The photographic systems employed in graphics vary, but 16mm and 35mm are the most popular sizes. Film is also available in 70mm and 105mm. The film must be high contrast, high resolution, and fine grained to be effective for reproducing drawings. Microfilm requires a minimum storage area. The data card which holds the microfilm can be retrieved automatically from storage. The automatic industrial office equipment used to store and reproduce prints from aperture cards is relatively expensive. The card may be used for printing a copy, or be used in a Microfilm Reader-Printer for study. Figure 15–15 shows the flow of the microfilm process. The reproduction from microfilm can be scheduled for full size, half size, or original intermediates by using projection film or paper. Half sizes are popular because the handling is easier, and paper costs are cut in half.

Microfilming Techniques for Quality Microproductions. Good drafting is the basis of good microfilm images and high-quality reproductions from microfilm. Microfilm standards have been a prime factor in producing excellent draftsmanship which is recognized by clear, clean line work, well-selected and well-positioned views, along with legible dimensions. The National Microfilm Association published Monograph No. 3, Modern Drafting Techniques for Quality Microreproductions, which is an educational guideline recommended for the preparation of drawings for microreproduction.

The major concerns in the preparation of high-quality original drawings for microfilming are listed as follows:

1. The selection of standard size drawing media of high-quality vellum or film with good surface texture.

2. The selection of opaque pencil

leads or ink that are proper for the medium, able to produce effective line contrasts, and yet can be erased easily.

3. Lettering which is legible, such as Microfont, a letter style established by NMA. The rules for height and spacing of letters are essential, as shown in Fig. 15-16.

4. The use of functional drafting with good spacing of views and a minimum of unnecessary detail.

These microfilm standards are important to industrial progress. Excellent camera work will complement only good draftsmanship. The reproduction on microfilm cannot improve the original drawing. Several systems have been developed which allow the preparation of secondary originals through economical techniques.

Systems of Preparing Intermediates. An intermediate is a duplicate production master of a drawing. It is often called a secondary original. The designer and draftsman can modify or revise the intermediate while leaving the original alone. The use of intermediates eliminates tracing from or redrawing on originals when modifications are required. The seven methods that follow are only the start of ways to save time on professional drafting assignments.

1. *Scissors drafting.* With scissors, knife, or razor blade, the unwanted sections of an intermediate are removed (Fig. 15-17). This is called editing, and then a new print or a second generation intermediate is produced. Additional data can then be drawn on this new original.

2. *Correction fluids.* After the intermediate is made from the original, unwanted data is removed by applying corrector fluid. New data can be drawn in the eradicated area. Prints

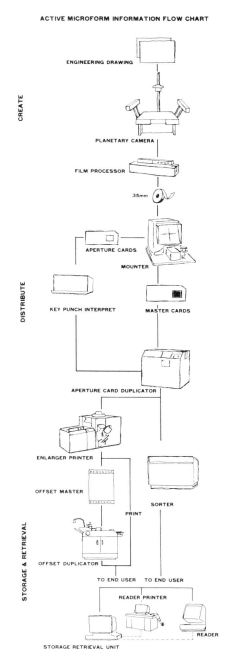

ACTIVE MICROFORM INFORMATION FLOW CHART

ENGINEERING DRAWING

CREATE

PLANETARY CAMERA

FILM PROCESSOR

35mm

APERTURE CARDS

MOUNTER

DISTRIBUTE

KEY PUNCH INTERPRET

MASTER CARDS

APERTURE CARD DUPLICATOR

ENLARGER PRINTER

OFFSET MASTER

PRINT

SORTER

STORAGE & RETRIEVAL

OFFSET DUPLICATOR

TO END USER   TO END USER

READER PRINTER

READER

STORAGE RETRIEVAL UNIT

Fig. 15-15 The microfilm process. (*Bruning Division, Addressograph Multigraph Corp.*)

Unit	Comments	Not This	Possible Error
B	Upper part small, but not too small.	B	8
H	Bar above center line.	H	———
M	Center portion extends below center line. Slight slant on uprights.	M or M	———
S	Lower part large, ends open. Slight angle on center bar.	S	8
T	Horizontal bar shall be full width of letter "E."	T	7
U	Full width.	V	V
V	Sharp point.	V	U
Z	All lines straight.	Z	2
I	Full height and heavy enough to be identified. No serifs.	1	7
2	Upper section curved with open hook. Bottom line straight.	2	8 or Z
3	Upper portion same as lower. Never flat on top.	3 or 3	8 or 5
4	Body open, ends extended.	4	7 or 9
5	Body large, curve dropped to keep large opening. Top fairly wide.	5 or 5 or 5	6 or 3 or S
6	Large body, stem curved and open.	6	8
8	Lower part larger than upper, full and round to avoid blur.	8 or 8	B
9	Large body, stem curved but open.	9	8

MICROFONT
ABCDEFGHIJKLMNOPQR
STUVWXYZ 1234567890

Fig. 15-16 Microfont, a lettering style developed by NMA. (*National Microfilm Association.*)

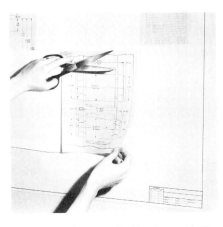

Fig. 15-17 Scissors drafting is used in the intermediate systems. (*Teledyne Post.*)

are produced from the corrected intermediates (Fig. 15-18).

3. *Erasable intermediates.* This kind of copy can be erased with ease. Corrections can be made without fluid. When redrawing, pencil or ink works equally well. The original is retained, and time is saved (Fig. 15-19).

4. *Block-out, or masking.* This system is recommended when time for

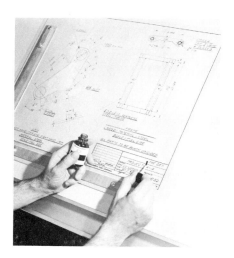

Fig. 15–18 Using correction fluids on an intermediate. (*Teledyne Post.*)

Fig. 15–19 Using an erasing machine on an intermediate. (*Teledyne Post.*)

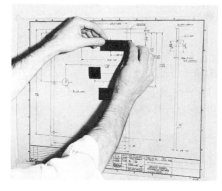

Fig. 15–20 Blocking out an area on an intermediate. (*Teledyne Post.*)

rework is limited. The area to be blocked out is covered with ink or pencil or opaque tape. The intermediate alteration can also be masked, and when unmasked after running through the white printer's ultraviolet light source, the exposed portion will burn out. Various techniques can be worked out, depending on the function of the new master intermediate (Fig. 15–20).

5. *Transparent tape.* Additional notes can be applied to paper, cloth or film, and intermediates and then taped in place. Dimensions, notes, or other data symbols can be added by this method. Varitype can be used to add notes to matte transparent press-on material which is then put in place (Fig. 15–21).

6. *Composite grouped intermediates.* When several drawings are required on one reproduction master, a composite grouping should be cut from intermediates and taped in the desired locations. A final intermediate is then made (Fig. 15–22).

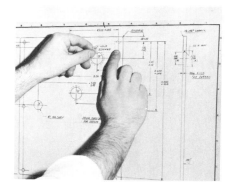

Fig. 15–21 Using transparent matte tape on an intermediate. (*Teledyne Post.*)

7. *Composite overlays.* Translucent original drawings can be combined into a composite intermediate by placing the originals in the desired location. The composite is then photographed. Photo positives or negatives can then be drawn upon and the intermediate master rerun for a second copy. Photo drafting is complementing drafting details with pictorials taken on film.

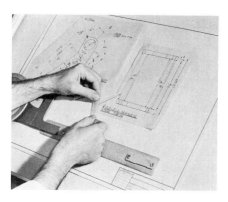

Fig. 15–22 Making a composite intermediate. (*Teledyne Post.*)

The draftsman often must decide which intermediate system described offers the highest quality secondary original at the least expense of time and materials. He must remember that the print reproduction quality of the system chosen must be appropriate to the task at hand. Along with intermediate systems affecting the reduced expense of the draftsman's time and materials, the com-

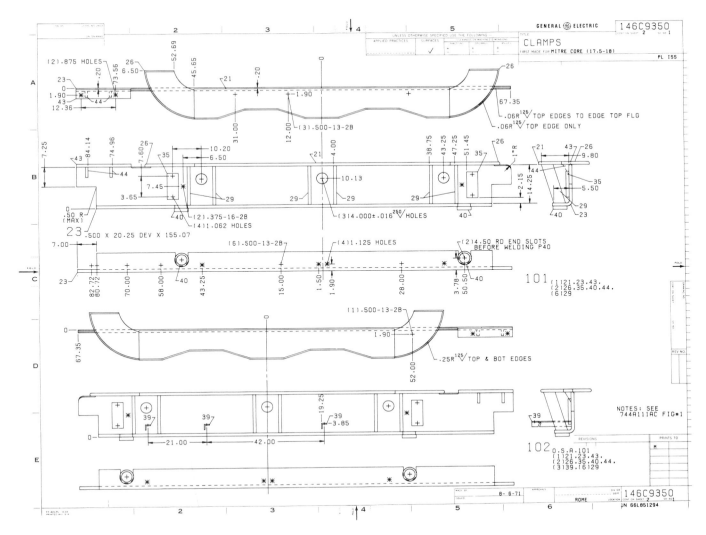

Fig. 15–23 A computer-aided drafting drawing. *(California Computer Products, Inc.)*

puter has become a working tool of the design team.

Computer-aided Graphics Systems. The automated drafting process is a digital plotting system which operates at a high speed. It delivers drawings from digital data obtained from a computer (Fig. 15–23). The computer is programmed to produce data from punched cards, from

punched tape, or from an alphanumeric keyboard. Computer-aided drafting (CAD) has the ability to transform XY coordinates into accurate multiview drawings. Using mathematical formulas and stored programs, CAD can also produce pictorial drawings including one-, two-, or three-point perspectives to any height or angle desired. Computer-aided drafting uses a programmed

drafting language as input. The drafting language consists of letters and numbers which are entered into the system to produce digital data. The corresponding drawing is produced on an incremental digital plotter (Fig. 15–24). The plotter may be either a drum or a flatbed type, and tracing paper, graph paper, vellum, or film can be used for plotting the drawings. The drum plotter is a pre-

Fig. 15–24 A flatbed plotter. (*California Computer Products, Inc.*)

cision instrument having a rotating drum as well as a traveling stylus mounted on a cross bar. The flatbed plotter uses a traveling beam on a carriage above the flatbed. The incremental motors that control the XY coordinate plotting move from two and one-half thousands of an inch per increment to five thousands of inch on slower plotters. The drawing head, or stylus, can use ball-point pens, capillary pens, or scribers. Figure 15–25 illustrates a pictorial drawing produced on an automated drafting machine.

The new systems for computer-aided drafting will assist in design and drafting routines. Professional advancements such as this enhance the role of the entire engineering design team. Draftsmen will be assured of

Fig. 15–25 A pictorial drawing done through computer-aided drafting. (*McDonnell Douglas.*)

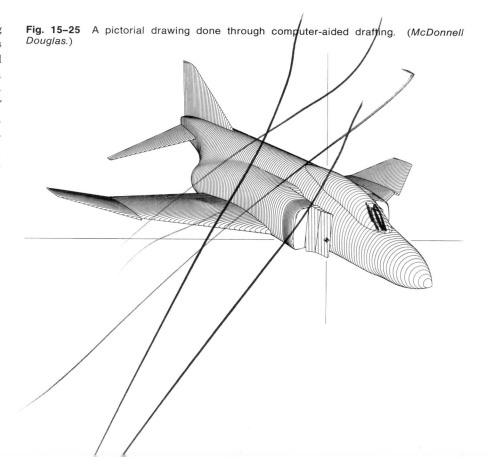

more work because of this advancement, but at the same time they may require a good education in science and mathematics. Figure 15–26 illustrates the cathode-ray tube (CRT) and the graphic plotting of images on the screen.

The systems discussed in this chapter are effectively aiding the design team. As new systems emerge, each member of the design team will be affected by technological change.

Fig. 15–26 A cathode-ray tube. (*IBM.*)

Review

1. Why are drafting standards important?

2. If a draftsman's job is to make drawings, why should he be required to provide information for various reports?

3. List several reasons why a draftsman should not check his own drawing.

4. Why is the proper care and control of drawings important?

5. List advantages and disadvantages of using intermediates.

6. The use of computers in drafting will reduce the number of jobs available for draftsmen. Agree/Disagree? Why?

Functional Drafting

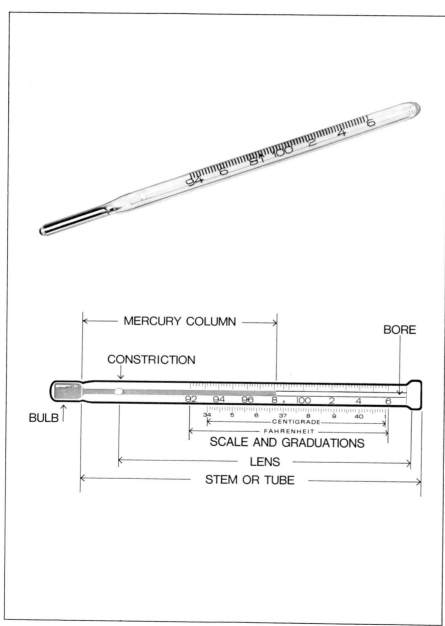

Chapter 16

MERCURY COLUMN

BORE

CONSTRICTION

BULB

92 94 96 8   100 2   4   6

34   5   6   37   8   9   40   1
CENTIGRADE
FAHRENHEIT

SCALE AND GRADUATIONS

LENS

STEM OR TUBE

(BECTON, DICKINSON AND COMPANY.)

The Meaning of Functional Drafting. The term *functional drafting* is not commonly used by draftsmen. However, the practice of functional drafting has been widely used for many years. *Simplified drafting,* a companion term, is more commonly used and is very often confused with the term functional drafting. Functional drafting is a family name for a variety of special drafting practices including simplified drafting. In other words, simplified drafting is one part of a larger category called functional drafting.

Drafting has been described as a means of communication through a graphic language. Every technical drawing has a particular function to perform. That function should be performed in the most effective way. It must be clear, have only one meaning, and be easily read and understood. In order to communicate properly, it must include only the necessary lines, views, notes, and dimensions to give a clear and complete description. *It must not leave anything to chance.*

Definition. *Functional drafting is a graphic means of describing exactly what is meant to be described, using the fewest views, lines, and details which will provide all necessary information in a way that can be easily read and understood by the user.* It is a clearly stated graphic description in the most direct and easily understood form.

Functional means practical, useful, and exactly suited to a purpose. A functional drawing must always be made with particular attention to the user. It suggests efficiency in the use of conventional drafting practices.

Classes of Functional Drawing. There are three types of functional drawings.

Class 1: In-company or local working drawings. These can be simplified in ways peculiar to that company's product, as the employees are familiar with the product. A minimum of detail and information is required. It must be remembered that such drawings, while functional for the company where used, would have little use in another company or for an unrelated product.

Class 2: For a field of industry or engineering. Various fields such as machine tool, aeronautical, automotive, electrical, etc. In a given industry, drawings can use simplification common to that industry. The various industries have developed simplified drawing practices which serve their own requirements. Some of these practices serve more than one industry and by modification have become general.

Class 3: General functional drawing. This is a common graphic language which is the basis for all uses of drawings. The *American National Standards Institute* is the accepted authority for the developing of functional drawings in general use and for certain fields of engineering.

Simplified Drafting. In order to save time and make technical drawing more functional, simplified drafting techniques have been developed. These shortcuts in conventional practices have been developed over many years. Early technical drawings were

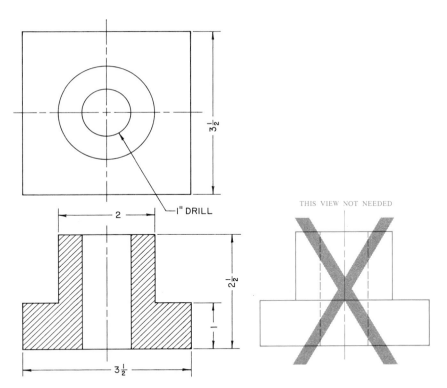

Fig. 16-1 The right-side view is unnecessary.

highly detailed works of art. They usually had more views than were necessary and were carefully shaded to give the views an almost photographic appearance. The lettering was fancy and notes were more wordy than necessary. In other words, drawings took a great deal of time to prepare, and attention to detail made them difficult to read and interpret.

As time went on, an increase in the number and complexity of technical drawings made it necessary to simplify the conventional methods being used at that time. The use of shading, unnecessary views, certain details, and special treatments were eliminated. Drawings have been made easier to read and understand by the use of simpler representation and symbols for familiar features. The two main reasons for simplification are to make a drawing more functional and to save drawing time.

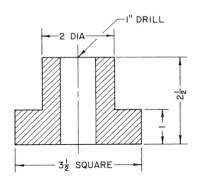

Fig. 16-2 A complete description can be given on one view.

Unnecessary views. Figure 16-1 is a three-view drawing of a flange. Since the front and right-side views are identical in shape and each shows exactly the same detail, it is obvious that one can be eliminated. In other words, the front and top views give as much information about the flange as do all three views. Figure 16-2 shows

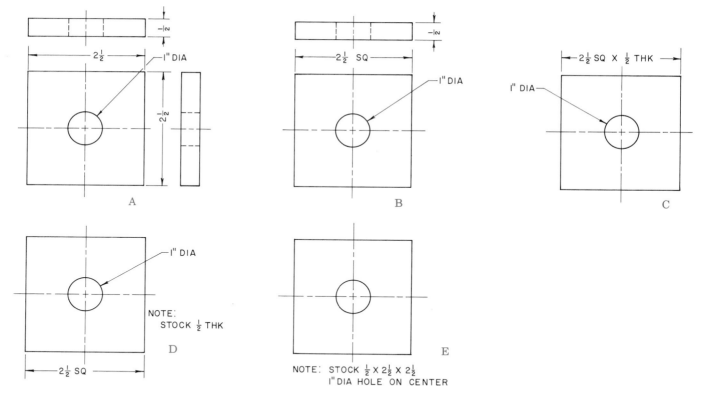

Fig. 16-3 Drawing of a flange in conventional and functional forms.

the same object with another view eliminated. Notice how the part has been dimensioned so that the complete size description is contained in the single view.

One view is usually sufficient to describe parts of uniform thickness. Figure 16-3 shows three views of a spacer block at A. Either the top or the right view may immediately be eliminated since their shapes are identical (Fig. 16-3 at B). Since the thickness is uniform throughout, the top view may also be eliminated and the thickness given in a dimension, as shown in Fig. 16-3 at C, or in a note, as in Fig. 16-3 at D. The same drawing could be further simplified as shown in Fig. 16-3 at E.

Fig. 16-4 The circular view is not necessary. Omit it.

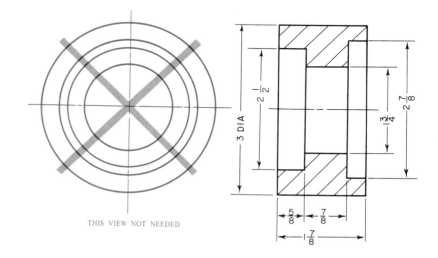

THIS VIEW NOT NEEDED

The circular view of many cylindrical parts does not aid in the description. In many cases, the principle view and a note will result in a more functional drawing (Fig. 16-4). Place your hand or a piece of paper over the circular view and notice that DIA, the abbreviation for diameter, quickly gives the same information as the circular view. Unnecessary views take the user's time to find out that it is not needed.

The same principle applies to a greater extent where two views serve better than three views. When a third view does not tell anything more than two views, do not draw it. Notice that the top view in Fig. 16-5 adds nothing to a clear description of the part.

Unnecessary detail. A great deal of detail can be eliminated by using notes to describe the size and shape of holes which are to be drilled, reamed, countersunk, etc. The location of the hole or holes can be shown by center lines, and information about the holes can be given in a note (Fig. 16-6). On a complex drawing this technique will save the draftsman a considerable amount of time.

When it is necessary to draw repeating detail, it may be drawn once and indicated by a note in repeated positions (Fig. 16-7 at A). Shown here is a simplified screw-thread symbol (not standard) which can be drawn quickly and which keeps an external thread, B, separate from an internal thread, C. Other simplified thread techniques are described in Chapter 10.

Other simplified techniques. Other ways of simplifying drawings to make them more functional have become regular practice and have been described in other chapters of this book. For example, the wide spacing for section lining, or just short lines

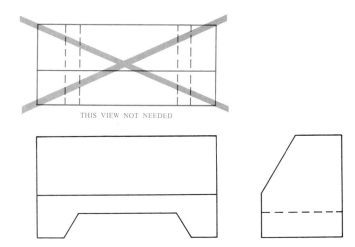

Fig. 16-5 In this case two views are better than three.

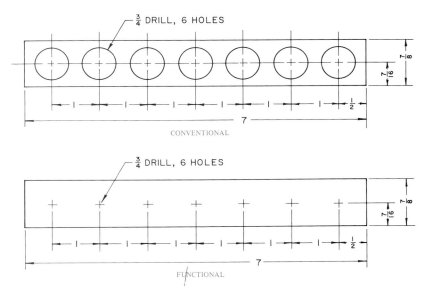

Fig. 16-6 Holes need not be drawn.

or shading around the outline, is described in Chapter 9. Examples of these techniques are shown in Fig. 16-8. The use of half views for symmetrical objects is another timesaving practice (Fig. 16-9).

Base-line dimensioning. The base-line system of dimensioning provides for another drafting economy (Fig. 16-10). In this case dimension

lines and arrowheads are eliminated. All dimensions are taken from the base lines (marked *0*).

Arrowheads. The conventional use of arrowheads can be time consuming and may add nothing to the clarity of the drawing. Methods ranging from no arrowhead at all to the use of a dot are in common practice (Fig. 16-11).

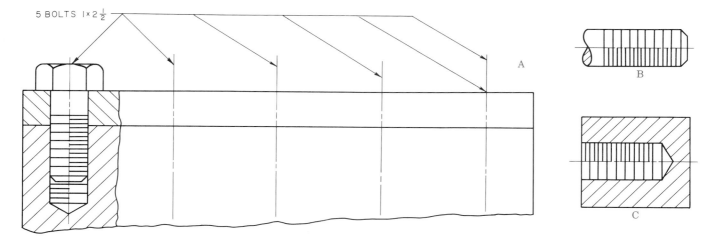

Fig. 16-7 Repeated details need not be drawn.

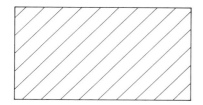

Fig. 16-8 Functional section-lining techniques.

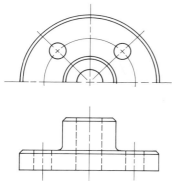

Fig. 16-9 The use of half views saves time and space.

Fig. 16-10 Substitutes for arrowheads.

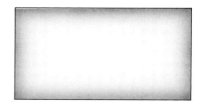

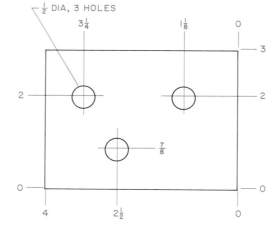

Fig. 16-11 Base-line dimensioning.

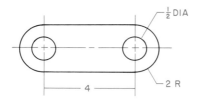

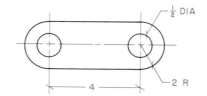

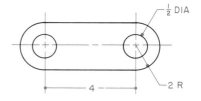

Freehand sketches. In many cases, simple freehand sketches are used as working drawings. This practice may use any or all of the simplified techniques explained earlier in this chapter. The use of sketches is most appropriate for parts of a less complex nature and when an exact scale is not necessary.

In-company standards. Many companies have their own standards for simplified practice. A few examples from Brown & Sharpe Manufacturing Company are shown in Fig. 16-12. This illustration shows the improvement by comparison. Some of the practices have come into wide use since they are general in character. When and where and how to use simplified drawings are three questions which require experience and thought to be sure of meeting the requirements of functional drawings.

Timesaving symbols. Nearly all fields of industry and engineering have various types of symbols common to that industry. For example, special symbols are widely used in architectural and structural drafting (Chapters 20 and 21), electrical and electronics drafting (Chapter 24), aerospace drafting (Chapter 25), welding drafting (Chapter 17), surface development (Chapter 18), etc. The use of symbols has become so widely accepted by industry that they have become standard practice through the American National Standards Institute (ANSI). Manuals regarding any of the above are available through the American National Standards Institute, Inc., 1430 Broadway, New York, NY 10018.

Templates. The development of a great variety of drafting templates has made the draftsman's job easier as well as timesaving. Such details as circles, squares, ellipses, electrical and

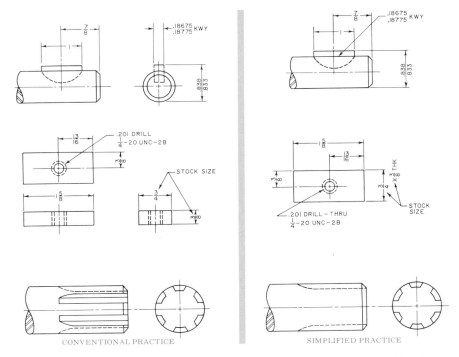

CONVENTIONAL PRACTICE   SIMPLIFIED PRACTICE

Fig. 16-12 Samples of in-company simplified practices. (*Brown & Sharpe Manufacturing Company.*)

electronics symbols, architectural symbols, and many others can very often be drawn faster and more accurately with templates. For example, Fig. 16-13 shows a four-center (approximate) ellipse (shown in red) with an ellipse of the same size (shown in black) drawn with a template. Notice that the template ellipse is a true ellipse, while the four-center ellipse is somewhat less accurate.

Figure 16-14 shows a variety of typical drafting templates. Most templates are made of thin plastic sheets. Many of them have reference marks and other helpful information. These features add to the timesaving and accuracy characteristics of the template.

In addition, many symbol templates are available for use with standard Leroy lettering equipment.

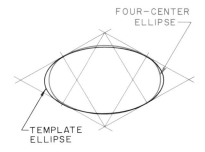

FOUR-CENTER ELLIPSE

TEMPLATE ELLIPSE

Fig. 16-13 Templates save time and may be more accurate than other drawing methods.

Figure 16-15 shows an electronics template in use. Figure 16-16 is an example of an electronics schematic drawing made with the Leroy template. Special typewritten copy has been added to give the values of the various components. Custom templates of the user's own design may be obtained through special order.

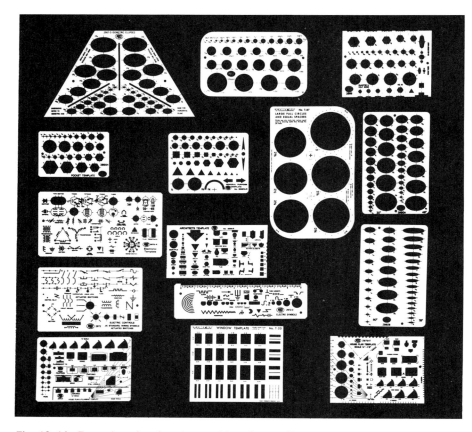

Fig. 16–14 Examples of various types of templates. (*Rapidesign, Inc.*)

Fig. 16–15 Leroy electronic symbol templates.

Pressure-Sensitive Overlays. Pressure-sensitive overlays may be printed or plain on transparent or translucent sheets with an adhesive backing. A throw-away carrier sheet against the adhesive back keeps the sheets from sticking together. The image sheet is peeled away from the carrier sheet, placed in position on the drawing, and rubbed briskly to make it adhere (Fig. 16–17).

Overlay sheets are available in a great variety of standard symbols or patterns (Fig. 16–18) and in blank (unprinted) sheets. A matte surface on the blank sheet will accept typewriter copy as well as pencil or ink lines. This material is often used for making corrections on drawings and for adding materials lists or detailed notes which can be typed faster than they can be hand lettered (Fig. 16–19).

Pressure-sensitive overlays are available in two basic types: cut-out and transfer. Cut-out overlays are applied by positioning the desired image in the correct position on the drawing, burnishing (rubbing) the image area, and cutting around it to remove the portion not wanted (Fig. 16–20). The transfer-type pressure-sensitive overlay works on a somewhat different principle. The carrier is removed from the translucent image sheet, and the area to be transferred is placed in position on the drawing. The image to be transferred is then rubbed over the top surface of the transfer sheet with a burnishing stick or other blunt instrument. When the transfer sheet is lifted, the image remains on the drawing sheet (Fig. 16–21). After the image has been transferred the carrier sheet is placed over the image and reburnished.

Pressure-sensitive overlays are not limited to sheets of material. Various patterns and colors of tape are also a part of the family (Fig. 16–22).

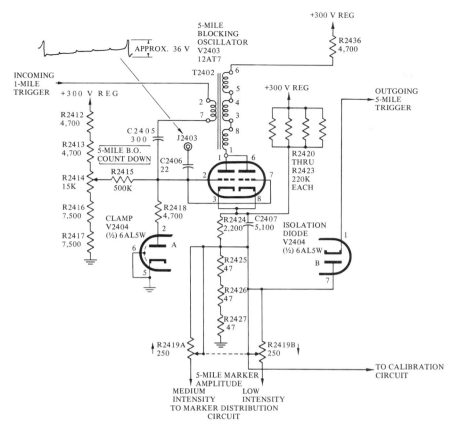

Fig. 16-16 Schematic drawing made with Leroy template.

Fig. 16-18 Examples of standard printed symbols and patterns available on pressure-sensitive overlays. (*Graphic Products Corp.*)

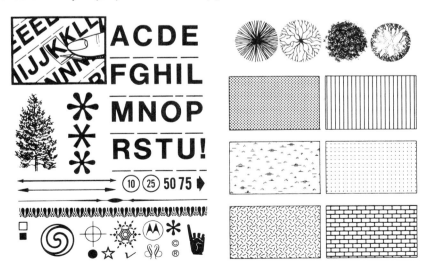

Fig. 16-17 The application of pressure-sensitive material. (*Artype, Inc.*)

Fig. 16-19 Type or draw an image on the sheet, cut out, and position on the drawing. (*Para-Tone, Inc.*)

Fig. 16-20 Method of applying cut-out-type pressure-sensitive overlays. (*Chartpak, A Division of Avery Products Corp.*)

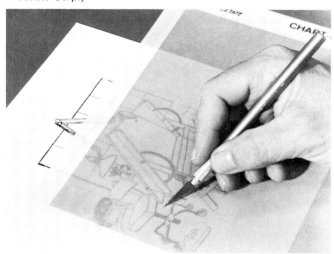

A. SCORE A SECTION SLIGHTLY LARGER THAN REQUIRED.

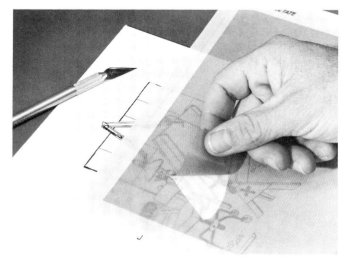

B. SLIDE POINT OF KNIFE UNDER FILM AND LIFT OUT.

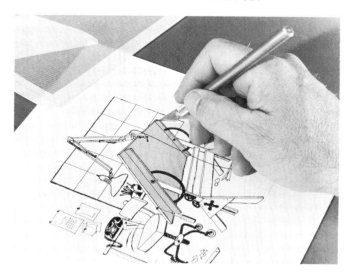

C. POSITION AND PRESS FIRMLY.

D. USING A SHARP KNIFE, TRIM AND PEEL OFF SURPLUS.

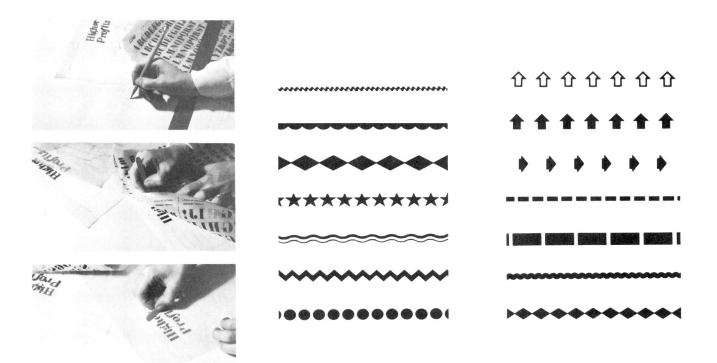

Fig. 16-21 Application of transfer-type pressure-sensitive overlay. (*Tactype, Inc.*)

Fig. 16-22 Some examples of pressure-sensitive tapes. These are available in a variety of colors and patterns. (*Chartpak, A Division of Avery Products Corp.*)

These tapes may be used for borders, standard lines, charts and graphs, and many other graphic applications. They may be applied by hand or with a tool called a *tape-pen* (Fig. 16–23).

Photo Drawing. Another type of functional drafting is *photo drawing*. This technique involves the use of photographs in various forms as an engineering shortcut. Photo drawings are generally used in one of two ways: (1) as a quick and inexpensive means of showing revisions, additions, or deletions to an existing apparatus, device, or machine, and (2) in the development of a product form the model or mock-up stage to the final product.

If an existing device or apparatus is to be altered, perhaps the most economical means of communicating and recording the desired change is to make a photograph and mark the changes on it (Fig. 16–24).

A photographic paper print may be used if only one copy is needed. However, it is often necessary to reproduce the photograph for filing and distribution purposes. In this case a special print, called a *reproducible intermediate* is made.

A reproducible intermediate is any medium (material) from which prints can be made. It is usually made by a photographic process on a transparent or translucent film. Lines can be added with a soft-lead pencil or drawing ink. In addition, the unwanted portion of the image can be removed with a chemical bleach or with a wet eraser. After alterations are made, conventional prints can be made in any quantity.

Engineers and other designers sometimes work with scale models and mock-ups during the preliminary design stage. Photographs may be used as a timesaving step in advancing the design. Minor changes in the shape and size of parts may be made on a photograph, thus eliminating the need for constructing several design models.

Scissors Drafting. Scissors drafting is a method of preparing a finished

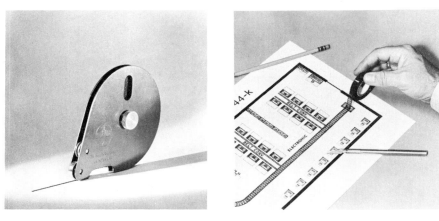

Fig. 16–23 Pressure-sensitive tape may be applied with a tape-pen or by hand. (*Chart-pak, A Division of Avery Products Corp.*)

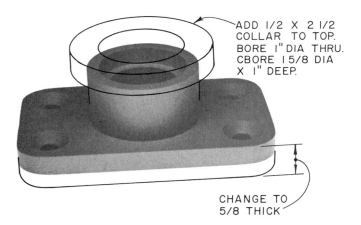

ADD 1/2 X 2 1/2 COLLAR TO TOP. BORE 1" DIA THRU. CBORE 1 5/8 DIA X 1" DEEP.

CHANGE TO 5/8 THICK

Fig. 16–24 A design change may be shown on a photograph.

working drawing by physically combining two or more existing drawings. In other words, the draftsman may have to draw some parts of an object but may add other parts by cutting and pasting parts from an existing drawing. A great deal of time can be saved by using this technique.

The advantage in scissors drafting is that it is not limited to the use of transparent or translucent materials. Catalog illustrations or pictures from magazines, books, or other printed material may be used if the proper copying equipment is available. Various pictures and detailed drawings are simply taped or pasted to an appropriate size sheet of white paper and reproduced by any copying equipment which uses reflected light. If the paste-up is made by combining drawings made on translucent material (tracing paper, cloth, or film), copies are made by using conventional methods such as blueprinting or the diazo method.

The paste-up may be made in a variety of ways. Rubber cement, paste, or other adhesive may be used to fasten an opaque sheet to the drawing. Since a reflected-light reproduction technique is necessary with opaque materials, the adhesive will not affect the final copy. If the components are on transparent or translucent material, they should be fastened together with a transparent tape. The best method is to place the drawing of the part to be added under the main drawing in the proper location, cut through both sheets with a knife or razor blade, remove the top cutout, and tape the inserted piece in place with transparent tape. It is usually better to place the tape on the underside of the drawing. This technique may also be used to remove details. A blank sheet is inserted and taped in place after a detail has been cut out.

Lettering. Functional lettering is a part of functional drawing, and this requires the use of one or more of the methods of lettering according to the means available and the purpose for which the lettering is to be used. Freehand lettering is the most widely used, but templates, guides and scribers, printed and typed adhesives, and special typewriters are used also to help make functional drawings. Long notes and parts lists can be typed to save the draftsman's time.

Review

1. Pressure-sensitive overlays are available in two basic types. Name them.

2. Name the functional drafting technique that involves the use of photographs.

3. What is the drafting term applied to the technique of preparing a drawing by cutting and pasting two or more existing drawings?

4. A graphic means of describing exactly what is meant to be described, using the fewest views, lines, and details which will provide all necessary information in a way that can be both easily read and understood by the user is called _____.

5. The elimination of unnecessary views, details, and arrowheads are techniques in _____ drafting practices.

6. Some companies have their own standards for simplified drawing practices. What are these called?

7. What is a reproducible intermediate?

8. Describe two uses for photo drawing.

9. Name the two main reasons for simplifying drawings.

10. Is it always necessary to make at least two views of a cylindrical part?

Problems for Chapter 16

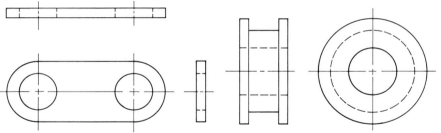

Fig. 16–25 Link. Take dimensions from the printed scale at the bottom of the page, using dividers. Make a working drawing of the link, using functional drafting techniques. Show only essential views and dimensions.

Fig. 16–26 Spool. Take dimensions from the printed scale at the bottom of the page, using dividers. Make a working drawing of the spool, using functional drafting techniques. Show only essential views and dimensions.

Fig. 16–27 V-block. Take dimensions from the printed scale at the bottom of the page, using dividers. Make a working drawing of the V-block, using functional drafting techniques. Eliminate unnecessary views. Dimension.

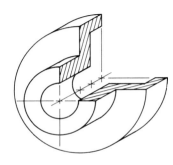

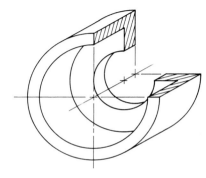

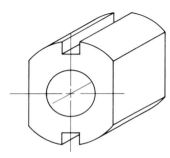

Fig. 16–28 Retainer. Take dimensions from the printed scale at the bottom of the page, using dividers. Make a working drawing of the retainer, using functional drafting techniques. Show only essential views and dimensions.

Fig. 16–29 Cap. Take dimensions from the printed scale at the bottom of the page, using dividers. Make a working drawing of the cap, using functional drafting techniques. Show only essential views and dimensions.

Fig. 16–30 Clutch block. Take dimensions from the bottom of the page, using dividers. Make a working drawing of the clutch block, using functional drafting techniques. Show only essential views and dimensions.

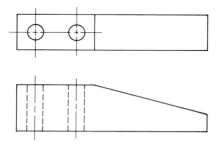

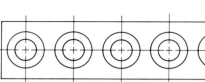

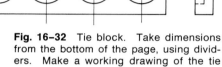

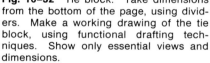

Fig. 16–31 Clamp jaw. Take dimensions from the bottom of the page, using dividers. Make a working drawing of the clamp jaw, using functional drafting techniques. Show only essential views and dimensions.

Fig. 16–32 Tie block. Take dimensions from the bottom of the page, using dividers. Make a working drawing of the tie block, using functional drafting techniques. Show only essential views and dimensions.

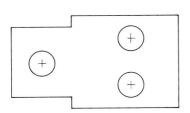

Fig. 16–33 Base plate. Take dimensions from the printed scale at the bottom of the page, using dividers. Make a working drawing of the base plate, using functional drafting techniques. Use base-line dimensioning. Indicate hole sizes in a note.

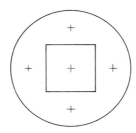

Fig. 16–34 Gasket. Take dimensions from the printed scale at the bottom of the page, using dividers. Make a working drawing of the gasket, using functional drafting techniques. Use base-line dimensioning. Indicate hole sizes in a note.

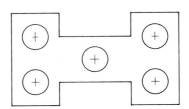

Fig. 16–35 Spacer. Take dimensions from the printed scale at the bottom of the page, using dividers. Make a working drawing of the spacer, using functional drafting techniques. Use base-line dimensioning. Indicate hole sizes in a note.

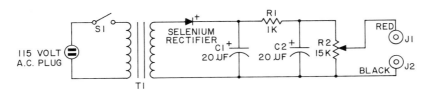

Fig. 16–36 Schematic drawing. Use an electronic schematic drawing symbols template to make a finished drawing of the schematic shown. Use a lettering template for all lettering. Estimate all sizes not given on templates.

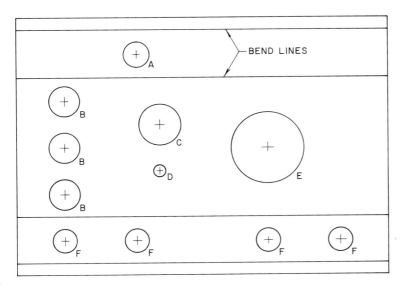

HOLE	DIA	QTY
A		
B		
C		
D		
E		
F		

Fig. 16–37 Chassis layout. Take dimensions from the printed scale at the bottom of the page, using dividers. Make a working drawing of the chassis, using functional drafting techniques. Use base-line dimensioning. Indicate hole sizes and quantities by completing the chart to the right of the chassis layout.

Welding Drafting | Chapter 17

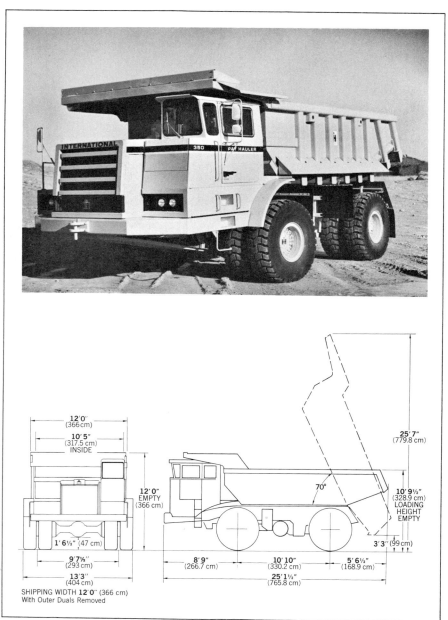

12'0" (366 cm)

10'5" (317.5 cm) INSIDE

12'0" EMPTY (366 cm)

1'6½" (47 cm)

9'7½" (293 cm)

13'3" (404 cm)

SHIPPING WIDTH **12'0"** (366 cm) With Outer Duals Removed

25'7" (779.8 cm)

70°

10'9½" (328.9 cm) LOADING HEIGHT EMPTY

3'3" (99 cm)

8'9" (266.7 cm)

10'10" (330.2 cm)

5'6½" (168.9 cm)

25'1½" (765.8 cm)

(INTERNATIONAL HARVESTER COMPANY.)

Joining Metals. Welding is an important industrial process used for joining machine parts and structural building components. The standard steel shapes, including plates and bars, can be welded together to make machine frames, bases, and mechanisms. Parts that formerly were fabricated by forging or casting can be designed with less weight and increased strength by using steel and welding techniques. A photograph and a drawing of a casting for a pulley housing are shown in Fig. 17–1. Compare the casting with the same part which was made by welding as shown in Fig. 17–2. The aircraft, automotive, shipbuilding, and building construction industries have developed welding as a major fabricating method for steel, aluminum, and magnesium (Fig. 17–3). Welding operations have increased the strength and durability of products along with improving their overall appearance. The art of welding is an ancient one. The prehistoric record of mankind reveals forged and welded jewelry, such as bracelets and rings. Welded materials are joined by heat or pressure or a combination of heat and pressure. The two *basic* processes are called *fusion* welding and *resistance* welding. Fusion welding uses only heat, and resistance welding uses a combination of heat and pressure.

Resistance welding was developed in 1857 by James Prescott Joule. Industry began to use resistance welding after electric power became abundant

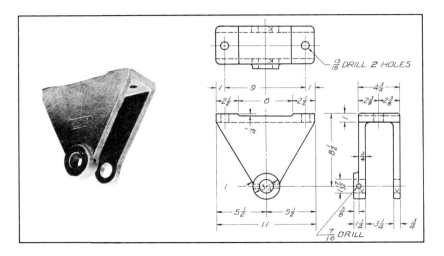

Fig. 17–1 Pulley housing made by casting. (*Wellman Engineering Co. and The Lincoln Electric Co.*)

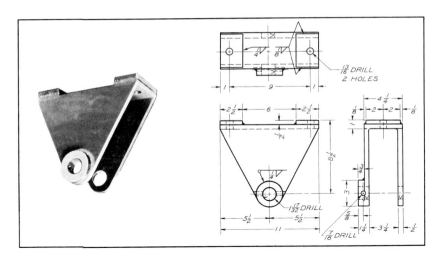

Fig. 17–2 Pulley housing made by welding. (*Wellman Engineering Co. and The Lincoln Electric Co.*)

in 1886. Gas welding (fusion) was effectively developed in 1895, when two gases—oxygen, from liquid air, and acetylene, from calcium carbide—were improved to a commercial quality. Arc welding, another fusion proc-

ess, was first performed in France by De Meritens in 1881.

Welding Processes. Welding processes include fusion, gas, arc, thermit, gas and shielded arc, and resistance

welding. Although called by their separate names, soldering and brazing are also a form of welding.

Fusion welding. Fusion welding uses a welding material in the form of a wire or rod which when heated to a melting point fills in a joint. These filler rods combine with the metal being welded. Gas or a carbon arc is used to create the heat so that the metal flows together.

Gas welding. Gas welding uses a flame of oxygen and acetylene, air and acetylene, or any gas combinations which can obtain welding heat (Fig. 17–4). The combustion temperature of these gases ranges between 5000 and 6500° F.

Arc welding. Arc welding is a process of forming an electric arc between the work and an electrode which will emit heat (Fig. 17–5). The heat in turn fuses a filler rod of suitable material for the work. The arc is produced by a dc generator or other power source. The intense heat brings the filler rod and the work to a melting point only at the tip of the electrode, where fusion or joining occurs.

Thermit welding. Thermit welding is based on the natural chemical reaction of aluminum with oxygen. A mixture, or charge, consisting of finely divided aluminum and iron oxide is ignited by a small quantity of special ignition powder. The very high temperature which results from the rapid burning process forms a molten metal which flows into molds and fuses mating members.

Gas and shielded arc welding. Aluminum, magnesium, low-alloy steels, carbon steels, stainless steel, copper, nickel, Monel, and titanium are some of the metals which can be welded with this kind of welding process. There are two forms of gas and shielded arc welding called *Tungsten-Inert-Gas* (TIG) and *Metallic-Inert-*

Fig. 17–3 Welding building components for Walt Disney World. (*United States Steel.*)

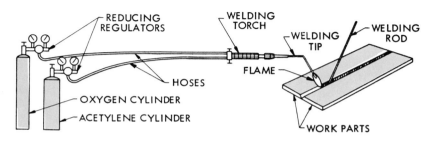

Fig. 17–4 The gas-welding process. (*General Motors Corp.*)

Fig. 17–5 The arc-welding process. (*Republic Steel Corp.*)

Gas (MIG). In TIG welding, a non-consumable tungsten electrode is used to provide the heated arc for welding. It does not provide a filler material, but only the arc for fusion. In MIG welding, a consumable metallic rod is used. It provides a filler material as well as the arc for fusion.

Resistance welding. Some metals can be fused together by heat and pressure. When two pieces of metal have electric current surging through the areas to be joined, the resistance to the charge causes that area of the material to become heated to a plastic state. With the proper electrical current and pressure the weld is completed. When current and pressure are confined to a small area between electrodes, a spot weld is produced.

Welding Drawings. Symbols for drawings have been standardized by the American Welding Society. They provide a means of giving specifications, type, location, and size of a weld. The many combinations available are illustrated in symbolic form

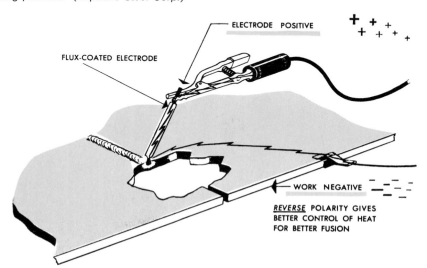

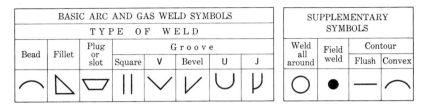

BASIC ARC AND GAS WELD SYMBOLS								SUPPLEMENTARY SYMBOLS			
TYPE OF WELD								Weld all around	Field weld	Contour	
Bead	Fillet	Plug or slot	Groove							Flush	Convex
			Square	V	Bevel	U	J				

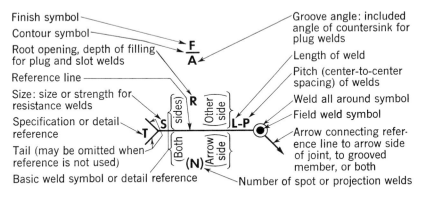

Finish symbol
Contour symbol
Root opening, depth of filling for plug and slot welds
Reference line
Size: size or strength for resistance welds
Specification or detail reference
Tail (may be omitted when reference is not used)
Basic weld symbol or detail reference

Groove angle: included angle of countersink for plug welds
Length of weld
Pitch (center-to-center spacing) of welds
Weld all around symbol
Field weld symbol
Arrow connecting reference line to arrow side of joint, to grooved member, or both
Number of spot or projection welds

Fig. 17–6   Location of welding information on welding symbols.

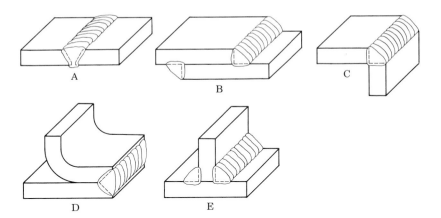

Fig. 17–7   Five basic types of joints.   A. Butt joint; B. lap joint; C. corner joint; D. edge joint; E. T-joint.

in Fig. 17–6. Every drafting room should be equipped with a copy of the latest edition of the American Welding Society Standard Welding Symbols. See the appendix for chart AWS A2.1-68.

Basic Welded Joints. There are five basic joints used in welding (Fig. 17–7). There can be many variations on these five basic joints. The many combinations and varieties of joints are used to meet the great number of

different conditions which occur in welding practice. Knowledge of materials, their conditions, and the possession of practical welding experience are necessary to make the proper selection of welded joints. A few factors to be considered in selecting a weld include the type of material, the equipment to be used, and the cost of preparation.

Basic Types of Welds. Figure 17–8 shows the basic types of grooved welds. Note that they may be in single or double form. These basic types of welds are shown only for the butt joint. They may be applied to all the basic types of joints shown in Fig. 17–7. Typical dimensions for a butt joint with a V-grooved weld are shown in Fig. 17–9. The dimensions for a T-joint with a bevel-grooved weld are shown in Fig. 17–10. The typical dimensions for a U-grooved weld on a butt joint are given in Fig. 17–11. Figure 17–12 shows the dimensions for a J-grooved weld on a T-joint.

Basic Arc and Gas Weld Symbols. Figure 17–13 shows the basic and supplementary welding symbols. Symbols may be selected to describe any desired weld. These symbols may be combined to describe the most simple or the most complicated joints.

The standard location of information on welding symbols is shown in Fig. 17–6. The notes indicate how symbols and data are placed in relation to the reference line. Note that the perpendicular leg of the fillet, bevel, and J-groove weld symbol (Fig. 17–14) is always placed to the left.

The welding symbol identifies the location and the particular type of weld to be used at a joint. The welding symbol may be placed on either side of the joint as space permits. When the type-of-weld part of the

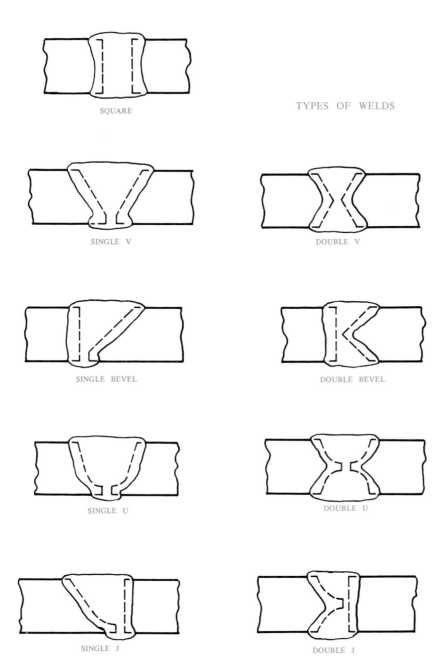

SQUARE

SINGLE V

DOUBLE V

SINGLE BEVEL

DOUBLE BEVEL

SINGLE U

DOUBLE U

SINGLE J

DOUBLE J

Fig. 17-8 Basic types of grooved welds as applied to a butt joint.

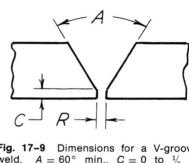

Fig. 17-9 Dimensions for a V-grooved weld. $A = 60°$ min., $C = 0$ to $\frac{1}{8}$ in., $R = \frac{1}{8}$ to $\frac{1}{4}$ in.

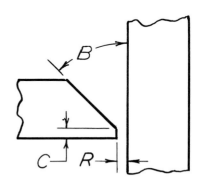

Fig. 17-10 Dimensions for a bevel-grooved weld. $B = 45°$ min., $C = 0$ to $\frac{1}{8}$ in., $R = \frac{1}{8}$ to $\frac{1}{4}$ in.

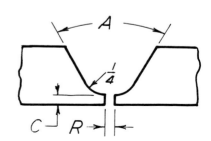

Fig. 17-11 Dimensions for a U-grooved weld. $A = 45°$ min., $C = \frac{1}{16}$ to $\frac{3}{16}$ in., $R = 0$ to $\frac{9}{16}$ in.

Fig. 17-12 Dimensions for a J-grooved weld. $B = 25°$ min., $C = \frac{1}{16}$ to $\frac{3}{16}$ in., $R = 0$ to $\frac{9}{16}$ in.

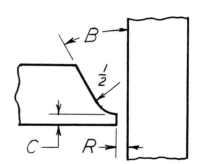

symbol lies below the horizontal line of the welding symbol, it indicates the weld is to be on the *arrow side* of the joint. Figure 17–15 at A and D shows the arrow-side symbol and the resulting weld. When the type-of-weld part of the symbol lies above the horizontal line of the welding symbol, it indicates the weld is to be on the *other side* of the joint. Figure 17–15 at B and E shows the other-side symbol and the resulting weld. If the weld is to be on both sides of a joint, the type of weld part is placed both above and below the horizontal line of the symbol, as at C.

With a J-groove weld the placement of the arrowhead of the welding symbol is important. In Fig. 17–16 at A, the meaning is not clear as to which member is to be grooved. The placement of the arrowhead at B clearly indicates that the vertical member is to be grooved (at C). At D the symbols on the reference line clearly indicate the J-groove weld is on the horizontal member on the arrow side. The fillet weld symbol indicates the fillet weld is to be on the other side. The weld would actually look as indicated at E.

In Fig. 17–6 the tail of the reference line is used for reference dimensions. At A in Fig. 17–17, *A*2 is placed in the tail, which indicates the dimensions described in connection with the weld. This is made clear in Fig. 17–17 at B.

The interpretation of Fig. 17–17B without the welding symbol would indicate that this joint can be described as follows:

A double filleted-welded, partially grooved, double-J-T-joint with incomplete penetration. The J-groove is of standard proportion. The radius (*R*) is ½ in. and the included angle is 20°. The penetration is ¾ in. for the other side and 1¼ in. deep for the arrow side. There is a continuous ⅜-in.

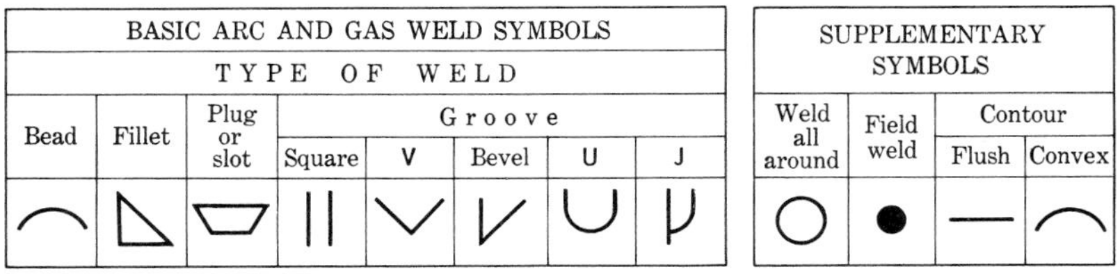

BASIC ARC AND GAS WELD SYMBOLS								SUPPLEMENTARY SYMBOLS			
TYPE OF WELD											
Bead	Fillet	Plug or slot	Groove					Weld all around	Field weld	Contour	
			Square	V	Bevel	U	J			Flush	Convex
⌒	◺	⏢	∥	∨	⌵	∪	∪	◯	●	—	⌒

Fig. 17–13 Basic and supplementary arc- and gas-weld symbols.

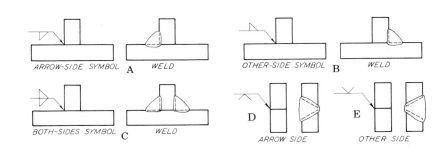

Fig. 17–14 The perpendicular leg on the weld symbol is always drawn to the left.

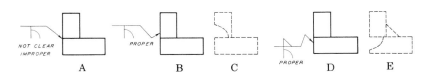

ARROW-SIDE SYMBOL A WELD   OTHER-SIDE SYMBOL B WELD

BOTH-SIDES SYMBOL C WELD   D ARROW SIDE   E OTHER SIDE

Fig. 17–15 Arrow side and other side.

NOT CLEAR IMPROPER   PROPER   PROPER
A   B   C   D   E

Fig. 17–16 J-groove weld indications.

Fig. 17–17 Reference specifications applied.

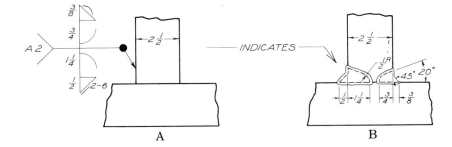

A   INDICATES   B

fillet weld on the other side, with a
½-in. fillet weld on the arrow side.
The fillet on the arrow side is 2 in.
long, and the pitch of 6 in. indicates
that it is spaced 6 in. center to center. All fillet welds are standard at
45°.

Supplementary Symbols. The
black, solid, dot on the reference line
at the elbow in Fig. 17–17 at A indicates that this weld is to be made in
the field or on the site of construction
rather than in the shop.

The typical specification *A2* placed
in the tail can be explained and expanded as follows. The work is to be
a metal-arc process, using a high-
grade, covered, mild-steel electrode;
the root is to be unchipped and the
welds unpeened, but the joint is to be
preheated.

In Fig. 17–17 at A the flush symbol
over the ½ fillet weld symbol indicates
that the contour shall be flat-faced
and unfinished. The ⅜ fillet weld on
the same reference line indicates that
the weld is to be finished to a convex
contour, since the convex-contour
symbol is added to the right of the
fillet symbol. Figure 17–13 indicates
supplementary symbols for the finished welding techniques employed in
Fig. 17–17 at A.

Typical Groove Welds. Figure 17–18
illustrates graphically the five typical
groove welds as applied to butt
joints. At A is a square groove joint,
at B is a squared both-sides joint, at
C is a V-groove joint, at D is a bevel-
groove joint, and at E is a U-groove
joint.

Typical Plug and Slot Welds. Examples of plug and slot welds are
shown in Fig. 17–19. The welding
symbol is the same for both types.

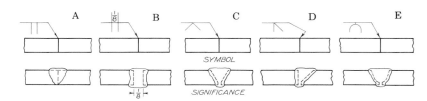

Fig. 17–18 Five typical groove welds.

Basic Resistance Welding. The
welding heat generated by the resistance of a part to the electric current
welds the parts together as pressure
is applied to the joint. Flash welding
occurs when the parts are placed in
very light contact or when a very
small air gap exists; the electric current flashes, or arcs, and melts the
ends of parts.

**Basic Resistance Welding Sym-
bols.** These are given in Fig. 17–20.
The basic reference line and arrow are
used with resistance welds as with arc-
and gas-welding symbols, but in general there is no arrow side or other
side. The same supplementary symbols apply as indicated in Fig. 17–20.
The four basic resistance symbols are
spot, projection, seam, and flash or
upset welds.

Typical Spot Welds. Resistance spot
welding is confined to a relatively
small portion of the area of lapped
parts. Figure 17–21 at A shows a reference symbol in the top view. The
arrow centers on the working center
line. The minimum diameter of each
weld at A is specified at 0.30 in. At
B the minimum shearing strength of
each weld is specified at 800 lb. At
C the reference data indicates that

Fig. 17–19 Plug and slot welds. The symbol shown is applied to both kinds.

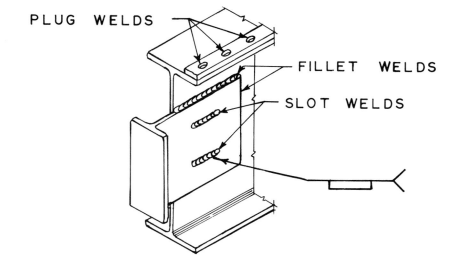

PLUG WELDS

FILLET WELDS

SLOT WELDS

BASIC RESISTANCE WELD SYMBOLS				SUPPLEMENTARY SYMBOLS			
TYPE OF WELD				Weld all around	Field weld	Contour	
Spot	Projection	Seam	Flash or upset			Flush	Convex
◯	◯	⊖	‖	◯	●	─	⌒

Fig. 17–20 Basic resistance-weld symbols.

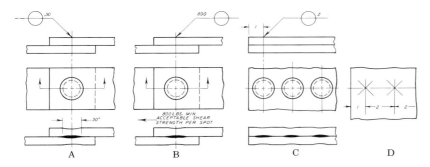

Fig. 17–21 Examples of spot-welding symbols and their meaning.

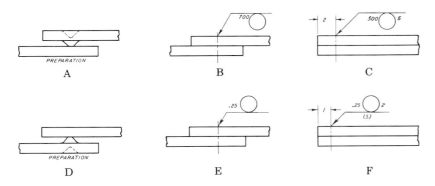

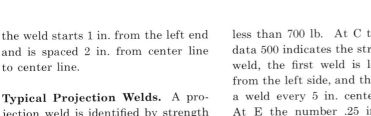

Fig. 17–22 Examples of projection-welding symbols.

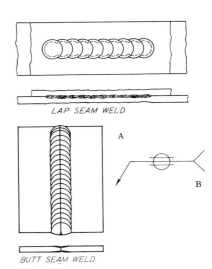

Fig. 17–23 Butt-seam and lap-seam welds and symbol.

the weld starts 1 in. from the left end and is spaced 2 in. from center line to center line.

Typical Projection Welds. A projection weld is identified by strength or size. Figure 17–22 at A and D shows typical preparation of parts with a boss projection. At B the reference 700 means that the acceptable shear strength per weld is to be not less than 700 lb. At C the reference data 500 indicates the strength of the weld, the first weld is located 2 in. from the left side, and the 5 indicates a weld every 5 in. center to center. At E the number .25 indicates the diameter of the weld. At F the diameter of the weld is .25, there is a weld every 2 in. beginning 1 in. in from the left side, and the (5) indicates that there are to be a total of five welds.

Notice the arrow-side and other-side indications.

Typical Seam Welds. Figure 17–23 at A illustrates butt-seam and lap-seam welds. The symbol for seam welding is shown in Figure 17–23 at B. The reference shows the two pieces positioned edge to edge for butt-seam welding and two pieces overlapping for tangent spot welds.

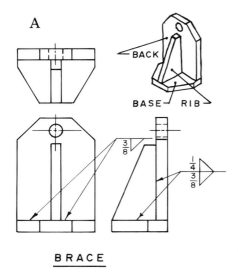

A

BRACE

Welding Symbols. Welding symbols can be used on a machine drawing, as shown in Fig. 17–24 at A, and on a structural drawing, Fig. 17–24B. Welding is required for fabricating steel in the shop. The mobile welding unit makes it possible to develop the framework for large multistory buildings and bridges.

Brazing. Brazing is a fusing process using nonferrous filler rods. It differs from welding in that the base metal is not melted. It is heated only to the extent that the nonferrous material will flow onto the joint. The brazing of small appliances which require silver, copper, or softer alloys can be done with a mobile welding unit.

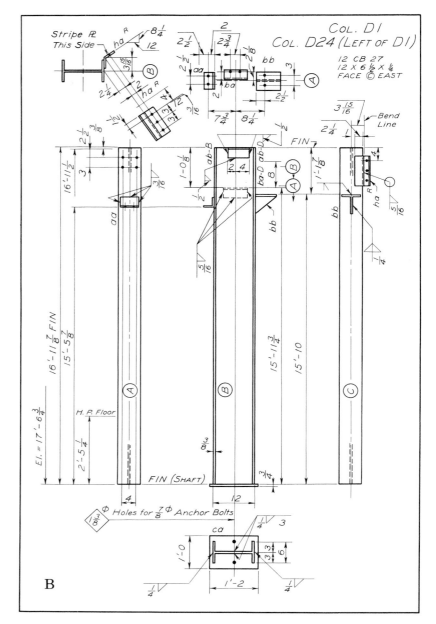

Fig. 17–24 The application of welding symbols. A. On a machine drawing; B. on a structural drawing.

Review

1. Which welding process was developed first, resistance or fusion?

2. The welding process that uses only heat is _____.

3. The welding process that uses heat and pressure in combination is _____.

4. Name the gases that are normally used in the gas welding process.

5. What is the basic principle of arc welding?

6. What are the meanings of MIG and TIG?

7. Name the basic welding joints.

8. Name the basic types of welds.

9. Identify the several parts of the welding symbol.

10. List and explain the meaning of several supplementary symbols.

Problems for Chapter 17

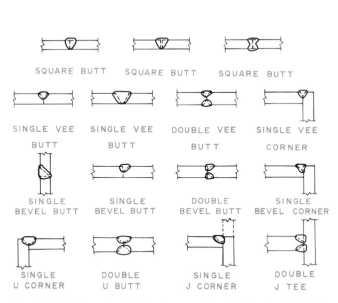

SQUARE BUTT   SQUARE BUTT   SQUARE BUTT

SINGLE VEE   SINGLE VEE   DOUBLE VEE   SINGLE VEE
BUTT   BUTT   BUTT   CORNER

SINGLE   SINGLE   DOUBLE   SINGLE
BEVEL BUTT   BEVEL BUTT   BEVEL BUTT   BEVEL CORNER

SINGLE   DOUBLE   SINGLE   DOUBLE
U CORNER   U BUTT   J CORNER   J TEE

Fig. 17–25 Draw the joints shown, using ½″ stock. Apply the proper weld symbol.

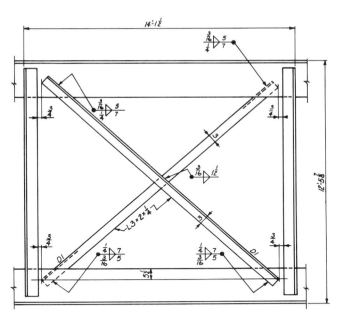

Fig. 17–26 Make a welding drawing of the double angle cross bracing. Draw the structural members at a suitable scale and note that the horizontal structural members are unidentified. Identify and label structural selections to complement this structure.

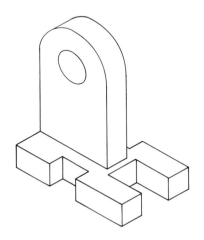

Fig. 17–27 Make a three-view welding drawing of the lever stand, assigning your own dimensions. Design a weld to support the upright member. Include dimensions and welding symbols on your drawing.

Fig. 17–28 Make a three-view welding drawing of the corner lug bracket. Apply weld symbols to indicate permanent assembly. Base = 3″ sq. × ¾″. Uprights = 3″ × 4″ × ½″ and 2″ × 4″ × ½″. Holes = 1″ drill located 1″ from each edge.

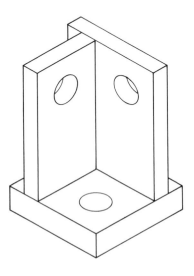

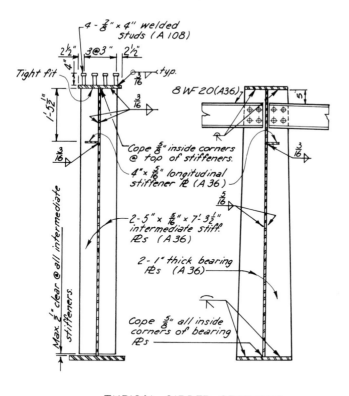

TYPICAL GIRDER SECTIONS
Scale: $\frac{3}{4}$" = 1'-0"

Fig. 17-29 Make a drawing of each section and place the weld symbols in appropriate locations. Identify the type of weld and letter the name close to the symbol.

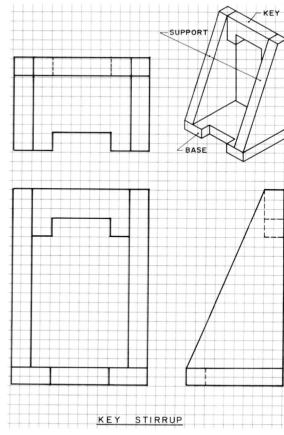

KEY STIRRUP

Fig. 17-31 Make a three-view welding drawing of the key stirrup. Choose appropriate weld symbols to join the four parts. Grid for dimensions = $\frac{1}{4}$".

Fig. 17-30 Develop a three-view drawing of the bearing support, using your own estimated dimensions. Dimension the drawing in decimals or millimeters as assigned. Apply the proper welding symbols to assemble the five parts. Prepare a separate parts list.

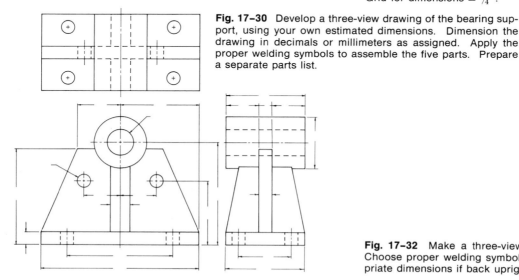

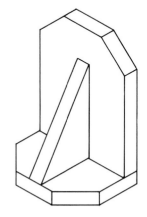

Fig. 17-32 Make a three-view welding drawing of the brace. Choose proper welding symbols for fabrication. Assign appropriate dimensions if back upright = $3\frac{1}{2} \times 4\frac{1}{4} \times \frac{1}{2}$.

Surface Developments and Intersections

Chapter 18

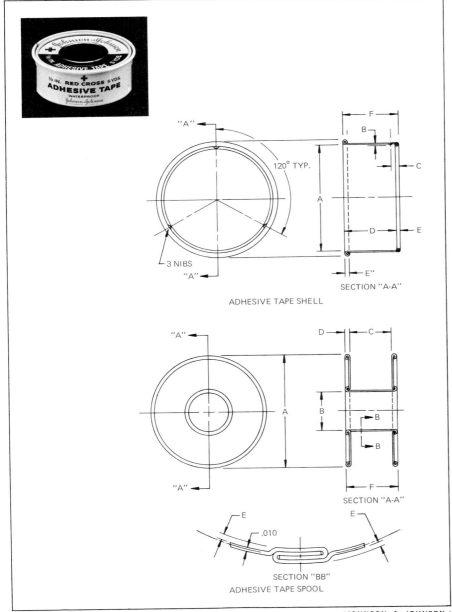

"A"

120° TYP.

3 NIBS

"A"

ADHESIVE TAPE SHELL

F

B

C

A

D

E

E"

SECTION "A-A"

"A"

"A"

A

B

D

C

B

B

B

F

SECTION "A-A"

E

.010

E

SECTION "BB"

ADHESIVE TAPE SPOOL

(JOHNSON & JOHNSON.)

Surface Development. The book cover shown in Fig. 18–1 is an example of a surface development. At A the cover is laid out flat. At B it has been wrapped around a book to serve as a protective covering. Notice that it fits neatly around all surfaces. This is accomplished by carefully measuring and laying out each part in its proper relationship to other parts. The layout is prepared full size and is made on a single flat plane. A surface development is also called a *stretchout,* a *pattern,* or simply a *development.*

Surface development is an important application of technical drawing in a great variety of industries. Some familiar items made from sheet materials include pipes, ducts for hot- or cold-air systems, parts of buildings, aircraft, automobiles, storage tanks, cabinets, office furniture, boxes and cartons, package containers for frozen foods, manufactured products, and countless other items.

Such uses require drawings made to provide patterns on flat sheets of material which can be folded, rolled, or otherwise formed to provide the required shape. Sheet materials used include paper, various cardboards, plastics and films, metals (such as steel, tin, copper, brass, aluminum), wood, fiberboard fabrics, etc.

The Packaging Industry. Packaging has become a very large industry which involves the principles of surface development. It combines

317

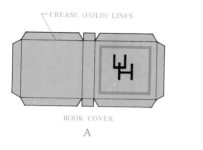

CREASE (FOLD) LINES

BOOK COVER

A

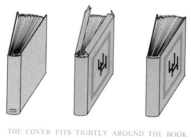

THE COVER FITS TIGHTLY AROUND THE BOOK

B

Fig. 18–1 A book cover is an example of a surface development.

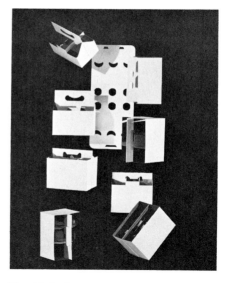

Fig. 18–2 Soft-drink cartons are cut in a flat pattern and then folded. (*Olin Packaging Division, Olin Mathieson Chemical Corporation.*)

engineering and fine arts for the design and manufacture of packages which are attractive and functional. These packages are designed for sales appeal and for protection of the contents in shipment. They may be intended for temporary or for permanent use.

Mass production of packages or containers at a reasonable cost is necessary for most items. The softdrink carrier is a familiar example (Fig. 18–2). It is made of strong kraftboard, white surfaced to permit attractive printing and design.

Flat patterns are printed, cut, creased, folded, glued (if necessary), and completed, all on machines designed especially for the purpose (Figs. 18–3 and 18–4).

Packages and cartons are made of many materials and in many thicknesses. Some are made of thin or medium-thickness paper stock (Figs. 18–5 and 18–6) which can be folded easily into the desired form. Some are designed so that no glue is required, and some may have glue on one or more tabs. The design of the pattern layout for a particular purpose and the lettering, color, and artwork are all a part of an important industry.

Packages made of cardboard, corrugated board, and many other materials require allowances for thickness. Boxes made in two parts, a container and a cover (Fig. 18–7), and a slide-in box (Fig. 18–8), are examples. Observe various cartons and packages to see many interesting and challenging problems in the development of surfaces.

Fig. 18–3 Twenty-up die on cylinder bed, showing makeready on cylinder. Dies are used to cut the sheet material for making packages. (*Olin Packaging Division, Olin Mathieson Chemical Corporation.*)

Fig. 18-4 Phase just prior to glue-lap contact. (*Olin Packaging Division, Olin Mathieson Chemical Corporation.*)

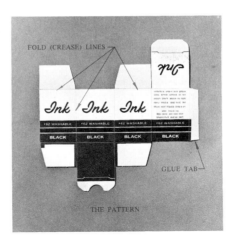

Fig. 18-5 A familiar container made by cutting and folding a flat sheet.

Fig. 18-6 Pattern for one-piece package with fold-down corner.

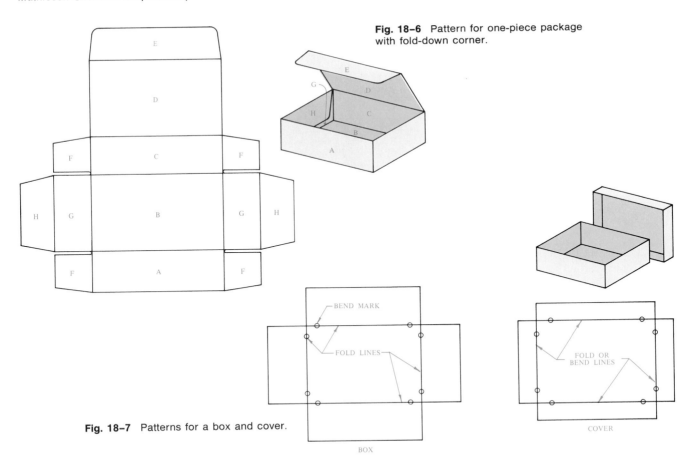

Fig. 18-7 Patterns for a box and cover.

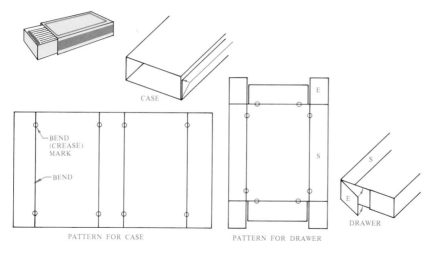

PATTERN FOR CASE

PATTERN FOR DRAWER

Fig. 18-8 A two-part package with a slide-in box.

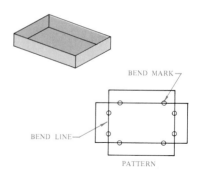

Fig. 18-9 Pictorial drawing and stretchout of a sheet-metal object.

Sheet-metal Pattern Drafting. A large variety of metal objects are made from sheets of metal which are laid out, cut, formed into the required shape, and fastened together. The shaping is done by bending, folding, or rolling. The fastening is done by riveting, seaming, soldering, or welding. Drawings for sheet-metal work consist of a graphic representation of the finished object and a drawing of the shape of the flat sheet which, when rolled or folded and fastened, will form the finished object (Fig. 18-9). This second drawing is called the *development,* or *pattern,* of the piece. Drawing the pattern is called *sheet-metal pattern drafting.*

A great many thin-metal objects without seams are formed by die stamping or by pressing a flat sheet into shape under heavy presses. Examples range from brass cartridge cases and household utensils to steel wheelbarrows and parts of automobiles and aircraft. Other kinds of thin-metal objects are made by spinning, for example, some brass- and aluminumware. In stamped and spun

work, the metal is stretched out of its original shape,

The operations for the preparation of sheet-metal work are performed on machines. These operations include cutting, folding, wiring, forming, turning, beading, and so forth. Some machines and operations are shown in Fig. 18-10. The machines shown are for hand operations. Large complex automatic equipment is generally used in industry for mass production.

Development. There are two general classes of surfaces: plane and curved. The six faces of a cube are plane surfaces. The bases of a cylinder are plane surfaces, whereas the lateral surface is curved (Fig. 18-11). Curved surfaces which can be rolled in contact with a plane surface, such as cylinders and cones, are called single-curved surfaces and can be exactly developed.

Another kind of curved surface is the double-curved surface, such as spheres and spheroids. They cannot be exactly developed; but approximate developments can be made.

It is possible to cut a piece of paper so that it can be folded into a cube, as in Fig. 18-12. The shape cut out is the pattern of the cube. There are five regular solids, and their patterns are made as shown in Fig. 18-13. A good understanding of the nature of developments may be gained by laying out these shapes on rather stiff drawing paper. These shapes can then be cut out and their patterns folded to form the figures. The joints can easily be secured with tape.

Thus, the pattern for any piece that has plane surfaces may be made by drawing each plane in its proper relationship to other planes.

Seams and Laps. While the basis of sheet-metal pattern drafting is development, it is also necessary to know the processes of wiring, seaming, and hemming. In addition, it is necessary to know the amount of material to be added for each. For example, open ends of articles are usually reinforced by enclosing a wire in the edge, as shown at A in Fig. 18-14. The amount added to the pattern may be

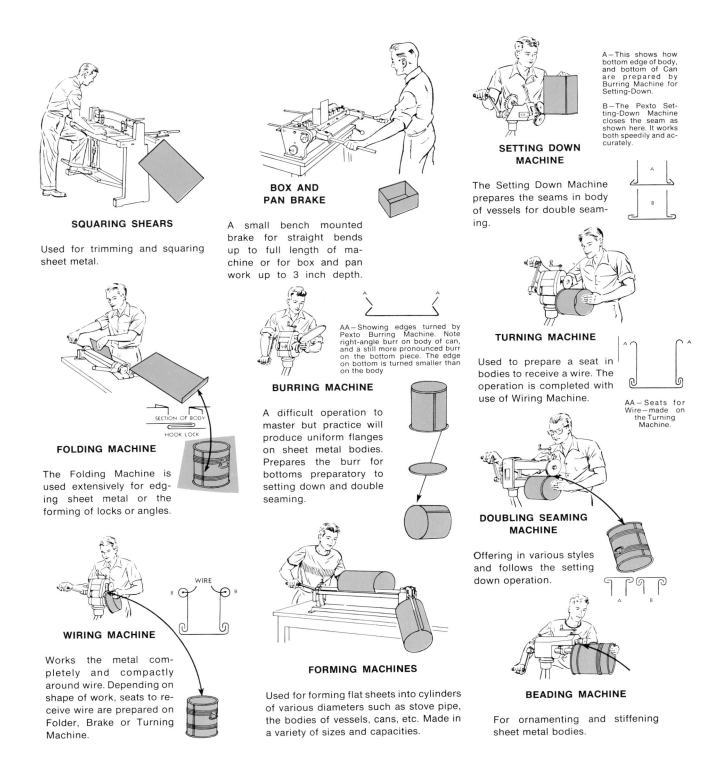

SQUARING SHEARS

Used for trimming and squaring sheet metal.

BOX AND PAN BRAKE

A small bench mounted brake for straight bends up to full length of machine or for box and pan work up to 3 inch depth.

SETTING DOWN MACHINE

The Setting Down Machine prepares the seams in body of vessels for double seaming.

A—This shows how bottom edge of body, and bottom of Can are prepared by Burring Machine for Setting-Down.

B—The Pexto Setting-Down Machine closes the seam as shown here. It works both speedily and accurately.

FOLDING MACHINE

SECTION OF BODY
HOOK LOCK

The Folding Machine is used extensively for edging sheet metal or the forming of locks or angles.

AA—Showing edges turned by Pexto Burring Machine. Note right-angle burr on body of can, and a still more pronounced burr on the bottom piece. The edge on bottom is turned smaller than on the body

BURRING MACHINE

A difficult operation to master but practice will produce uniform flanges on sheet metal bodies. Prepares the burr for bottoms preparatory to setting down and double seaming.

TURNING MACHINE

Used to prepare a seat in bodies to receive a wire. The operation is completed with use of Wiring Machine.

AA—Seats for Wire—made on the Turning Machine.

WIRING MACHINE

WIRE

Works the metal completely and compactly around wire. Depending on shape of work, seats to receive wire are prepared on Folder, Brake or Turning Machine.

DOUBLING SEAMING MACHINE

Offering in various styles and follows the setting down operation.

FORMING MACHINES

Used for forming flat sheets into cylinders of various diameters such as stove pipe, the bodies of vessels, cans, etc. Made in a variety of sizes and capacities.

BEADING MACHINE

For ornamenting and stiffening sheet metal bodies.

Fig. 18–10 Some of the machines used in sheet-metal working. (*The Peck, Stow, and Wilcox Company.*)

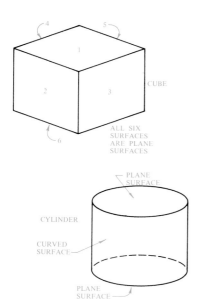

Fig. 18–11 Plane and curved surfaces.

taken as $2\frac{1}{2}$ times the diameter of the wire. Edges may also be stiffened by hemming. Single- and double-hemmed edges are shown at B and C. Edges are fastened by soldering on lap seams (D), flat lock seams (E), or grooved seams (F). Other types of seams and laps are also shown in Fig. 18–14. Each has its own general or specific use. The amount of material

allowed in each case will vary according to the thickness of material, method of fastening, and application. In most cases the corners of the lap are notched in order to produce a neater joint.

Parallel-line Development. In parallel-line development the stretch-out is made by means of parallel lines which represent segments of the object. Figures 18–12 and 18–15 are examples. The development of the rectangular prism shown in Fig. 18–16 is illustrated in Fig. 18–17. The complete drawing is made by following steps A through E.

A. Draw the front and top views full size and label the points as shown (18–17A).

B. Draw the stretchout line SL, and on it lay off distances equal to sides 1-2, 2-3, 3-4, and 4-1 obtained from the top view (Fig. 18–17B).

C. At points 1, 2, 3, 4, and 1 on the stretchout line, draw vertical crease (fold) lines equal in length to the height of the prism (Fig. 18–17C). The crease lines are also referred to as measuring lines.

D. Draw the top line of the stretch-out parallel to SL and projecting from

the top of the front view. Darken all outlines using thick black lines. A small circle or × may be used to identify a fold line (Fig. 18–17D).

E. The top and bottom may be added by transferring distances 1-4 and 2-3 from the top view to the stretchout, as shown in Fig. 18–17E. Laps may be added for assembly.

A slight variation is the truncated prism shown in Fig. 18–18. A front, top, and auxiliary view are first drawn full size and points are labeled as shown. The next two steps are the same as steps B and C in Fig. 18–17. Next, project horizontal lines from points *A-B* and *C-D* on the front view to establish points on the stretchout. Connect the points to complete the top line of the stretchout. Add the top and bottom as shown.

Cylinders. The developed surface of a cylinder is illustrated in Fig. 18–19, where the cylinder has been rolled out on a plane surface.

For a cylinder the stretchout line is a straight line equal in length to the circumference of the cylinder. If the base of the cylinder is perpendicular to the axis, it will roll out into a straight line and form the stretchout line. If the prism or cylinder does not

Fig. 18–12 Pattern for a cube.

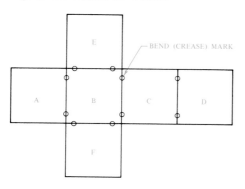

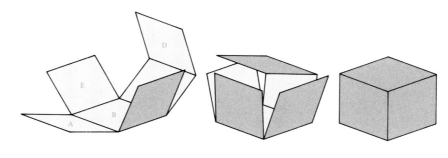

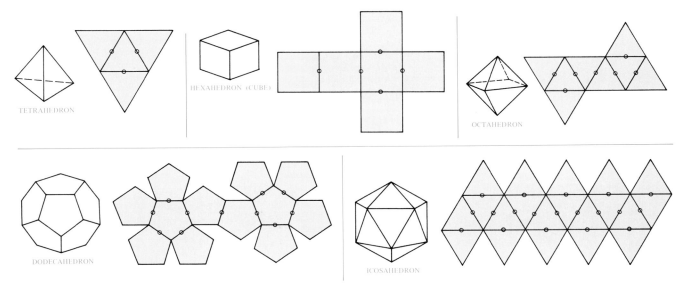

Fig. 18-13 The five regular solids.

TETRAHEDRON

HEXAHEDRON (CUBE)

OCTAHEDRON

DODECAHEDRON

ICOSAHEDRON

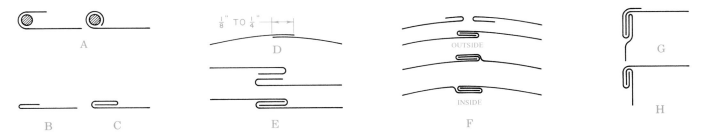

Fig. 18-14 Wiring, seaming, and hemming.

$\frac{1}{8}$" TO $\frac{1}{4}$"

OUTSIDE

INSIDE

A   B   C   D   E   F   G   H

Fig. 18-15 A pattern for a prism, showing stretchout line and lap.

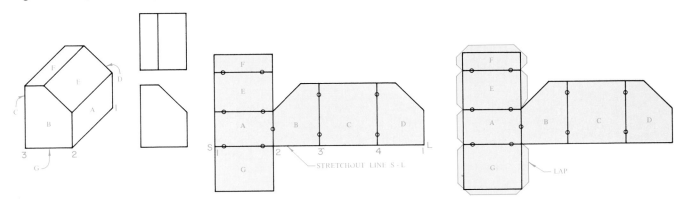

STRETCHOUT LINE S-L

LAP

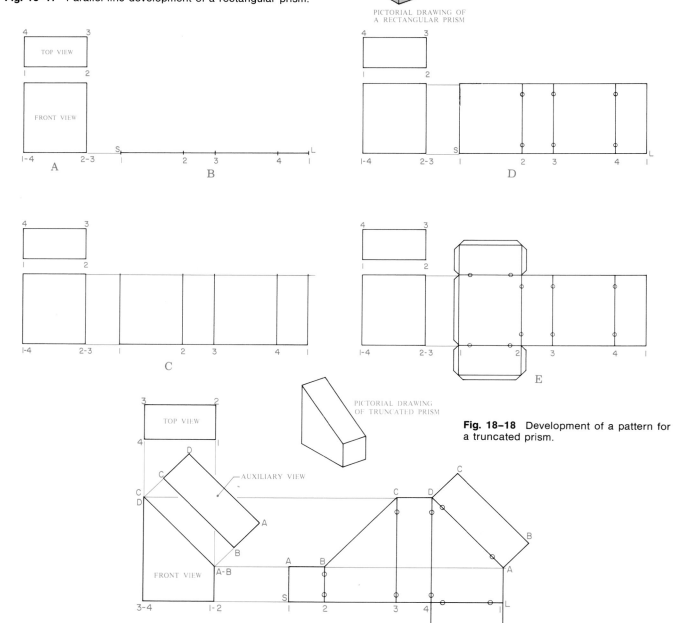

Fig. 18–16 Pictorial drawing of a rectangular prism.

PICTORIAL DRAWING OF
A RECTANGULAR PRISM

Fig. 18–17 Parallel-line development of a rectangular prism.

TOP VIEW

FRONT VIEW

A

B

D

C

E

PICTORIAL DRAWING
OF TRUNCATED PRISM

Fig. 18–18 Development of a pattern for a truncated prism.

TOP VIEW

AUXILIARY VIEW

FRONT VIEW

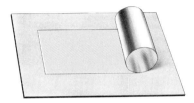

Fig. 18–19 Developed surface of a right circular cylinder.

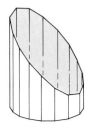

Fig. 18–20 Pictorial drawing of a truncated right cylinder.

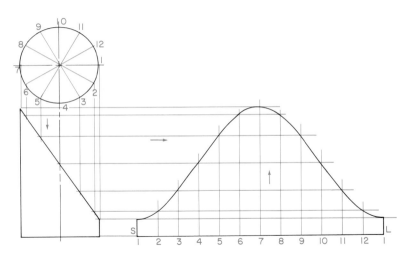

Fig. 18–21 Development of a pattern for a cylinder.

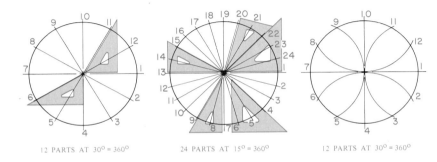

12 PARTS AT 30° = 360°   24 PARTS AT 15° = 360°   12 PARTS AT 30° = 360°

Fig. 18–22 Dividing a circle.

have a base perpendicular to the axis, a right section must be taken to obtain the stretchout line.

Development of Cylinders. A cylinder may be thought of as being a many-sided prism. Each of these sides forms an edge called an element. Technically, the elements are of an infinite number since the lateral surface of a cylinder is a smooth curve. In other words, the closer the elements, the smoother the curved surface. However, for ease of preparing the stretchout, it is necessary to space the elements equally at a convenient working distance apart. This makes it possible to obtain the length of the stretchout by stepping off the same number of equal spaces along the stretchout line.

The development of a truncated right cylinder (Fig. 18–20) is shown in Fig. 18–21. The complete drawing is made by following steps A through G.

A. Draw the front and top views full size. Divide the top views into a convenient number of equal parts (12 in this case). Various methods for dividing a circle are shown in Fig. 18–22.

B. Draw the stretchout line (*SL*). Its actual length will later be determined when the elements are stepped off.

C. Use dividers to step off the same number of spaces along *SL* as there are on the top view and label each as shown in Fig. 18–21. Draw vertical lines from each point to the approximate height of the front view.

D. Project lines downward from the elements on the top view to the front view. Label their points of intersection on the front view.

E. Project horizontal lines from points 7, 6-8, 5-9, 10-4, 11-3, 12-2, and 1 to the development.

F. Locate corresponding points of intersecting on the development and draw a smooth curve.

G. Darken outlines and add laps as necessary. The red arrows are used to show the direction in which the various lines are projected.

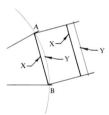

Fig. 18-23 A straight line is the shortest distance between two points.

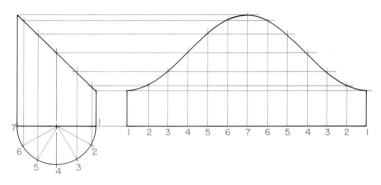

Fig. 18-24 Development of a pattern for a cylinder using a front and half-bottom view.

Since the surface of a cylinder is a smooth curve, the stretchout as obtained is only approximate. The inaccuracy results from measuring point to point on a straight line rather than on a curved line, as shown in Fig. 18-23. Notice that when the curved line is measured accurately and stretched out into a straight line, it is slightly longer than the straight line taken from point A to point B. In most cases, however, the difference is so slight that the inaccuracy is not a critical amount. The actual difference may be tested by figuring the actual length, using the formula circumference $= \pi D$, and measuring this distance along the stretchout line.

A slightly different method for developing a cylinder is shown in Fig. 18-24. In this case a front and a half-bottom view are used. Attaching the half-bottom view to the front view saves time and increases accuracy. Notice that both methods produce the same development.

To Draw the Pattern for a Two-piece, or Square, Elbow. Since this elbow consists of two cylinders cut off at 45°, only one part needs to be developed for a pattern, as shown in Fig. 18-25. Lap is allowed as indicated,

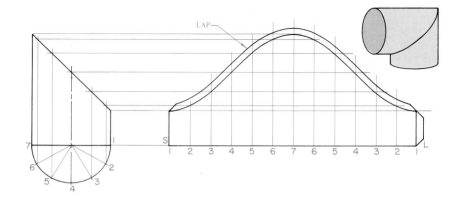

Fig. 18-25 Pattern for a square elbow.

depending upon the type of seams to be made. If no lap allowance is necessary on the curved edge, the two parts may be developed as shown in Fig. 18-26. Notice that the seam on part A is on the short side, while on part B it is on the long side. In Fig. 18-25 the seam on both pieces would be on the short side. In most cases this is not critical.

The Development of a Four-piece Elbow. The development of a four-piece elbow is shown in Fig. 18-27. The pattern is developed by following steps A through C.

A. Draw arcs having the desired inner and outer radii, as shown at A. Divide the outer circle into six equal parts. Draw radial lines from points 1, 3, and 5 to locate the joints (seams). Draw tangents to the arcs through points 2 and 4 on both arcs to complete the front view.

B. Draw a half-bottom view and divide it as shown at B. Project the assumed edges (elements) from the partial view to the front view.

C. Develop the pattern for a part A as described in Figs. 18-25 and 18-26. The patterns for the four pieces may be cut from one rectangular piece if

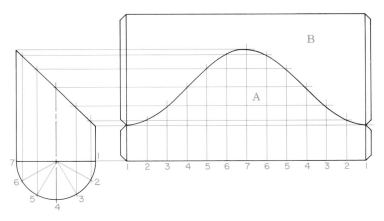

Fig. 18–26 Both parts of the pattern may be made on one stretchout.

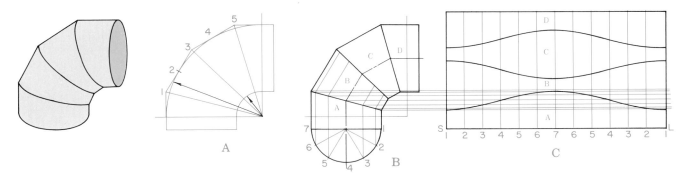

Fig. 18–27 Pattern for a four-piece elbow.

the seams are made alternately on the inside (throat) and the outside.

To draw the pattern for section B, extend the measuring lines of section A, and with the dividers take off the lengths of the assumed edges on the front view, starting with the longest one. Sections C and D are made in a similar way. Since the curve is the same for all sections, only one need be plotted, and that one can be used as a template (pattern) for the other curves.

Radial-line Development. In the case of prisms and cylinders, the stretchout line is straight, with the measuring lines perpendicular to it and parallel to each other. Thus, the name parallel-line development. However, since the edges of cones and pyramids are not parallel, the measuring lines will not be parallel, and the stretchout line will not be a continuous straight line. The measuring lines (elements) will radiate (project) from a point and, therefore, will not be parallel to one another. This type of development is called *radial-line development*.

Cones. The curved surface of a cone may be thought of as being made up of an infinite number of triangles. The development of the curved surface might be obtained by placing each triangle, in order, in contact with a plane surface. The result would be a sector of a circle having a radius equal to an element of the cone and an arc equal in length to the circumference of the base of the cone. The developed surface of a cone is illustrated in Fig. 18–28.

Fig. 18–28 Developed surface of a cone.

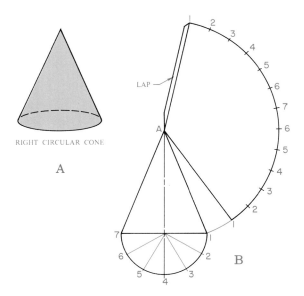

Fig. 18–29 Development of a pattern for a cone and a frustum of a cone.

To Draw the Pattern for a Right Circular Cone. A right circular cone is one in which the base is a true circle and the apex is a point directly over the center of the base (Fig. 18–29A). A *frustum* of a cone is produced when a cut is made through the cone and parallel to the base (Fig. 18–29C). To draw a pattern for a cone, follow steps A through E.

A. Draw front and half-bottom view to the desired size.

B. Divide the half-bottom view into several equal spaces and label the points, as shown in Fig. 18–29B.

C. With the slant height taken from the front view as a radius (A1 in Fig. 18–29B), draw an arc of indefinite length as a measuring arc. Draw a line from the apex (A) to the measuring arc at any location a short distance from the front view.

D. Use dividers to step off spaces 1-2, 2-3, 3-4, etc., along the measuring arc. Label the points to be sure none have

been missed. Complete the development by drawing line A1 at the far end.

E. Add the laps for the seam as desired. The width of the allowance for the seam depends upon the size of the development and the type of joint to be made.

The development for a frustum of a cone is shown in Fig. 18–29D. The procedure is the same as in Fig. 18–29B, except that a second arc AB′ is drawn from point B on the front view.

The Pattern for a Truncated Circular Cone. A *truncated circular cone* is one in which a cut not parallel to the base has been made through a circular cone (Fig. 18–30A). Follow steps A through E to develop the pattern.

A. Draw the front and top or bottom views (Fig. 18–30B). A half-top or bottom view may be used if desired.

B. Proceed as in Fig. 18–29 to develop the overall layout for the pattern.

C. Project points 1 through 6 from the bottom view to the front view and then to the apex. Label the points where they intersect the miter (cut) line to avoid mistakes. The elements as drawn on the front view are not true-length lines and must be projected to line A1 or A6 to determine their true length.

D. Draw the elements on the development from the apex to the points on the stretchout arc.

E. Draw arcs from the front view to the development and mark corresponding intersections. Draw a smooth curve and add an allowance (lap) to complete the pattern.

To Find the True Length of an Edge of a Pyramid or an Element of a Cone. The pyramid at A in Fig. 18–31 is shown by top and front views at B. The edge OA does not show in its true length in either view; if we draw the pyramid in the position shown at C, however, the true length of *OA* is shown in the front view. At C the pyramid has been revolved from the position of B about a vertical axis until the line *OA* is parallel to the vertical plane. At D the line *OA* is shown before and after revolving; thus, the construction for finding the true length of a line is as follows: In the top view, with radius *OA* and center *O*, revolve the top view of *OA* until it is horizontal. Project the end of the line down to meet a horizontal line through the front view of A. Join this point of intersection with the front view of *O*. The true length is shown at *OA'*.

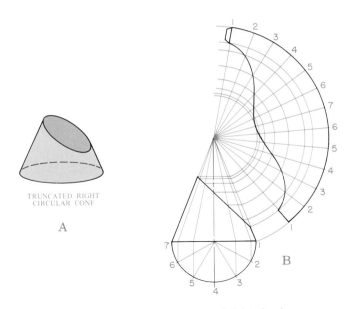

Fig. 18–30 Development of a pattern for a truncated right circular cone.

Fig. 18–31 To find the true length of a line.

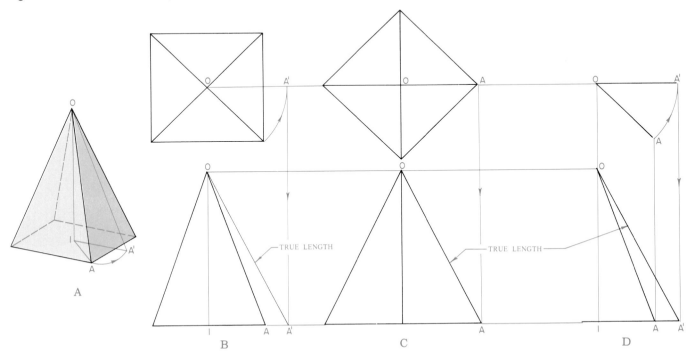

To Draw the Pattern for a Right Rectangular Pyramid. Follow steps A through D to develop a pattern for a right rectangular pyramid.

A. Find the true length of one of the edges (O1 in this case) by swinging it around until it is parallel to the vertical plane (O1′) and proceed as shown in Fig. 18–32.

B. With the true length as a radius, draw an arc of indefinite length for a measuring arc (curved stretchout).

C. On the measuring arc, mark off as chords the four edges of the base 1–2, 2–3, 3–4, and 4–1.

D. Connect the points and draw crease lines. Mark the crease lines if desired.

To Develop the Pattern for an Oblique Pyramid or a Truncated Oblique Pyramid. The development of an oblique pyramid (Fig. 18–33) is made by following steps A through C.

A. Find the true lengths (TL) of the lateral edges by revolving them parallel to the vertical plane as shown for edges O2 and O1. Edge O2 is revolved

in the top view and projected to O2′ in the front view, where it shows the true length of edge O2. In like manner find the true length of edge O1. Edge O2 = edge O3. Edge O1 = edge O4.

B. Start the development by laying off 2-3. Locate point O by intersecting arcs with centers 2 and 3 and radius true length of O2′ from front view.

C. Construct triangles O-3-4, O-4-1, and O-1-2 with the true lengths of the sides to complete the development as shown.

For a truncated pyramid (Fig. 18–34) with the inclined surface ABCD, find the true lengths of OA, OB, OC, and OD. For this pyramid OA = OD and OB = OC. The true lengths of OB and OC are shown at OB′ in the front view and of OA and OD at OA′. Lay off these true lengths on the corresponding edges of the development and join them to find the development of the lateral surface of the frustum. The inclined surface is shown in the auxiliary view and could

be attached to the lateral surface as indicated.

Triangulation. This is a convenient method for the approximate development of surfaces which cannot be exactly developed. It consists of dividing the surface into triangles, finding the true lengths of the sides, and then constructing the triangles in regular order on a flat surface. By using triangles with one short side, the plane triangles will approximate the curved surface. Triangulation is sometimes used for single curved surfaces.

The development of part of an oblique conical surface in Fig. 18–35 illustrates the use of triangulation. Divide the surface into triangles. For a closer approximation more triangles should be used than are shown here. Find the true lengths of the elements by revolving them parallel to the vertical plane or construct a true-length diagram as at C. At C, lay off O1, O2, etc., equal to O1, O2, etc., taken from the top view at A and draw O1, O2, etc., at C, which will be the true lengths of the elements. To

Fig. 18–32 Development of a pattern for a right rectangular pyramid.

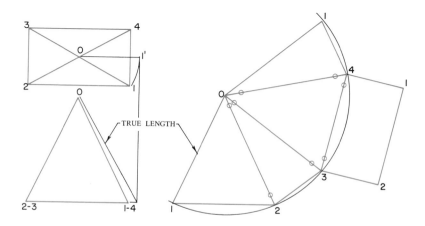

Fig. 18-33 To develop a pattern for an oblique pyramid.

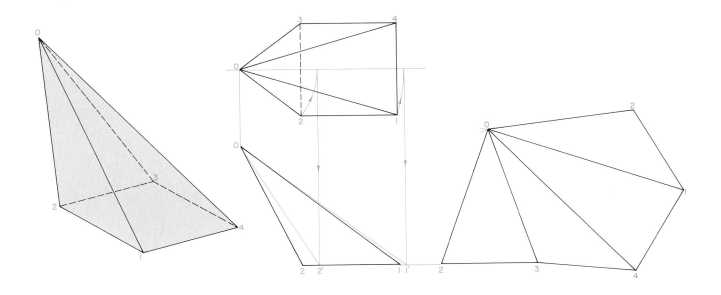

Fig. 18-34 To develop a pattern for a truncated oblique pyramid.

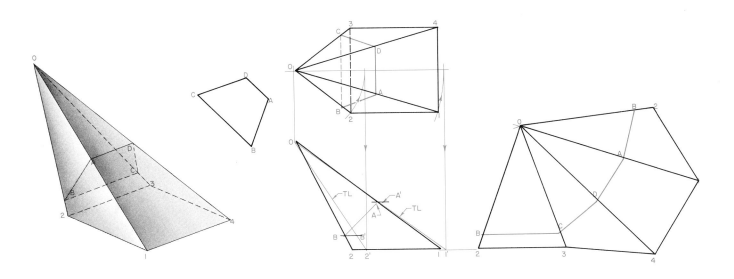

lay out the pattern at D, construct the triangles in the order in which they occur, taking distances 1-2, 2-3, etc., from the top view and O1, O2, etc., from the true-length diagram.

Transition Pieces. These are used to connect pipes or openings of different shapes, sizes, or positions. Transition pieces have a surface made up of parts of surfaces which may be plane, or curved, or both. A few transition pieces are shown in Fig. 18–36.

A square-to-round transition piece (Fig. 18–37) being formed in the plant of Dreis & Krump Manufacturing Company is described as follows: "A square-to-round transition is easily made in two pieces. Illustrated is the positioning of one of the pieces for making the last of the 8 partial bends on the second conical corner. This conical bending is done by moving one end of the plate out the proper distance for each partial bend while the point for the square corner remains fixed. When completed, the round end is 12″ in diameter and the opposite end 4′ by 5′ and the height is 6′. The material is ¼″ steel plate. The floor-to-floor time for the complete

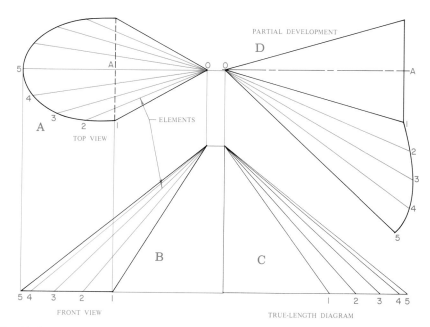

Fig. 18–35 Triangulation used in developing an oblique cone.

bending of each section as shown in the machine is approximately 30 minutes."

Development of a Transition Piece. As stated earlier, transition pieces are made up of parts of surfaces, and they are developed by tri-

angulation. This consists of dividing a surface into triangles (exact or approximate) and laying them out on the developed pattern in regular order.

The example shown in Fig. 18–38 connects two square ducts, one of which is at 45° with the other. This

Fig. 18–36 Some transition pieces.

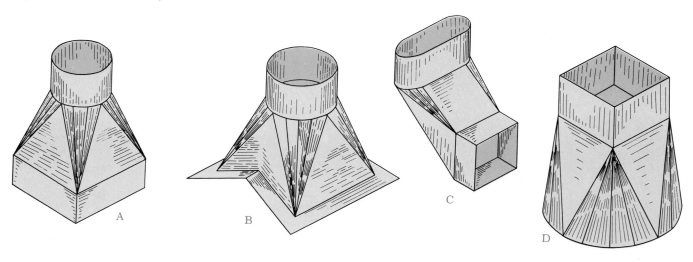

Fig. 18–37 Forming a square-to-round transition piece. (*Dreis & Krump Manufacturing Company.*)

piece is made up of eight triangles, four of one size and four of another.

To draw the developed surface follow steps A through F.

A. Find the true size of each triangle and place them in the proper order. Lines 1–2, 2–3, 3–4, and 4–1 show in their true size, as do lines AB, BC, CD, and DA.

B. Find the true length of one of the other lines as for $A4$. Revolve it parallel to the vertical plane, and the true length shows at $4A'$ in the front view.

C. Start the development by drawing line DA.

D. Then with centers at A and D and radius $4A'$ taken from the front view (true lengths of $A4 = D4 = D3$, etc.),

draw intersecting arcs to locate point 4 on the development.

E. With D as a center, draw an arc with radius $4A'$ and intersect it with an arc of radius 4-3 and center 4 to locate point 3.

F. Proceed to lay off the remaining triangles until the transition piece is completed.

The example shown in Fig. 18-39 connects a round pipe with a rectangular one. This piece is formed by four triangles. Between these triangles are four conical parts with apexes at the corners of the rectangular opening and bases each one-quarter of the round opening. To draw the development follow steps A through J.

A. Starting with the cone whose apex is at *A*, divide its base 1-4 into a number of equal parts, as 2, 3, and draw the lines *A*2, *A*3 to give triangles approximating the cone.

B. Find the true length of each of these lines. This is done in practical work by constructing a separate diagram, diagram I. The construction is based on the fact that the true length of each line is the hypotenuse of a triangle whose altitude is the altitude of the cone and whose base is the length of the top view of the line.

C. On the front view, draw the vertical line *AE* as the altitude of the cone. On the base *EF* lay off the distances *A*1, *A*2, and so forth, taken from the top view. This is done in the figure by swinging each distance about the point *A* in the top view and dropping perpendiculars of *EF*.

D. Connect the points thus found with the point *A* in diagram I to obtain the desired true lengths. Diagram II, constructed in the same way, gives the true lengths of lines *B*4, *B*5, and so forth, of the cone whose apex is at *B*.

E. After the true-length diagrams are constructed, start the development with the seam at *A*1. Draw a line *A*1 equal to the true length of *A*1.

F. With 1 as a center and radius 1-2 taken from the top view, draw an arc. Intersect this arc with an arc from center *A* and radius equal to the true length of *A*2, thus locating the point 2 on the development.

G. With 2 as center and radius 2-3, draw an arc and intersect it by an arc with center *A* and radius of the true length of *A*3.

H. Proceed similarly with point 4 and draw a smooth curve through the points 1, 2, 3, and 4 thus found.

I. Then attach the true size of the triangle *A*4*B*, locating point *B* on the development by intersecting arcs from *A* with radius *AB* taken from the top view, and from 4 with the radius the true length of *B*4.

J. Continue until the piece is completed.

Intersections. When a line pierces a plane, the point where the line passes through the plane is called the *point of intersection* (Fig. 18-40). When two plane surfaces meet, the line where they come together, or where one passes through the other, is called the *line of intersection* (Fig. 18-41). When a plane surface meets a curved surface, or where two curved surfaces meet, there is a line of intersection which may be either a straight line or a curved line, depending upon the surfaces and/or their relative positions.

It is necessary for the package designer, the sheet-metal worker, and the machine designer to be able to locate the point at which a line pierces a surface and to locate the line of intersection between two surfaces.

Fig. 18-38 Development of a square-to-square transition piece.

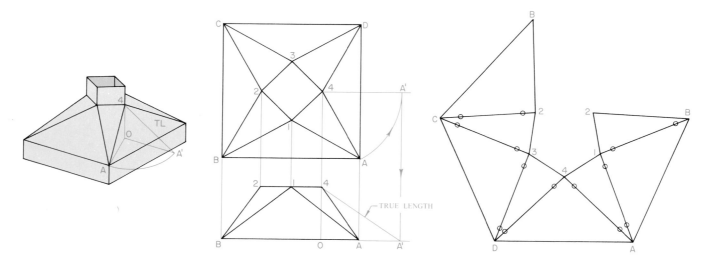

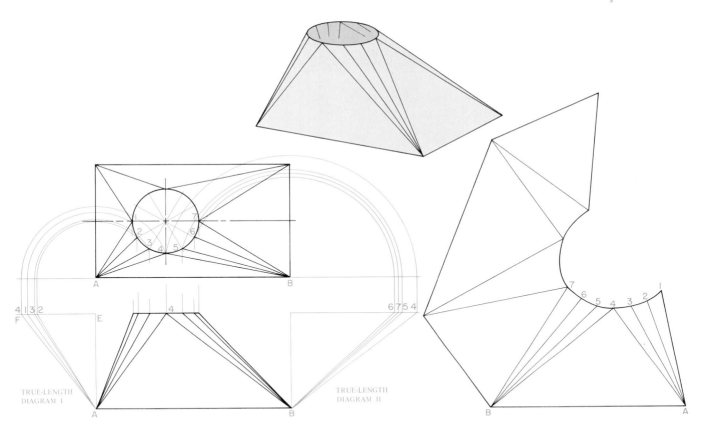

Fig. 18–39 Development of a rectangular-to-round transition piece.

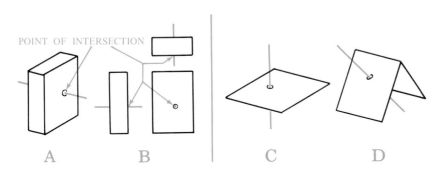

Fig. 18–40 The intersection of a line and a plane is a point.

Fig. 18–41 The intersection of two planes is a line. The arrow points to the line of intersection.

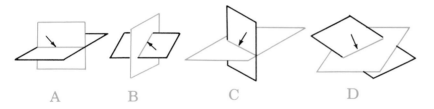

Intersecting Prisms. Several examples of intersecting surfaces are illustrated in Figs. 18–42 and 18–43.

To draw the line of intersection of two prisms, first start the orthographic views. In Fig. 18–44 a square prism passes through a hexagonal prism. Through the front edge of the square prism, pass a plane parallel to the vertical plane. The top view of this plane appears as a line *AA*. The intersection of the plane *AA* with one of the faces of the vertical prism shows in the front view as line *aa* and is crossed by the front edge of the square prism at point 1. Point 1 is a point on both prisms and, therefore, a point in the desired line of intersection. Plane *BB* is parallel to plane *AA* and contains an edge of the inclined prism, which meets at point 2 in the front view. Points 1 and 2 are in both planes; therefore, a line joining them will be in both planes and a part of the line of intersection. Plane *BB* also determines point 3.

These planes are called *cutting planes,* and they may be used for the solution of most problems in intersections. For intersecting prisms, pass planes through all the edges of both prisms within the limits of the line of intersection. Where the lines that are cut from both prisms by the same plane cross, there is a point on the required line of intersection. In Fig. 18–45 four cutting planes are required. The limiting planes are *AA* and *DD*, as a plane in front of *AA* or in back of *DD* would cut only one of the prisms.

Intersecting Cylinders. To draw the line of intersection of two cylinders (Fig. 18–46), since there are no edges on the cylinders, it will be necessary to assume positions for the cutting planes. Plane *AA* contains

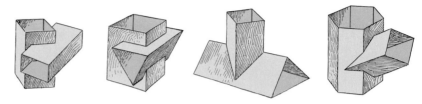

Fig. 18–42 Some lines of intersection.

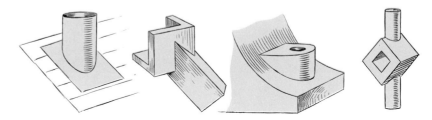

Fig. 18–43 Some lines of intersection.

the front line (element) of the vertical cylinder and cuts a line (element) from the horizontal cylinder. Where these two lines intersect in the front view, there is a point on the required curve. Each plane cuts lines from both cylinders which intersect at points common to both cylinders. The development of the vertical cylinder is shown in the figure.

The solution for an inclined cylinder is given in Fig. 18–47, where the

Fig. 18–44 Intersecting prisms.

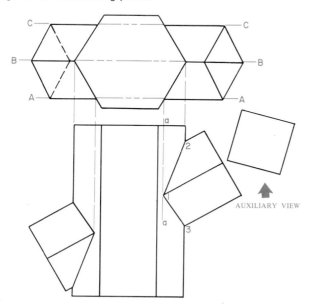

AUXILIARY VIEW

Fig. 18–45 Intersection of prisms.

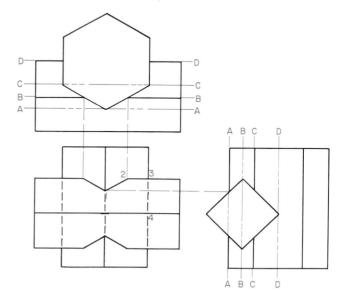

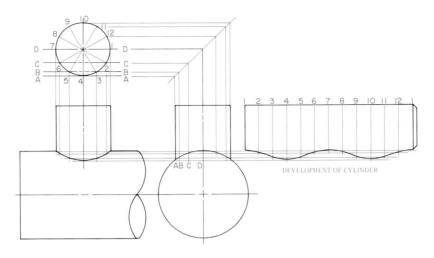

Fig. 18-46 Intersection of cylinders at a right angle.

positions of the cutting planes are located by an auxiliary view. In the development of the inclined cylinder the auxiliary view is used to get the length of the stretchout. If the cutting planes have been chosen so that the circumference of the auxiliary view is divided into equal parts, the measuring lines will be equally spaced along the stretchout line. Project the lengths from the front view and join ends with a smooth curve.

Intersection of Cylinders and Prisms. The intersection of a cylinder and a prism is found by the use of cutting planes, as already described. In Fig. 18-48 a triangular prism intersects a cylinder. The planes A, B, C, and D cut lines from the prism and lines from the cylinder which cross in the front view and determine the curve of intersection as shown. The development of the triangular prism is found by taking the length of the stretchout line from the top view and the lengths of the measuring lines from the front view. Note that one plane of the triangular prism

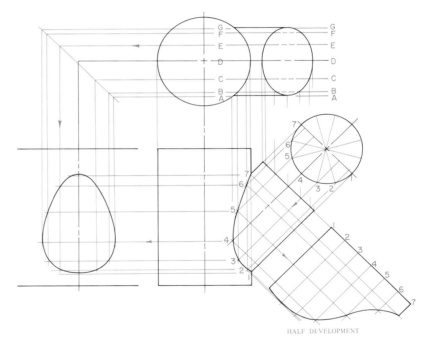

Fig. 18-47 Intersection of cylinders at an angle.

(line 1-5 in the top view) is perpendicular to the axis of the cylinder. The curve of intersection on that face is the radius of the cylinder.

Intersection of Cylinders and Cones. The intersection of a cylinder and a cone may be found by passing planes parallel to the horizontal

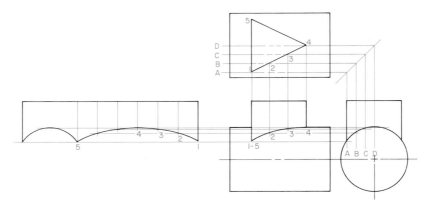

Fig. 18–48 Intersection of a prism and a cylinder and development of the prism.

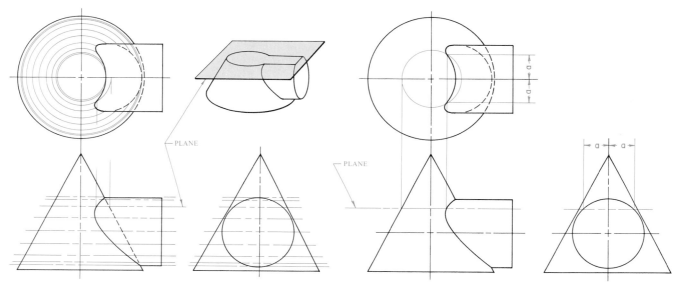

Fig. 18–49 Intersection of a cylinder and a cone.

Fig. 18–50 A cutting plane.

plane, as shown in Fig. 18-49. Each plane will cut a circle from the cone and two straight lines from the cylinder. The straight lines of the cylinder cross the circle of the cone in the top view at points on the curve of intersection. These lines are then projected to the front view, as in Fig. 18-50, where the construction is shown for a single plane. Use as many planes as are necessary to obtain a smooth curve.

Intersection of Planes and Curved Surfaces. The line of intersection of a cone cut by a plane, *MM*, as in Fig. 18-51, may be found by horizontal cutting planes *A*, *B*, *C*, and *D*. Each plane cuts a circle from the cone and a straight line from the plane *MM*. Thus, points common to both the plane *MM* and the cone are located as shown in the top view. These points, when projected to the front view, give the curve of intersection.

The intersection at the end of a connecting rod is found by passing planes perpendicular to the axis which cut circles as shown in the end view of Fig. 18-52. The points at which these circles cut the "flat" are projected back as points on the curve.

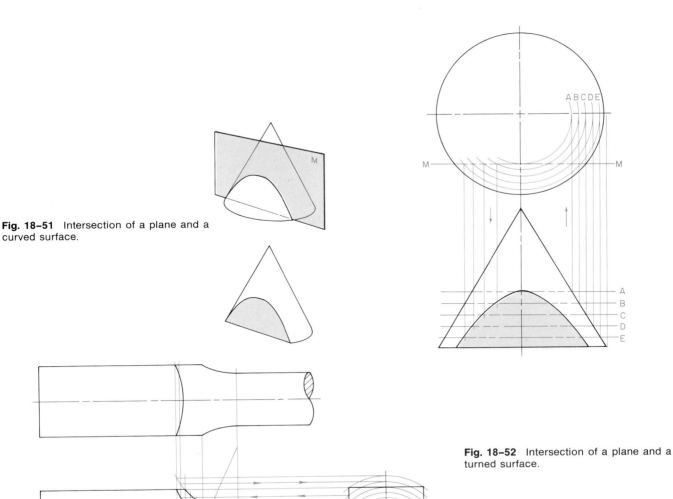

Fig. 18–51 Intersection of a plane and a curved surface.

Fig. 18–52 Intersection of a plane and a turned surface.

Review

1. Name the three basic types of surface developments.

2. Name the two general classes of surfaces.

3. A sphere has a _____-curved surface.

4. The drawing of the shape of a flat sheet which, when folded and fastened, will form the object is called the _____ of the piece.

5. Lines on a stretchout which indicate where a fold is to be made are called _____ lines.

6. Using a series of triangles to develop a pattern is called _____ .

7. Parts used to connect openings of different shapes, sizes, or positions are called _____ .

8. A point of intersection is a point where a line passes through a _____ .

9. When two plane surfaces meet, the line where they come together is called the _____ .

10. Sheet materials are usually fastened together by _____ or _____ .

Problems for Chapter 18

Fig. 18–53 Developments. Scale: full size. Problems *A* through *L* are planned to fit on an 11″ × 17″ or 12″ × 18″ drawing sheet. Draw the front and top views of the problem assigned. Develop the stretchout (pattern) as shown in the example at the right. For problems *A* through *F*, add the top in the position it would be drawn for fabrication. Include dimensions and numbers if instructed to do so. Patterns may be cut out and assembled.

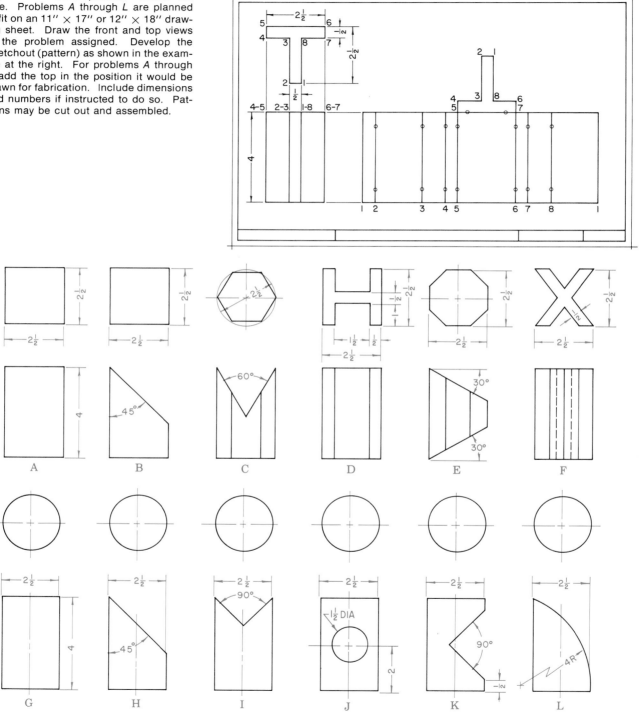

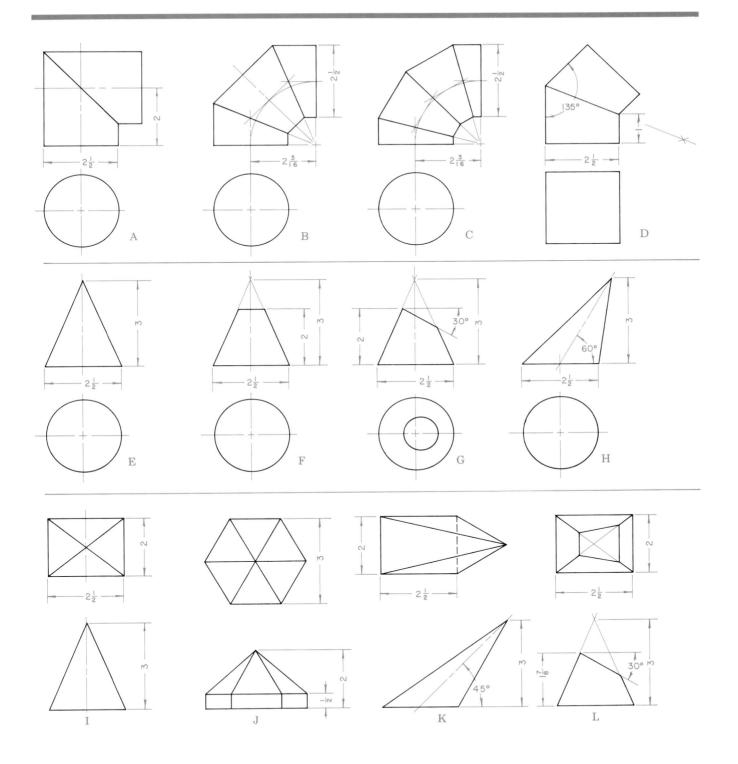

Fig. 18–54 Make two views of the problem assigned and develop the pattern. Scale: full size.

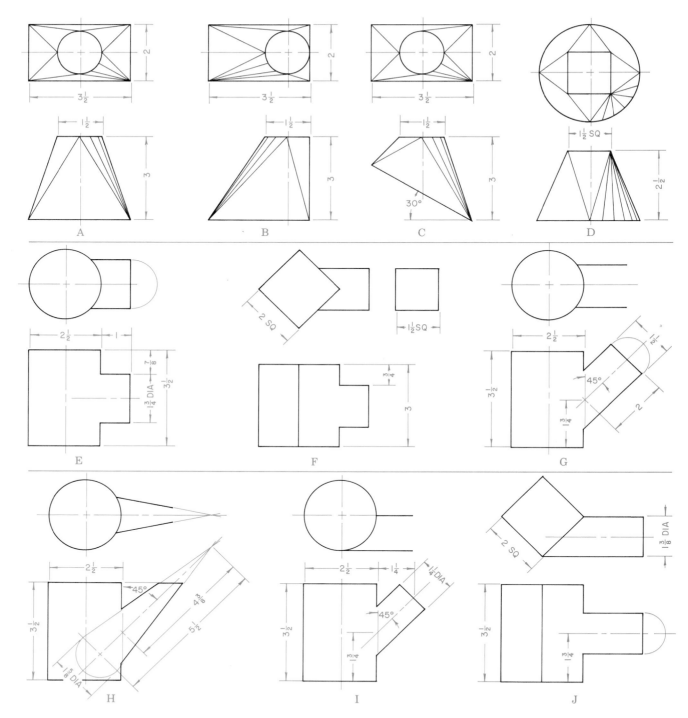

Fig. 18–55 Make two views of the problem assigned. In problems *A* through *D*, complete the top view and develop the pattern. In problems *E* through *J*, complete views where necessary by developing the line of intersection and completing the top view in *G*, *H*, and *I*. Develop patterns for both parts in problems *E* through *J*.

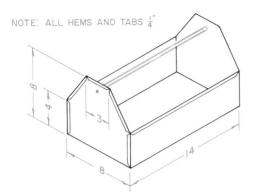

NOTE: ALL HEMS AND TABS $\frac{1}{4}$"

Fig. 18-56 Tool tray.

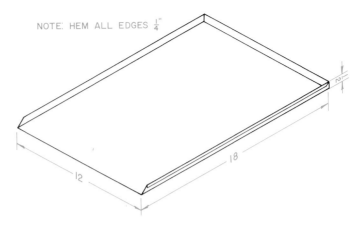

NOTE: HEM ALL EDGES $\frac{1}{4}$"

Fig. 18-57 Cookie sheet.

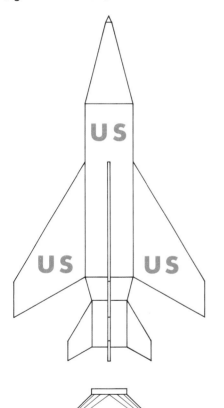

US
US   US

Fig. 18-60 Rocket with launching pad.

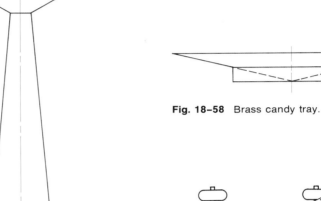

Fig. 18-59 Candle stick.

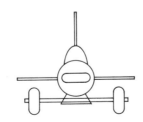

Fig. 18-58 Brass candy tray.

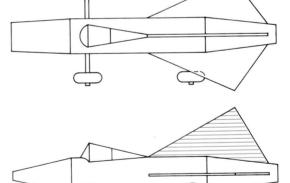

Fig. 18-61 Model racer.

Make pattern drawings for Figs. 18-56 and 18-57. No other views are necessary. Make complete working drawings of Figs. 18-58 through 18-61, including all necessary views and patterns. Take dimensions from the printed scale at the bottom of the page for Figs. 18-58 through 18-61.

1 0 1 2 3 4 5 6 7 8 9

Cams and Gears | Chapter 19

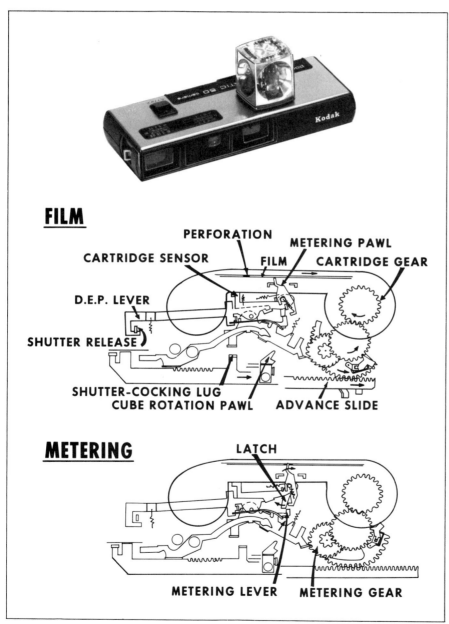

FILM

PERFORATION
CARTRIDGE SENSOR
FILM   METERING PAWL
CARTRIDGE GEAR
D.E.P. LEVER
SHUTTER RELEASE
SHUTTER-COCKING LUG
CUBE ROTATION PAWL   ADVANCE SLIDE

METERING

LATCH
METERING LEVER   METERING GEAR

(EASTMAN KODAK COMPANY.)

Cams and Gears. Cams and gears are machine parts which perform specific functions. Machine designers combine cams and gears into groups, called mechanisms, to form machines which equip industry for manufacturing (Fig. 19-1). A mechanism is a combination of stationary and movable parts which are shaped and connected in such a way that a definite motion of one part will cause other parts to have definite motion. Cams and gears can transmit motion, change the direction of motion, and change the speed of motion within a machine. The ability to draw cams and gears and to understand their function is a first step in becoming a machine designer. This chapter will introduce the basic techniques in drawing and understanding these important machine parts.

Cams. A cam is a machine part usually having an irregular curved outline or a curved groove which, when rotated, gives a specified motion to another part, called the follower. The plate cam illustrated in Fig. 19-2 would have a follower which moves up and down as the cam revolves on the shaft. The cylindrical cam in Fig. 19-3 revolves on a shaft, while its follower moves back and forth, parallel to the axis of the shaft, by following the groove. The grooved cam in Fig. 19-4 would revolve on a shaft, while its follower, placed in the groove, provides an irregular pattern of motion. A cam mechanism generally consists

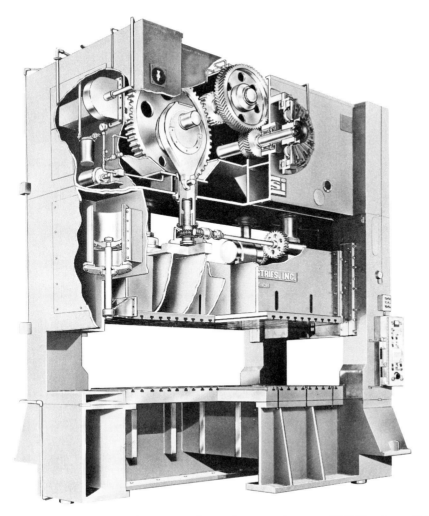

Fig. 19-1 Machine mechanisms are made up of cams and gears. (*USI-Clearing, a U.S. Industries Company.*)

Fig. 19-2 A plate cam. (*Camco.*)

Fig. 19-3 A cylindrical cam. (*Camco.*)

Fig. 19-4 A grooved cam. (*Camco.*)

of two elements: the cam and its follower. The cam usually drives the follower and the design, either plate or cylindrical, provides for continuous rotation common to automatic machinery. The importance of the cam to automatic control and accurate timing makes cams common to many classes of machinery. The many applications of cams are not limited by shape, for it is the unlimited variety of shapes which make it an indispensable tool of the designer.

Cam Followers. Three types of cam followers for plate cams are shown in Fig. 19-5. The roller follower is used to reduce friction to a minimum and transmits greater forces at higher speeds. The flat-surface follower and the point follower usually are hardened surfaces which provide durability. They generally are used on slower cams where friction is overcome with a transfer of force. Plate cams usually require a follower which must be spring loaded in order to permit contact through a full revolution. The spring pushes the follower against the cam. The lifting (rise) of the follower by the cam will be assured through direct contact, but uniform contact during a fall or at rest cannot be assured unless brought about by an external force such as a spring.

Cam Terms. Some terms are graphically expressed in Fig. 19-6, which illustrates how the cam acts. The stroke, or rise, occurs within one-half a revolution, or 180°, and repeats every 360°, or one full revolution.

Kinds of Cams. A cam for operating the valve of an automobile engine is illustrated in Fig. 19-7. This cam has a flat follower which rests against the face of the plate cam. In Fig. 19-8 the slider cam at A moves the follower up and down as the cam moves back and forth from right to left. An offset plate cam with a point follower off center is shown at B. A pivoted roller follower cam is shown at C. The cylindrical edge cam with a swinging follower is shown at D. All cams can be classified as simple inclines with a wedged surface that produces predetermined motion.

Cam Layout. In the layout of a cam mechanism it is necessary to determine which shape of cam profile will produce the proper motion in the follower. The displacement diagram in Fig. 19-9 indicates the desired motion through one revolution. The length represents one revolution of 360°, and the height of the diagram represents the total displacement stroke, 1.875, of the follower from its lowest position. The cam is assumed to rotate with a constant speed so that the length or time of a revolution is divided into parts proportional to the number of degrees for each action, called time

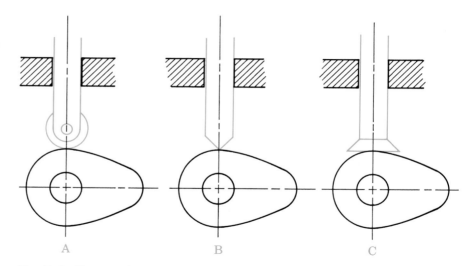

Fig. 19-5   Plate cam followers: A, roller; B, point; C, flat-surface. (*Camco.*)

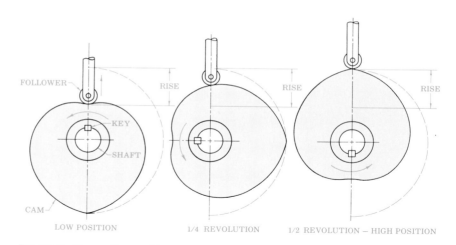

Fig. 19-6   Cam action and terms.

periods. These proportional parts are identified A, 1^1, 2^1, 3^1, B^1, C, 4^1, 5^1, D^1, E, 6^1, 7^1, A, as shown. Each unit represents a 30° angular division of the base circle.

The Problem. Given point O, the center of the cam shaft, and point A, the lowest position of the center of the roller follower. It is necessary to raise the center of the roller follower 1.875 with uniform motion during the first 120° of a revolution of the shaft; to remain at rest (dwell) for 30°, to drop 1.250 for 90°, to dwell for another 30°, and to drop 0.625 during the remaining 90°. The shaft is assumed to revolve uniformly.

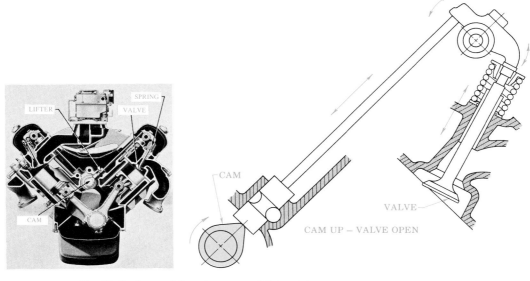

Fig. 19-7 Automobile valve cam. (*Oldsmobile Div., General Motors Corp.*)

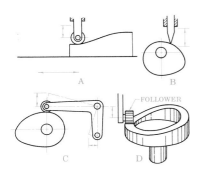

Fig. 19-8 Kinds of cams: A, slider; B, offset plate; C, pivoted roller; D, cylindrical roller.

Fig. 19-9 Cam displacement diagram.

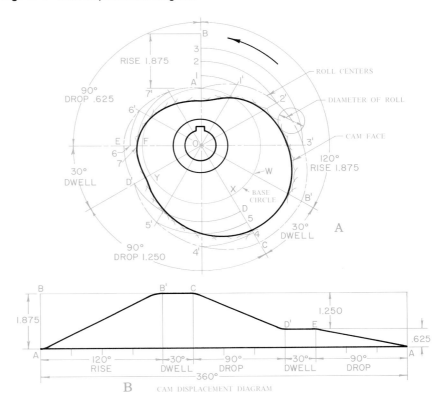

Displacement Diagram. The displacement diagram at B in Fig. 19-9 illustrates the solution to the problem. Points A, B^1, C, D^1, and E relate to the travel pattern which will occur in one revolution.

To Draw the Profile of the Cam. The profile of the cam shown at A in Fig. 19-9 has five significant features.

A. *Rise.* Divide the rise AB, or 1.875″, into a number of equal parts. Four parts are used, but eight parts could

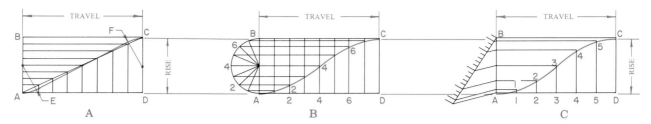

Fig. 19–10 Kinds of cam motion: A, uniform; B, harmonic; C, uniformly accelerated or decelerated.

increase the accuracy of the layout. The rise occurs from *A* to *W* (120°) and is divided into the same number of equal parts as the rise (four at 30°) with radial lines from *O*. With center *O*, draw arcs with radii *O*1, *O*2, *O*3, and *O*B until they locate 1^1, 2^1, 3^1, and B^1 on the four radial lines. Using an irregular curve, draw a smooth line through these four points.

B. *Dwell.* Draw an arc B^1C (30°) with radius O^1B^1. This will allow the follower to be at rest, for it will be at a constant distance from center *O*.

C. *Drop.* At *C* lay off *CD*, 1.250″, on a radial line from *O* and divide it into a number of equal parts (three are shown). Divide the arc *XY* (90°) on the base circle onto the same number of equal parts (three) and draw three radial lines from center *O* at 30° increments. Draw arcs with center *O* and radii *O*4, *O*5, and *O*D to locate points 4^1, 5^1, and D^1 on the three radial lines. Draw a smooth curve with the irregular curve as a guide through points 4^1, 5^1, and D^1.

D. *Dwell.* Draw an arc D^1E (30°) with radius *OD* to provide for the constant distance from center *O* while roller follower is at rest.

E. *Drop.* In the last 90° of a full revolution the roller will return to point *A*, which is a distance *EF*, or 0.625″. Divide *EF* into a number of equal parts (three as shown). Divide arc *FA* into the same number of equal

parts (three) and draw radial lines at 30° increments. Draw arcs with radii *O*6 and *O*7 to locate points 6^1 and 7^1. Using the irregular curve as a guide, draw a smooth curve through points *E*, 6^1, 7^1, and *A*. This finishes the roll centers.

With centers on the line-of-roll centers, draw successive arcs with the radius of the roller, as indicated. Then using an irregular curve, draw the cam profile, which is a smooth curve tangent to the arcs that represent the roller, as shown.

Motion. A cam may be designed so that the followers can have three types of motion. The displacement is used to plot the variations.

Fig. 19–11 A drawing of a plate cam.

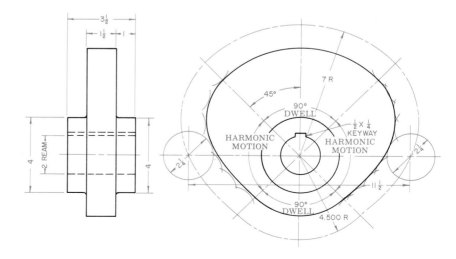

Uniform-motion cams. These are suitable for high-speed operations. In Fig. 19–10 at A, the thin line represents uniform motion. Equal distances on the rise (eight units) are made for equal distances on the travel (equal intervals of time, eight units). To avoid a sudden jar at the beginning and end of motion, arcs are used to modify the motion, as shown by the heavy line formed by arcs *E* and *F*.

Harmonic motion. The method for plotting a harmonic curve is shown at B. Draw a semicircle with the rise as the diameter. Divide the semicircle into eight equal parts using radial lines and the travel into the same number of equal parts. Project the eight points horizontally from the semicircle until they intersect the corresponding vertical projections as shown. Draw a smooth curve through all eight points with an irregular curve.

Uniformly accelerated and decelerated motion. At C we are plotting a parabolic curve. A cam designed on this basis has a uniform acceleration and deceleration curve. The rise is divided into parts proportional to 1, 3, 5, . . ., 5, 3, 1 (six parts not equal). The travel is divided into the same number of parts (equal parts). Project the six points horizontally from the rise until they intersect the corresponding vertical projections as shown. Draw a smooth curve as shown through all the points of intersection.

Cam Drawings. A drawing for a face (plate) cam is shown in Fig. 19–11. Note that the amount of movement (rise) is given by indicating the radii for the dwells, a 4.5-in. radius and a 7.0-in. radius. Harmonic motion is used, and there are apparently two rolls working on this cam. In Fig. 19–12 a drawing for a barrel (cylindrical) cam is shown with a displacement diagram. The diagram shows two dwells and two kinds of motion. Note that the distance traveled from center to center is 1.5 in.

There are many types of cams which are designed for modern machinery. This introduction involves principles of travel and time which apply to all types of cams regardless of their complexity.

Gears. One of the most practical and dependable machine elements for transmitting rotary motion from one shaft to another is the spur gear. There are many kinds of gears, a few of which are illustrated in Fig. 19–13. The operation of simple spur gears

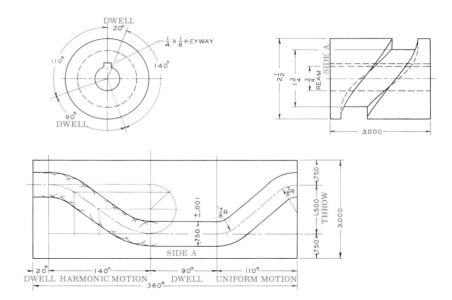

Fig. 19–12 A drawing of a cylindrical cam.

Fig. 19–13 Several gears which are used as typical machine elements: A, B, and C, spur gears; D, a bevel gear; E, a pinion (spur gear); F, a rack (E and F together are called a rack and pinion); G, an internal gear. (*The Fellows Gear Shaper Company.*)

may be explained as follows: If two wheels are in contact, as in Fig. 19-14, both will revolve if one is turned. If the small friction-wheel disk is two-thirds the diameter of the larger wheel, it will make one and one-half revolutions for one revolution of the larger wheel if no slipping occurs. When the load on the driven member is increased and the wheel is hard to turn, slipping begins to occur. Friction wheels are not reliable for a uniform transfer of rotary motion.

Spur Gear. Teeth are added to the wheels in Fig. 19-15 to form spur gears. The shape of the teeth added to the wheels can cause the same kind of motion as rolling friction wheels, but they do this without slipping. The spur gear and pinion gear with parallel shafts (the small spur gear is the pinion, illustrated in Fig. 19-15) are typical of the involute system of gearing. The involute curve unique to gears is illustrated in Fig. 19-19. An involute can be thought of as a curve formed by a taut string as it unwinds from around a cylinder (circle).

Gear Teeth. The basic forms used for gear teeth are involute and cycloidal curves. A cycloidal curve can be thought of as the path of a curve formed by a point on a rolling circle (Fig. 19-16). The information given in this chapter is for the $14\frac{1}{2}°$ involute system. It can be used for a 20° system as well since the only practical difference is the number of degrees of the pressure angle. The $14\frac{1}{2}°$ or 20° refers to the pressure angle (Fig. 19-17). The pressure angle and the distance between the centers of mating spur gears will determine the diameters of the base circles (Fig. 19-17). Note that the gears are related by the touching (point of tangency) pitch diameters which corre-

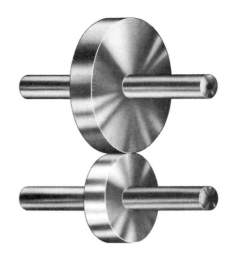
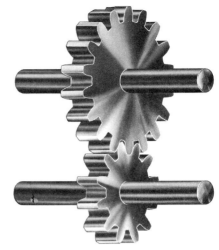

Fig. 19-14 Friction wheels are a simple means of transmitting rotary motion from one shaft to another.

Fig. 19-15 The spur gear. Teeth added to friction wheels provide a more efficient means of transmitting rotary motion.

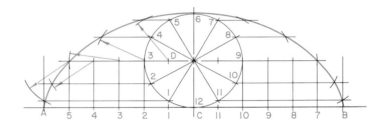

Fig. 19-16 A cycloidal curve.

Fig. 19-17 The pressure angle. Note that the center distance indicates the distance between shafts of mating gears.

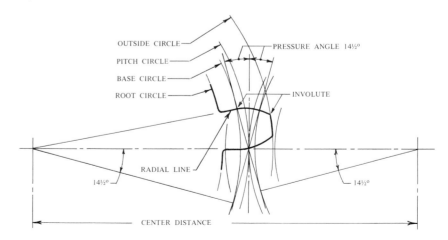

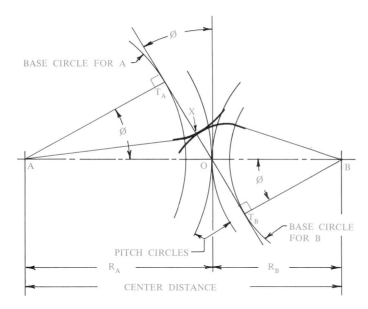

Fig. 19-18 Gear tooth interaction; the rolling nature of surface contact.

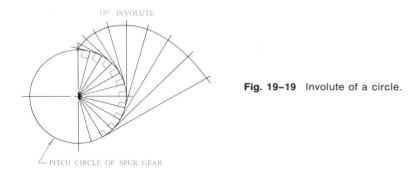

Fig. 19-19 Involute of a circle.

Fig. 19-20 Simplified method of drawing a gear tooth.

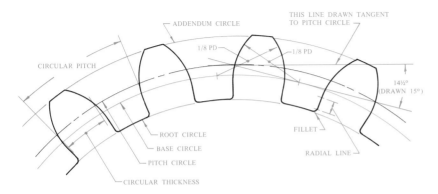

spond to the diameters of the rolling friction wheels that are replaced by the gears. The base circle from which the involute is derived is smaller than the pitch circle.

In Fig. 19-18, R_A is the radius of the pitch circle of the gear with center at A, and R_B is the radius of the pitch circle of the pinion with the center B. The distance between gear centers is $R_A + R_B$. The line of pressure $T_A T_B$ is drawn through O (which is the point of tangency of the pitch circles) and makes the pressure angle ϕ (Greek letter phi) with the perpendicular to the line of centers. This angle is $14\frac{1}{2}°$. Note that lines $A T_A$ and $B T_B$ are drawn from centers A and B perpendicular to the pressure-angle line $T_A T_B$. A point X on a cord (line of pressure $T_A T_B$) will describe the points which form the involute curve as the cord winds and unwinds to represent the outlines of gear teeth outside the base circles. The profile of the gear tooth inside the base circle is a radial line. Figure 19-19 illustrates the cord unwinding off the surface of the base circle. To simplify the drawing of the involute curve, note that the radius in Figure 19-20 is one-eighth the pitch diameter.

Spur-gear Terms and Formulas. The names of some parts of a spur gear are illustrated in Fig. 19-21. The following terms and formulas are for use in finding the required dimensions for standard $14\frac{1}{2}°$ involute spur-gear problems.

Spur-gear terms

N = number of teeth
  N_G = number of teeth of gear
  N_P = number of teeth of pinion

D = pitch diameter—diameter of pitch circle
  D_G = pitch diameter of gear
  D_P = pitch diameter of pinion

P = diametral pitch—number of teeth per inch of pitch diameter

a = addendum—radial distance the tooth extends above pitch circle

b = dedendum—radial distance the tooth extends below pitch circle

D_O = outside diameter—pitch diameter plus twice the addendum gives the overall gear size

D_R = root diameter—pitch diameter minus twice the dedendum gives the root diameter

h_t = whole depth—radial distance from the root diameter to the outside diameter, equal to addendum plus dedendum

p = circular pitch—distance from a point on one tooth to the same point on the next tooth measured along the pitch circle; the distance of one tooth and one space

t_c = circular thickness—thickness of a tooth measured along the pitch circle

c = clearance—difference between the addendum and the dedendum; i.e., the distance between the top of a tooth and the bottom of the mating space

h_K = working depth—the mating distance a tooth projects into the mating space; twice the radial distance of the addendum

o = pressure angle—the direction of pressure between teeth at point of contact

D_b = base-circle diameter—the circle from which the involute profile evolves

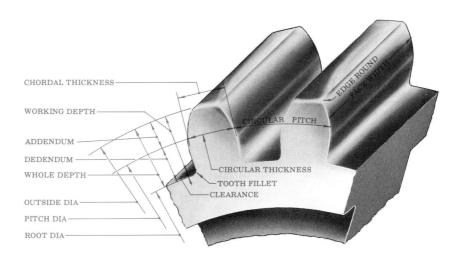

Fig. 19–21 Gear terms illustrated.

Spur-gear formulas

Number of teeth:
$$N = PD \text{ or } N = \frac{\pi D}{P}$$

Pitch diameter:
$$D = \frac{N}{P} \text{ or } D = \frac{N_p}{\pi}$$

Diametral pitch:
$$P = \frac{\pi}{p} \text{ or } P = \frac{N}{D}$$

Addendum: $a = \dfrac{1}{P}$ or $a = \dfrac{\pi}{p}$

Dedendum:
$$b = \frac{1.157}{P} \text{ or } b = \frac{1.157}{\pi}p$$

Outside diameter:
$$D_O = D + 2a \text{ or } D_O = \frac{N+2}{P}$$
$$\text{or } D_O = \frac{(N+2)p}{\pi}$$

Root diameter:
$$D_R = D - 2b$$
$$\text{or } D_R = D_O - 2(a + b)$$

Whole depth:
$$h_t = a + b \text{ or } h_t = \frac{2.157}{P}$$
$$\text{or } h_t = \frac{2.157p}{\pi}$$

Circular pitch:
$$p = \frac{\pi D}{N} \text{ or } p = \frac{\pi}{P}$$

Circular thickness:
$$t = \frac{p}{2} \text{ or } t = \frac{\pi}{2P}$$

Clearance:
$$c = \frac{0.157}{P} \text{ or } c = \frac{0.157p}{\pi}$$

Working depth: $h_K = 2a$ or $h_K = \dfrac{2}{P}$

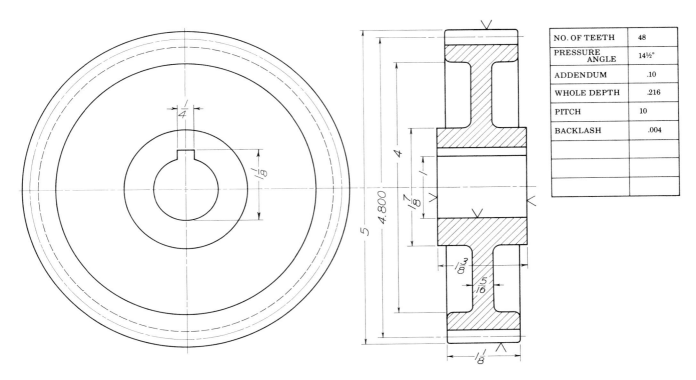

NO. OF TEETH	48
PRESSURE ANGLE	14½°
ADDENDUM	.10
WHOLE DEPTH	.216
PITCH	10
BACKLASH	.004

Fig. 19–22 Simplified profile and cross section of a spur-gear working drawing.

Application of Gear Formulas. Before the formulas can be applied, written specifications must be given. Typical information given to the draftsman is as follows. A pair of involute gears are to be drawn according to the following specifications.

The *pressure angle* will be 14½°.
The distance between parallel shaft centers will be 12 in.
The driving shaft will turn at 800 rpm clockwise.
The driven shaft will turn at 400 rpm.
The *diametral pitch* equals 4 (number of teeth per inch of pitch diameter).

Typical computations. The following computations are necessary before the formulas can be applied.

1. Pitch radius of the pinion (smaller gear)

$$R_p = \frac{400}{400 + 800} \times 12''$$
$$= 4\text{-in. radius}$$

2. Pitch radius of the spur gear

$$R_s = \frac{800}{400 + 800} \times 12''$$
$$= 8\text{-in. radius}$$

Velocity ratio ½ (4 in. : 8 in.)

The first two computations are based on the ratio of the two cylinders' velocity. One cylinder drives the other, and so the ratio is obtained by dividing the velocity (rpm) of the driver by the velocity (rpm) of the driven member.

3. Number of teeth on the pinion = $N_p = DP = 4 \times 8 = 32$

4. Number of teeth on the spur gear = $N_s = DP = 4 \times 16 = 64$

5. Addendum = $a = \dfrac{1}{P} = \dfrac{1}{4}$ in.

6. Dedendum = $b = \dfrac{1.157}{p} = \dfrac{1.157}{4}$
$$= 0.289 \text{ in.}$$

Standard Involute Gears. Involute gears are interchangeable when they have set conditions which will allow them to run properly together. The four conditions for interchanging involute gears are: the same diametral pitch, the same pressure angle, the same addendum, and the same dedendum.

Gear Drawings. It is not necessary to show the teeth on typical gear drawings. A drawing for a cut spur gear is shown in Fig. 19–22. The gear blank should be drawn with dimensions for making the pattern and for the machining operations. The spur-gear drawing should include data

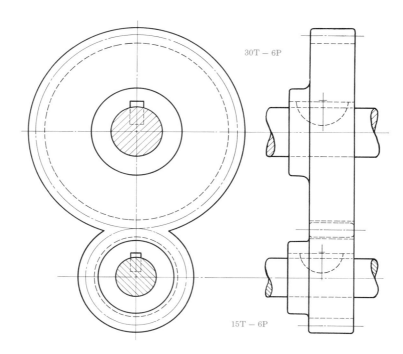

Fig. 19-23 Simplified drawing of mating spur gears. Note the tangent pitch circles, the number of teeth, and the pitch.

for cutting the teeth, data for the tolerances required, and a notation of the material to be used. On assembly drawings the representation may be simplified as shown in Fig. 19-23. Even though the drawing may be simplified, it should include all necessary notes for making the gear.

Involute rack and pinion. A rack and pinion is shown in Fig. 19-24. A rack is simply a gear with a straight pitch line instead of a circular pitch line. The tooth profiles become straight lines, and these lines are perpendicular to the line of action.

Worm and wheel. Figure 19-25 shows how the worm and wheel mesh at right angles. The worm gear is similar to a screw and may have single or multiple threads. This system is used to transmit motion between perpendicular nonintersecting shafts, and the wheel is similar to a spur gear in design except that the teeth must be curved to engage the worm gear.

Fig. 19-24 A rack and pinion. (*Brad Foote Gear Works, Inc.*)

Bevel gears. When two gear shafts intersect, bevel gears are used to transfer motion. These are at times referred to as miter gears when the gears are the same size and the shafts are at right angles. In Fig. 19-26 the illustration shows four rolling cones. Bevel gears may be thought of as replacing the friction cones, just as the spur gear replaced the circular friction

Fig. 19-25 Application of a worm and wheel. (*Industrial Drives Division, Eaton Corp.*)

wheels. Figure 19-27 shows mating beveled gears. The smaller gear is called the pinion.

Bevel-gear terms. Some basic information about bevel gears is given in Fig. 19-29. There are three Greek letters used in the following list: α = alpha, δ = delta, and Γ = gamma. Examine Fig. 19-28 for the design similarities common to spur and bevel

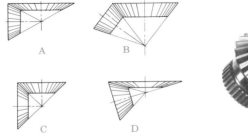

Fig. 19–26 Rolling cones which represent bevel gearing.

Fig. 19–27 Mating bevel gears. (*Arrow Gear Co.*)

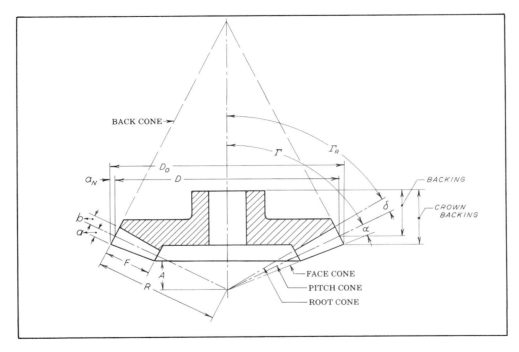

Fig. 19–28 Bevel gear terms. Note the cone shape.

gears. Note the differences. The pitch diameter is the diameter of the pitch cone in the bevel-gear design. The circular pitch and the diametral pitch are based on the pitch diameter just as in spur gears. The important features, such as the angles, are listed here.

α = addendum angle
δ = dedendum angle
Γ = pitch angle
Γ_R = root angle
Γ_o = face angle

a = addendum
b = dedendum
a_n = angular dedendum
A = cone distance
F = face
D = pitch diameter
D_0 = outside diameter
N = number of teeth
P = diametral pitch
R = pitch radius

Bevel-gear drawing. Working drawings of bevel gears require the dimensions for machining the blank, together with the necessary gear data. An example of a working drawing for a cut bevel gear is shown in Fig. 19–29.

Gear information. The American National Standards Institute has established standards for satisfactorily detailing gear drawings. ANSI Y14.7, Section Seven, and B6.1 and B6.5 are standard references that can be used for further study of the subject of gear detailing.

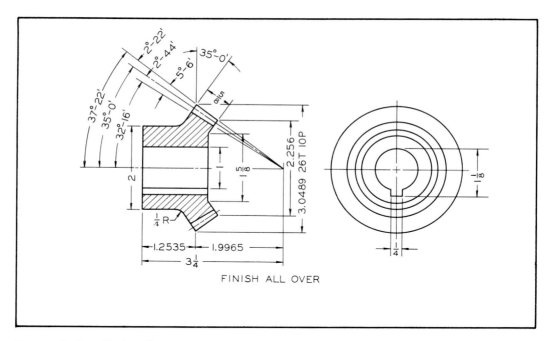

Fig. 19-29 Simplified profile and detailed section of a bevel gear.

Review

1. If time is an important factor in the design of a cam, what is the other essential factor?

2. What is the purpose of a displacement diagram?

3. List three types of cams and three types of followers.

4. List three applications of a cam.

5. Describe harmonic motion used in cam development.

6. How is the curve on a spur-gear tooth developed?

7. What technical phrase describes the ratio of the number of teeth to the pitch diameter?

8. Rolling cones are used to describe the interaction of what kind of gears?

9. What two bits of information are required to find circular pitch?

10. List two applications of bevel gears.

11. Name the four circles used by the draftsman in drawing gears.

Problems for Chapter 19

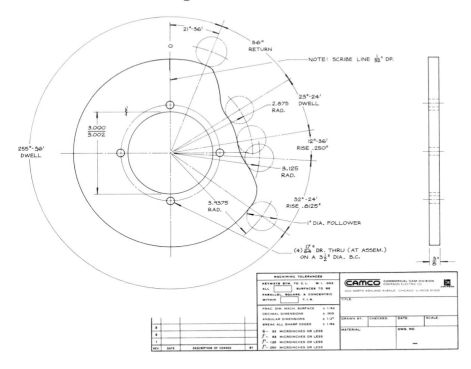

Fig. 19–31 Prepare a displacement diagram to illustrate the specified motion travel patterns from the original base circle. Make a profile drawing of the radial plate cam. (*Camco*.)

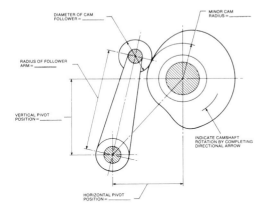

Note:

A. Locate the position of a pertinent timing feature (i.e. keyway, dowel hole, split line, etc.) assuming that the follower as shown is at zero degrees of camshaft rotation.

B. Complete the graph and table below indicating all critical points, transitional motions and timing features.

Fig. 19–32 Complete the data with assistance from the instructor, and prepare a displacement diagram for the follower arm cam. Develop a profile of the cam and a left-side view showing the cam and follower. Choose appropriate dimensions. (*Camco*.)

GRAPH OF FOLLOWER DISPLACEMENT VS. CAMSHAFT ROTATION

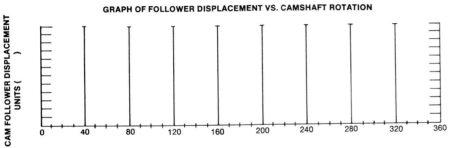

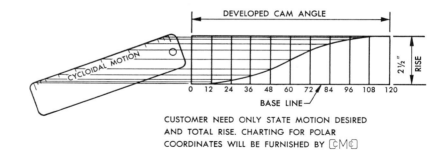

TYPICAL EXAMPLE: CONSTRUCT A 2½″ CYCLOIDAL RISE IN 120° CAM ROTATION.

(1) DRAW TWO PARALLEL HORIZONTAL LINES 2½″ APART REPRESENTING THE RISE.

(2) DIVIDE THE 120° CAM ANGLE INTO 10 EQUAL PARTS (12°-24°-36° ETC.).

(3) LAY CAM SCALE BETWEEN BASE LINE & 2½″ RISE. TRANSFER POINTS FROM CAM SCALE TO PAPER.

(4) PROJECT THESE POINTS HORIZONTALLY TO THEIR RESPECTIVE CAM DIVISIONS AND DRAW THE CURVE.

COPYRIGHT 1962 COMMERCIAL CAM AND MACHINE CO.

CUSTOMER NEED ONLY STATE MOTION DESIRED AND TOTAL RISE. CHARTING FOR POLAR COORDINATES WILL BE FURNISHED BY [CMC]

Fig. 19–33 Examine the displacement diagram carefully, and then make a drawing along a base line as shown, with a rise of 2½″ in 120°, a 90° dwell, a drop of 2½″ in 120°, and a dwell of 30°. Prepare a profile of this plate cam, using a base circle of 3½. (*Camco.*)

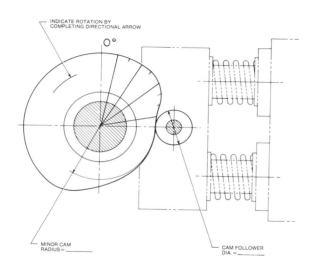

Note:

A. Locate the position of a pertinent timing feature (i.e. keyway, dowel hole, split line, etc.) assuming that the follower as shown is at zero degrees of camshaft rotation.

B. Complete the graph and table below indicating all critical points, transitional motions and timing features.

GRAPH OF FOLLOWER DISPLACEMENT VS. CAMSHAFT ROTATION

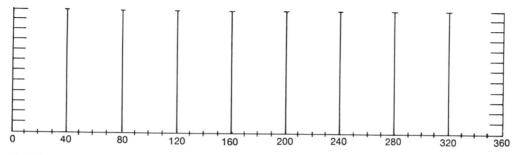

Fig. 19–34 Develop a displacement diagram for the radial cam. Start at 0° and work clockwise to determine the travel pattern at 10° intervals. Choose the size of the minor cam radius and the cam follower diameter. Complete the drawing with a cam profile, using dimensions of your choice or as supplied by the instructor. (*Camco.*)

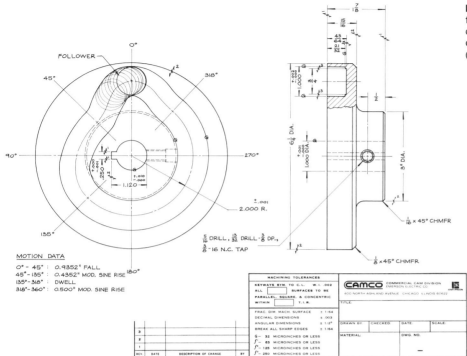

Fig. 19-35 Develop a working drawing of the grooved cam as shown. Prepare a displacement diagram illustrating the counterclockwise motion of this cam. (Note the motion of the follower.) (*Camco.*)

FOLLOWER

MOTION DATA

0° - 45° : 0.9352" FALL
45° - 135° : 0.4352" MOD. SINE RISE
135° - 318° : DWELL
318° - 360° : 0.500" MOD. SINE RISE

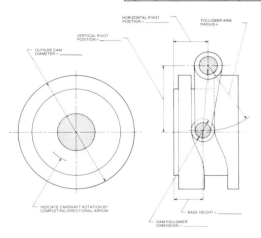

HORIZONTAL PIVOT POSITION =
FOLLOWER ARM RADIUS =
VERTICAL PIVOT POSITION =
OUTSIDE CAM DIAMETER =
INDICATE CAMSHAFT ROTATION BY COMPLETING DIRECTIONAL ARROW
BASE HEIGHT =
CAM FOLLOWER DIMENSION =

Fig. 19-36 As design detailer, determine the data necessary for drawing a displacement diagram of the barrel cam. Prepare the cam profile with adequate detail dimensions. (*Camco.*)

Note:

A. Locate the position of a pertinent timing feature (i.e. keyway, dowel hole, split line, etc.) assuming that the follower as shown is at zero degrees of camshaft rotation.

B. Complete the graph and table below indicating all critical points, transitional motions and timing features.

GRAPH OF FOLLOWER DISPLACEMENT VS. CAMSHAFT ROTATION

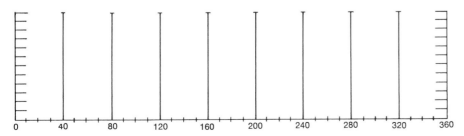

0   40   80   120   160   200   240   280   320   360

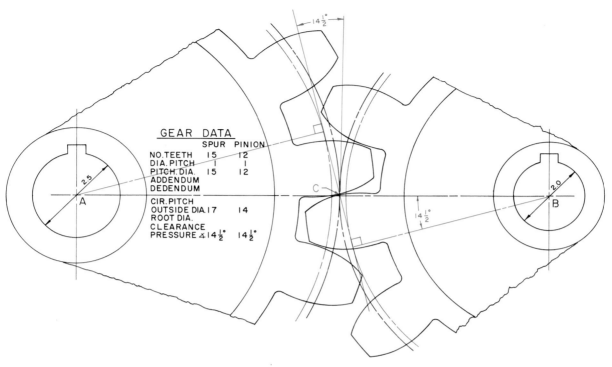

GEAR DATA		
	SPUR	PINION
NO. TEETH	15	12
DIA. PITCH	1	1
PITCH. DIA.	15	12
ADDENDUM		
DEDENDUM		
CIR. PITCH		
OUTSIDE DIA.	17	14
ROOT DIA.		
CLEARANCE		
PRESSURE	$14\frac{1}{2}°$	$14\frac{1}{2}°$

Fig. 19-37 Complete the gear data table, using formulas from Chapter 19. Make an enlarged drawing of a mating spur and pinion as shown. Select a suitable scale. Use an involute to draw the gear teeth or $\frac{1}{8}$ PD as directed by the instructor.

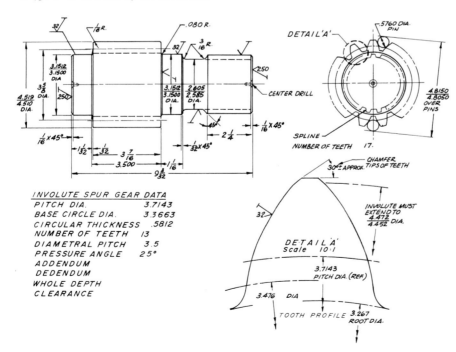

INVOLUTE SPUR GEAR DATA	
PITCH DIA.	3.7143
BASE CIRCLE DIA.	3.3663
CIRCULAR THICKNESS	.5812
NUMBER OF TEETH	13
DIAMETRAL PITCH	3.5
PRESSURE ANGLE	25°
ADDENDUM	
DEDENDUM	
WHOLE DEPTH	
CLEARANCE	

Fig. 19-38 Examine the spur gear data and drawing carefully before making the following revisions. The 3.500 and the $9\frac{3}{32}$ dimensions are to be reduced by 1.500″. Select a suitable scale and make a gear drawing showing two teeth, bottom and top, in profile. Complete the gear data where necessary and list on the drawing.

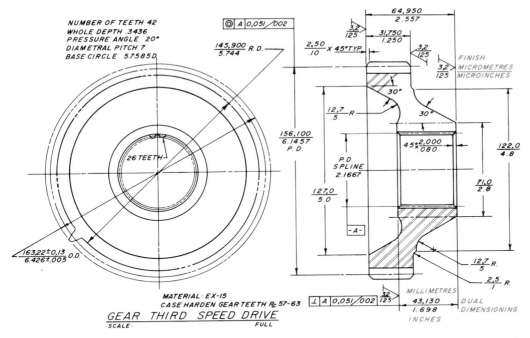

Fig. 19-39 Prepare a spur gear detail drawing, using the data given to determine necessary dimensions. Note that the gear is designed in metric dimensions (using a comma in place of a decimal point) and is dual dimensioned with decimal inches. (*International Harvester Co.*)

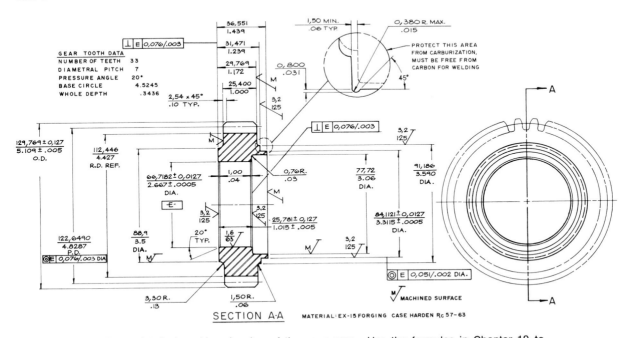

Fig. 19-40 Develop a detailed working drawing of the spur gear. Use the formulas in Chapter 19 to complete the gear data. Dimension in millimeters, decimal inches, or dually, as directed by the instructor. Note the datum used for accuracy in machine finish dimensioning. (*International Harvester Co.*)

Architectural Drafting | Chapter 20

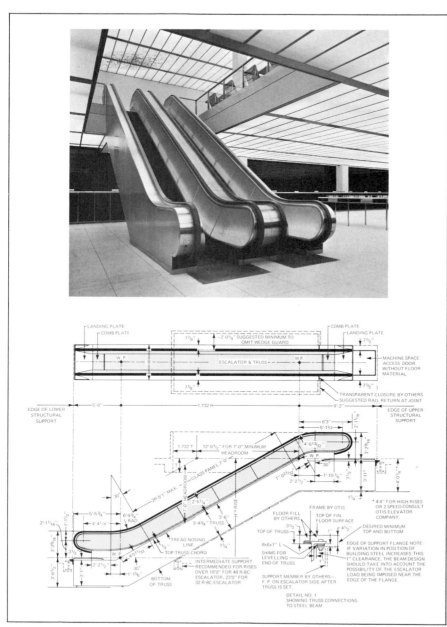

(OTIS ELEVATOR COMPANY.)

Today's Environment. Architects make important contributions to our patterns of living. The wide range of life styles and activities which take place every day within a community are related to some physical environment designed by an architect. Architectural drawings have to do with the planning of a three-dimensional form called *architecture*. The architect is the designer of buildings and the spaces around them which create urban or suburban communities. The job of the architect in its simplest form is to plan, design, and supervise the construction of buildings which will encourage and heighten all human activity (Fig. 20–1).

Career Opportunities. We enter an era in history which has a good reason for recreating man's environment. The methods of construction and the materials available provide the designer the means to do so. An individual trained in architecture may find opportunities to design, but each firm needs staff which can contact clients, program design development, write specifications, select materials, determine structural and mechanical requirements, analyze cost factors, and supervise construction. The career opportunities are developed in a professional publication available from the following organizations:

1. Architectural Careers
   American Institute of Architects
   (AIA)

Fig. 20-1 Architectural planning of a three-dimensional form. (*J. E. Barclay, Jr., & Associates.*)

1735 New York Avenue N.W.
Washington, D.C. 20006

2. Urban Planning
   American Institute of Planners
   907 15th Street, N.W.
   Washington, D.C. 20056

3. Landscape Architects
   American Society of Landscape
      Architects
   2000 K Street, N.W.
   Washington, D.C. 20006

4. Construction Contracting
   Associated General Contractors of
      America, Inc.
   1957 E. Street N.W.
   Washington, D.C. 20006

5. Careers in Building Trades
   Building and Construction Trades
      Department
   AFL—CIO
   815 16th Street, N.W.
   Washington, D.C. 20005

These references are only a few of the organizations which can provide specific information about careers related to man's environment. How well it is planned today affects the health, safety, and stability of entire com-

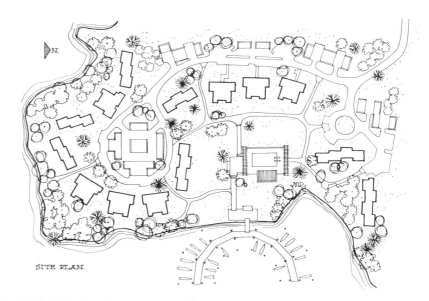

SITE PLAN

Fig. 20-2 Site plan of a new community.

munities planned for tomorrow (Fig. 20-2).

Community Architecture. The effect of the architect's planning can be examined by observing any urban community. The architectural styles of established residential communities

unfold dramatically with an overall look. As the styles are examined, the building materials should also be reviewed. There may be wooden (frame) cottages and brick bungalows. The larger residential buildings are the rented apartment buildings and the condominiums, which are liv-

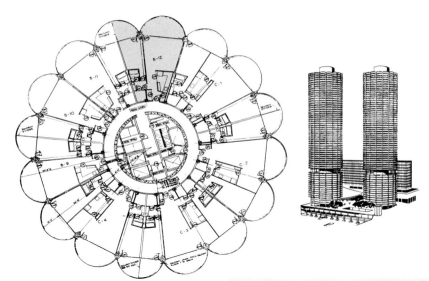

Fig. 20-4 Several styles of houses. (*Orrin Dressler, Designer-Contractor.*)

ing units individually owned within a large complex dwelling (Fig. 20-3). The buildings which are planned for multistory dwellings are generally constructed of such materials as brick, stone, heavy timber, steel, and concrete, which emphasize strength and durability. Architectural styles are usually identified by size, roof shape, and the materials employed. Several houses are illustrated in Fig. 20-4 to assist the student in identifying styles. Contemporary Ranch, French Contemporary, Colonial, French Chateau, Georgian, and English Tudor are some of the significant residential styles.

Fig. 20-4 (Cont.)

Neighborhood Arrangements. The residential structures are arranged in quiet settings formed by geometric patterns. The pattern is generally rectangular and is called a block. The church is reverently located on an important street in or near the center of the neighborhood. Many churches are located on the corner of a block (Fig. 20-5). Noisy parks or recreation centers are spotted conveniently throughout each neighborhood. Figure 20-6 illustrates the contemporary recreation center which emerges in the neighborhood that can afford the space required. The more important busy streets passing through communities form the neighborhood boundary and are lined with small stores, offices, and apartments.

Urban Development. Cities of various sizes are composed of neighborhoods. The main environmental force at work in the world today is the gathering of people to the city and the suburbs. The organization of these communities requires municipal buildings which serve the people. Municipal buildings often include the city hall and facilities for services such as a police station, fire station, bureau of streets, and perhaps water and sanitation departments. Cultural and health centers are also vital to the growth patterns of communities and may include theaters, libraries, museums, performing arts centers, medical offices, and hospitals. Architectural services are needed as communities are planned and developers seek to provide the environmental needs. Figure 20-7 lists typical architectural services.

Traffic Patterns. The mobility and flexibility of a community are reflected in the planned transportation systems, expressways, and arterial interchanges which serve the com-

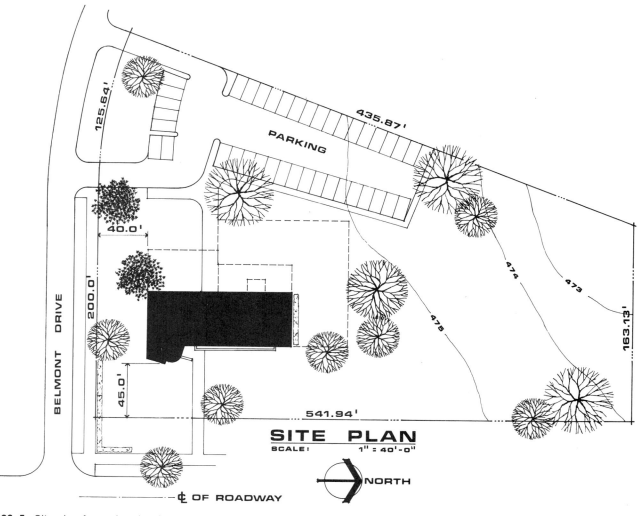

Fig. 20-5 Site plan for a church. (*Robert C. Taylor, AIA, and Associates.*)

munity. The geometric blocks or weaving circular lanes with cul-de-sacs which form a community are surrounded by major arteries that have along their ways large shopping centers and office buildings. Multistory housing, banking facilities, and hospitals are generally located on or near key arteries which make them easily accessible. Building and zoning commissions appropriately specify areas to be dedicated for commercial, social, and educational structures. Land for an industrial complex may be located in the heart of town or on the perimeter, and the traffic patterns will normally allow easy access. Industrial planning may be the center of a successful community plan.

Architecture Defined. Architecture is not just a matter of buildings. It is basically a matter of people—how they live, work, play, and worship. Good architectural environments can encourage and enhance any human activity. Architecture then is considered to be any physical environment. It is the sprawling suburb, the conspicuous ghetto, the glitter of a Park Avenue, or the elegance of a metropolitan cultural center.

Architecture Evaluated. Architecture affects everyone's life and can be examined in three significant ways by those who study buildings (Fig. 20-8).

First, it has to satisfy a social purpose, that is, to exhibit functional patterns for human activity. Begin with the study of the movement of people through the plan to see how the various spaces relate to the needs of people (Fig. 20-9).

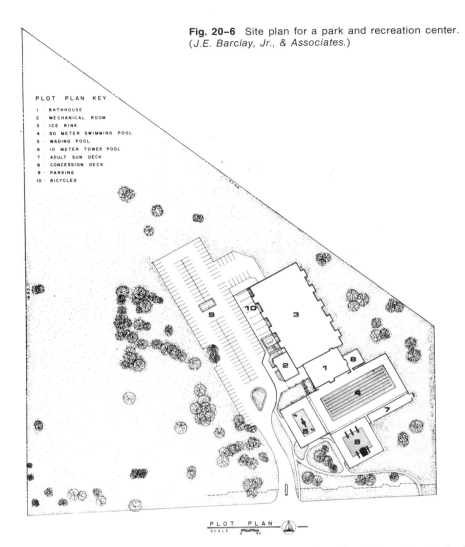

PLOT PLAN KEY
1 · BATHHOUSE
2 · MECHANICAL ROOM
3 · ICE RINK
4 · 50 METER SWIMMING POOL
5 · WADING POOL
6 · 10 METER TOWER POOL
7 · ADULT SUN DECK
8 · CONCESSION DECK
9 · PARKING
10 · BICYCLES

PLOT PLAN
SCALE 0 40

Second, it must be well engineered. It must be made of a good selection of materials and structural members that are well constructed (Fig. 20–10).

Third, it must have beauty or aesthetic value. A pleasant form which has appealing design qualities is essential to successful architecture (Fig. 20–11).

Architectural form is easily evaluated by trained architects. The design form of human environment is based on the merging of function, structure, and beauty.

The Architect's Office. The typical architectural firm is a partnership of two to four principal partners employing six to twelve persons, although some may consist of one or as many as one hundred or more. The large firm may offer all services— architectural design, structural design, mechanical engineering, civil engineering, landscape design, interior design, and the urban planning required for entire cities or a single structure. Architectural firms may concentrate on one service and contract with other design firms on a consultation basis for specialized serv-

ices. Some firms are organized to operate as a proprietorship, a partnership, or a corporation, depending upon the wishes of the principal architect. An architectural office is, in effect, a team on which each member has a position. The project director has team leaders who can assist in finding clients and contracting the office's talent for particular building projects. The National Council of Architecture Licensing Boards in Washington, D.C., promotes professional registration of architects within each state.

The Basic Drawings. For centuries architects have been preparing four basic drawings which are considered essential for communicating the details for construction. They are plan, section, elevation, and perspective drawings. The drawings require the same general technical projection methods of representation used in mechanical drawing.

1. *The plan.* This is a drawing which shows the horizontal plane in which spaces are arranged for human use. The three principal areas of the plan for residential design are living, sleeping, and service (Fig. 20–12). The plan is a section that cuts through walls showing structural opening.

2. *The cross-detail section.* In choosing a vertical plane to cut a section for construction details, typical material details will be labeled. A section across the entire structure is cut to assist in interpreting the relationship of the important spaces to be used. Proportions for construction and living can be examined more closely through a full section (Fig. 20–13).

3. *The elevation.* The elevation, or facade, defines the structural form and architectural style. The roof shape, sides, and side openings are

Schedule of Normal Services

An outline of professional services performed by Robert C. Taylor and Associates for their clients in the approximate sequence in which they are normally carried out:

1. PROGRAMMING OR SCHEMATIC DESIGN PHASE
   a. Conferences, research and study to determine the scope of the project, its relation to the site and its surroundings; development of the general program plan, suitable structural systems, and probable cost.
   b. Study of the pertinent building codes, ordinances, zoning requirements, standards, and recommendations of insurance underwriters.
   c. Preparation of a written statement of the requirements and limitations of the project.
   d. Preparation of schematic or diagramatic sketches of the building and its relation to the site, and budget estimates of its probable cost. Presentation of these studies to the client.

2. BASIC OR PRELIMINARY DESIGN PHASE
   a. Upon approval of the preferred schematic study, preparation of a preliminary site plan, floor plan, elevations, and details to indicate type of construction, materials, mechanical, and electrical equipment, and preparation of perspective studies of the building.
   b. Preparation of brief outline specifications establishing the quality of materials, workmanship, and mechanical and electrical systems to be employed.
   c. Preparation of a quantitative preliminary cost estimate based upon these preliminary drawings.

3. CONSTRUCTION DOCUMENT OR WORKING DRAWING PHASE
   a. Upon acceptance and approval of the preliminary study, the preparation of construction documents consisting of working drawings, detailed specifications, bidding information, and proposal forms based on the preliminary drawings, and indicating in detail the design, type, extent, and location of materials, structure, finishes, mechanical and special equipment.
   b. Preparation and coordination of structural, mechanical, electrical, and special equipment drawings and collaboration with engineers, technicians and specialists to facilitate proper integration of the structural, heating, ventilating, air conditioning, sanitary, electrical, fire prevention, and control systems.
   c. Adjustment of preliminary cost estimates as required.

4. SUPERVISION OR CONSTRUCTION PHASE
   a. Upon acceptance of the construction documents and the decision to build, the interviewing or investigating of contracting organizations, the negotiation of one of the various types of fixed-fee, or cost-plus agreements, or the issuance of invitations to qualified contractors to submit lump-sum construction proposals.
   b. Assistance in the analysis of the proposals received and recommendations for their acceptance, and in the preparation of the agreements, lending agency forms, and permit applications.
   c. Checking of subcontractors' drawings, data sheets, manufacturer's specifications, samples, and tests to determine their compliance with the construction documents. Examination of contractors' progress schedules and investigation and approval of subcontractors.
   d. Making periodic inspections of the construction in progress at intervals deemed necessary by the architect to ascertain that the work is being executed in substantial conformity to the contract documents. On large projects, arranging for the employment by the owner of a fulltime resident superintendent when required.
   e. Checking contractors' applications for payments, affadavits and waivers of lien, and issuing certificates to authorize proper payments.
   f. Upon completion of the construction, making an inspection of the work with the owner, verifying that guarantees and waivers are delivered, and issuing a certification that the terms of the construction contract have been satisfied.

This certification completes the normal services. The client will retain for his files sets of the documents used during construction, but all such documents are instruments of service and as such are the property of the architect and may not be used on other work except by specific agreement and consent of the architect.

Phase I of the services outlined represents about ten per cent of the total services offered. Phase II is about twenty per cent and Phases III and IV represent approximately fifty and twenty percent, respectively.

Special services available for an additional compensation for those clients who require them include: site selection and evaluation prior to initiation of the building program; design of furniture and custom or built in units; assisting in the selection of furnishings and equipment so that colors and materials are coordinated throughout the building; preparation of perspective drawings, diagrams, layouts and other materials suitable for use in fund raising or informational brochures; full time supervision; and administration of construction when the owner desires to award separate contracts.

ROBERT C. TAYLOR AND ASSOCIATES

ARCHITECTS ▼ ENGINEERS

212 SOUTH MARION STREET / OAK PARK, ILLINOIS 60302 / VILLAGE 8-8577

Fig. 20-7 Typical architectural services. (*Robert C. Taylor, AIA, and Associates.*)

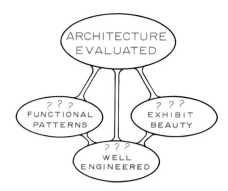

Fig. 20-8 The basis for evaluating architecture.

The four levels of communication discussed in Chapter 1 apply to architectural drafting (Fig. 20-16).

Level one—Creative communication. The preliminary architectural studies are sketched or laid out lightly for self-evaluation.

Level two—Technical communications. The sketches of refined ideas used for related professional approval.

Level three—Market communications. The pictorial drawings and colored perspective-presentation drawing designed for client interpretation and approval.

Level four—Construction communication. These are detailed plans, elevations, and sections along with specifications for builders' interpretation and approval. Working drawings by the supplier of materials are complementary to the architect's details.

Architectural Drafting Techniques. A line technique is the most important skill which can be developed for architectural design and detail. In Fig. 20-17 a series of lines are shown intersecting and distinctly stopping after a flared corner has been formed. The visual weights applied to

determined by examining required styles of the plan. Alterations can be made to provide balance or symmetry. The elevation studies are limited to the two-dimensional effect, which restricts in-depth study (Fig. 20-14).

4. *Perspective drawing.* The pictorial-presentation drawing is the easiest way to examine architectural form. The three-dimensional space enclosed in structural form can be described with one-, two-, or three-point perspective. Interior spaces are also studied and examined in perspective. Architectural perspective studies are not required to be mechanically accurate; therefore, freehand techniques along with good proportions are the essential objectives of establishing a realistic form to see in three dimensions (Fig. 20-15).

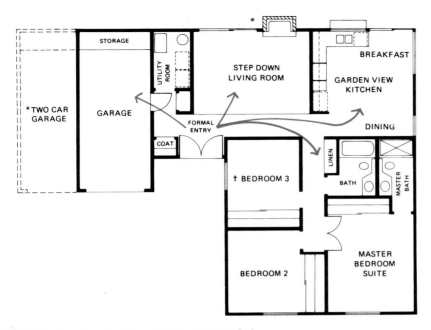

Fig. 20-9 Functional traffic patterns. (*Larwin Co.*)

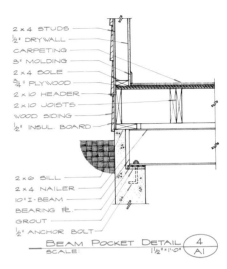

Fig. 20-10 Structural detail which indicates good engineering.

Fig. 20-11 Residential design in a contemporary form. (*Larwin Co.*)

lines are means of expressing spatial relationships. Good lines are required for reproduction, developing appeal for the design and discerning depth on a two-dimensional media as illustrated in Fig. 20-18. Line tone and texture are obvious complimentary traits of well-prepared architectural details.

Lettering. Architects have traditional and contemporary styles of lettering which are common and acceptable practice. The traditional lettering is based upon the Old Roman alphabet, as illustrated in Fig. 20-19. These Roman letters have serifs, and the examples used here are pressure-sensitive graphic aids which can be used on all media. Elaborate and important titles may be designed with Old Roman letters, but plain single-stroke Gothic is preferred on architectural working drawings (Fig. 20-20). The condensed and extended styles illustrated are typical of experienced professionals who can insert

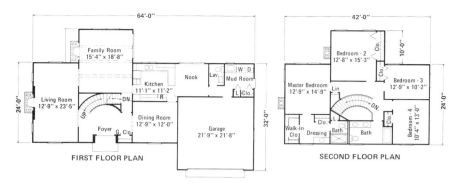

Fig. 20-12 A plan, the first of the four basic drawings. (*Inland Steel Company.*)

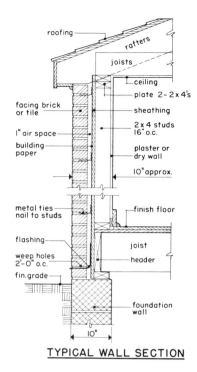

TYPICAL WALL SECTION

Fig. 20-13 A cross-detail section, the second of the four basic drawings.

SOUTHERN COLONIAL

Fig. 20-14 The elevation, the third of the four basic drawings. (*Inland Steel Company.*)

Fig. 20-15 A perspective, the fourth of the four basic drawings. (*Inland Steel Company.*)

brief notes or balance the drawing with either style.

Preparing title blocks. The image of an architectural firm may be quickly identified with a logo, as shown in Fig. 20-21. Some firms prefer preprinted forms which have the border, title, and revision blocks ready to be filled in with black ink. The typical title blocks shown in Fig. 20-22 were created by students preparing to play professional roles. Information required generally consists of owner, owner's address, type of structure, name and address of architect, title of the sheet and sheet number, date, scale, initials of detailing draftsman, initials of supervisor, and revision block.

Fig. 20-16 The four levels of graphic communication.

ACCEPTED   PREFERRED

Fig. 20-17 Architectural line technique.

Fig. 20-18. Techniques which feature the major proportions. (*Justus Company, Inc.*)

OLDTOWN PUBLIC LIBRARY
NINTH AVENUE AND STANTON STREET
OLDTOWN, TEXAS

DRAWN BY	DATE	JAMES R. JOHNSTON	REVISED BY	DATE
TRACED BY	DATE	ARCHITECT	JOB NO	
CHECKED BY	DATE	LEA BLDG. – OLDTOWN, TEX	SHEET **2** OF 8	

Fig. 20-19 Old Roman lettering applied in an architectural style.

ABCDEFGHIJKLMNOPQRSTUVWXYZ

FIRST FLOOR PLAN

SCALE   1/4" = 1'-0"

ABCDEFGHIJKLMNOPQRSTUVWXYZ

TYPICAL WALL SECTION

SCALE: 1/2" = 1'-0"

ABCDEFGHIJKLMNOPQRSTUVWXYZ

GENERAL NOTES AND MATERIALS

Fig. 20-20 Single-stroke Gothic lettering in condensed, extended, and the general Gothic stroke.

Fig. 20-21 Title block including the architectural firm's logo.

PROPOSED ADDITION AND ALTERATIONS
TO THE RESIDENCE OF
MR. & MRS. THOMAS ALLAN
145 PARK AVE.   GLEN ELLYN, ILLINOIS

DOUGLASS STROM AND ASSOCIATES
ARCHITECTURAL DESIGNERS
ONE SALT CREEK ROAD   HINSDALE, ILLINOIS

| DRAWN: DAS | CHECKED: | APPROVED: | SCALE: 1/4"=1'-0" | DATE: 4 MAY 1976 |

SHEET NO:
A1
OF TWELVE

Student-created title blocks

	CLIFFORD MARSHALL AND ASSOCIATES SEARS TOWER, CHICAGO	
	DRAWN: *LM* DATE: 10-16-72	JOB NO.
	CHECKED: SCALE: ¼"=1'-0"	A-10-72

	GARY STAIGER & ASSOCIATES: ARCHITECTS STANDARD OIL BUILDING, CHICAGO, ILLINOIS	
	DR BY GS DECEMBER 14, 1975	JOB NO.
	OK BY GRADE:	127514

	J. NEALE SCOTTY AND ASSOCIATES 3476 NORTH BEACON ST.– BOSTON, MASSACHUSETTES	
	DRAWN: J.N.S. DATE– OCT. 13, 1972	SHEET
	CHECKED: SCALE– ⅛"=1'-0"	A-33

	MICHAEL J. LEARY AND ASSOCIATES ARCHITECTURAL DESIGNERS JOHN HANCOCK CHICAGO, ILLINOIS	
	DRAWN: DATE:	SHEET NO.
	CHECKED: SCALE: ½"=1'-0"	A-3

Fig. 20-22 Student-created title blocks.

MATERIAL INDICATION

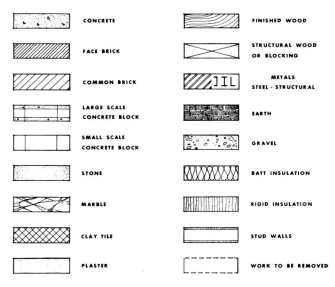

CONCRETE		FINISHED WOOD	
FACE BRICK		STRUCTURAL WOOD OR BLOCKING	
COMMON BRICK		METALS STEEL - STRUCTURAL	
LARGE SCALE CONCRETE BLOCK		EARTH	
SMALL SCALE CONCRETE BLOCK		GRAVEL	
STONE		BATT INSULATION	
MARBLE		RIGID INSULATION	
CLAY TILE		STUD WALLS	
PLASTER		WORK TO BE REMOVED	

Fig. 20-23 Architectural material symbols.

Fig. 20-24 Examples of architectural pressure-sensitive symbols. (*Para-tone, Inc.*)

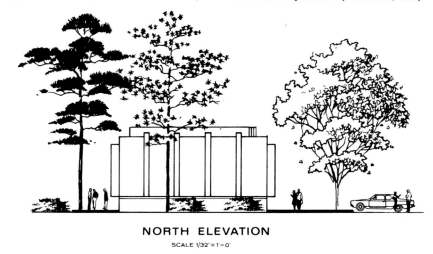

NORTH ELEVATION
SCALE 1/32"=1'-0"

Material symbols. Architectural symbols are relative to the materials used in building and are shown in Fig. 20-23. The techniques shown apply to plan, section, and elevation.

Pressure-sensitive symbols. Many graphic aids have been developed to meet professional techniques on both the plan and elevation. An example is Para-Tipe pressure symbols (Fig. 20-24), which are heat resistant and pressure sensitive. A large selection of landscape symbols, such as shrubs, hedges, and trees, are available in plan and elevation. Door swings, furniture, and plumbing and electrical symbols are only a few of the useful items available for updated presentation and detail drawings. They are transferred to vellum, film, or paper by burnishing with a smooth, blunt instrument.

Architectural wall symbols. The details prepared for the plan with dimensions have graphic meaning for all those whose work is related to the building industry. Typical residential wall symbols are illustrated in Fig. 20-25. The cross-hatched masonry wall at A is dimensioned to the out- side face of the wall (this includes solid brick, concrete, and concrete block walls). The frame wall at B is shown dimensioned to the face of the stud. The brick veneer wall at C

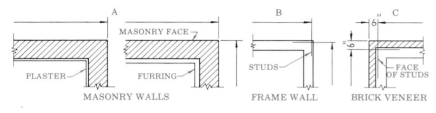

Fig. 20–25 Wall symbols.

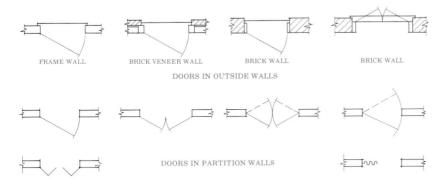

Fig. 20–26 Door symbols.

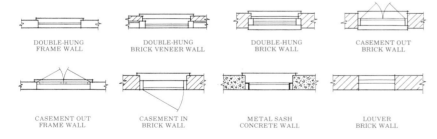

Fig. 20–27 Window symbols.

Fig. 20–28 House plan and plumbing template. (*Teledyne Post.*)

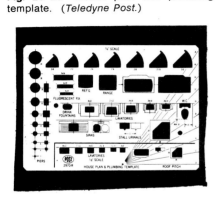

Fig. 20–29 Window template. (*Teledyne Post.*)

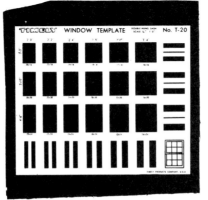

showing cross-hatched brick and a sole plate is dimensioned to the face of the stud. The remaining material is assumed to be exterior wall-facing material.

Door and window openings in symbol. Some door variations are illustrated in Fig. 20–26. The plan shows the outside door ajar in a single line, with an arc indicating the direction of the swing. The doors are also shown in a variety of ways for interior partitions. Typical windows are illustrated in Fig. 20–27. Symbols will assist in interpreting the type of window movement and type of wall material framing the window.

Templates for design and detail. The large variety of templates available today have complemented the quality of drawings developed in the architect's office. They are generally a light-weight plastic with openings which symbolize structural qualities or fixtures. The pencil or pen follows the outline of the opening to provide the graphic image, as illustrated in Fig. 20–28.

Figure 20–29 illustrates one of the many special-purpose templates that are available with door, window, landscape, furniture, structural-steel-shape, plumbing, and electrical symbols.

Dimensioning Techniques. The frame and brick veneer structures use similar methods for dimensioning structural openings in walls on the plan drawing. General dimensioning techniques apply as discussed in Chapter 6, except that the dimension line is unbroken and the dimensions appear above the dimension line, as shown in Fig. 20–30. Dimensions may be terminated by the conventional arrowhead or several other symbols, as illustrated in Fig. 20–31. Note that dimension lines may cross when determining sizes of interior rooms.

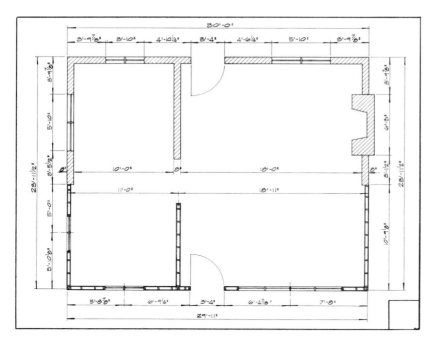

Fig. 20–30 Architectural style of dimensioning.

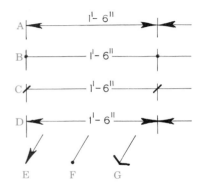

Fig. 20–31 Dimension-line arrowheads and alternate terminating symbols.

The plan carries only width and depth dimensions, as shown. Any required heights are carried as floor to ceiling dimensions on one of the elevations or on a pertinent wall section.

The building industry and the American National Standards Institute established Modular Coordination under A62.1, which was sponsored by the American Institute of Architects. The basic module is 4 in. for all United States building materials and products. The International Standards has established 10 cm (almost 4 in.) for countries using the metric system. Modular components are typically accommodated on a 4-in. center line. The multiple of 4 in. would include the planning for all buildings currently designed for modular characteristics. Figure 20–32 shows Crown Hall School of Architecture at the Illinois Institute of Technology. This building was designed by Mies Van der Rohe using a modular grid. Note the grid lines on the floor plan at B. Schematic line drawings of elevations at C are drawn to a modular grid. Modular planning reduces building costs with standard components and reduces excessive cutting and fixing time.

The dimensions used on architectural drawings may be termed construction or finish details. The elevations generally carry the expected finished dimensions, whereas the floor plan and sectional details may carry construction dimensions or finish dimensions, as shown in Fig. 20–33.

Dimension lines are placed about $\frac{3}{8}$ in. apart and are needed for finished overall building and room dimensions. The structural openings and wall offsets are defined with detailed construction dimensions. Arrows generally terminate the distances indicated by dimension lines between center lines for modules or extension lines for general design. Dimensions placed above the line are given in feet and inches as shown. Note the variations (Fig. 20–34) which are considered acceptable practice on architectural drawings.

Scales and Dimensioning Relationship. Residential design and details are usually developed with the aid of an architect's scale, which is discussed in Chapter 3. Normally the $\frac{1}{4}'' = 1'$-0 scale is best suited for house plans and small buildings. The usual scale for larger buildings is $\frac{1}{8}'' = 1'$-0. Plot plans may be drawn at $\frac{1}{10}'' = 1'$-0, $\frac{1}{32}'' = 1'$-0, or preferably with an engineer's scale at $1'' = 20'$, $1'' = 30'$ and $1'' = 40'$. Other scales used for enlarged details may be developed at $1'' = 1'$, $1\frac{1}{2}'' = 1'$ or sectional details may be well defined at $\frac{1}{2}'' = 1'$, $\frac{3}{4}'' = 1'$, and some details may require full size.

Materials Design and Detail. The architect and draftsman create environments which take on new forms

A

Fig. 20-32 A. Crown Hall School of Architecture; B. Modular floor plan of Crown
Hall; C. Schematic line elevations of Crown
Hall. (*The office of Mies Van der Rohe
and Hedrich—Blessing.*)

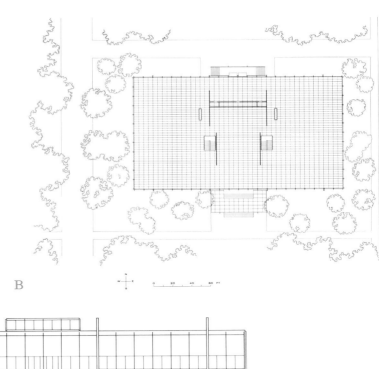

B

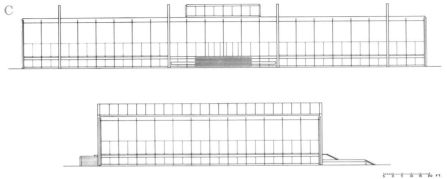

C

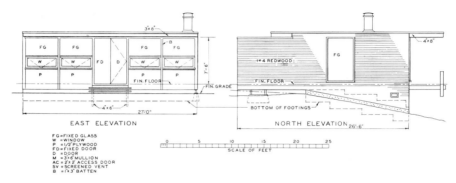

Fig. 20-33 Dimensions applied to an elevation. (*Adapted from drawings of Henry Hill, architect, AIA, and John W. Kruse, AIA, associate, San Francisco, CA.*)

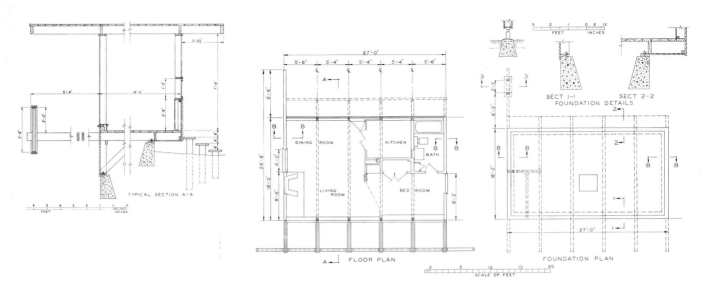

Fig. 20-34 Dimensions which apply to plan and section drawings. (*Adapted from drawings of Henry Hill, architect, AIA, and John W. Kruse, AIA, associate, San Francisco, CA.*)

with materials that must be extensively detailed for fabrication and construction. The Sweets Architectural Catalog for manufacturers of materials is one of the most important tools of the design team. A great variety of materials are available as standard units for the building trades, and many others are custom design. A few standards are considered here.

Lumber. Lumber may be specified by nominal dimensions, which differ from the actual dimensions of the surfaced wood. For example, most of the boards and lumber for residential construction are surfaced on four sides; the finished or dressed sizes are noted in Fig. 20-35. The standards for dressed lumber have been changed recently and the tables reflect these new standards.

Masonry. Figure 20-36 gives the sizes of brick building materials. The common brick has modular dimensions. Brick, block, stone, and stucco may be used for exterior and interior walls and floors. Properly bonded, most of these materials also serve as structural load-bearing walls for supporting floor and roof loads. They generally can be bonded (overlap units to develop strength) for structural unity, as illustrated in Fig. 20-37, and many decorative patterns formed. Limited upkeep and the various colors and forms complement architectural design.

Concrete. Concrete may be considered a structural unit for footing,

LUMBER				
Nominal Size	2×4	2×6	2×8	2×10	2×12
Dressed Size	$1\frac{1}{2}'' \times 3\frac{1}{2}''$	$1\frac{1}{2}'' \times 5\frac{1}{2}''$	$1\frac{1}{2}'' \times 7\frac{1}{2}''$	$1\frac{1}{2}'' \times 9\frac{1}{2}''$	$1\frac{1}{2}'' \times 11\frac{1}{2}''$
Nominal Size	4×6	4×8	4×10	6×6	6×8
Dressed Size	$3\frac{9}{16}'' \times 5\frac{1}{2}''$	$3\frac{9}{16}'' \times 7\frac{1}{2}''$	$3\frac{9}{16}'' \times 9\frac{1}{2}''$	$5\frac{1}{2}'' \times 5\frac{1}{2}''$	$5\frac{1}{2}'' \times 7\frac{1}{2}''$
Nominal Size	6×10	8×8	8×10		
Dressed Size	$5\frac{1}{2}'' \times 9\frac{1}{2}''$	$7\frac{1}{2}'' \times 7\frac{1}{2}''$	$7\frac{1}{2}'' \times 9\frac{1}{2}''$		

BOARDS				
Nominal Size	1×4	1×6	1×8	1×10	1×12
Actual Size, Common Boards	$\frac{3}{4}'' \times 3\frac{9}{16}''$	$\frac{3}{4}'' \times 5\frac{9}{16}''$	$\frac{3}{4}'' \times 7\frac{1}{2}''$	$\frac{3}{4}'' \times 9\frac{1}{2}''$	$\frac{3}{4}'' \times 11\frac{1}{2}''$
Actual Size, Shiplap	$\frac{3}{4}'' \times 3''$	$\frac{3}{4}'' \times 4\frac{15}{16}''$	$\frac{3}{4}'' \times 6\frac{7}{8}''$	$\frac{3}{4}'' \times 8\frac{7}{8}''$	$\frac{3}{4}'' \times 10\frac{7}{8}''$
Actual Size, Tongue-and-Groove	$\frac{3}{4}'' \times 3\frac{1}{4}''$	$\frac{3}{4}'' \times 5\frac{3}{16}''$	$\frac{3}{4}'' \times 7\frac{1}{8}''$	$\frac{3}{4}'' \times 9\frac{1}{8}''$	$\frac{3}{4}'' \times 11\frac{1}{8}''$

Fig. 20–35 New standard sizes for lumber and boards.

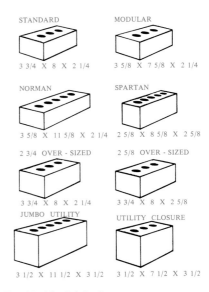

Fig. 20–36 Brick sizes.

foundation walls, and poured-in forms so as to support floor loads. Walls, both interior and exterior, may be formed and be precast for architectural finishes. Color, texture, and pattern may be interrelated to obtain the desirable wall detail.

Parts of a House. The essential parts of a house are illustrated in Fig. 20–38. All these parts do not appear in every house, and different materials may be used for some of the parts. Typical wood framing of an exterior wall is illustrated in Fig. 20–39. The framing begins on the foundation wall with the sill, header, and floor plate. After the stud wall is erected and sheathing, plywood, or insulation board is applied, many types of facing are possible. Horizontal or vertical siding is typical, but shakes or shingles are also common today along with brick veneer, illustrated in Fig. 20–40.

Housing frame. The framework of a building must be strong and rigid to ensure low maintenance costs over a long period of years. A prefabricated home and a custom designed home may vary in the construction procedures, but some features are typical.

Western framing. Western framing, or platform framing, at best has each floor framed separately (Fig. 20–41). The first floor is built on top of the foundation wall as a platform. Studs are one story in height and are used to develop the framework for the second story along with load-bearing interior walls.

Balloon framing. Balloon framing has studs extended for two full stories from sill to plate, as in Fig. 20–42. The second floor joists are carried on a false girt inserted in the stud wall. A box sill is used, and diagonal bracing brings rigidity to the corners. This system is not common in the construction industry today. It is necessary to know when remodeling older homes.

Plank and beam framing. This system employs heavier posts and beams to carry the continuous planking. The architectural effect can be an open ceiling with added height and fewer members which generally affect labor and savings (Fig. 20–43).

Sill construction. Types of sill and wall construction are shown in Fig.

8" ALL ROLOK WALL COMMON BOND

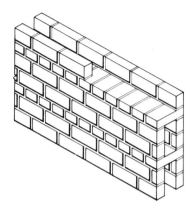

12" ALL ROLOK WALL FLEMISH BOND

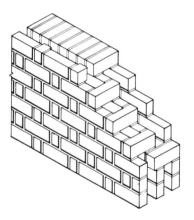

Fig. 20-37 Common and flemish bonds. (*Structural Clay Products Institute.*)

Fig. 20-38 Essential parts of a house. (*From National Bureau of Standards Circular 489.*)

1. Gable end
2. Louver
3. Interior trim
4. Shingles
5. Chimney cap
6. Flue linings
7. Flashing
8. Roofing felt
9. Roof sheathing
10. Ridge board
11. Rafters
12. Roof valley
13. Dormer window
14. Interior wall finish
15. Studs
16. Insulation
17. Diagonal sheathing
18. Sheathing paper
19. Window frame and sash
20. Corner board
21. Siding
22. Shutters
23. Exterior trim
24. Waterproofing
25. Foundation wall
26. Column
27. Joists
28. Basement floor
29. Gravel fill
30. Heating plant
31. Footing
32. Drain tile
33. Girder
34. Stairway
35. Subfloor
36. Hearth
37. Building paper
38. Finish floor
39. Fireplace
40. Downspout
41. Gutter
42. Bridging

20-44. At A the frame wall is set up on box-sill construction. Note the metal shield called for in sectioned detail. At B the brick veneer is started below the floor line on a stepped foundation wall. At C the slab construction is obviously reinforced with a wire mesh. The detail at D is a monolithic slab and foundation wall because it is laid in a continuous form. As detailed sections at E, F, G, H, and I are interpreted, obvious changes can be identified by the symbols and notes used by the architect.

Corner studs and sheathing. Some typical corner bracing is shown in Fig. 20-45. Diagonal sheathing was formerly typical rigid bracing, but horizontal sheathing and plywood along with modular insulation board are prevalent in today's labor-saving system. Plywood is used not only on exterior walls but for interior decking (subflooring) and roof sheathing. Anchor bolts are used to secure the superstructure to the substructure. They are normally $\frac{1}{2}$- to $\frac{3}{4}$-in. bolts placed about 8 ft apart and 18 in. into the concrete.

Roof designs. Some basic roof types are illustrated and named in Fig. 20-46. The style of architecture is noted many times through the shape of the roof. The common shapes are the gable, hip, flat, and shed roof.

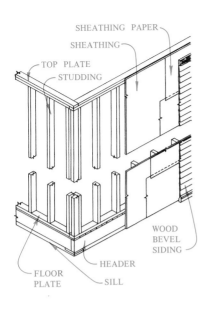

SHEATHING PAPER
SHEATHING
TOP PLATE
STUDDING

WOOD
BEVEL
SIDING

FLOOR
PLATE

HEADER
SILL

Fig. 20-39 Frame wall with wood siding.

Fig. 20-40 Frame wall with brick veneer.

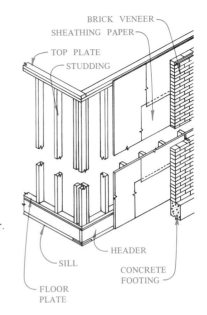

BRICK VENEER
SHEATHING PAPER
TOP PLATE
STUDDING

HEADER

SILL

CONCRETE
FOOTING

FLOOR
PLATE

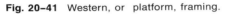

Fig. 20-41 Western, or platform, framing.

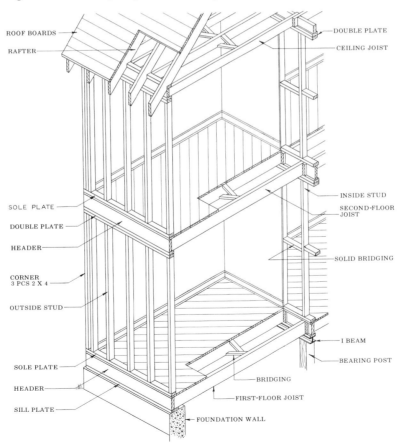

ROOF BOARDS
RAFTER

SOLE PLATE
DOUBLE PLATE
HEADER
CORNER
3 PCS 2 X 4
OUTSIDE STUD

SOLE PLATE
HEADER
SILL PLATE

DOUBLE PLATE
CEILING JOIST

INSIDE STUD
SECOND-FLOOR
JOIST

SOLID BRIDGING

I BEAM
BEARING POST

BRIDGING
FIRST-FLOOR JOIST

FOUNDATION WALL

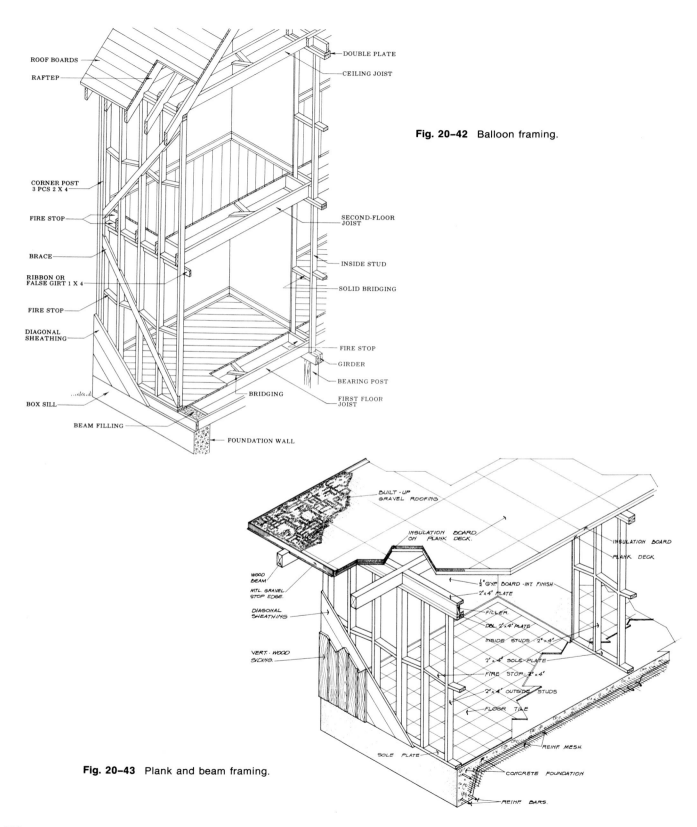

Fig. 20–42 Balloon framing.

ROOF BOARDS
RAFTEP
CORNER POST 3 PCS 2 X 4
FIRE STOP
BRACE
RIBBON OR FALSE GIRT 1 X 4
FIRE STOP
DIAGONAL SHEATHING
BOX SILL
BEAM FILLING
FOUNDATION WALL
BRIDGING

DOUBLE PLATE
CEILING JOIST
SECOND-FLOOR JOIST
INSIDE STUD
SOLID BRIDGING
FIRE STOP
GIRDER
BEARING POST
FIRST FLOOR JOIST

Fig. 20–43 Plank and beam framing.

BUILT-UP GRAVEL ROOFING
INSULATION BOARD ON PLANK DECK.
INSULATION BOARD PLANK DECK
WOOD BEAM
MTL. GRAVEL STOP EDGE.
DIAGONAL SHEATHING
VERT. WOOD SIDING.
½" GYP. BOARD-INT. FINISH
2"x 4" PLATE
FILLER
DBL. 2"x 4" PLATE
INSIDE STUDS 2"x 4"
2"x 4" SOLE PLATE
FIRE STOP 2"x 4"
2"x 4" OUTSIDE STUDS
FLOOR TILE
SOLE PLATE
REINF. MESH.
CONCRETE FOUNDATION
REINF. BARS.

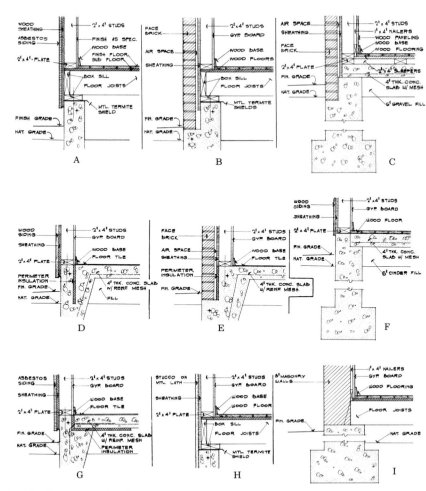

Labels in figure A: WOOD SHEATHING, ASBESTOS SIDING, WOOD BASE, 2'x4' PLATE, 2'x4' STUDS, FINISH AS SPEC., FINISH FLOOR, SUB FLOOR, BOX SILL, FLOOR JOISTS, MTL. TERMITE SHIELD, FINISH GRADE, NAT. GRADE

Labels in figure B: FACE BRICK, AIR SPACE, SHEATHING, 2'x4' STUDS, GYP BOARD, WOOD BASE, WOOD FLOORS, BOX SILL, FLOOR JOISTS, MTL. TERMITE SHIELDS, FIN. GRADE, NAT. GRADE

Labels in figure C: AIR SPACE, SHEATHING, FACE BRICK, 2'x4' PLATE, FIN. GRADE, NAT. GRADE, 2'x4' STUDS, WOOD PANELING, WOOD BASE, WOOD FLOORING, 1'x4' SLEEPERS, 4' THK. CONC. SLAB W/ MESH, 6' GRAVEL FILL

Labels in figure D: WOOD SIDING, SHEATHING, 2'x4' PLATE, PERIMETER INSULATION, FIN. GRADE, NAT. GRADE, 2'x4' STUDS, GYP BOARD, WOOD BASE, FLOOR TILE, 4' THK. CONC. SLAB W/ REINF MESH, FILL

Labels in figure E: FACE BRICK, AIR SPACE, SHEATHING, PERIMETER INSULATION, FIN. GRADE, 2'x4' STUDS, GYP BOARD, WOOD BASE, FLOOR TILE, 4' THK. CONC. SLAB W/ REINF. MESH

Labels in figure F: WOOD SIDING, SHEATHING, 2'x4' PLATE, FIN. GRADE, NAT. GRADE, 2'x4' STUDS, GYP BOARD, WOOD FLOOR, 4' THK. CONC. SLAB W/ MESH, 6' CINDER FILL

Labels in figure G: ASBESTOS SIDING, SHEATHING, 2'x4' PLATE, FIN. GRADE, NAT. GRADE, 2'x4' STUDS, GYP BOARD, WOOD BASE, FLOOR TILE, 4' THK. CONC. SLAB W/ REINF. MESH, PERIMETER INSULATION

Labels in figure H: STUCCO ON MTL. LATH, SHEATHING, 2'x4' PLATE, 2'x4' STUDS, GYP BOARD, WOOD BASE, WOOD FLOOR, BOX SILL, FLOOR JOISTS, FIN. GRADE, MTL. TERMITE SHIELD

Labels in figure I: 8' MASONRY WALLS, 1'x4' NAILERS, GYP BOARD, WOOD FLOORING, FLOOR JOISTS, FIN. GRADE, NAT. GRADE

Fig. 20-44 Sill constructions.

The mansard, gambrel, butterfly combination, clerestory, and A-frame are a little more unique to specific design styles. The terms which are common are the rise and run, which determine what is called a roof pitch.

Roof-framing sections. Figure 20–47 illustrates some typical cornice details. Note the terms which apply to the members that finish off the joints between the wall and roof. The illustrations also include open rafter ends and boxed rafters, along with built-up flat roofing for plank and beam construction. Aluminum and galvanized gutters are still common ways of controlling rain water, and one detail shows a built-in metal-lined gutter, which is a costly detail common on formal designs. The note "COND @ BRICK" (condition at brick) indicates a possible addition of brick veneer construction.

Stairway framing and detail. Three types of stairs are (1) the straight run, (2) the platform, and (3) the circular. The *rise* of a flight of stairs is the height measured from the top of one floor to the top of the next floor (Fig. 20–48). The *run* of a flight of stairs is the horizontal distance from the face of the first riser to the

Fig. 20-45 Corner studs and sheathing.

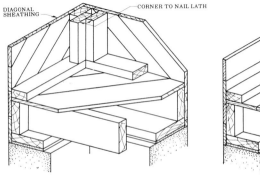

DIAGONAL SHEATHING, CORNER TO NAIL LATH

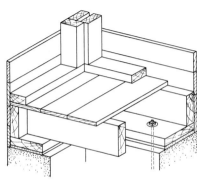

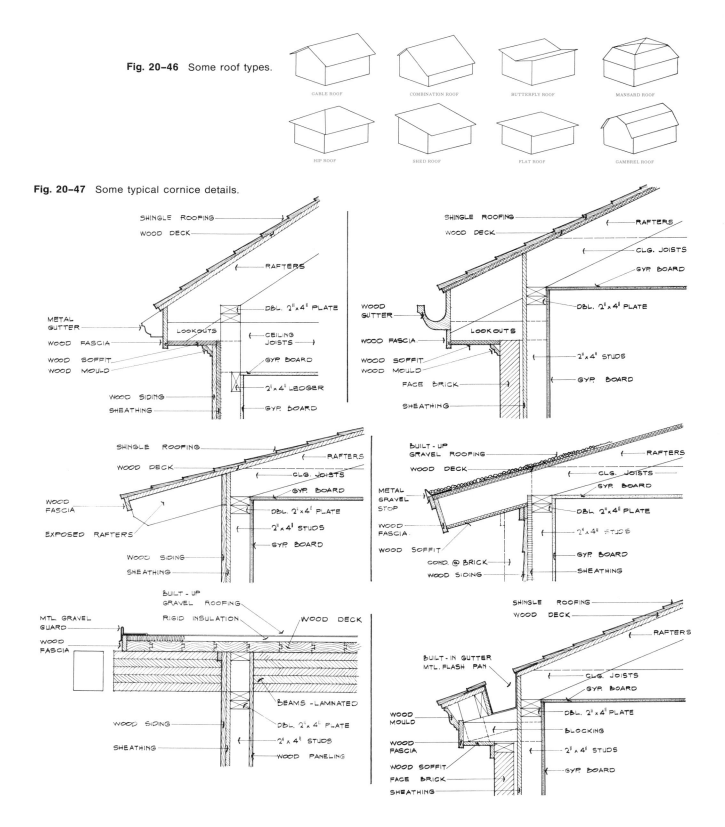

Fig. 20-46 Some roof types.

GABLE ROOF   COMBINATION ROOF   BUTTERFLY ROOF   MANSARD ROOF

HIP ROOF   SHED ROOF   FLAT ROOF   GAMBREL ROOF

Fig. 20-47 Some typical cornice details.

SHINGLE ROOFING
WOOD DECK
RAFTERS
METAL GUTTER
LOOKOUTS
DBL. 2" x 4" PLATE
WOOD FASCIA
CEILING JOISTS
WOOD SOFFIT
WOOD MOULD
GYP. BOARD
WOOD SIDING
2" x 4" LEDGER
SHEATHING
GYP. BOARD

SHINGLE ROOFING
WOOD DECK
RAFTERS
CLG. JOISTS
GYP. BOARD
WOOD GUTTER
DBL. 2" x 4" PLATE
WOOD FASCIA
LOOKOUTS
WOOD SOFFIT
WOOD MOULD
2" x 4" STUDS
FACE BRICK
GYP. BOARD
SHEATHING

SHINGLE ROOFING
WOOD DECK
RAFTERS
CLG. JOISTS
GYP. BOARD
WOOD FASCIA
DBL. 2" x 4" PLATE
2" x 4" STUDS
EXPOSED RAFTERS
GYP. BOARD
WOOD SIDING
SHEATHING

BUILT-UP GRAVEL ROOFING
WOOD DECK
RAFTERS
CLG. JOISTS
GYP. BOARD
METAL GRAVEL STOP
DBL. 2" x 4" PLATE
WOOD FASCIA
2" x 4" STUDS
WOOD SOFFIT
GYP. BOARD
COND. @ BRICK
WOOD SIDING
SHEATHING

BUILT-UP GRAVEL ROOFING
RIGID INSULATION
WOOD DECK
MTL. GRAVEL GUARD
WOOD FASCIA
BEAMS -LAMINATED
WOOD SIDING
DBL. 2" x 4" PLATE
SHEATHING
2" x 4" STUDS
WOOD PANELING

SHINGLE ROOFING
WOOD DECK
RAFTERS
BUILT-IN GUTTER MTL. FLASH PAN
CLG. JOISTS
GYP. BOARD
WOOD MOULD
DBL. 2" x 4" PLATE
BLOCKING
WOOD FASCIA
2" x 4" STUDS
WOOD SOFFIT
GYP. BOARD
FACE BRICK
SHEATHING

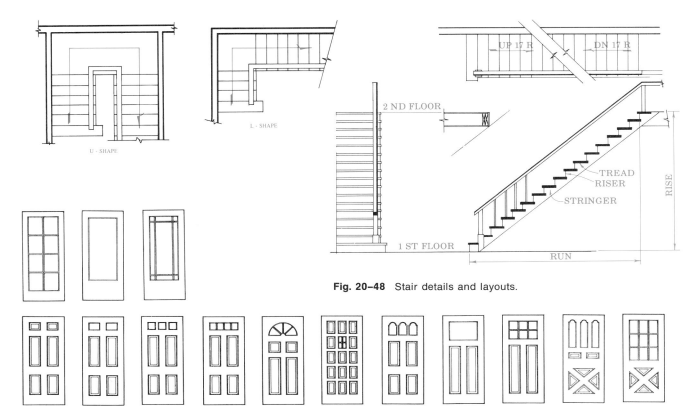

U - SHAPE

L - SHAPE

UP 17 R   DN 17 R

2 ND FLOOR

1 ST FLOOR

TREAD
RISER

STRINGER

RISE

RUN

Fig. 20–48 Stair details and layouts.

Fig. 20–49 Door patterns.

face of the last riser in one stairwell, which equals the sum of the width of the treads. The riser, or height from one step to the next, is generally $6\frac{1}{2}$ to $7\frac{1}{2}$ in. The width of the tread is such that the sum of one riser and one tread is about 17 to 18 in. (a 7-in. riser and an 11-in. tread is considered a general standard). Note the use of the scale to divide the floor-to-floor height into the number of risers. On working drawings it is not customary to draw entire flights of stairs, and so a break line is used to show what is on the floor under it. A good stairway should be comfortable for ascending and descending with a feeling of security. A simple rule for determining safe rise is to keep the angle of incline between 28 and 35 degrees. Note the typical terms applied to the illustration.

Doors. Usual heights of doors are 6 ft 8 in. and 7 ft 0 in. The width may vary for some purposes from 2 ft 0 in. to 3 ft 0 in., but the usual widths are 2 ft 8 in. and 3 ft 0 in. The thicknesses vary from $1\frac{3}{8}$ in. to $1\frac{3}{4}$ in. for interior doors and from $1\frac{3}{4}$ in. to $2\frac{1}{2}$ in. for exterior doors. The head, jamb, and sill details may vary with the swinging, sliding, or folding door used for residential functions. The fit must be close to keep out wind, rain, and snow and yet open and close easily (Fig. 20–49).

Exterior doors are usually larger to allow access for furniture and traffic (3 ft 0 in.). Bedroom doors are usually 2 ft 6 in. and bathroom doors run 2 ft 4 in. The bifold and folding accordian doors have special features which need the manufacturer's individual specifications for assembly.

Details for a formal entrance are illustrated in Fig. 20–50.

Windows. Residential design will prescribe the style of window to be used and how it is to be placed. Double-hung and casement windows are practical for most residential design, although horizontal sliding types may be considered contemporary and very popular.

The casement window is popular for French and English designs, and the double-hung is commonly used for Colonial and American styles. Figure 20–51 illustrates the various types available to the architect. Casement windows are hinged at the sides to swing open (in or out). Projected windows are hinged at the top or bottom. The one hinged at the bottom is considered a hopper type. The one hinged at the top is termed an awning

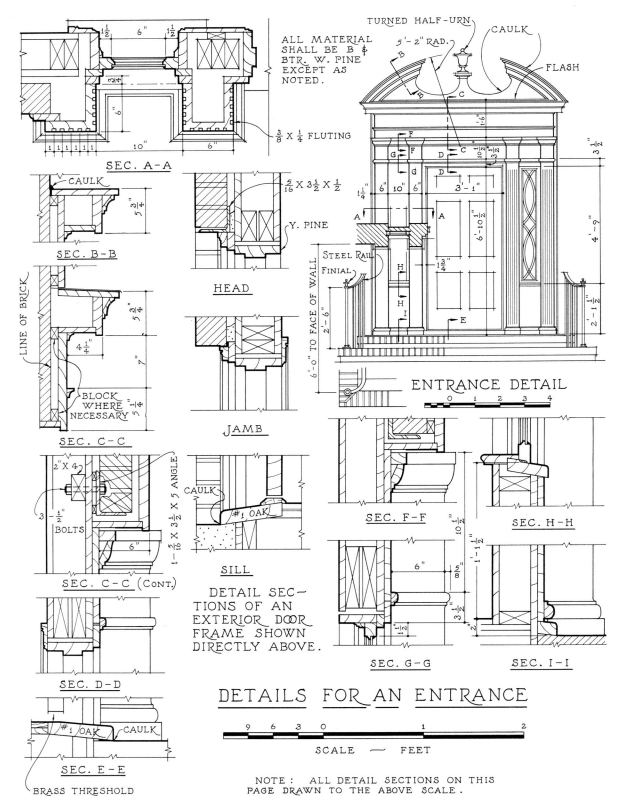

ALL MATERIAL SHALL BE B & BTR. W. PINE EXCEPT AS NOTED.

$\frac{3}{8}$ X $\frac{1}{4}$ FLUTING

SEC. A-A

CAULK

SEC. B-B

LINE OF BRICK

BLOCK WHERE NECESSARY

SEC. C-C

2" X 4"

3 - $\frac{1}{2}$" BOLTS

SEC. C-C (CONT.)

SEC. D-D

#1 OAK   CAULK

SEC. E-E

BRASS THRESHOLD

$\frac{5}{16}$ X $3\frac{1}{2}$ X $\frac{1}{2}$

Y. PINE

HEAD

JAMB

1 - $\frac{5}{16}$ X $3\frac{1}{2}$ X 5 ANGLE

CAULK

#1 OAK

SILL

DETAIL SECTIONS OF AN EXTERIOR DOOR FRAME SHOWN DIRECTLY ABOVE.

TURNED HALF-URN   CAULK

5' - 2" RAD.

FLASH

STEEL RAIL FINIAL

6'-0" TO FACE OF WALL

2'-6"

ENTRANCE DETAIL

0 1 2 3 4

SEC. F-F

SEC. H-H

SEC. G-G

SEC. I-I

DETAILS FOR AN ENTRANCE

9   6   3   0   1   2

SCALE — FEET

NOTE: ALL DETAIL SECTIONS ON THIS PAGE DRAWN TO THE ABOVE SCALE.

Fig. 20–50 Details for a formal entrance.

type. Sliding windows move sidewise instead of up and down, as ordinary double-hung windows do. The double-hung contains two independent sashes that can move in a vertical track. A counterbalance holds them at any position desired, and the new style allows a press-in spring-loaded track which is very convenient. Fixed windows and jalousie are special featured windows. Some common terms are illustrated in Fig. 20–52. Normally, windows are placed on the elevation so that they are aligned with the top of the door. Sectional details in Fig. 20-53 are for interpretation.

Working Drawings. These form the most important class of drawings and include plans, elevations, sections, schedules, schematics, and details. When these are read with the specifications of materials and finish, they give information for the erection of the building. Working drawings for a ranch house developed from a preliminary design (Fig. 20–54) are in Figs. 20–55 to 20–60. They include a complete set of plans.

Plans. The basement plan in Fig. 20–55 must be completely dimensioned because the foundation construction of the house is begun with this plan. It should be checked with the first-floor plan and may be traced from it. Note the foundations for the porch and garage. Windows should be located for structural advantage.

The first-floor plan in Fig. 20–56 is a horizontal section taken above the floor and prepared to show all walls, doors, windows, and other structural features. All the fixed features should include cabinets, stairways, heating and plumbing, lighting outlets in walls, and ceilings. Frame walls are drawn 6 in. thick. After locating windows and doors, draw them as conventional symbols. Sizes of doors and windows are found on the schedules.

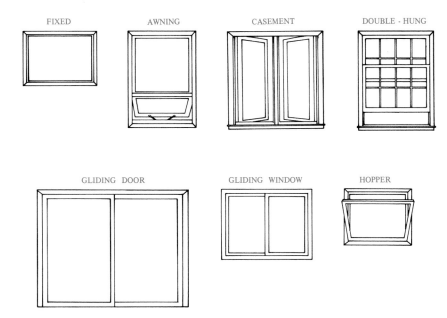

Fig. 20–51 Types of windows.

Fig. 20–52 Window in a frame wall with sectional details and technical terms.

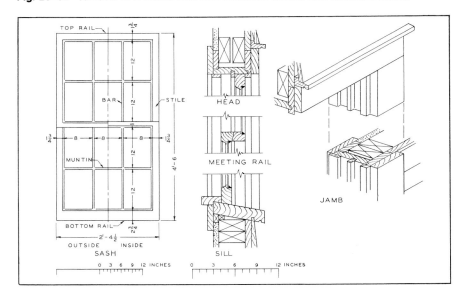

Elevations. Front and side elevations are shown in Fig. 20–57. Four elevations should be included in a complete set of plans. The elevations show the exterior appearance of the house, floor and ceiling heights, openings for windows, doors, roof pitch, and selected materials. To draw an elevation, start with the grade line; then locate the center lines which indicate the finish working dimensions. The elevation openings can be developed by working over a print of the floor plan.

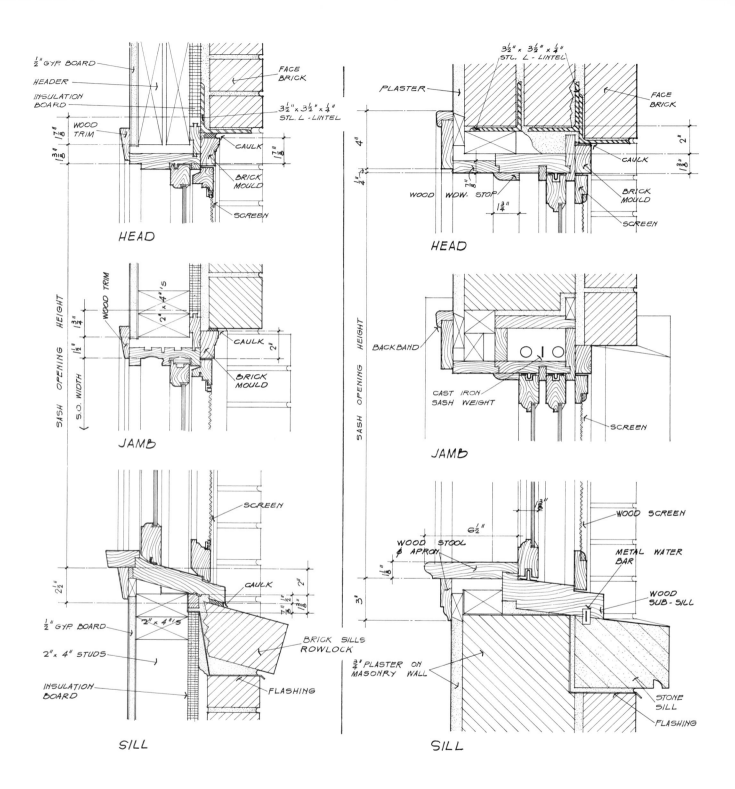

Fig. 20–53 Detail sections of a double-hung window in a brick veneer and a masonry wall.

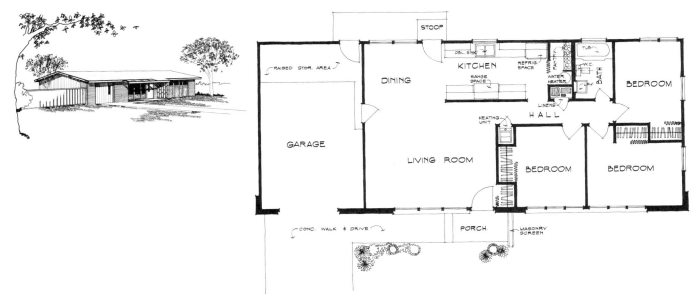

Fig. 20–54 A preliminary study developed for client approval.

Fig. 20–55 Foundation plan.

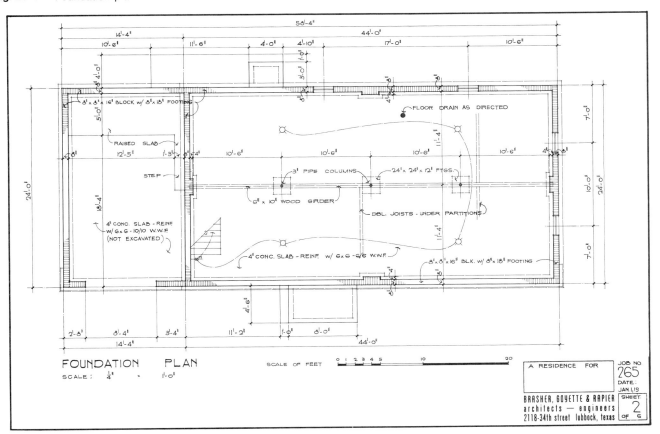

FOUNDATION   PLAN

SCALE: ¼" = 1'-0"

SCALE OF FEET   0 1 2 3 4 5   10   20

A RESIDENCE FOR

BRASHER, GOYETTE & RAPIER
architects — engineers
2118-34th street lubbock, texas

JOB NO 265
DATE: JAN 1, 19
SHEET 2 OF 6

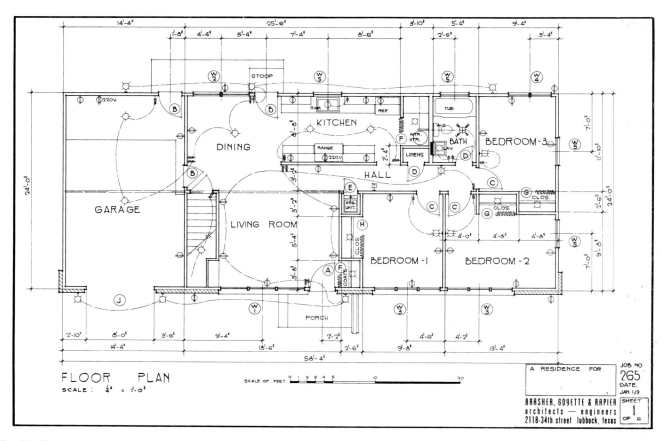

Fig. 20-56 Floor plan.

Sections. A typical wall section from substructure to superstructure is detailed in Fig. 20-58. The scale is larger, and clarifies and describes the ¼-in. scale of the elevation. Door and window details are also developed as the building progresses so that the millwork and custom framework will assemble readily.

Electrical planning. Electrical wiring circuits are shown in Fig. 20-59. It shows electric outlets and switches located for the major appliances. A list of symbols is shown to identify the ones on the layout. Circuits for 110-V and 220-V service are shown. Additional data would be listed in the formal specifications.

Site plan and schedules. The site illustrates an ordinary urban lot and locates the house on the lot (Fig. 20-60). It should give complete and accurate dimensions and indicate all driveways, sidewalks, and pertinent information required by the building inspector. Note the roof plan with a center ridge which represents a gable roof. The north symbol should be noted after studying the solar orientation best suited for this contemporary ranch. What scale would be suitable for the development of this site plan?

The schedules relate five types of windows and nine types of doors which are used. The typical finish

required for the floors, walls, and ceiling are perhaps the important details to examine.

Plumbing symbols. Some of the standard symbols are illustrated on architectural drawings. There are schematics and diagrams which should be examined by referring to the latest edition of *Architectural Graphic Standards,* which are periodically revised by the American Institute of Architects.

Electrical symbols. The many standard symbols are shown in Fig. 20-61. These are typically used on house plans to indicate the basic circuit layout required for normal activity.

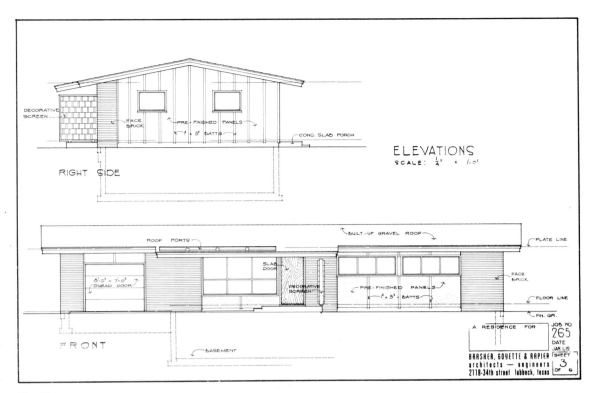

Fig. 20–57 Elevations.

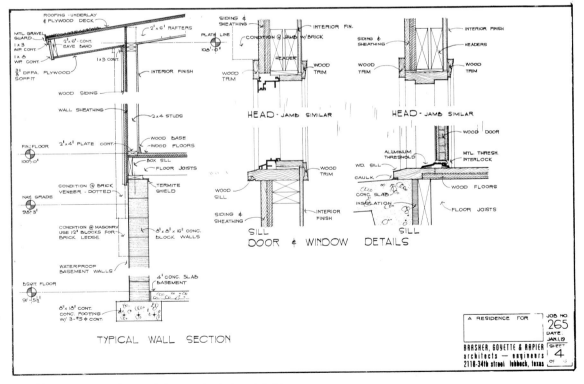

Fig. 20–58 Wall section and door and window details.

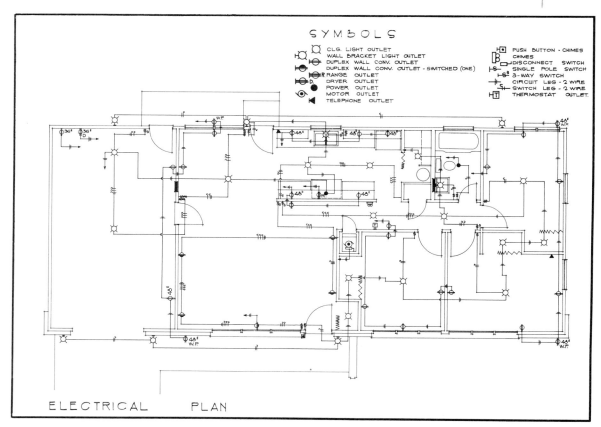

Fig. 20–59 Electrical plan.

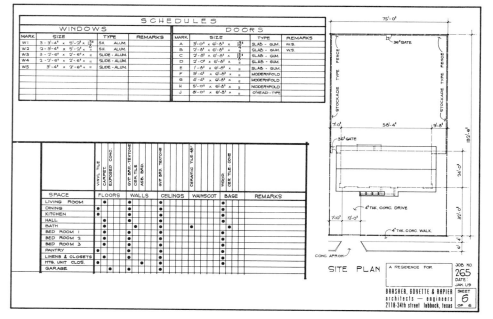

Fig. 20–60 Schedules, site, and roof plan.

Fig. 20–61 Electrical symbols.

GENERAL OUTLETS

Ceiling Wall

Ceiling	Wall	Description
◯	—◯	Outlet
Ⓑ	—Ⓑ	Blanked Outlet
Ⓓ		Drop Cord
Ⓔ	—Ⓔ	Electric Outlet For use only when circle used alone might be confused with columns, plumbing symbols, etc.
Ⓕ	—Ⓕ	Fan Outlet
Ⓙ	—Ⓙ	Junction Box
Ⓛ	—Ⓛ	Lamp Holder
Ⓛ$_{PS}$	—Ⓛ$_{PS}$	Lamp Holder with Pull Switch
Ⓢ	—Ⓢ	Pull Switch
Ⓥ	—Ⓥ	Outlet for Vapor Discharge Lamp
Ⓧ	—Ⓧ	Exit-light Outlet
Ⓒ	—Ⓒ	Clock Outlet (Specify Voltage)

CONVENIENCE OUTLETS

Duplex Convenience Outlet

Convenience Outlet other than Duplex$_{1,3}$
1=Single, 3=Triplex, etc.

Weatherproof Convenience Outlet $_{WP}$

Range Outlet $_{R}$

Switch and Convenience Outlet $_{S}$

Radio and Convenience Outlet [R]

Special Purpose Outlet
(Des. in Spec.)

Floor Outlet

SWITCH OUTLETS

S — Single-pole Switch

S_2 — Double-pole Switch

S_3 — Three-way Switch

S_4 — Four-way Switch

S_D — Automatic Door Switch

S_E — Electrolier Switch

S_K — Key-operated Switch

S_P — Switch and Pilot Lamp

S_{CB} — Circuit Breaker

S_{WCB} — Weatherproof Circuit Breaker

S_{MC} — Momentary Contact Switch

S_{RC} — Remote-control Switch

S_{WP} — Weatherproof Switch

S_F — Fused Switch

S_{WF} — Weatherproof Fused Switch

SPECIAL OUTLETS

⊖$_{a,b,c,etc.}$ $S_{a,b,c,etc.}$

Any standard symbol as given above with the addition of a lower-case subscript letter may be used to designate some special variation of standard equipment of particular interest in a specific set of architectural plans.

When used they must be listed in the Key of Symbols on each drawing and if necessary further described in the specifications.

AUXILIARY SYSTEMS

⦿ — Pushbutton

⊡ — Buzzer

⊐ — Bell

◇ — Annunciator

◀ — Outside Telephone

◁ — Interconnecting Telephone

◁| — Telephone Switchboard

Ⓣ — Bell-ringing Transformer

D — Electric Door Opener

F — Fire-alarm Bell

F — Fire-alarm Station

✖ — City Fire-alarm Station

FA — Fire-Alarm Central Station

FS — Automatic Fire-alarm Device

W — Watchman's Station

[W] — Watchman's Central Station

H — Horn

N — Nurse's Signal Plug

M — Maid's Signal Plug

R — Radio Outlet

[SC] — Signal Central Station

▢ — Interconnection Box

┇┇┇┇ — Battery

—·—·— Auxiliary System Circuits

Note: Any line without further designation indicates a 2-wire system. For a greater number of wires designate with numerals in manner similar to —·—
12-No. 18W-3/4"C., or designate by number corresponding to listing in Schedule.

Fig. 20-61 (*Cont.*)

☐ a,b,c   Special Auxiliary Outlets
Subscript letters refer to notes on plans
or detailed description in specifications.

S   Single-pole switch

S$_P$ Single-pole switch and pilot light

S$_3$ Three-way switch

S$_D$ Door-operated switch

⏀   Duplex convenience outlet

⏀   Duplex convenience outlet for
2 circuit installation

R⏀   Electric-range outlet

WP   Weatherproof

Ⓜ   Vent-hood fan motor

TV△   Television-antenna outlet

▬   Lighting panel

▲   Telephone outlet

J▲   Telephone jack

▣   Push button

◻   Door bell

○   Ceiling or pendent mounted
incandescent-lighting fixture

○−   Wall mounted incandescent-
lighting fixture

⊢—⊣   Surface mounted flourescent-
lighting fixture

▢   Recessed incandescent-lighting
fixture

▭○▭   Recessed flourescent-lighting
fixture

———   Conduit run in ceiling and walls

- - - - Conduit run under floors and in walls

5,6
⊬⊬⊬→ { Arrow indicates home run to
lighting panel
Hash lines indicate number of wires
Numerals indicate circuit numbers

A   Letters indicate type of lighting
fixture

(See lighting-fixture schedule)

PANELS, CIRCUITS, AND MISCELLANEOUS

▬   Lighting Panel

▨   Power Panel

———   Branch Circuit; Concealed in
Ceiling or Wall

—·—   Branch Circuit; Concealed in Floor

- - - - Branch Circuit; Exposed

→   Home Run to Panel Board. Indicate
number of circuits by number of
arrows.
Note: Any circuit without further
designation indicates a two-wire
circuit. For a greater number of
wires indicate as follows: —⧸⧸⧸
(3 wires) —⧸⧸⧸⧸ (4 wires), etc.

———   Feeders. Note: Use heavy lines and
designate by number corresponding
to listing in Feeder Schedule.

⧈   Underfloor Duct and Junction Box—
Triple System
Note: For a double or single systems
eliminate one or two lines. This
symbol is equally adaptable to
auxiliary system layouts.

Ⓖ   Generator

Ⓜ   Motor

Ⓘ   Instrument

Ⓣ   Power Transformer
(Or draw to scale.)

⊠   Controller

▭   Isolating Switch

Contemporary House Plans. Contemporary house plans may be used for study and comparison. The house should be examined for the merging of function, structure, and beauty, as previously described.

Split-level design. The town house in Fig. 20–62 at A, B, and C is described as a split-level residence. It is of masonry and frame construction with exterior walls of textured cement and patio garden walls surrounding the patio-viewed living room. The red clay tile roof has copper flashing.

The traffic pattern is functional and contemporary in the openness exhibited in the living-room level. The main-stair traffic pattern is centrally located for efficient use.

A

B

C

Fig. 20–62 A split-level town house. A. Perspective; B. floor plan; and C. interior. (*Larwin Co.*)

A

B

C

The California-styled House. The house in Fig. 20–63 at A has the dramatic open spaciousness illustrated in the interior photograph, Fig. 20–63 at B. The garden-view kitchen with breakfast and dining is compact and functional (Fig. 20–63 at C). The exterior materials blend the textured vertical boarding with a highly textured roof line. The contemporary form that includes a spacious master suite is shown in the plan at D.

D

Fig. 20–63 The California-styled house. A. Exterior; B. interior; C. interior showing kitchen; D. floor plan. (*Larwin Co.*)

The Vacation Home. Figure 20–64 describes the retreat lodge which is ideal for a rolling and wooded lakeside. The home is a prefabricated design with a few variations. Natural Western Red Cedar causes the home to be warmer in the winter and cooler in the summer.

The Contemporary House of God. Figures 20–65A, B, C, and D aptly describe the contemporary form often exhibited in the community church. The plot plan at A and the floor plan at B are typical preliminary studies. The interior and exterior perspectives at C and D give a feeling of materials, reverence perhaps, and function.

Fig. 20–64 A vacation home. (*Justus Co.*)

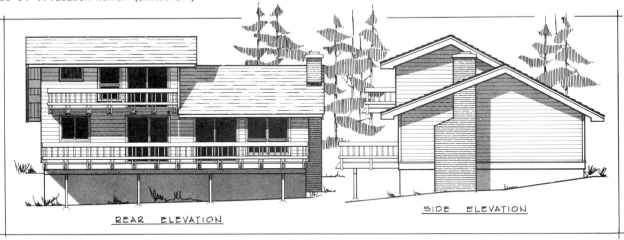

REAR ELEVATION

SIDE ELEVATION

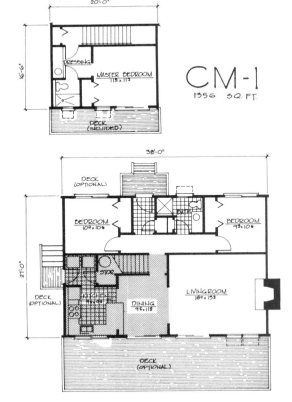

CM-1
1356 SQ. FT.

CM-2
1148 SQ. FT.

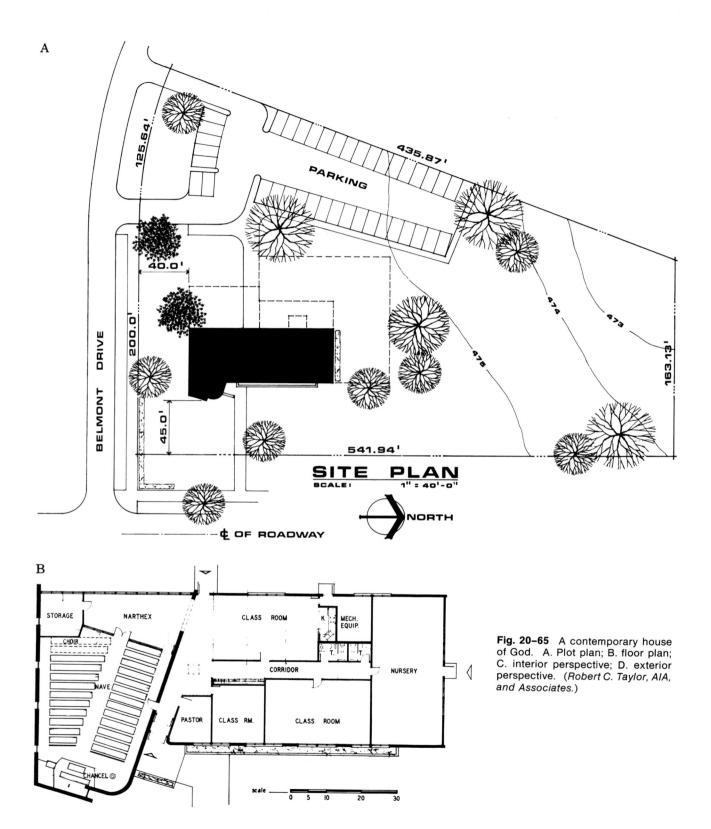

A

125.64'

PARKING

435.87'

40.0'

BELMONT DRIVE

200.0'

45.0'

474

473

478

163.13'

541.94'

SITE PLAN
SCALE: 1" = 40'-0"

NORTH

⊄ OF ROADWAY

B

STORAGE   NARTHEX

CHOIR

NAVE

PASTOR   CLASS RM.

CHANCEL ⊙

CLASS ROOM   K   MECH. EQUIP.

CORRIDOR

NURSERY

CLASS ROOM

scale   0 5 10   20   30

Fig. 20-65 A contemporary house of God. A. Plot plan; B. floor plan; C. interior perspective; D. exterior perspective. (*Robert C. Taylor, AIA, and Associates.*)

C

ROBERT C. TAYLOR & ASSOCIATES
ARCHITECTS & ENGINEERS
212 So. MARION ST.   OAK PARK ILL.   60302

D

Fig. 20–65 *(Cont.)*

Working Drawings for a Contemporary Multi-Level Residence. The French mansard roof which is shaped to fit and hug a residence is shown in Fig. 20–66. It is textured so as to give a feeling of warmth in natural wood shades. The site plan is contoured to reveal the gentle rolling pattern of the land (Fig. 20–67). The family-room (lower-level), main-level (Fig. 20–68), and upper-level (Fig. 20–69) plans are gracefully brought together with a central traffic pattern. The front elevation (Fig. 20–70) illustrates the rolling contour of the site. Note the line weights on the cedar shakes, which add depth to the roof forms. The rear

Fig. 20–66 Contemporary, French-mansard styled residence. (*Orrin Dressler, designer-contractor.*)

Fig. 20–67 Site plan. (*Orrin Dressler, designer-contractor.*)

SHEET:	DRAWING DESCRIPTION
A1	SITE PLAN & DRAWING INDEX
A2	MAIN FLOOR PLAN
A2ᵤ	UPPER LEVEL FLOOR PLAN
A2ₗ	LOWER LEVEL FLOOR PLAN
A3	FRONT ELEVATION
A4	LEFT ELEVATION
A5	REAR ELEVATION
A6	RIGHT ELEVATION
A7	TYPICAL WALL SECTIONS
A8	DOOR & WINDOW SCHEDULES
A9	LONGITUDINAL SECTION "X-X"
—	SCHEDULES - SCHEMATICS

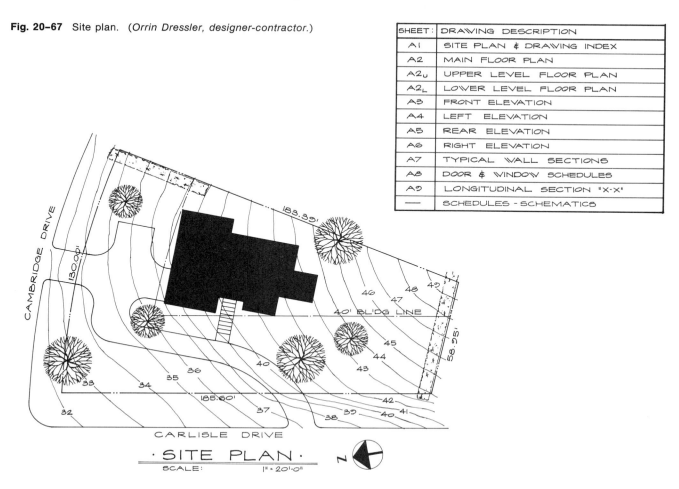

· SITE PLAN ·
SCALE: 1" = 20'-0"

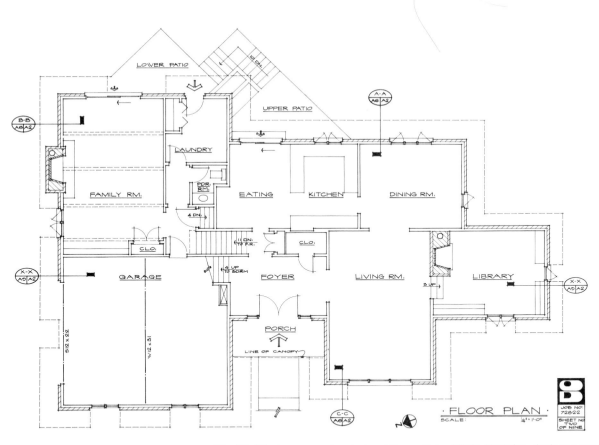

Fig. 20–68 Lower- and main-level floor plan. (*Orrin Dressler, designer-contractor.*)

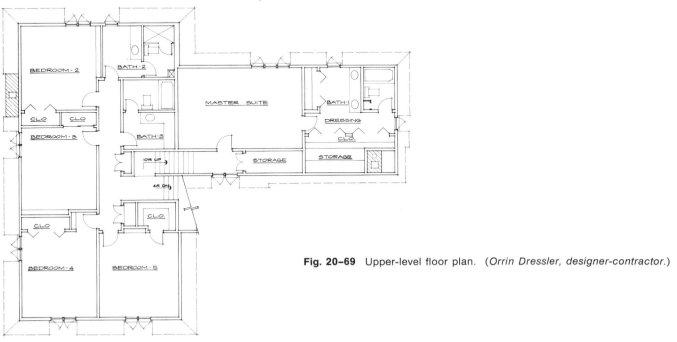

Fig. 20–69 Upper-level floor plan. (*Orrin Dressler, designer-contractor.*)

Fig. 20-70 Front elevation. (*Orrin Dressler, designer-contractor.*)

Fig. 20-71 Rear elevation. (*Orrin Dressler, designer-contractor.*)

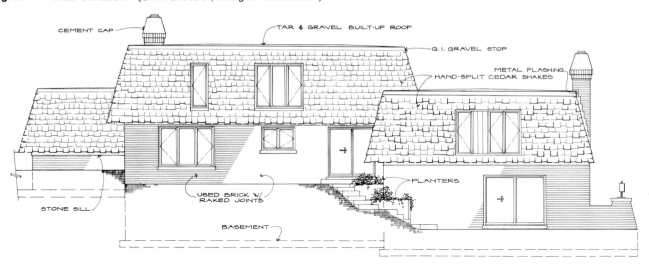

elevation (Fig. 20-71) shows contoured stairs between the main and lower levels. The right and left elevations are shown in Figs. 20-72 and 20-73. The longitudinal section (Fig. 20-74) explains the arrangement of levels and respective rooms. The sections illustrated in Fig. 20-75 are identified on the main floor plan (Fig. 20-68).

Architectural specifications are essential for turning a set of drawings into a building. They note the general conditions of the site to be developed, and the material descriptions are described for the client's and contractor's mutual agreement. Guidelines for specifications have been established by the American Institute of Architects and are available through local state chapters of the AIA. A set of formal written specifications for this contemporary, multilevel house appears in the appendix.

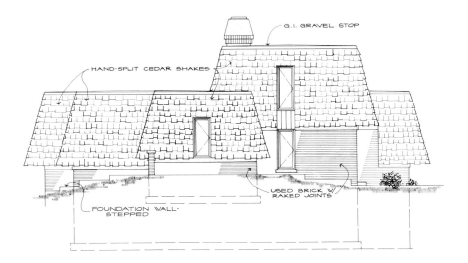

Fig. 20–72 Right elevation. (*Orrin Dressler, designer-contractor.*)

G.I. GRAVEL STOP

HAND-SPLIT CEDAR SHAKES

USED BRICK W/ RAKED JOINTS

FOUNDATION WALL-STEPPED

Fig. 20–73 Left elevation. (*Orrin Dressler, designer-contractor.*)

G.I. GRAVEL STOP

TAR & GRAVEL BUILT-UP ROOF

USED BRICK-RAKED JOINTS

TOP OF FOUND.

ROUGH-SAWN VERT. CEDAR SID.

16'-0" x 7'-0" FLUSH OVERHEAD DOOR

Fig. 20–74 Longitudinal section. (*Orrin Dressler, designer-contractor.*)

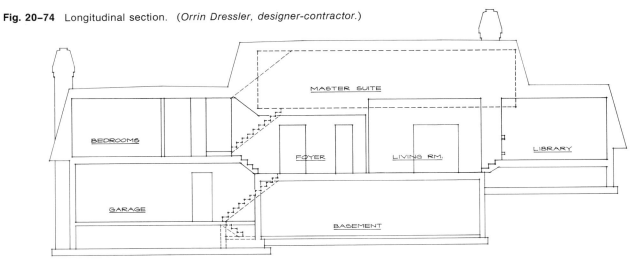

MASTER SUITE

BEDROOMS

FOYER

LIVING RM.

LIBRARY

GARAGE

BASEMENT

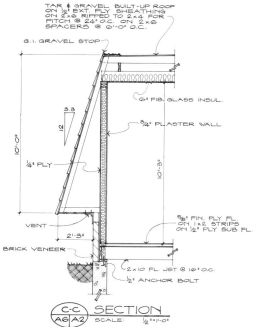

TAR & GRAVEL BUILT-UP ROOF
ON ½" EXT. PLY SHEATHING
ON 2x6 RIPPED TO 2x4 FOR
PITCH @ 24" O.C. ON 2x6
SPACERS @ 6'-0" O.C.

G.I. GRAVEL STOP

6" FIB. GLASS INSUL.

¾" PLASTER WALL

3.3
12

10'-0"

10'-3"

¼" PLY

VENT

2'-3"

⅝" FIN. PLY FL.
ON 1x2 STRIPS
ON ½" PLY SUB FL.

BRICK VENEER

2x10 FL. JST @ 16" O.C.

½" ANCHOR BOLT

C-C SECTION
A6 A2   SCALE: ½"=1'-0"

Fig. 20–75 Section views. (*Orrin Dressler, designer-contractor.*)

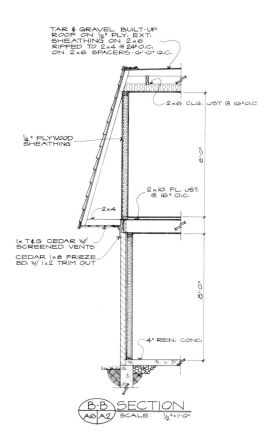

TAR & GRAVEL BUILT-UP
ROOF ON ½" PLY. EXT.
SHEATHING ON 2x6
RIPPED TO 2x4 @ 24" O.C.
ON 2x6 SPACERS 6'-0" O.C.

¼" PLYWOOD
SHEATHING

2x6 CLG. JST @ 16" O.C.

8'-0"

2x4

2x10 FL. JST.
@ 16" O.C.

1x T&G CEDAR W/
SCREENED VENTS

CEDAR 1x8 FRIEZE
BD. W/ 1x2 TRIM OUT

8'-0"

4" REIN. CONC.

B-B SECTION
A6 A2   SCALE: ½"=1'-0"

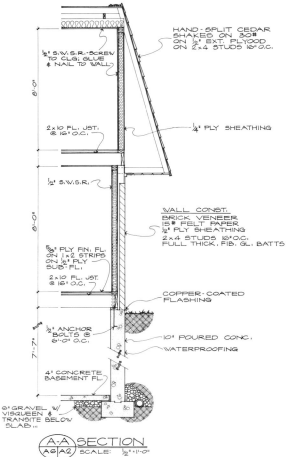

HAND-SPLIT CEDAR
SHAKES ON 30#
ON ½" EXT. PLYWOOD
ON 2x4 STUDS 16" O.C.

½" S.W.S.R.-SCREW
TO CLG; GLUE
& NAIL TO WALL

8'-0"

2x10 FL. JST.
@ 16" O.C.

¼" PLY. SHEATHING

½" S.W.S.R.

8'-0"

WALL CONST.
BRICK VENEER
15# FELT PAPER
½" PLY SHEATHING
2x4 STUDS 16" O.C.
FULL THICK. FIB. GL. BATTS

⅝" PLY FIN. FL.
ON 1x2 STRIPS
ON ½" PLY
SUB-FL.

2x10 FL. JST.
@ 16" O.C.

COPPER-COATED
FLASHING

½" ANCHOR
BOLTS @
6'-0" O.C.

7'-7"

10" POURED CONC.

WATERPROOFING

4" CONCRETE
BASEMENT FL.

6" GRAVEL W/
VISQUEEN &
TRANSITE BELOW
SLAB

A-A SECTION
A6 A2   SCALE: ½"=1'-0"

The Farnsworth residence, in Plano, Illinois, was designed by the famous American architect, Ludwig Mies Van der Rohe (Fig. 20–76). This design of structural steel has received much acclaim. The modular grid on the floor plan (Fig. 20–77) and the elevation section (Fig. 20–78) illustrate the pure form of the design. The stair detail (Fig. 20–79) is shown in its actual form in Fig. 20–80. The structural steel wall section in Fig. 20–81 illustrates the steel column and channel at work. Figure 20–82 reveals the proportions of the glass and steel curtain wall which dominates the contemporary skyline of many large American cities (Fig. 20–83).

Fig. 20–76 Farnsworth residence, Plano, IL. (*The office of Mies Van der Rohe and Hedrich—Blessing.*)

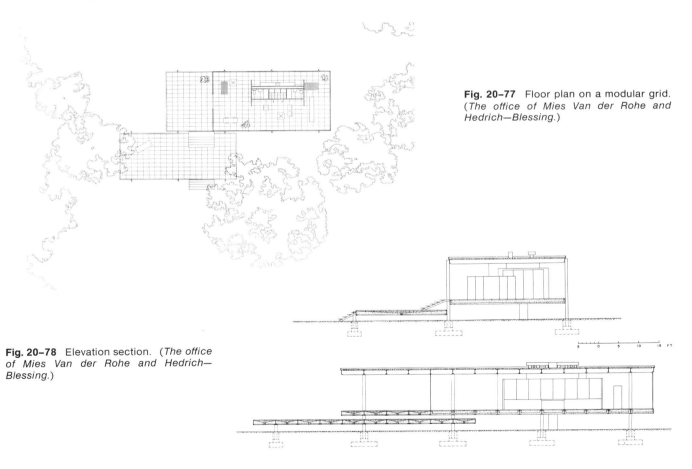

Fig. 20–77 Floor plan on a modular grid. (*The office of Mies Van der Rohe and Hedrich—Blessing.*)

Fig. 20–78 Elevation section. (*The office of Mies Van der Rohe and Hedrich—Blessing.*)

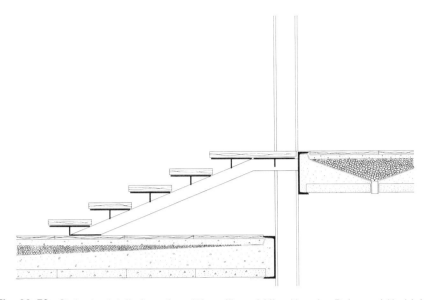

Fig. 20–79 Stairs in detailed section. (*The office of Mies Van der Rohe and Hedrich—Blessing.*)

Fig. 20–80 The finished stairs complement structural form. (*The office of Mies Van der Rohe and Hedrich—Blessing.*)

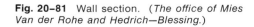

Fig. 20–81 Wall section. (*The office of Mies Van der Rohe and Hedrich—Blessing.*)

Fig. 20–82 Steel and glass for open design. (*The office of Mies Van der Rohe and Hedrich—Blessing.*)

Fig. 20–83 Steel and glass help form a skyline. (*The office of Mies Van der Rohe and Ron Vickers, Ltd.*)

Review

1. Name the two styles of lettering used in architecture.

2. List four residential house styles.

3. What are the three significant ways in which we can evaluate architecture?

4. What are the four basic drawings which are developed by the architect?

5. What are working drawings?

6. What are the major services rendered by an architect?

7. Define architecture in your own words.

8. Illustrate four material symbols used by the architect.

9. Illustrate a line technique important to architectural style.

10. What are pressure-sensitive symbols?

11. List six types of windows available for residential design.

12. What are the preferred scales for plan and elevation drawings of small buildings?

13. What scale is preferred for enlarging details?

14. List three types of wall construction common in residential design.

15. What is modular coordination?

Problems for Chapter 20

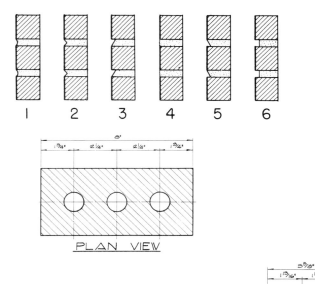

INDEX TO DRAWINGS

ARCHITECTURAL

A1 — SITE PLAN, INDEX TO DRAWINGS, ARCHITECTS SYMBOLS
A2 — GROUND FLOOR PLAN
A3 — FIRST FLOOR PLAN
A4 — TYPICAL FLOOR PLAN (2ND THRU 10TH FLRS.)
A5 — PENTHOUSE & ROOF PLANS
A6 — LARGE SCALE CORE PLANS, DETAILS, SECTIONS
A7 — LARGE SCALE APARTMENT PLANS, ELEVATIONS, DETAILS
A8 — EXTERIOR ELEVATIONS
A9 — EXTERIOR ELEVATIONS, STAIR ENTRANCE, DETAILS
A10 — WALL SECTIONS, CROSS SECTION, MISCELLANEOUS DETAILS
A11 — ENTRANCE SECTIONS, MISCELLANEOUS DETAILS

STRUCTURAL

S1 — FOUNDATION & GROUND FLOOR PLAN
S2 — GENERAL NOTES, SECTIONS, DETAILS
S3 — FIRST FLOOR FRAMING PLAN
S4 — TYPICAL FLOOR FRAMING PLAN (2ND THRU 10TH FLRS.)
S5 — ROOF & PENTHOUSE FRAMING PLAN
S6 — COLUMN SCHEDULE & DETAILS
S7 — BEAM SCHEDULE & DETAILS
S8 — STAIR SECTIONS & DETAILS
S9 — PARKING DECK GRADES, BEAM & SLAB DETAILS

Fig. 20–84 Practice lettering the architectural or structural index to drawings as assigned by the instructor. Use a bold extended style. The letter sizes should be $\frac{1}{4}$, $\frac{3}{16}$, and $\frac{1}{8}''$. Use guide lines.

PLAN VIEW

SIDE VIEW

END VIEW

DR'WG SCALE: 6" = 1'-0"

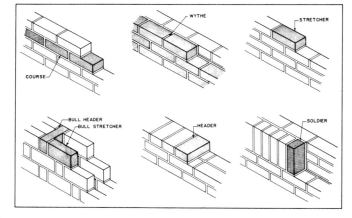

Fig. 20–85 Make a three-view drawing of a brick, showing dimensions. Label the appropriate surface with "header," "stretcher," and "face." Make a pictorial drawing of the brick bonds as assigned. The nominal dimensions are $2\frac{1}{4}$, 4, and 8'', with $\frac{1}{2}''$ for mortar joints. Use a scale of $1\frac{1}{2}'' = 1'\text{-}0''$. Draw six sectional views of joint shapes: (1) concave, (2) v-joint, (3) weathered, (4) flush, (5) struck, and (6) raked.

Fig. 20–86 Prepare an isometric view of a wall 4′ long, showing the section illustrated. The brick bonding may be shown, or it may appear as horizontal lines in pictorial. Scale: 1½″ = 1′-0″. (*Doug Strom.*)

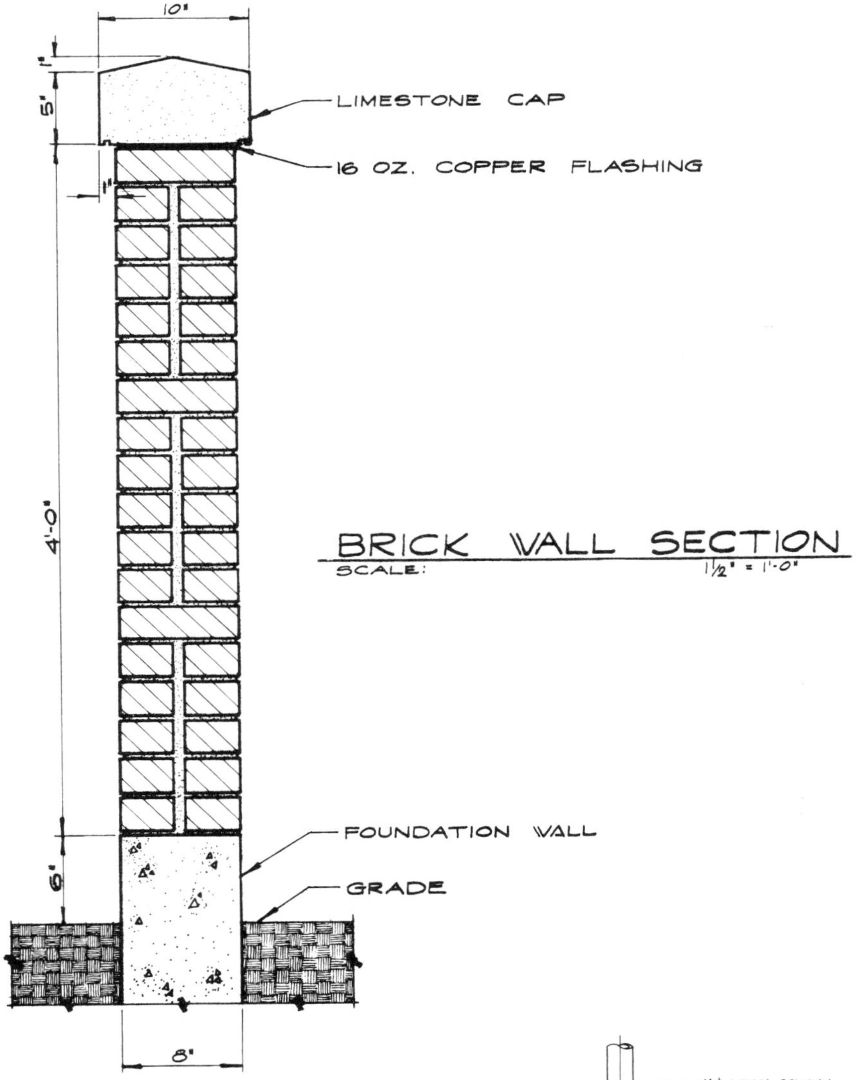

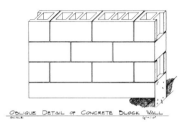

OBLIQUE DETAIL OF CONCRETE BLOCK WALL
SCALE:

Fig. 20–87 Prepare an isometric detail of a concrete block wall. Investigate and use the nominal sizes of concrete block. Scale: 1½″ = 1′-0″. (*Doug Strom.*)

Fig. 20–88 Draw a footing detail as shown. Use sectional symbols common to architectural detailing. Design your footing to be on load-bearing soil at a depth below a local frost line or as established by building codes. (*Doug Strom.*)

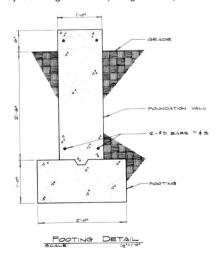

GRADE

FOUNDATION WALL

2-#5 BARS T & B

FOOTING

FOOTING DETAIL
SCALE: 1½″ = 1′-0″

10″

1″

5′

1″

4′-0″

6″

8″

LIMESTONE CAP

16 OZ. COPPER FLASHING

BRICK WALL SECTION
SCALE: 1½″ = 1′-0″

FOUNDATION WALL

GRADE

Fig. 20–89 Prepare an isometric view of a typical column footing using the given sizes of the structural items. (*Doug Strom.*)

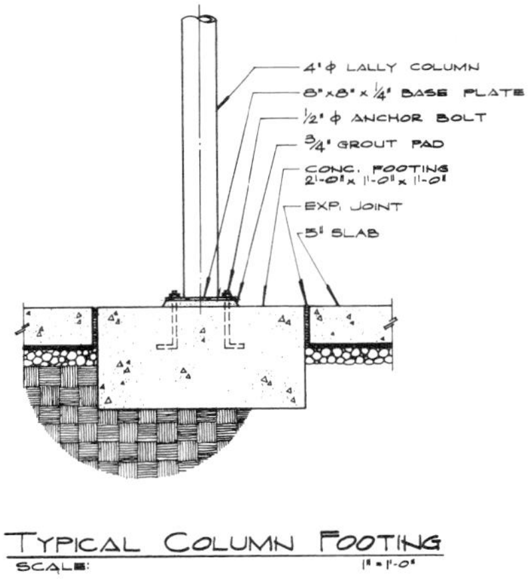

4″ ⌀ LALLY COLUMN
8″×8″×¼″ BASE PLATE
½″ ⌀ ANCHOR BOLT
¾″ GROUT PAD
CONC. FOOTING 2′-0″ × 1′-0″ × 1′-0″
EXP. JOINT
5″ SLAB

TYPICAL COLUMN FOOTING
SCALE: 1″ = 1′-0″

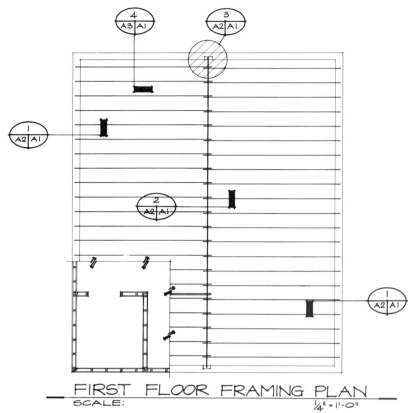

FIRST FLOOR FRAMING PLAN

SCALE: ¼" = 1'-0"

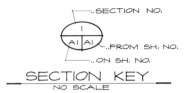

SECTION KEY
NO SCALE

Figs. 20–90 through 20–95 are all part of the same structure. When preparing the details (Figs. 20–92 through 20–95), refer back to the framing plan (Fig. 20–91). (*Doug Strom.*)

Fig. 20–90 Prepare a sectional key diagram using an ellipse, circle, or rectangle as suits your style. Note how it applies to the framing plan and details.

Fig. 20–91 Prepare a framing plan as shown. Outside building dimensions are 24'-0" by 30'-0". Joists are 16" on center. Include your sectional key diagrams for the enlarged details to be drawn.

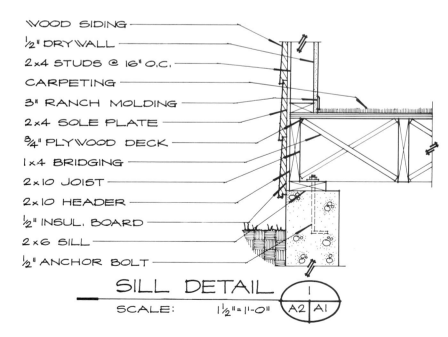

WOOD SIDING
½" DRYWALL
2×4 STUDS @ 16" O.C.
CARPETING
3" RANCH MOLDING
2×4 SOLE PLATE
¾" PLYWOOD DECK
1×4 BRIDGING
2×10 JOIST
2×10 HEADER
½" INSUL. BOARD
2×6 SILL
½" ANCHOR BOLT

SILL DETAIL

SCALE: 1½" = 1'-0"

Fig. 20–92 Using the nominal dimensions of the materials listed, construct a sill detail in section. The lapped, 1" wood siding has an 8" exposure. Scale: 1½" = 1'-0".

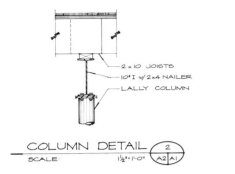

COLUMN DETAIL 2
SCALE: 1½"=1'-0" A2 A1

Fig. 20-93 Prepare a partial column-girder detail. The column is 4″ round with a 6″ plate welded to the top. The 10″ I-beam is bolted to the plate. The 2 × 4 rests on the 4½″ flange of the I-beam, supporting the overlapped floor joists.

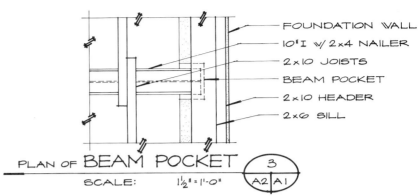

— FOUNDATION WALL
— 10" I W/ 2x4 NAILER
— 2x10 JOISTS
— BEAM POCKET
— 2x10 HEADER
— 2x6 SILL

PLAN OF BEAM POCKET 3
SCALE: 1½"=1'-0" A2 A1

Fig. 20-94 Prepare a plan of the 6″ beam pocket. The beam is to have a 4″ bearing on the concrete wall.

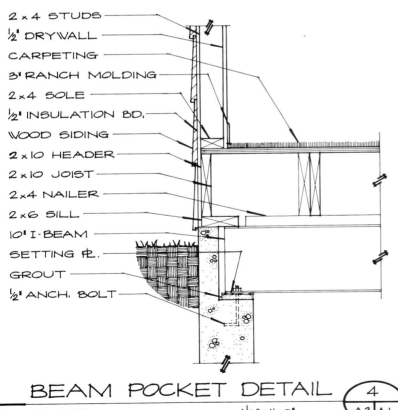

2 x 4 STUDS
½" DRYWALL
CARPETING
3' RANCH MOLDING
2x4 SOLE
½" INSULATION BD,
WOOD SIDING
2x10 HEADER
2x10 JOIST
2x4 NAILER
2x6 SILL
10' I-BEAM
SETTING ₽.
GROUT
½" ANCH, BOLT

BEAM POCKET DETAIL 4
SCALE: 1½"=1'-0" A3 A1

Fig. 20-95 Prepare a sectional beam pocket detail showing the 10″ I-beam on the stepped foundation wall. Use nominal dimensions for all the materials listed. Allow the flange of the I-beam to show as a nominal ½″ on the top and bottom of the beam.

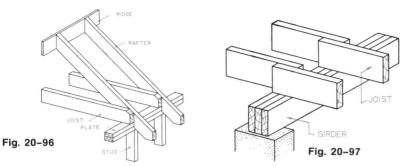

Fig. 20–96

Fig. 20–97

Figs. 20–96 and 20–97 Prepare a pictorial of typical roof framing. Studs, ceiling joists, and rafters are 16″ on center. Joist and rafter size: 2″ by 6″. Studs and top plate size: 2″ by 4″. Ridge size: 1″ by 8″. Use the nominal sizes at a scale of $1\frac{1}{2}″ = 1'\text{-}0″$. Draw a pictorial of the built-up girder to support the floor joist. Girder and joist size: 2″ by 10″.

Figs. 20–98 through 20–102 are a partial set of plans for a two-story residence. Make drawings as assigned by the instructor. (*A. W. Wendell & Sons, designer-contractor.*)

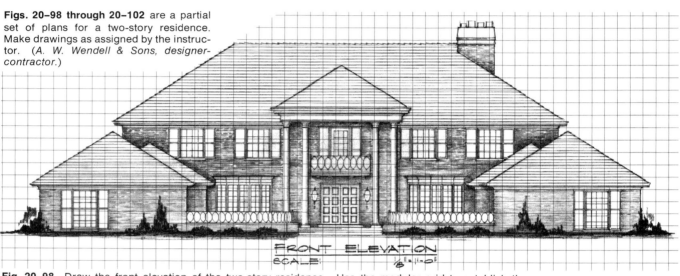

FRONT ELEVATION
SCALE: 1/8″=1'-0″

Fig. 20–98 Draw the front elevation of the two-story residence. Use the modular grid to establish the rectangular proportions, roof slope, and structural openings. Scale: $\frac{1}{8}″ = 1'\text{-}0″$. Illustrate the brick, masonry walls, and asphalt roofing with horizontal lines. Add window shutters, window panes, and decorative appointments to suit your own design taste.

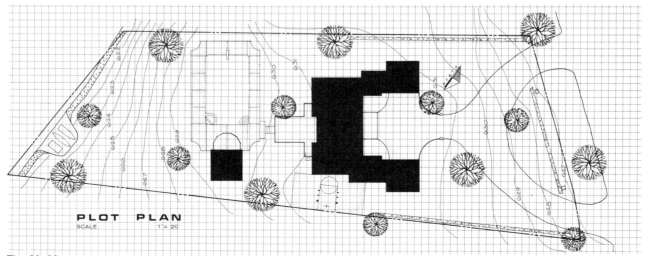

PLOT PLAN
SCALE 1″= 20'

Fig. 20–99

Fig. 20–99 Examine the gridded plot plan, and develop the boundary lines at a scale of $1'' = 20'$-$0''$, on a C-size sheet. Dimension the length of each boundary line. Complete the plot plan with the plan of the house and your own landscaping design. All radii on the drive must be 18' minimum. Lay out the approximate contour lines, noting the change of terrain. The basketball court and cabana are optional design appointments. If instructed, prepare a simple plan and elevation of a cabana to fit the space provided.

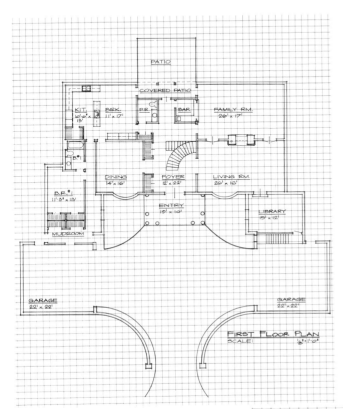

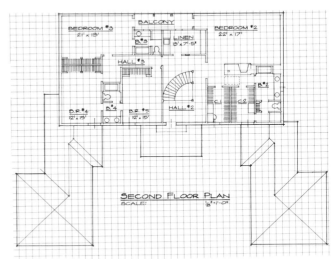

Fig. 20–101 Prepare a second floor plan to include the roof plan of the single-story area. Scale: $\frac{1}{8}'' = 1'$-$0''$. Locate the missing windows according to the front and right-side elevations. Locate the windows you believe necessary on the left and rear elevation walls. Label the rooms and their approximate sizes. Note the hip roof design and the intersecting inclined planes forming the roof.

Fig. 20–100 Draw a first floor plan at a scale of $\frac{1}{8}'' = 1'$-$0''$. By examining the modular grid, locate and include missing windows on your drawing. Label the rooms and their approximate sizes. If instructed, plot the functional traffic patterns on an overlay to the floor plan.

Fig. 20–102 Draw a right-side elevation by examining the proportions on the modular grid. Note the center lines which locate the finished floors and ceilings. Find the common roof pitch and label it on the elevation. Draw in the materials that describe this elevation. Prepare a left-side and rear elevation as assigned by the instructor.

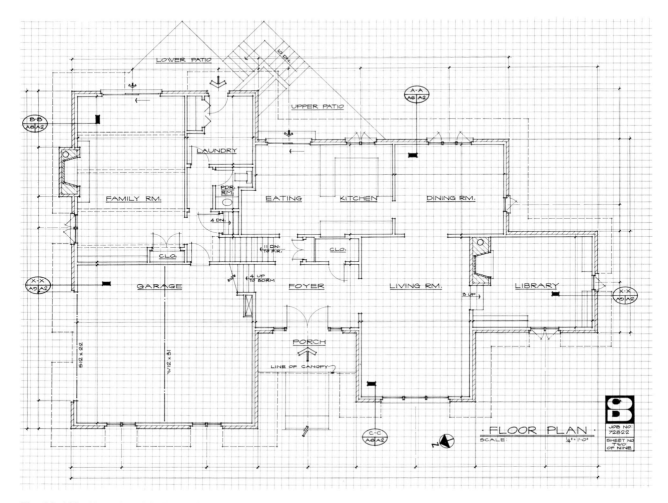

Fig. 20–103 Note that this floor plan is Fig. 20–68 in the text. Prepare the first floor plan at a scale of $\frac{1}{4}'' = 1'\text{-}0''$ on a C-size sheet. Add the dimensions to the plan. Examine the $\frac{1}{8}''$ modular grid to establish the room sizes, window opening sizes, and the overall building size. (*Orrin Dressler.*)

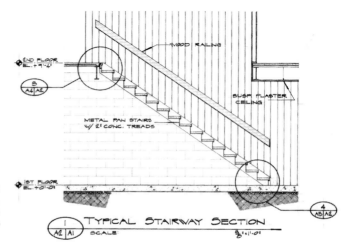

Fig. 20–104 Prepare a sectional elevation of a typical stairway. Riser 7″, tread 11″, nosing 1″, and stringer 10″. Note the elevational readings from finished floor to finished floor. Scale: $\frac{3}{8}'' = 1'\text{-}0''$.

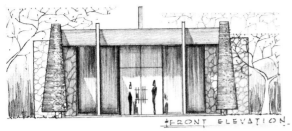
Fig. 20–105

Fig. 20–106

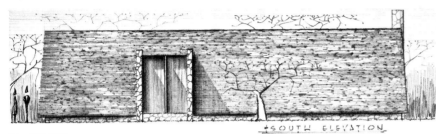

Figs. 20–105 through 20–108 Examine the front, west, and south elevations of the church in relationship to the floor plan. Draw a floor plan and lay out the narthex, nave, and chancel at a scale of $\frac{1}{8}'' = 1'$-0''. Use dividers and the scale on the floor plan for sizes. Render the floor plan and add the site appointments. Draw elevations as directed by the instructor. Elevations are 18' at the height of the shingled roof. Apply textured materials for the roof and stone walls. Add landscaping appointments. (*Doug Strom.*)

Fig. 20–107

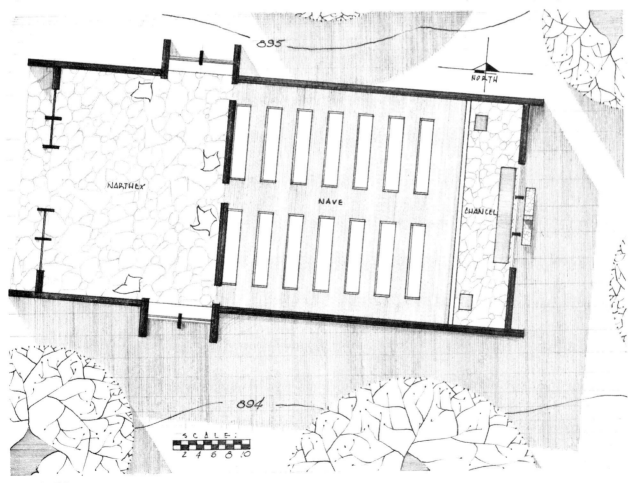

Fig. 20–108

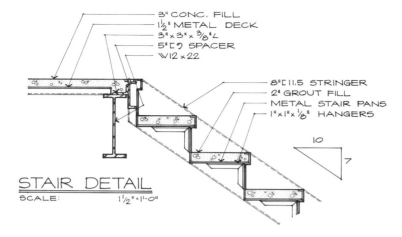

STAIR DETAIL

SCALE: 1½" = 1'-0"

- 3" CONC. FILL
- 1½" METAL DECK
- 3" x 3" x ⅜" L
- 5" [9 SPACER
- W12 x 22
- 8" [11.5 STRINGER
- 2" GROUT FILL
- METAL STAIR PANS
- 1" x 1" x ⅛" HANGERS

10 / 7

Fig. 20–109 Problem 1: Draw stair details for a straight run rise of 7' on the outside of a building. Scale: 1" = 1'-0". Include necessary dimensions. Problem 2: Design a run of stairs with a rise of 14' from a sunken plaza in front of a building to street level, with a U or L turn as assigned by the instructor. Include all dimensions and notes.

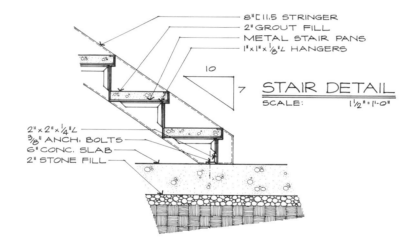

- 8" [11.5 STRINGER
- 2" GROUT FILL
- METAL STAIR PANS
- 1" x 1" x ⅛" L HANGERS
- 2" x 2" x ¼" L
- ⅜" ANCH. BOLTS
- 6" CONC. SLAB
- 2" STONE FILL

10 / 7

STAIR DETAIL

SCALE: 1½" = 1'-0"

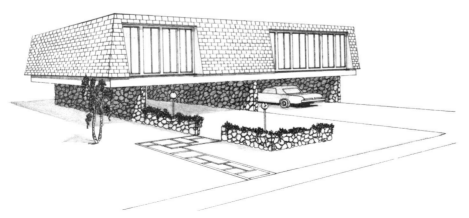

Fig. 20–110 Prepare a front elevation of the professional office shown in the perspective view. Design a floor plan and office layout for a profession of your choice. Select overall sizes and a suitable scale.

Fig. 20-111 The twin towers contain both office space and apartments. The outside dimensions are 60 feet square. Prepare typical floor plans for an office floor and for an apartment floor, designed around a central service core containing elevators, stairwells, and so forth. Mix the kinds of apartments—from one-room efficiencies to two- or three-bedroom apartments.

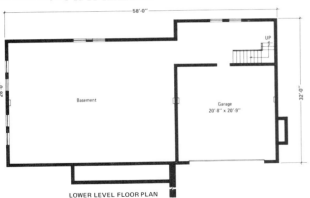

LOWER LEVEL FLOOR PLAN

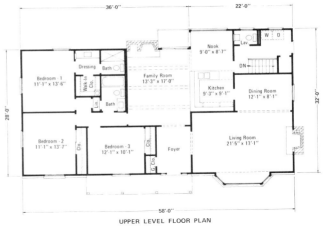

UPPER LEVEL FLOOR PLAN

Fig. 20-112 Draw a front- and left-side elevation of the split-level house from the floor plans and perspective given. Choose an appropriate scale. Prepare a plot plan with contour lines for this house. Scale: $\frac{1}{8}'' = 1'\text{-}0''$.

Structural Drafting | Chapter 21

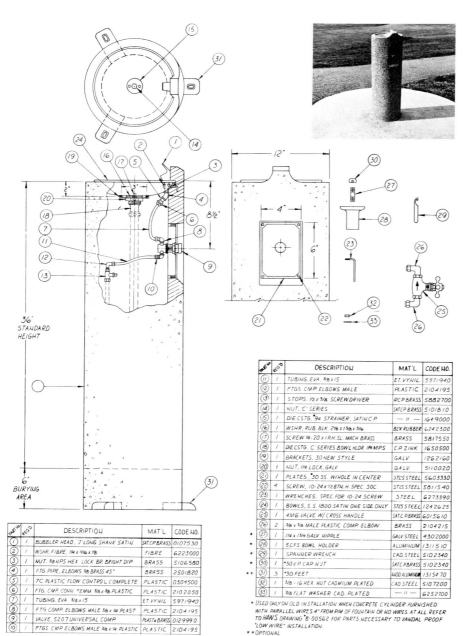

The Structural Draftsman. Detailed structural drawings are required for framing and support of buildings, bridges, dams, storage tanks, communication towers, and many of the other structures built by America's construction industry. The thousands of structural drawings are generally prepared by structural draftsmen. The structural draftsman is usually a member of an engineering team, often working in cooperation with other teams under the direction of a project manager or a job superintendent. An example of teamwork can be the combined efforts of structural and architectural designers. The structural designer designs the frame of a building in accordance with the functional design forms of an architectural designer (Fig. 21–1). The detailed work of the structural draftsman is important to the engineering profession, and construction of buildings, bridges, and other structures is dependent upon detailed instructions provided in structural drawings (Fig. 21–2).

The type of work performed by a structural draftsman is found in five specific applications:

1. As a detailer in an architect's or engineer's office.

2. Preparing construction details for a contractor (construction company's shop drawings).

3. Drafting the structural details for a manufacturer of structural materials.

PART NO.	REQ'D	DESCRIPTION	MAT'L	CODE NO.
1	1	BUBBLER HEAD, 3" LONG SHANK SATIN	SAT.CP.BRASS	0107530
2	1	WSHR. FIBRE, 1¼ x ¹¹/₁₆ x ⅛	FIBRE	6223000
3	1	NUT, ⅜ NPS HEX LOCK BR. BRIGHT DIP	BRASS	5106580
4	1	FTG. PIPE, ELBOWS ⅜ BRASS 45°	BRASS	2501820
5	1	7C PLASTIC FLOW CONTROL COMPLETE	PLASTIC	0304500
6	1	FTG. CMP. CONN. FEMM. ⅜ x ⅜ PLASTIC	PLASTIC	2102050
7	1	TUBING, EVA ⅜ x 15	ET. VYNIL	5971940
8	1	FTG. COMP. ELBOWS MALE ⅜ x ¼ PLAST	PLASTIC	2104195
9	1	VALVE, S207 UNIVERSAL COMP	PLAST & BRASS	0129990
10	1	FTGS CMP ELBOWS MALE ⅜ x ¼ PLASTIC	PLASTIC	2104195

PART NO.	REQ'D	DESCRIPTION	MAT'L	CODE NO.
11	1	TUBING, EVA. ⅜ x 15	ET. VYNIL	5971940
12	1	FTGS CMP ELBOWS MALE	PLASTIC	2104195
13	1	STOPS, ½ x ⅜ SCREWDRIVER	RCP.BRASS	5882700
14	1	NUT, 'C' SERIES	SATCP BRASS	5101810
15	1	DIE CSTG. '94 STRAINER, SATIN C.P.	— " —	1649000
16	1	WSHR. RUB. BLK. 2⅛ x 1⅜ x ³/₁₆	BLK.RUBBER	6242300
17	1	SCREW ¼-20 x 1 RH SL. MACH BRASS	BRASS	5817550
18	1	DIE CSTG. 'C' SERIES BOWL HLDR 1¼ MPS	C.P. ZINK	1650500
19	1	BRACKETS, 30 NEW STYLE	GALV.	1262160
20	1	NUT, 1¼ LOCK, GALV.	GALV.	5110020
21	1	PLATES, '30 SS. WHOLE IN CENTER	STC'S STEEL	5603330
22	4	SCREW, 10-24 x ½ BT.H. SPEC. 50C	STC'S STEEL	5811540
23	1	WRENCHES, SPEC. FOR 10-24 SCREW	STEEL	6273390
24	1	BOWLS, S.S. 1800 SATIN ONE SIDE ONLY	STC'S STEEL	1242625
25	1	4MG VALVE W/ CROSS HANDLE	SAT.C.P.BRASS	6015610
26	2	⅜ x ⅜ MALE PLASTIC COMP. ELBOW	BRASS	2104215
* 27	1	1¼ x 1¾ GALV NIPPLE	GALV STEEL	4302000
* 28	1	5 CFS BOWL HOLDER	ALUMINUM	1311510
* 29	1	SPANNER WRENCH	CAD. STEEL	5102340
* 30	1	'30 V.P. CAP N.JT	SAT.C.P.BRASS	5102340
* * 31	3	'30 FEET	ANOD ALUMINUM	1315470
32	1	⅜ -16 HEX NUT CADMIUM PLATED	CAD STEEL	5107200
33	1	⅜ FLAT WASHER CAD. PLATED	— " —	6252700

* USED ONLY ON OLD INSTALLATION WHEN CONCRETE CYLINDER FURNISHED WITH PARALLEL WIRES 4" FROM RIM OF FOUNTAIN OR NO WIRES AT ALL. REFER TO HAWS DRAWING 'B-0056' FOR PARTS NECESSARY TO VANDAL PROOF 'LOW WIRE' INSTALLATION.

* * OPTIONAL

(HAWS DRINKING FAUCET COMPANY.)

36" STANDARD HEIGHT

6" BURYING AREA

12"

4"

6"

8½"

Fig. 21–1 The Deere & Company administrative center designed by Eero Saarinen and Associates required many large sheets of structural details. (*John Deere & Co.*)

Fig. 21–2 A structural detail of the Deere building. (*John Deere & Co.*)

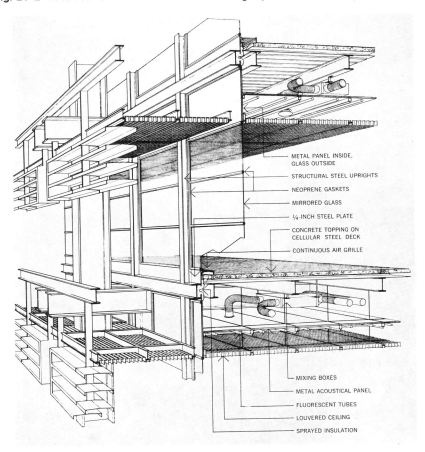

METAL PANEL INSIDE, GLASS OUTSIDE

STRUCTURAL STEEL UPRIGHTS

NEOPRENE GASKETS

MIRRORED GLASS

¼-INCH STEEL PLATE

CONCRETE TOPPING ON CELLULAR STEEL DECK

CONTINUOUS AIR GRILLE

MIXING BOXES

METAL ACOUSTICAL PANEL

FLUORESCENT TUBES

LOUVERED CEILING

SPRAYED INSULATION

4. Working for the engineering department of a plant which maintains its own engineering operations.

5. Preparing drawings for government and agencies which control construction and design of public buildings, bridges, dams, and other structures.

Career opportunities for the structural draftsman are generally related to his practical experience gained as a junior member of an engineering team. Some of the typical job titles which may be achieved in a promotion are structural detail checker, estimator, chief draftsman, construction supervisor, or building inspector (Fig. 21–3).

Additional career opportunities for structural draftsmen can be related to a continuing education program. Formal training at a junior college or university will help the structural draftsman to advance. Among the most important skills required for promotion are an ability to communicate and mechanical aptitude. The high school or college student interested in the construction industry should direct his interests to language skills, mathematics, science, and graphic communication.

Structural Materials. As an engineering team works together they must learn about new materials and systems which become available to the construction specialists. The designer and detail draftsman must be acquainted with a great many materials. All structural materials have special methods or systems for fastening and erection which must be considered in the accurate drafting details prepared by the engineering team.

The basic structural materials which are used in today's contemporary structures are steel, wood, concrete, structural clay products, and

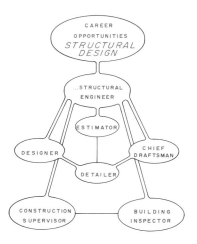

Fig. 21–3 Structural design career opportunities.

stone masonry. The characteristics of materials vary, but it is the designer's job to bring any combination of these together so that the forces (compression, tension, and shear) interacting in a structure's materials are safely at rest.

Steel as a Structural Material. Steel lends itself to construction because of the shapes formed at the mill. Steel shapes are produced in rolling mills and then shipped to fabrication shops where they are cut to specific lengths and connections are prepared. Some of the basic shapes include the wide flanged W-shape used as a beam or column; the S-shape, formerly known as an I-beam; the C-shape, formerly known as the channel; and the L-shape, formerly known as the angle.

These steel shapes have framed the skyscrapers of our cities for nearly three-quarters of a century. The American Institute of Steel Construction maintains regional offices from coast to coast to asist in maintaining guide lines for the design and construction of substructures and superstructures.

A steel-framed village. The structural steel form of the Tempo Bay Resort Hotel and Polynesian Village in Walt Disney World, Florida, is an interesting example of contemporary engineering and construction. Figure 21–4 shows the A-shape structure as it begins to rise. Structural steel is rising around the concrete elevator core. In the left foreground, complete modular steel-framed rooms are stacked three high. All the modular rooms are 9 ft high and 15 ft wide. The length of each unit varies from 32 to 39 ft, including a balcony.

The basic building blocks shown in Fig. 21–5 are trucked to the site alongside the structural support for the monorail transportation system. The two monorail tracks and the platform between them are poured-in-place concrete spanning 66 ft. The monorail passes through the building but is structurally free from it to avoid any transfer of vibrations.

The tubular structural columns of a basic building block are welded together in Fig. 21–6. Workmen shown installing the wall panels are working with 25-gage galvanized steel runners and studs.

The illustration of the completed structural A-frame form is shown in Fig. 21–7. Each steel A-frame is approximately 220 ft across at the base, 135 ft at the top, and 15 ft across the vertical bents. It is constructed of tubes, wide-flanged sections, and chords made of 18- by 26-in. tubes. Each of the A-frames is assembled in the field and erected in five pieces. All the steel members are connected with high-strength bolts (HSB).

The 13 A-frames were designed with the aid of a computer for all kinds of loading combinations (wind loads, temperature changes, and the vertical dead and live loads). The wind loads are based on hurricanes with wind speeds of up to 100 mph. The forces

Fig. 21–4 Structural steel begins to rise at Walt Disney World. (*United States Steel Corp.*)

Fig. 21-5 Basic building blocks are trucked to the site. (*United States Steel Corp.*)

Fig. 21-6 Welding the basic building block. (*United States Steel Corp.*)

Fig. 21-8 A steel-framed skyscraper, the John Hancock Building, Chicago. (*Skidmore, Owings & Merrill; Hedrich-Blessing.*)

Fig. 21-7 The completed A-frame form. (*United States Steel Corp.*)

are carried through the complete structure to the ground. All structures must be designed for a given geographical location, the prevailing weather pattern, their function, and the size and shape of the form determined by the architectural designer.

A steel-framed skyscraper. The framework of the 100-story John Hancock Center in Chicago is a fine example of a new design and system developed economically. Big John, as it is called, tapers from a 265 × 165 ft area at the base to a 160 × 100 ft area at the top floor. The towering giant presses upward to a little over 1100 ft in height. The walls are stiffened by diagonals (Fig. 21-8) in the form of huge 18-story-high steel "X's." This unusual design saved about 27,000 tons of structural steel and $15,000,000. The firm of Skidmore, Owings and Merrill and their designer Bruce Graham used the computer in refining this building's structural form.

The John Hancock project includes 700 apartment units ranging from one-room efficiencies to four-bedroom luxury apartments. Of the 2,800,000 square feet, there are 812,000 square feet for offices and 300,000 square feet for commercial entrances, lobbies, and major traffic patterns. The complementary facilities include parking within the structure, restaurants, health clubs, a swimming pool, and an ice-skating rink. The three acres available for this site helped in creating the character of this steel superstructure.

A steel-framed bridge. Rigid structural framing of high-strength steel fastened with high-strength bolts and welds describes the bridge designed by engineers in New Mexico. One of New Mexico's recent bridges spans a deep gorge of the Rio Grande River near Taos, New Mexico (Fig. 21–9). The bridge contains over 1900 tons of structural steel. The center span is 600 feet long, and the two side spans are each 300 feet long. The distance from the canyon floor to the bridge floor is 600 feet (Fig. 21–10).

A typical welded-steel member of this bridge appears in Fig. 21–11. The structural detail in Fig. 21–12 is bolted in the field as a framed unit according to the construction diagram.

Steel Systems. Steel has been used to design new structural systems. The Unistrut Space Frame consists of four or five modular units which have a geometric pattern, as shown in Fig. 21–13. The basic unit has been developed with four or five parts which are bolted together. The system is primarily designed for canopies and roof structures.

Figure 21–14 illustrates the erection of a roof system after it was assembled on the ground. The characteristics of a steel roof framed in geometric detail are shown in Fig. 21–15 as the interior of a mall in the model city of Columbia, Maryland.

Fig. 21–9 Rio Grande Gorge Bridge. (*New Mexico State Highway Department.*)

Fig. 21–10 Schematic of bridge framing. (*United States Steel Corp.*)

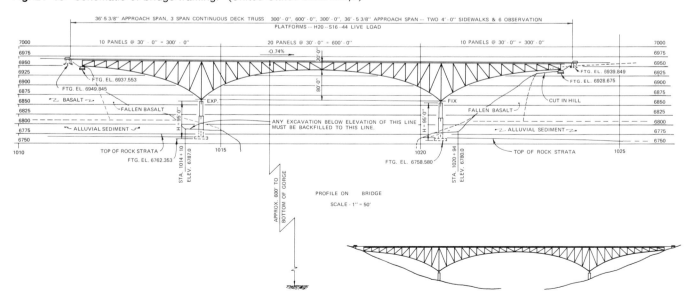

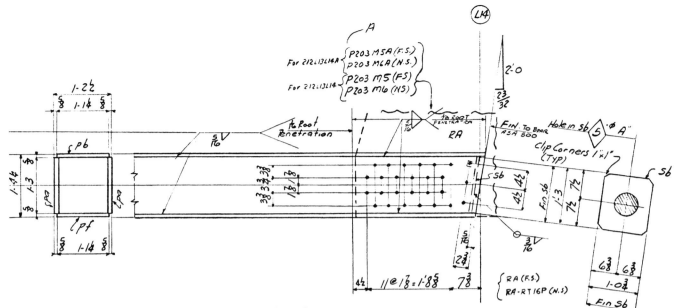

Fig. 21–11 Welded bridge member. (*United States Steel Corp.*)

Fig. 21–12 Bolted bridge member. (*United States Steel Corp.*)

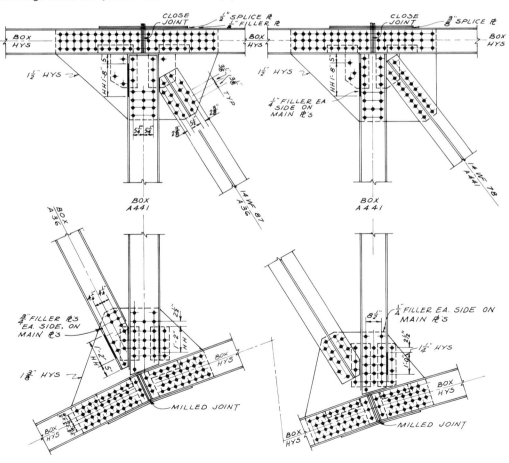

Fig. 21-13 Unistrut geometric form. (*Unistrut Corp.*)

Fig. 21-14 The erection of the unistrut roof. (*Unistrut Corp.*)

Fig. 21-16 Dome system for a theater. (*Temcor.*)

Fig. 21-15 Finished interior steel roof frame. (*Unistrut Corp.*)

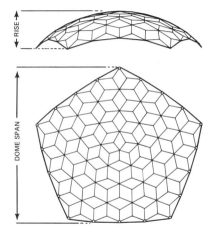

Fig. 21-17 Diagram of dome. (*Temcor.*)

Structural Dome Systems. The dome system illustrated in Fig. 21-16 shows the dome erected for a theatre in Reno, Nevada. The geometry used in this structure is called *geodesic.* It is different from other dome geometry in that its strength is in all directions (omnidirectional). The three-dimensional (Fig. 21-17) triangulated framing of the dome makes it exceptionally strong while using less material than other dome frames. Dr. R. Buckminster Fuller is the inventor of geodesic geometry. Figure 21–18 illustrates the erection of the flexible-skin dome.

Structural Storage System. The water tank, called a water spheroid because of its shape, shown under construction in Figs. 21–19A through D, is typical of welded fastening. The parts are preformed in the shop and are welded in the field. This structural geometric solution to storage is clever and calls for ingenious erection techniques.

Structural Drafting of Steel Shapes. Some of the basic structural shapes were identified earlier in this chapter. The seventh edition of the *Manual of Steel Construction*[1] or any handbook published by a major steel company will list all the major shapes of steel available and the great variety of sizes and their weights. The *AISC Manual* contains tables for designing and detailing the various shapes in any combination. Figure 21–20 shows the cross section of the plain material which is grouped by AISC as follows:

1. American Standard beams (S).

2. American Standard channels (C).

3. Miscellaneous channels (MC). These include special-purpose channels which are other than standard.

4. Wide-flange shapes (W). These are used as both beams and columns.

5. Miscellaneous shapes (M). These light-weight shapes are similar in cross-sectional profile to W shapes.

6. Structural tees (ST, WT, MT). These are made by splitting S, W, and M shapes, usually along the mid-depth of their webs.

7. Angles (L). These consist of two legs of equal or unequal widths. The

[1] *Manual of Steel Construction,* or, as it is called, *AISC Handbook,* is published by the American Institute of Steel Construction, 101 Park Avenue, New York, NY 10017.

Fig. 21–18 Erection of dome. (*Temcor.*)

Fig. 21–19A Cylinder erection. (*Chicago Bridge & Iron Co.*)

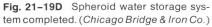

Fig. 21–19B Beginning erection of spheroid. (*Chicago Bridge & Iron Co.*)

Fig. 21–19C Partially completed spheroid. (*Chicago Bridge & Iron Co.*)

Fig. 21–19D Spheroid water storage system completed. (*Chicago Bridge & Iron Co.*)

legs are set at right angles to each other.

8. Plates (PL) and flat bars (Bar). These are rectangular in cross section.

These plain forms are basic to structural detailing and must be understood in order to prepare adequate drawings.

Scales. Structural details are prepared at a scale of $1'' = 1'\text{-}0''$ for beams up to 21 in. in depth. When beams are over 21 in., a $\frac{3}{4}'' = 1'\text{-}0''$ scale is preferred. The overall length of structural members can be drawn shortened if the details are adequately shown. Very small dimensions such as a clearance may be exaggerated to clarify views.

Typical Steel Details on the Design Drawings. Design drawings prepared by the designer will always have all the information necessary for the preparation of shop drawings. Figure 21–21 represents a small part of a structurally framed floor plan. The view of the framed system is from above. A shop drawing of the wide-flanged beam (W) can be prepared from the notes and dimensions on the designer's plan.

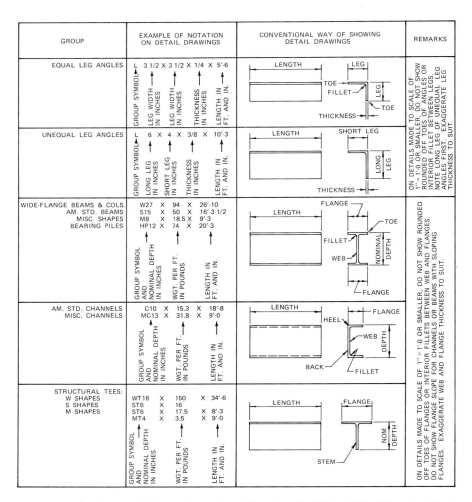

Fig. 21–20 Steel material shapes in section. (*American Institute of Steel Construction.*)

Fig. 21–21 Small part of designed floor plan. (*Adapted from* AISC Handbook, *with permission of publisher*)

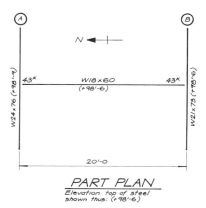

Fig. 21–22 Typical beam detail. (*Adapted from* AISC Handbook, *with permission of the publisher.*)

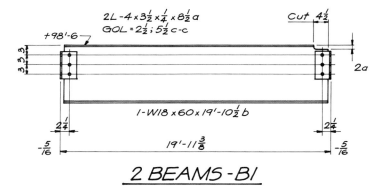

The 20-ft dimension is presumed to be the structural bay, or distance from *A* to *B,* and the structural members are at right angles to one another unless noted. The height recorded on the line diagram of a beam is significant. These height elevations are assumed to be level at the figure given. Figure 21–21 shows two elevations, 98′-6″ and 98′-9″.

Shop Drawings. In order to fabricate a beam, the typical details in Fig. 21–22 are essential. The beam detail seldom shows the detailed connections of mating parts as they are shown in Fig. 21–23. The beam will only have the features (e.g., connection angles) which will have a bearing on the erection of this individual unit.

In preparing the structural details, the draftsman will make reference to the handbook and the dimensions for detailing. Figure 21–24 shows both framed and seated connections.

Riveting. Beam connections may be riveted together. The standard symbols for rivets are given in Fig. 21–25 at A. A typical buttonhead rivet is shown in Fig. 21–25 at B. When rivet assembly takes place in the shop, the rivet is shown on the drawing by open circles of the diameter of the rivet head. Field riveting which takes place on the site of construction is shown on drawings by blacked-in circles of the diameter of the rivet hole, as shown in Fig. 21–25 at C. Lines on which rivets are spaced are called *gage lines.* The distance between rivet centers on the gage lines is called the *pitch.*

Structural Bolting. High-strength-steel bolts are rated by the American

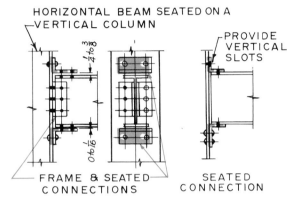

Fig. 21–24 Frame and seated connection. (*Adapted from* AISC Handbook, *with the permission of the publisher.*)

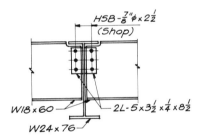

Fig. 21–23 Typical connection of mating parts. (*Adapted from* AISC Handbook, *with the permission of the publisher.*)

Fig. 21–25 Rivet symbols and buttonhead rivet.

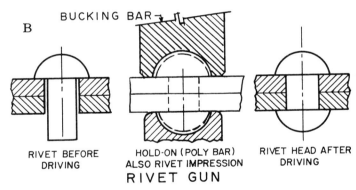

Society for Testing and Materials (ASTM). The bolt is applied in the field or in the shop. The hole prepared for this connector is normally $\frac{1}{16}$ in. larger than the bolt. Figure 21–26 illustrates both bolted frame and seated connections. The bolt transmits the force of the beam load to the column; therefore, the stress in the bolt is shear (cutting). The force transmitted into the column is a compressive (pushing together) stress.

Welding Structural Members. Structural steel members are welded with the metal-arc process by most fabricators of steel. The fillet weld is typically the most common on structural connectors. The symbols for welding standards were reviewed in Chapter 17.

Dimensioning. Dimensions should be given to the center lines of beams, to the backs of angles, and normally to the backs of channels. The vertical dimensions should be given to the tops or bottoms of beams and channels. In general the edges of flanges and toes of angles are not to be dimensioned for detail purposes. Continuous and unbroken dimension lines are used on structural drawings. The dimensions are placed above the dimension lines. Dimensions are given primarily to working points.

When dimensions are in feet and inches, the foot symbol is used but not the inch symbol.

Structural drawings. In Fig. 21–27 a small steel roof truss which is symmetrical is detailed about the left of a center line. A close study of the drawing will show that each member is completely dimensioned or described and that dimensions adequately relate the fixed location of each structural member.

Roof trusses. Diagrams for some roof trusses and bridge trusses are shown and named in Fig. 21–28. The ones shown are only a few of those available and may be modified to carry designed loads.

Wood for Construction. The construction of homes and other small structures has been commonly framed out in wood. The National Forest Products Association has provided industry with typical construction details, and these are now a standardized method of construction.

The wood forms manufactured by wood laminators are providing the construction industry with "factory grown" timbers to any size or shape (Fig. 21–29). The American Institute of Timber Construction (AITC) has provided guidelines for manufacturers of structural glued laminated timber. Some of the forms available include tudor arches, radial arches, parabolic arches, A frames, and tapered beams (Fig. 21–30).

Figure 21–31 illustrates some of the construction details necessary for building systems around structural timber. The details pictorially describe how to anchor the structural members to foundations and how two or more timbers are connected together.

Systems for timber construction. Structural timber for construction is graded according to allowable stresses and the types of connections which will allow the transfer of stresses from one member to another. Wood commonly used for building includes pine, ash, birch, cedar, cypress, elm, oak, and redwood. Design handbooks list the specific grades and related structural properties.

Figure 21–32 illustrates the erection of engineered timber construction. In Fig. 21–33 the templates are laid out for a specific predesigned arch. The construction in Fig. 21–32 shows connections and wrappings used for protection.

Concrete Systems. Concrete can be reinforced or prestressed. Many of

Fig. 21–26 Structural bolt and bolted connection.

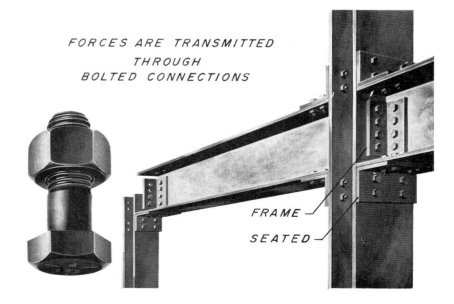

FORCES ARE TRANSMITTED THROUGH BOLTED CONNECTIONS

FRAME

SEATED

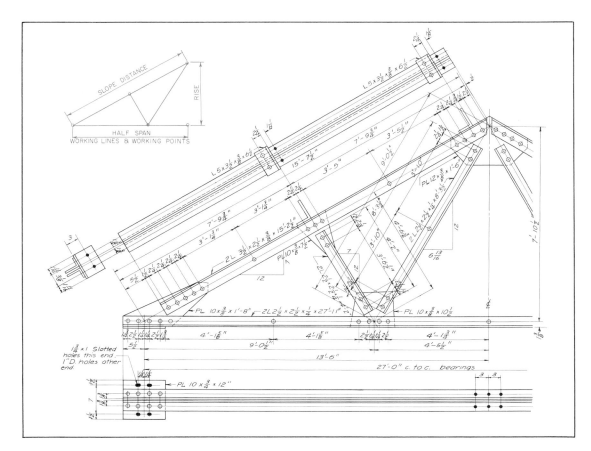

Fig. 21-27 Roof truss detail.

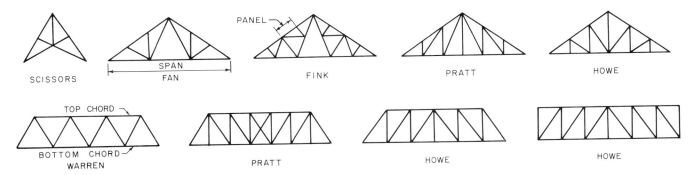

Fig. 21-28 Roof and bridge truss diagrams.

our buildings, bridges, and dams require the special properties of concrete. The American Concrete Institute is a national organization which has prepared a manual of standard practice for concrete structures.

Concrete is a building material that has limited strength without special preparation. Gravel, sand, and portland cement mixed with water in predetermined proportion produce various grades of concrete. Reinforced

concrete has steel bars embedded in the concrete. The reinforced bars are arranged to resist the stresses imposed by the working loads within the structure which concrete could not support by itself. When concrete and steel are

Fig. 21-29 Laminated wood forms take any size or shape. (*American Institute of Timber Construction.*)

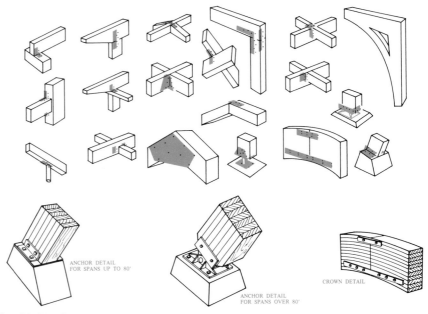

ANCHOR DETAIL
FOR SPANS UP TO 80'

ANCHOR DETAIL
FOR SPANS OVER 80'

CROWN DETAIL

Fig. 21-31 Construction details for timber construction. (*American Institute of Timber Construction.*)

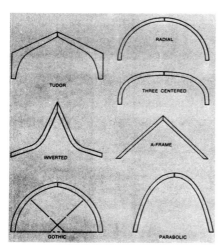

RADIAL

TUDOR

THREE CENTERED

A-FRAME

INVERTED

GOTHIC

PARABOLIC

Fig. 21-30 Typical shapes of laminated wood forms. (*American Institute of Timber Construction.*)

Fig. 21-32 Erection of laminated wood forms. (*American Institute of Timber Construction.*)

Fig. 21-33 Preparing templates for glued laminated wood arches. (*American Institute of Timber Construction.*)

combined they may be used in the monolithic form shown in Fig. 21-34. A typical reinforced concrete detail is illustrated in Fig. 21-35.

Prestressed concrete is a term used when the reinforcing bars are stretched before the concrete is poured into the designed form. The prestressed form will then accept a predesigned load. This combination of material (concrete and stretched steel) will provide a structural system superior to plain concrete or reinforced concrete.

Detailed drawings. Concrete forms which are designed by the structural engineer are drawn for the manufacturer's use only. Construction drawings are prepared for the construction contractor by the manufacturer. The purpose of the construction drawing is to show location, placement, and connection. See Fig. 21-36 for typical details.

Structural Clay Systems. The solid brick wall is the oldest type of masonry construction known. Bricks are produced from different types of clay in many shapes, forms, and colors. They are assembled into walls (exterior and interior) which carry the loads of the floor and roof systems. In order to carry the loads of these horizontal systems, the clay brick used in the vertical wall must have high compressive strength.

Structural bonding. Brick masonry units are interlocked or tied together so as to act as a single structural unit. Various bonding patterns are formed by masonry units and mortar joints. The overlapping (interlocking) of these masonry units embedded in connecting mortar joints produces a structural assembly. A structural bond is illustrated in Fig. 21-36. Note the terms applied to the brick. The common brick size is $2\frac{1}{4} \times$

Fig. 21-34 Monolithic concrete form. (*Ceco Steel Products Corp.*)

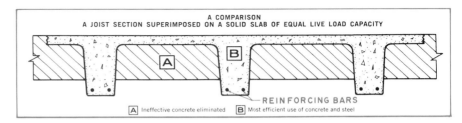

Fig. 21-35 Reinforced concrete detail. (*Ceco Steel Products Corp.*)

Fig. 21-36 Concrete placement drawing. (*Ceco Steel Products Corp.*)

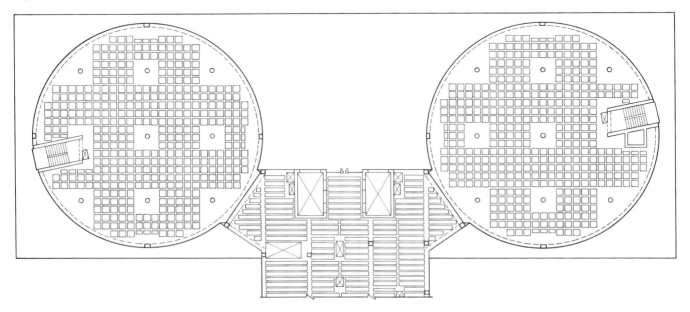

$3\frac{3}{4} \times 8''$; some of the common bonds are illustrated in Fig. 21–37.

Structural clay products are generally dependent upon the strength of the mortar joints. The limited strength of mortar joints on some designs requires the designer to specify reinforcing for added strength.

Steel rods or wire embedded in the mortar of masonry construction is called *reinforced masonry*. Masonry in the form of brick or concrete is used to enclose structural steel framework. This provides adequate fireproofing to meet standard building codes.

Natural stone and manufactured stone are used in masonry construction, but very few structural designs today include these materials for structural purposes. Manufactured stone, along with marble, granite, limestone, and sandstone, is used for ornamental treatment on buildings.

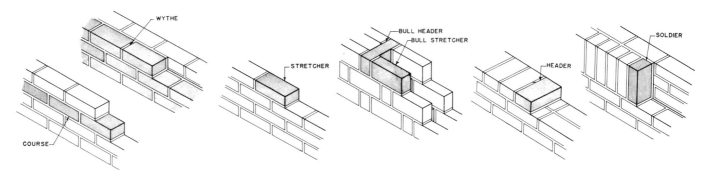

Fig. 21–37 Bonds and structural patterns of common brick. (*Structural Clay Products Institute.*)

Review

1. Name five places of employment for a structural draftsman.

2. Name five basic structural steel shapes.

3. What is the major advantage of a geodesic dome?

4. How are beams, angles, and channels dimensioned?

5. How are dimension lines drawn on structural drawings?

6. What is the difference between reinforced and prestressed concrete?

7. What symbols are used in structural dimensions, and when are they used?

8. What is the limiting factor in structural clay systems which must carry a load?

Problems for Chapter 21

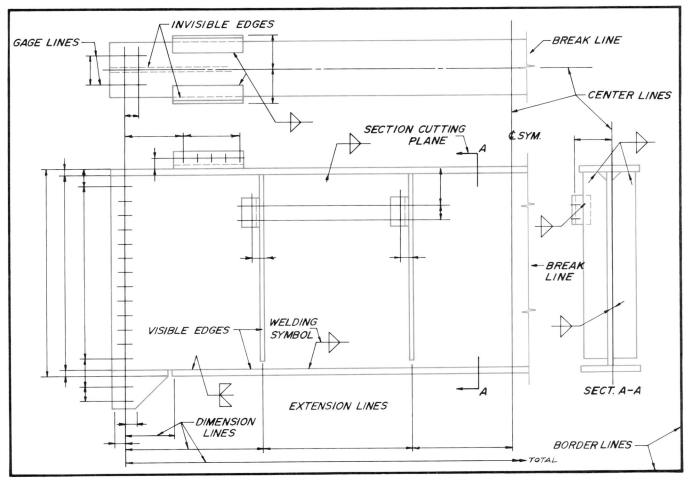

Fig. 21-38 Prepare a detail of a 22″ high girder with a 6″ flange. Develop the girder, showing conventional lines and their relative weights. Identify angles, welds, stiffeners, and field bolts. Fill in missing dimensions with appropriate selection. Scale: $1\frac{1}{2}″ = 1'\text{-}0″$.

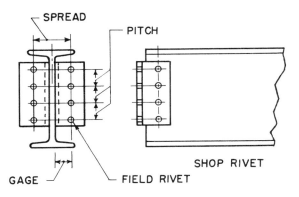

Fig. 21-39 Draw the detail of a framed connection on an S-beam which is 8″ by 18.4 lbs. The flange width is 4″ and the web is .270″. The 3″ by 3″ connection angle is 6″ long. Pitch = $1\frac{1}{2}″$. Gage = $1\frac{3}{4}″$. Rivet = $\frac{1}{2}″$. Prepare details at half scale.

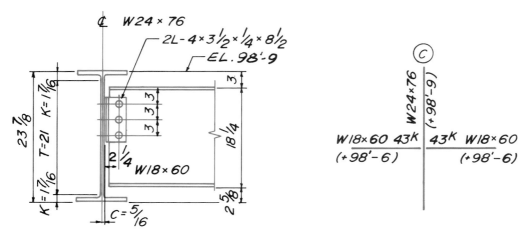

Fig. 21–40 Prepare a partial detail of the framed connection of the 18″-wide flanged beam with a 24″-wide flanged beam. Scale: 3″ = 1′-0″. Prepare a simple erection diagram, no scale assigned.

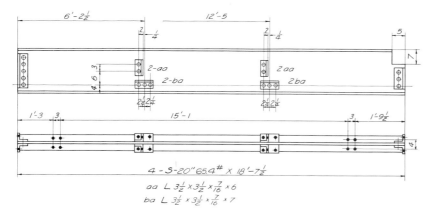

Fig. 21–41 Prepare a detail drawing of a standard S-beam, 20″ by 65.4 lbs. Flange = 6¼″ wide with web thickness of ½″. Scale: ¾″ = 1′-0″.

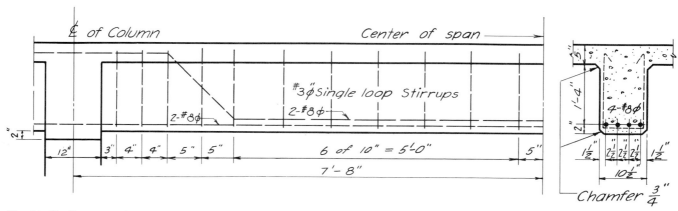

Fig. 21–42 Prepare a drawing of the reinforced concrete detail shown. Scale: 1½″ = 1′-0″.

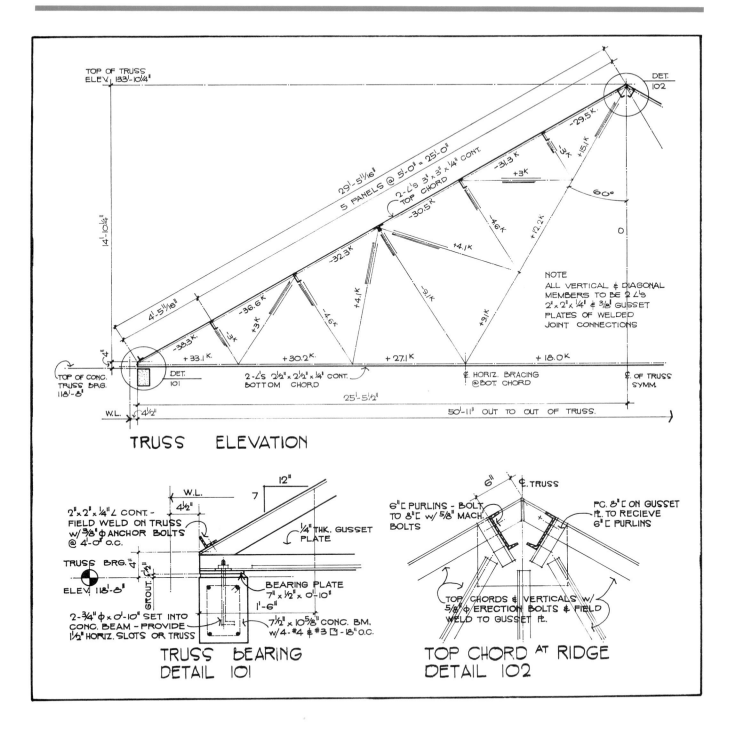

Fig. 21–43 Prepare the schematic diagram for the welded truss construction. Scale: $\frac{1}{2}'' = 1'$-$0''$. Details 101 and 102 may be prepared at a suitable scale to enlarge the detail: $1\frac{1}{2}'' = 1'$-$0''$ is suggested.

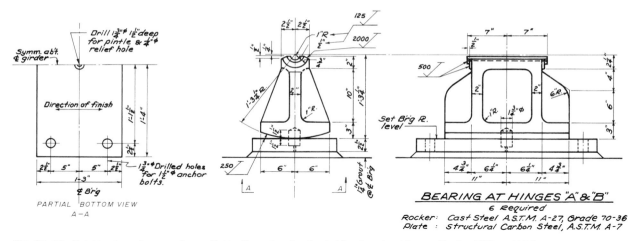

Fig. 21-44 Prepare two views and a partial bottom view for the bridge bearing hinge. Scale: 1½" = 1'-0".

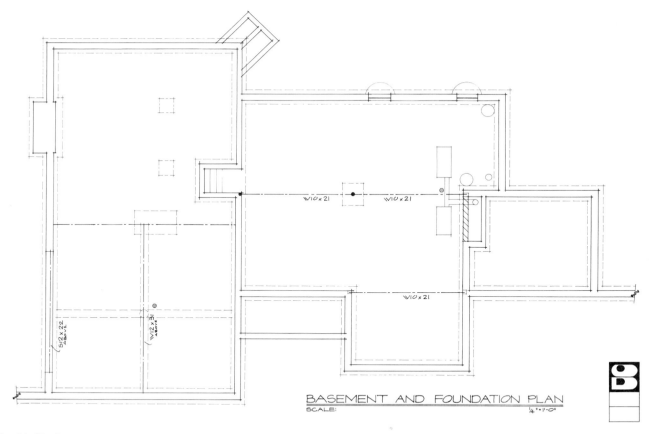

BASEMENT AND FOUNDATION PLAN
SCALE: ¼" = 1'-0"

Fig. 21-45 Prepare a structural deck plan for the French Provincial residence. Examine Fig. 20–103 for dimensions. Show structural wood framing, 16" on center, placed on the appropriate structural support. Refer to the framing plan (Fig. 20–90) for the placement of a joist on a structural member. Scale: ¼" = 1'-0". (*Orrin Dressler.*)

Map Drafting | Chapter 22

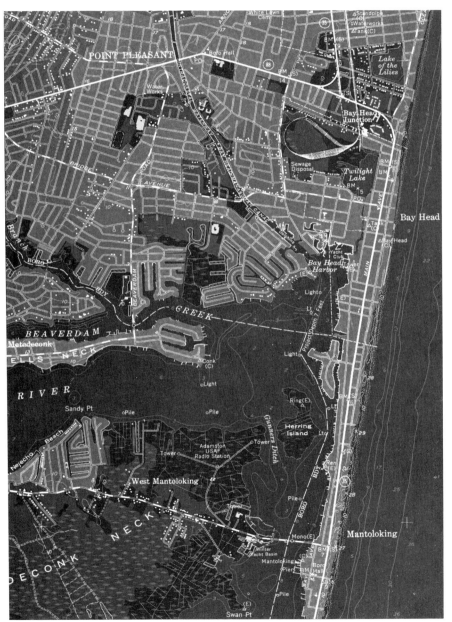

(U.S. COAST AND GEODETIC SURVEY.)

Mapping—A Changing Industry. Cartographers are designers who have been trained to gather data and prepare maps. The skilled map draftsman prepares maps in detail. He is closely linked to the civil engineer, scientist, geographer, and geologist. Map making is the process of representing facts about the surface of a sphere, such as the earth or other bodies in the solar system (Fig. 22-1). Missiles fired from Cape Kennedy (Fig. 22-2), whether to the moon or to Mars as in the Mariner program, carry the necessary equipment for mapping these distant bodies. The moon and other planets have been mapped by *photogrammetry* (aerial photography), a technique used in the modern mapping industry. Based on the speed of light and microwaves, electronic devices now can *survey* (measure) great distances in short periods of time with a great degree of accuracy. Surveying or tracking has become highly sophisticated.

Maps were used for early transportation systems when streams and rivers were avenues of commerce. Many early examples of maps were generally artistic representations, but they lacked the accuracy found in present day maps (Fig. 22-3). Modern surveying has assisted in developing maps which connect cities, states, and nations with the arteries of travel and commerce. Surveying also establishes the property lines of the nation's homes. The uses of maps have multiplied greatly, and many professional

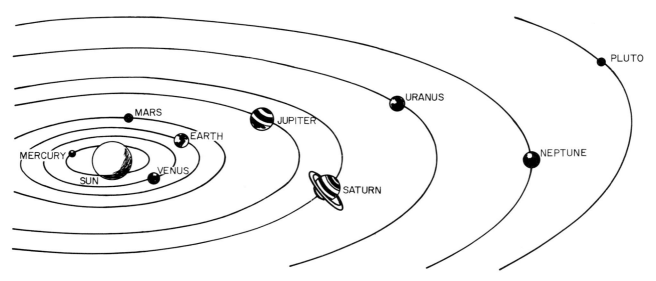

Fig. 22-1 The solar system. The Sun is a star located in the center of the system with nine planets revolving around it in noncircular orbits. All the planets move in the same direction around the Sun and are located in nearly the same plane. The Earth is 90 million miles (average 92.8) from the Sun. The planet Pluto is $3\frac{1}{2}$ billion miles from the Sun.

opportunities are available for those qualified in assisting in their preparation.

Career Opportunities. The field of civil engineering is always expanding. The planning of railroads, highways, harbor facilities, airports, and space stations only begin to define the map planning in this broad industry.

The draftsman may prepare maps and charts under the supervision of the design engineer, or cartographer. Advancement opportunities may include photogrammetry, surveying with laser beams, or research and development projects with the geographer. Additional information is available from:

Association of American
  Geographers
1146 16th Street N.W.
Washington, D.C. 20036

American Society of Certified
  Engineering Technicians

2029 K Street N.W.
Washington, D.C. 20036

American Geographical Union
1145 19th Street N.W.
Washington, D.C. 20036

Map Sizes. As the world has entered the space age, there are maps showing enormous distances. The *scale* (size) is relatively small on maps which cover large distances. These maps may not always be accurate but may have accurate dimensions applied to them. Some maps having to do with ownership of property, such as city plats, must be extremely accurate. They may be drawn to a large scale so as to note all the data of physical property. Geographic maps of states or countries, which show boundary lines, streams, lakes, or coastlines, may use a scale of several miles to the inch.

Scales Used on Maps. The civil engineer's scale is used for map drawings. Distances are given in decimals

of a foot or meter, such as tenths, hundredths, and so forth. (See the maps in your geography and history books.) Practically all the countries of the world have adopted the metric system, with the kilometer in general use instead of miles. The Earth's

Fig. 22-2 Missiles can survey our land. (*Official U.S. Air Force photo.*)

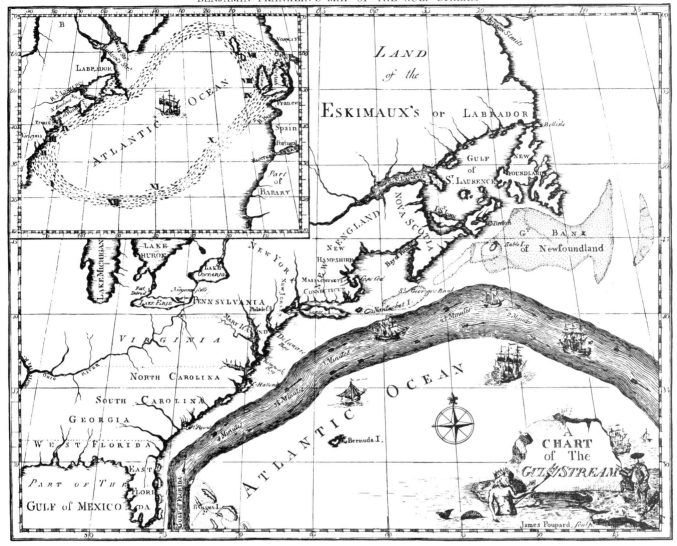

Fig. 22–3 Benjamin Franklin's map of the Gulf Stream. (*U.S. Coast and Geodetic Survey.*)

spherical measurements provided the French with a basis for originating the metric scale. The meter is one ten-millionth of the distance between the equator and the pole. This unit of measure was adopted by the French in 1791 as their official measure. The metric system is based on multiples of ten. See the appendix for a discussion of the metric system.

The scale of a map is generally noted as 1 in. equals 500 ft, or 1 part equals 6000 parts, noted as 1:6000. The scale of 1 in. equals 1 mile can also be shown as 1:63,360. Graphic scales should be shown on maps as part of their basic information.

Plats of a Survey. A map used to record the boundaries of a tract of land and to identify it is called a *plat*. The amount and kind of information presented will depend upon the purpose for which the map is required. The plat of a plane survey which was made to accompany the legal description of a property is shown in Fig. 22–4. Accuracy of data is all-important and must correspond to the legal description.

City Plat. Maps of cities are made for many purposes, such as to maintain a record of street improvements, to show the location of utilities, and to record sizes and location of property for tax assessments. A part of such a city plat is shown in Fig. 22-5. Notice the numbering of the lots and the location of streets, sidewalks, and various other details found in a city.

Operations Maps. Engineering and management groups, or government agencies, frequently have projects or programs under discussion which are greatly aided by the use of operations maps. These maps show the relationship of various components to a general plan. Such a map is found in Fig. 22-6. A good presentation will assist greatly in the selling of a program.

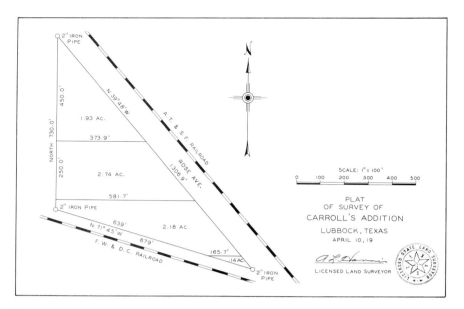

Fig. 22-4 Plat of a survey. Note parts which make up the plat: the acreage of each part, the iron pipe locating the corners of the graphics scale, the signature of the surveyor, and the official seal.

Fig. 22-5 Part of a city map drawn with the aid of a computer. (*City of Las Vegas Department of Public Works, and Calcomp.*)

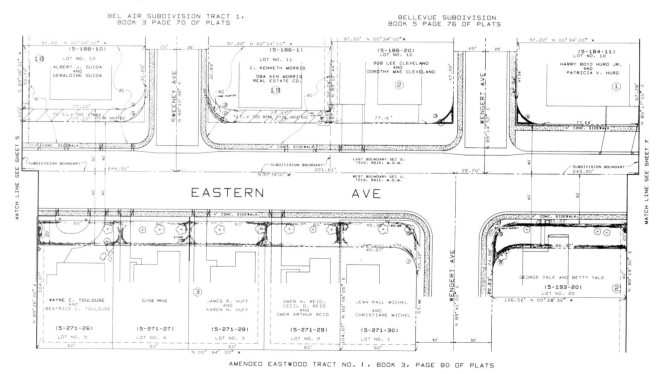

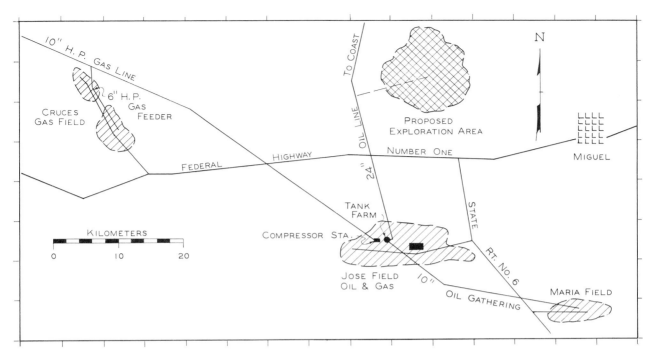

Fig. 22-6 An oil-field operations map. Note scale of kilometers. One kilometer = 0.621 statute mile.

Contours. Since maps are one-view drawings, vertical distances and variations in ground level do not show. They can, however, be indicated by lines of constant level called *contours.* This is illustrated in Fig. 22-7, where the contours show the location of lines on the ground which are at stated heights above the ocean (sea level). Contours which are close together indicate a steeper slope than lines that are farther apart. This can be seen by projecting the intersections of the horizontal level lines with the profile section, as shown in Fig. 22-7.

It will be observed that the contour map and the profile correspond to the plan and section of an ordinary drawing. Note the horizontal line, *A-A*, or cutting plane, on the contour map which shows the position, or line, on which the profile is taken. Notice how the profile would change if the

cutting plane were moved toward the ocean or to some other new position.

Spacing of contour intervals will vary according to requirements and the scale of maps. A 10-ft interval may be quite satisfactory, reserving a 5-ft interval for use on larger-scale maps. For close detail work, such as an irrigation project, the contour intervals may be reduced to 0.5 ft, 1 ft, or 2 ft. In case of small-scale maps with a high degree of relief, intervals may be increased from 20 to 200 ft, or even more. As an aid to interpretation, every fifth contour is generally accentuated by drawing a much heavier line (refer to Fig. 22-11).

A technical pen is satisfactory for inking contour lines. It is replacing the contour pen (Fig. 22-8), which has blades that swivel and so can easily follow contour lines.

There are many forms of flexible

curves and splines which can be bent to match curves to be drawn. The rubber-covered curve in Fig. 22-9 has a core composed of a strip of lead and two strips of steel. The spline in Fig. 22-10 is a metal strip which can be bent to fit a desired curve and is held in place by ducks (metal weights), as shown.

A contour map is shown in Fig. 22-11; it uses a contour interval of 20 ft (vertical distance between contour lines). The elevation in feet is marked in a break in each contour line. Notice that the drainage is indicated as intermittent streams.

Before a contour map can be drawn, elevations must be obtained in the field for various key points controlling the construction of the contours. Various methods are used: a grid system where all intersection elevations are obtained, along with critical ele-

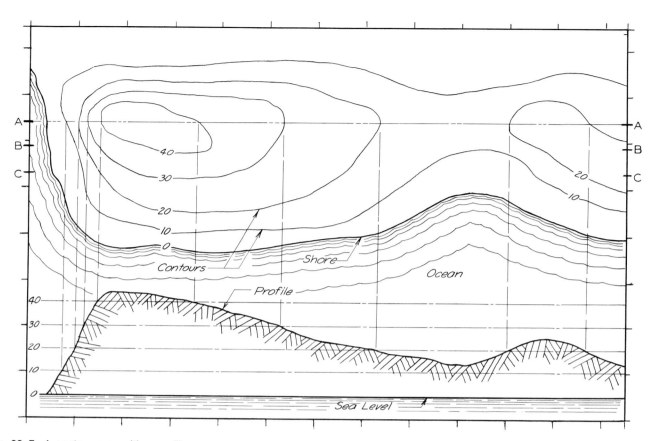

Fig. 22-7 A contour map with a profile.

Fig. 22-8 A contour pen.

Fig. 22-9 Part of a rubber-covered flexible curve.

Fig. 22-10 A metal spline held in place by "ducks."

vations on grid lines; points located by transit (Fig. 22-12) and stadia rod, with the corresponding elevations calculated by plane table; and aerial photographic surveys (Fig. 22-13). The finished map in Fig. 22-13 was produced by photogrammetric methods. The equipment for making photogrammetric maps is shown in Fig. 22-14. The actual drawing of the map is usually done with the scribing technique on a coated film. An example of scribing on film is shown in Fig. 22-15.

All these methods require experience in surveying, an important part of civil engineering.

Topographic maps are made to give rather complete descriptions of the

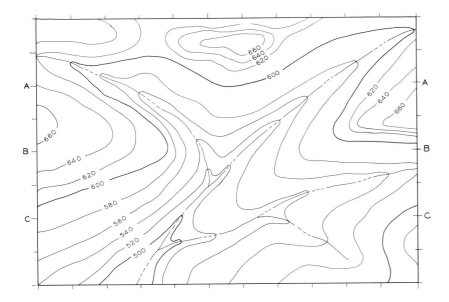

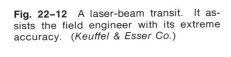
Fig. 22-11 A contour map.

Fig. 22-12 A laser-beam transit. It assists the field engineer with its extreme accuracy. (*Keuffel & Esser Co.*)

Fig. 22-13 Portions of an aerial photo and a topographic map of Concepcion, Chile, compiled by photogrammetric methods. The photo and map are part of a Chilean-OAS program to speed reconstruction of earthquake-damaged areas and to advance abroad planning for national growth. Actual map sheets were done at a scale of 1:2000 with 1-meter contours. Photos from which maps were prepared are at a scale of 1:10,000. This project is being carried out by a group of companies under the direction of Aero Service Corporation, Philadelphia, a division of Litton Industries.

areas shown (Fig. 22-16). This includes such information as boundaries, natural features, the works of man, vegetation, and relief (elevations and depressions). Symbols are used for many of the features shown on topographic maps, some of which are given in Fig. 22-17. Maps using topographic symbols can be obtained at nominal cost from the Director, U.S. Geological Survey, Department of the Interior, or from the U.S. Coast and Geodetic Survey, Department of Commerce, Washington, D.C. Naval charts (maps) come from the Hydrographic Office, Bureau of Navigation, Department of the Navy.

Aeronautical maps (Fig. 22-18) make use of special symbols which need to be understood in order to read them. Some symbols from the U.S. Coast and Geodetic Survey are shown in Fig. 22-19.

Fig. 22-14 Photogrammetric equipment with computerized storage. (*Keuffel & Esser Co.*)

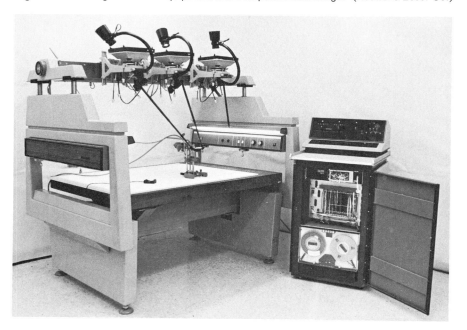

Fig. 22-15 Scribing with a jewel point on coated film provides a very accurate and durable map. (*Keuffel & Esser Co.*)

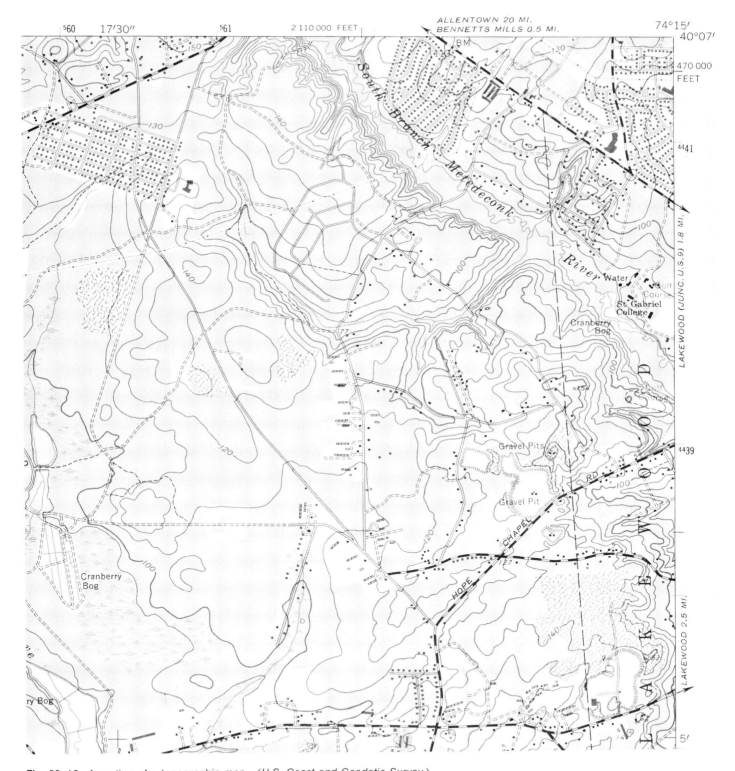

Fig. 22-16 A portion of a topographic map. (*U.S. Coast and Geodetic Survey.*)

TOPOGRAPHIC MAP SYMBOLS

Hard surface, heavy-duty road ....
Hard surface, medium-duty road ....
Improved light-duty road ....
Unimproved dirt road ....
Trail ....
Railroad: single track ....
Railroad: multiple track ....
Bridge ....
Drawbridge ....
Tunnel ....
Footbridge ....
Overpass—Underpass ....
Power transmission line with located tower ....
Landmark line (labeled as to type) .... TELEPHONE

Dam with lock ....
Canal with lock ....
Large dam ....
Small dam: masonry — earth ....
Buildings (dwelling, place of employment, etc.) .
School—Church—Cemeteries .... Cem
Buildings (barn, warehouse, etc.) ....
Tanks; oil, water, etc. (labeled only if water) ... Water Tank
Wells other than water (labeled as to type) ...... ○ Oil ○ Gas
U.S. mineral or location monument — Prospect ...
Quarry — Gravel pit ....
Mine shaft—Tunnel or cave entrance ....
Campsite — Picnic area ....
Located or landmark object—Windmill ....
Exposed wreck ....
Rock or coral reef ....
Foreshore flat ....
Rock: bare or awash ....

Horizontal control station .... △
Vertical control station .... BM ×671 ×672
Road fork — Section corner with elevation ...... ×429 +58
Checked spot elevation .... × 5970
Unchecked spot elevation .... × 5970

Boundary: national ....
State ....
county, parish, municipio ....
civil township, precinct, town, barrio ....
incorporated city, village, town, hamlet ....
reservation, national or state ....
small park, cemetery, airport, etc. ....
land grant ....
Township or range line, U.S. land survey ....
Section line, U.S. land survey ....
Township line, not U.S. land survey ....
Section line, not U.S. land survey ....
Fence line or field line ....
Section corner: found—indicated .... + +
Boundary monument: land grant—other .... □ □

Index contour .... Intermediate contour
Supplementary cont. .... Depression contours
Cut — Fill .... Levee
Mine dump .... Large wash
Dune area .... Tailings pond
Sand area .... Distorted surface
Tailings .... Gravel beach

Glacier .... Intermittent streams
Perennial streams .... Aqueduct tunnel
Water well—Spring .... Falls
Rapids .... Intermittent lake
Channel .... Small wash
Sounding—Depth curve ...... Marsh (swamp)
Dry lake bed .... Inundated area

Woodland .... Mangrove
Submerged marsh .... Scrub
Orchard .... Wooded marsh
Vineyard .... Bldg. omission area

Fig. 22–17 Some conventional symbols used on maps. (*U.S. Coast and Geodetic Survey.*)

Block Diagrams. Consideration has been given to mapping in the horizontal plane and the vertical plane by profiles, or sections. To aid in the visual understanding of this three-dimensional problem a *block diagram* is also employed. It is a three-dimensional projection using the isometric view (Fig. 22–20). This block diagram has been developed from Fig. 22–11. Keep in mind that each contour represents a level plane, similar to a card in a deck of cards. True lengths are measured on the isometric axes.

Geological Mapping. Geology is the science dealing with the makeup and structure of the earth's surface and interior depths. As an aid to understanding, the methods of pictorial representation discussed in this book are used in modified form to meet the special needs of this science.

The crust of the earth is made up of three groups of rock: igneous (crystalline), sedimentary, and metamorphic. Crystalline rocks, for pur-

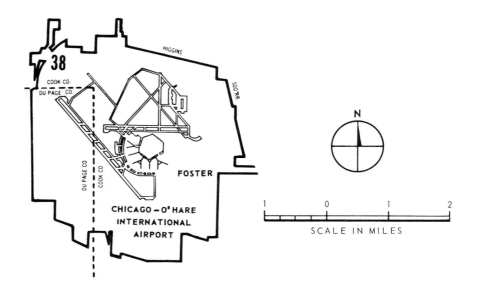

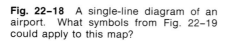

Fig. 22–18 A single-line diagram of an airport. What symbols from Fig. 22–19 could apply to this map?

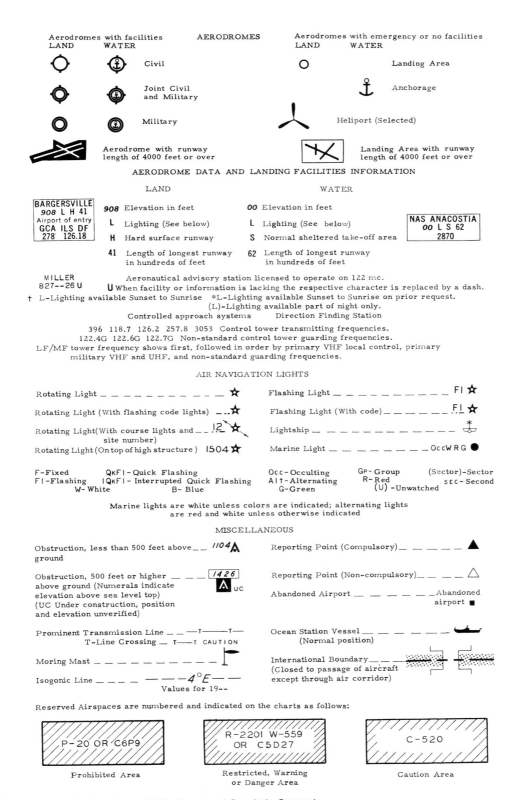

Aerodromes with facilities
LAND   WATER

AERODROMES

Aerodromes with emergency or no facilities
LAND   WATER

Civil

Joint Civil and Military

Military

Landing Area

Anchorage

Heliport (Selected)

Aerodrome with runway length of 4000 feet or over

Landing Area with runway length of 4000 feet or over

AERODROME DATA AND LANDING FACILITIES INFORMATION

LAND

WATER

BARGERSVILLE
908 L H 41
Airport of entry
GCA ILS DF
278' 126.18

908 Elevation in feet
L   Lighting (See below)
H   Hard surface runway
41 Length of longest runway in hundreds of feet

00 Elevation in feet
L Lighting (See below)
S Normal sheltered take-off area
62 Length of longest runway in hundreds of feet

NAS ANACOSTIA
00 L S 62
2870

MILLER
827--26 U

Aeronautical advisory station licensed to operate on 122 mc.
U When facility or information is lacking the respective character is replaced by a dash.
† L-Lighting available Sunset to Sunrise   *L-Lighting available Sunset to Sunrise on prior request.
(L)-Lighting available part of night only.
Controlled approach systems   Direction Finding Station
396 118.7 126.2 257.8 3053 Control tower transmitting frequencies.
122.4G 122.6G 122.7G Non-standard control tower guarding frequencies.
LF/MF tower frequency shows first, followed in order by primary VHF local control, primary military VHF and UHF, and non-standard guarding frequencies.

AIR NAVIGATION LIGHTS

Rotating Light ☆

Flashing Light Fl ☆

Rotating Light (With flashing code lights) ☆

Flashing Light (With code) Fl ☆

Rotating Light (With course lights and site number) 12 ☆

Lightship

Rotating Light (On top of high structure) 1504 ☆

Marine Light Occ W R G ●

F-Fixed   QkFl-Quick Flashing   Occ-Occulting   Gp-Group   (Sector)-Sector
Fl-Flashing   IQkFl-Interrupted Quick Flashing   Alt-Alternating   R-Red   sec-Second
   W-White   B-Blue   G-Green   (U)-Unwatched

Marine lights are white unless colors are indicated; alternating lights
are red and white unless otherwise indicated

MISCELLANEOUS

Obstruction, less than 500 feet above ground 1104 ▲

Reporting Point (Compulsory) ▲

Obstruction, 500 feet or higher above ground (Numerals indicate elevation above sea level top) (UC Under construction, position and elevation unverified) 1426 ▲ UC

Reporting Point (Non-compulsory) △

Abandoned Airport Abandoned airport ■

Prominent Transmission Line
T-Line Crossing CAUTION

Ocean Station Vessel (Normal position)

Moring Mast

Isogonic Line 4°E
Values for 19--

International Boundary (Closed to passage of aircraft except through air corridor)

Reserved Airspaces are numbered and indicated on the charts as follows:

P-20 OR C6P9

R-2201 W-559 OR C5D27

C-520

Prohibited Area

Restricted, Warning or Danger Area

Caution Area

Fig. 22–19 Some aeronautical symbols. (*U.S. Coast and Geodetic Survey.*)

poses of this general discussion, are the basic materials making up the earth's crustal ring. This rock was once molten; it has cooled, but it has not been eroded nor has its makeup changed. Sedimentary rocks, as a rule, are deposited in water in layers of varying thicknesses similar to the layers of an onion. If the onion is cut perpendicular to its axis, a series of concentric rings will be noted. In a slice of the earth's crust made in a sedimentary area, a similar pattern could be seen; a series of layers identifiable by texture, color, and material is visible (refer to pictures of the Grand Canyon). Metamorphic rocks are generally considered to be sedimentary rocks which have been deeply buried, subjected to high temperatures, and so recomposed that they are no longer identifiable as sedimentary rocks.

Nature, being ever-changing, folds, tips, and slices these sedimentary layers in an infinite number of ways. It pierces these layers with intrusions of crystalline rock; it allows sedimentary layers to be formed on top of crystalline rocks and then tips the whole mass, perhaps pushing it for miles, raising it high above sea level or dropping it thousands of feet. The geologist making investigations has the problem of representing what has happened or what a particular area looks like.

Figure 22–21 is a part of a geological surface map. The red lines represent the line of surface exposure of the contact between two formations (the line of two contacting layers of a cut onion). The geologist locates this in the field and notes his observation point with the "tee" symbol. The direction of this contact (strike) is shown by the top of the tee, and the slope of the contact (dip) is indicated by the figures at the tee, such as 23° on the tee and the right-hand side and

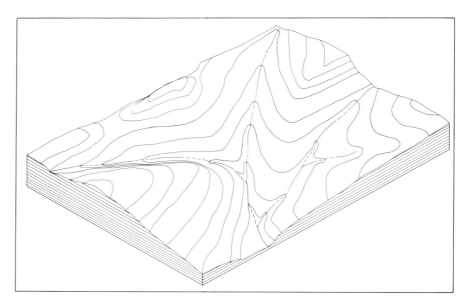

Fig. 22–20 Block diagram shows a block of earth in an isometric view. This picture was made from Fig. 22–11.

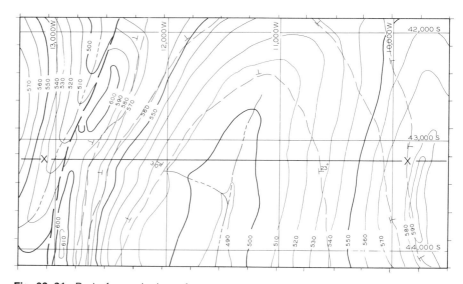

Fig. 22–21 Part of a geologic surface map.

below the section line X–X. Since the stem of the tee, in this case, is pointing to the east, or to the right, the dip is 23° to the east; in other words, this formation slopes 23° below the horizontal and to the east. Another tee symbol on the left side shows a 30° dip to the west. The heavy, broken line near the left edge repre-

sents a fault trace (the line along which the layer broke).

Geological Sections. These supplement the surface map, and an example (Fig. 22–22) shows in an idealized way what the geologist believes the area below the surface to be like. This is a section along line X–X of Fig.

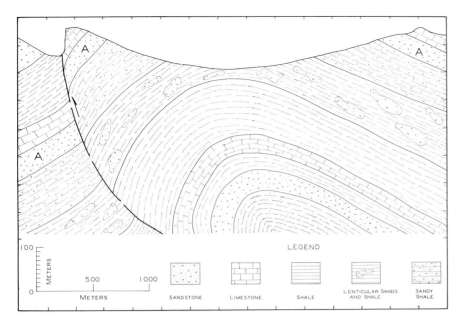

Fig. 22-22 A geologic section along line *X-X* of Fig. 22–21.

Fig. 22-23 A structural map showing strata details below the surface.

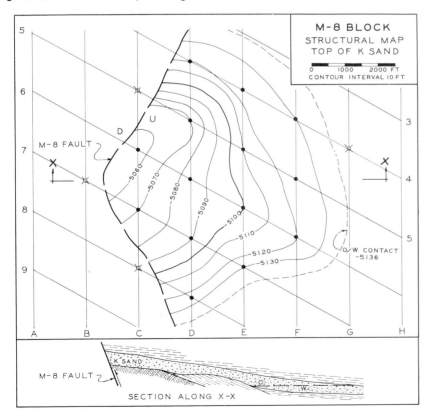

22–21. The dips which the geologist noted are used in developing the curvature of the folds. The geologist determines, too, by means of type section of the region, the various normal thicknesses of each formation, or stratum, and applies these values in making this section. The fault, as indicated, shows the area to the right to be upthrown. The relative displacement is apparent by comparing the position of formation *A* on either side of the fault.

Subsurface Mapping. This is a means of showing details of strata lying below the surface, such as the top or bottom of a given formation or possibly an assumed horizon. Data for constructing such a map is obtained from many sources, such as core holes, electrically recorded logs, seismograph surveys, and so forth. An example is shown in Fig. 22-23, where data were obtained from electrically recorded logs taken in a series of oil wells. The wells are located on a grid pattern, with producing wells indicated by a solid black circle and dry holes by an open circle with outward-extending rays. The top of a producing sand which is cut by a fault on the west is shown. Notice that the contours are numbered with negative values, or subsea depths; the larger the subsea value, the deeper the point below sea level. Section *X–X* shows thickness of sand and the level of the *oil-water contact*.

Geological maps, sections, and so forth, are greatly improved by the use of colors. In Fig. 22-21 colors are applied to each of the formations exposed between the red formation-contact lines. This can also be done in Fig. 22-22, applying the same colors to the corresponding formations. This device aids in bringing out the three-dimensional relationships of the surface and the shape of the structure, and it helps greatly to understand the geology of the area. Paper prints of the tracing are colored and then rubbed carefully to give smooth, even color texture. Examples of color use may be seen on a U.S. Geological Survey map.

The making of geological maps and drawings is an important part of the extractive minerals industry, particularly for petroleum. With the aid of maps it is possible to maintain proper records and information to ensure continued activity in this economic field. Between 40,000 and 50,000 oil wells are drilled each year in the United States, and all this information, obtained and properly recorded, aids in keeping this industry moving forward. Standards for records vary from company to company, but generally they are well covered in technical literature, such as publications of the AIME, petroleum branch, the AAPG, U.S. Geological Survey, U.S. Bureau of Mines, and others.

Notes and Definitions. Maps are important in the understanding of the news, in the study of geography and history, in making auto trips, and in surveying, geology, civil engineering, petroleum engineering, and space exploration.

The earth's surface is in the general form of a sphere. The surface of sphere cannot be exactly developed; however, systems have been devised for making maps of varying degrees of accuracy to meet the requirements for different purposes. The subject of map projection and the various systems which have been worked out is a rather extensive study and is beyond the scope of this book.

Brief definitions for some of the terms used are listed here. For more complete descriptions refer to books on surveying and geology.

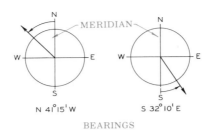

BEARINGS

1. *Bearing:* The direction of a line as shown by its angle with a north-south line (meridian).

2. *Meridian:* North-south line.

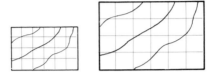

3. *Grid:* Generally, a series of uniformly spaced horizontal and perpendicular lines used to locate points by coordinates, or for enlarging or reducing a figure.

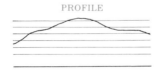

PROFILE

4. *Profile:* A section of the earth on a vertical plane showing the intersection of the surface of the earth and the plane (see Fig. 22–7).

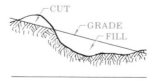

5. *Cut:* Earth to be removed to prepare for construction as for a desired level or slope for a road.

6. *Fill:* Earth to be supplied and put in place to prepare for a construction in order to obtain a desired level or slope.

7. *Grade:* A particular level or slope, as a downgrade.

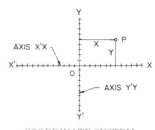

COORDINATE SYSTEM

8. *Coordinate system:* A system for locating points by reference to lines, generally at right angles.

9. *Contour:* A line of constant level showing where a level plane cuts through the surface of the earth (see Fig. 22–7).

10. *Block diagram:* A pictorial drawing (generally isometric) of a block of earth showing profiles and contours (see Fig. 22–20).

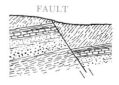

FAULT

11. *Fault:* A break in the earth's crust with a movement of one side of the break parallel to the line of the break.

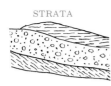
STRATA

12. *Stratum:* A layer of rock, earth, sand, and the like, horizontal or inclined, arranged in flat form distinct from the matter next to it. (Plural is strata.)

FOLD

13. *Fold:* A bend in a layer or layers of rock resulting from forces acting upon the rock after it has been formed.

DIP   STRIKE

14. *Dip:* The angle that an inclined stratum, or like geological feature, makes with a horizontal plane. The direction of dip is perpendicular to the strike.

15. *Strike:* The direction at the surface of the intersection of a stratum with a horizontal plane.

Review

1. If a map covers a large area the scale is _____.

2. If a map covers a small area the scale is _____.

3. Define a plat.

4. Define a contour line.

5. Name some reasons for making and using a contour map.

6. What is a topographic map?

7. What is a geological map?

8. What is a cartographer?

Problems for Chapter 22

Fig. 22-24 Plot the map of Chicago on a C-size sheet. Scale: $\frac{3}{4}'' = 1$ mile. Calculate the approximate number of square miles or the number of acres that make up this metropolitan center.

Fig. 22-25 On the elliptical projection of the earth, the isotherms indicate the average July temperature in degrees Fahrenheit around the world. Using dividers and an irregular or flexible curve, plot the temperature zones at a scale of your choice.

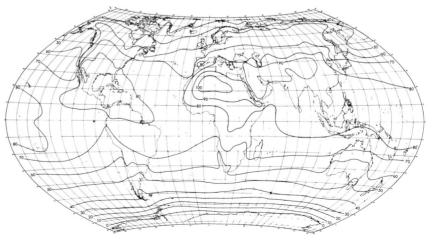

Fig. 22-26 Problem 1: Make a city map, using the data provided in Fig. 22-5. Problem 2: Enlarge the city map, Fig. 22-5, to include five complete blocks. Begin with the data provided and complete the map with data of your choice. Name all lots and streets added and provide dimensions on the drawing.

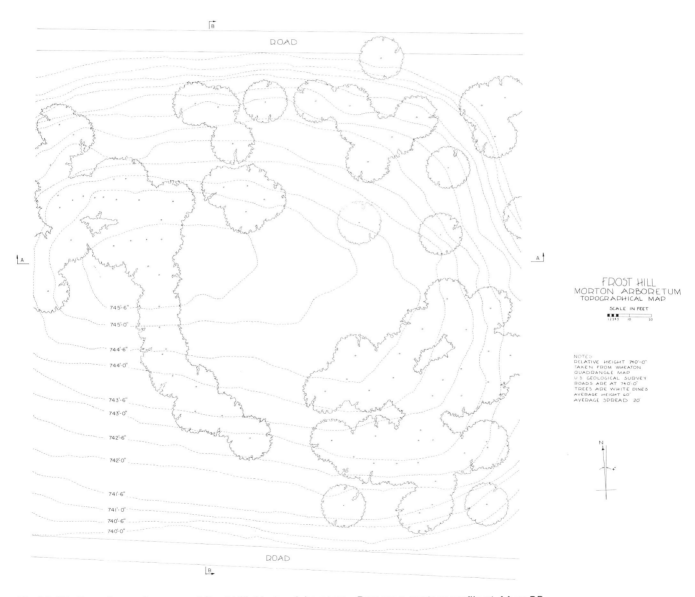

FROST HILL
MORTON ARBORETUM
TOPOGRAPHICAL MAP

SCALE IN FEET

NOTES
RELATIVE HEIGHT 740'-0"
TAKEN FROM WHEATON
QUADRANGLE MAP
U.S. GEOLOGICAL SURVEY
ROADS ARE AT 740'-0"
TREES ARE WHITE PINES
AVERAGE HEIGHT 60'
AVERAGE SPREAD 20'

N

ROAD

745'-6"
745'-0"
744'-6"
744'-0"
743'-6"
743'-0"
742'-6"
742'-0"
741'-6"
741'-0"
740'-6"
740'-0"

ROAD

Fig. 22-27 Draw the contour map of Frost Hill, Morton Arboretum. Prepare a contour profile at *AA* or *BB* as instructed. Choose a suitable scale. Locate all trees that appear within 20 feet of the profile. (*Joseph DeSalvo.*)

Graphic Charts and Diagrams

Chapter 23

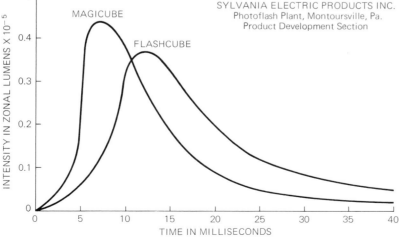

Fig. 1

Lighting Characteristics

	MAGICUBE	FLASHCUBE
Time to Peak (Milsec.)	7.0	12.0
Peak Intensity (Zonal-Lumens X 10^{-5})	0.44	0.38
Zonal Lumen Output (0-25 Milsec.)	460 (90% of Total)	460 (75% of Total)
Rise Time to 10,000 Zon-Lum. (Milsec.)	4.0	6.5
Decay Time to 10,000 Zon-Lum. (Milsec.)	19.0	27.0

SYLVANIA ELECTRIC PRODUCTS INC.
Photoflash Plant, Montoursville, Pa.
Product Development Section

(COURTESY OF GTE SYLVANIA INC.)

Importance of Graphic Charts and Diagrams.

Graphic charts and diagrams are an important part of technical drawing. They are important to scientists, engineers, mathematicians, and nearly everyone else in everyday life.

Scientists use charts and diagrams to record and study the results of research. Engineers use them to record information about materials and conditions. Mathematicians use them to record facts and trends of numerical information. Doctors use charts to record body temperature, heart action, and other body functions. Most people use and read charts and diagrams to see about the weather, the stock market, where the tax dollar goes, and for many other purposes. Because Fig. 23–1 is in chart form, it can be seen at a glance that driver reaction distance and automobile braking distance increase as speed is increased. Figure 23–2 has the same information but requires a great deal more study time to obtain it. In addition, the graphic (pictorial) aspect of Fig. 23–1 makes the relationship of speed and distance more easily understood.

Definitions.

Graphic charts and diagrams are used as a graphic (pictorial) way to show information that is numerical. They show trends, such as whether the cost of living and wages are rising or falling over a period of time or whether one is rising and the other is falling.

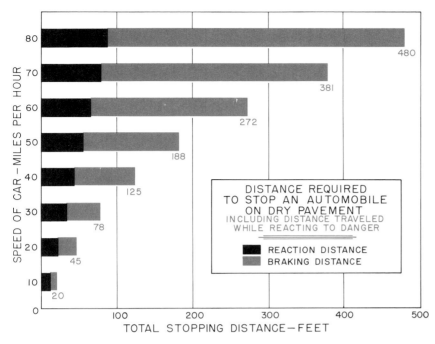

Fig. 23-1 Chart showing stopping distances at different speeds for automobiles.

DISTANCE REQUIRED TO STOP AN AUTOMOBILE ON DRY PAVEMENT			
MILES PER HOUR	REACTION DISTANCE FT.	BRAKING DISTANCE FT.	TOTAL STOPPING DISTANCE FT.
10	11	9	20
20	22	23	45
30	30	45	78
40	44	81	125
50	55	133	188
60	66	206	272
70	77	304	381

Fig. 23-2 The same information takes longer to read and understand in this form.

Charts can show ratios, such as speed versus distance in Fig. 23–1, or they may show percentages of a whole, as in a bar chart or a pie chart. Charts can also be used to explain information which is not numerical. A flow chart shows sequential information, that is, which operation is first, second, third, and so on.

Graphic charts may also be used to solve various kinds of mathematical problems. The following problems can easily be solved by studying the chart in Fig. 23–3:

1. What is the normal water level in the lake?

2. What is the maximum rise?

3. When does the highest level occur?

You probably had no trouble answering the questions because charts of this type are easily understood without much instruction.

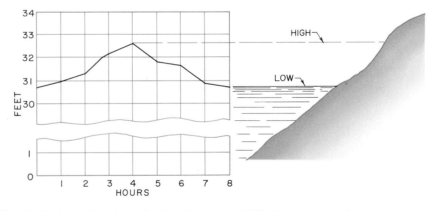

Fig. 23-3 A graphic chart showing the rise and fall of water on a lake.

The curve on a chart is not necessarily a curved line. As shown in Fig. 23–4, a curve on a chart may be a straight line, a curved line, a broken line, a stepped line, or a straight line or curved line adjusted to plotted points.

The selection of the proper scales (vertical and horizontal squares) is

important. The vertical and horizontal scales must be such that a true pictorial impression is given by the angle of slope of the curve. In Fig. 23–5, notice the different impressions given by the three charts. At A it appears as though there is a very abrupt change; at B, a normal change; and at C, a very slow or gradual

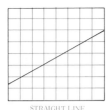 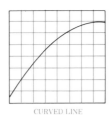 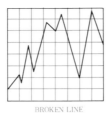  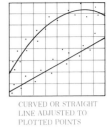

STRAIGHT LINE   CURVED LINE   BROKEN LINE   STEPPED LINE   CURVED OR STRAIGHT LINE ADJUSTED TO PLOTTED POINTS

Fig. 23-4 Curves on graphs may have different forms.

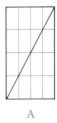 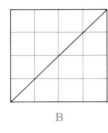 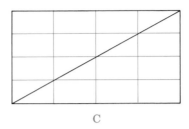

A   B   C

Fig. 23-5 A false impression may result if vertical and horizontal scales are not properly selected.

change. The scale chosen should be the one that gives the most accurate pictorial impression.

Printed grid or graph paper is available in many forms. It may be purchased with lines ruled (drawn or printed) 4, 5, 8, 10, 16, and 20 to the inch and in many other forms (Fig. 23-6). Graph paper may also have certain lines printed heavier to make it easier to plot points and to read the finished chart. Figure 23-6 shows every tenth line printed heavier. In this case, the heavy lines are 1 in. apart. The lines on grid paper may form squares or rectangles, as shown in Fig. 23-5.

Line Charts. Line charts are most often used to show *trends* or changes. For example, changes in the weather, ups and downs in sales, or trends in population growth may be plotted and shown graphically on a line chart. A line chart may contain one or several curves. A conversion chart with one curve (Fig. 23-7) is often convenient

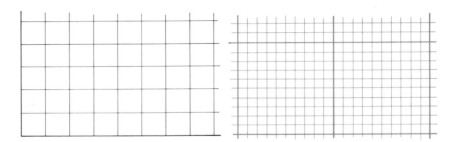

Fig. 23-6 Graph paper is available in many forms and is printed in different colors.

Fig. 23-7 A conversion chart.

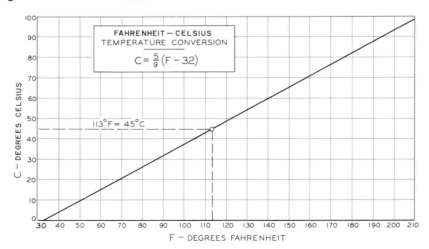

FAHRENHEIT — CELSIUS
TEMPERATURE CONVERSION

$$C = \frac{5}{9}(F - 32)$$

113°F = 45°C

C — DEGREES CELSIUS

F — DEGREES FAHRENHEIT

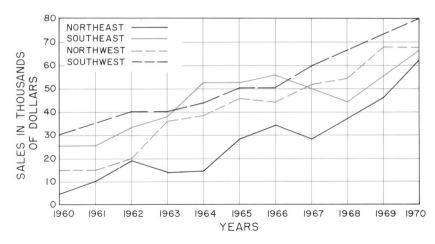

Fig. 23-8 A multiline chart.

SCORING INFORMATION EASTERN HIGH SCHOOL	
GAME NUMBER	POINTS SCORED
1	38
2	20
3	50
4	40
5	40
6	30
7	10
8	40
9	55
10	45

Fig. 23-9 Information to be presented in a line chart.

for changing from one value to another. This one is used to convert Fahrenheit to Celsius. Figure 23-8 contains four curves comparing sales in various parts of the country.

To Draw a Line Chart. A line chart is made by following steps A through I.

A. Prepare and list the information to be presented (Fig. 23-9).

B. Select plain or ready-ruled graph paper.

C. Select a suitable size and proportion to give the desired result.

D. Select the proper scale.

E. If graph paper is not used, lay off and draw thin horizontal (called X axis, or abscissa) lines and vertical (called Y axis, or ordinate) lines, as shown in Fig. 23-10A. The intersection of X and Y is zero.

F. Lay off the scale divisions on the X axis and the Y axis (Fig. 23-10B).

G. Mark the scale values on the X and Y axes (Fig. 23-10B).

H. Plot the points accurately from the listed information (Fig. 23-9). It is usually better to use small open circles, triangles, or squares rather than crosses or solid dots for plotting purposes (Fig. 23-10C).

I. Connect the points to complete the line chart (Fig. 23-10D).

If more than one curve is drawn on a chart, use different types of lines or different colors for each curve (Fig. 23-11). Use a full, continuous line of the brightest color for the most important curve. In general, the scales and other identifying notes, or captions, are placed below the X axis and to the left of the Y axis. For large charts it is sometimes desirable to show the Y-axis scales at both the right and the left and the X-axis scales at the top and bottom for convenience in reading.

Engineering Charts. Experimental data may be plotted from tests and used to obtain an unknown value. In Fig. 23-12 the results of tests have been plotted and show as a straight-line curve when drawn in an adjusted position. (Ω is the Greek letter

Fig. 23-10 Steps in drawing a line chart.

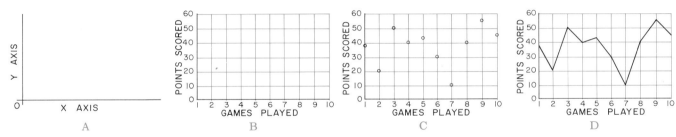

A   B   C   D

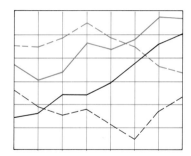

Fig. 23–11 Use different types of lines and different colors to distinguish curves on a multiline chart.

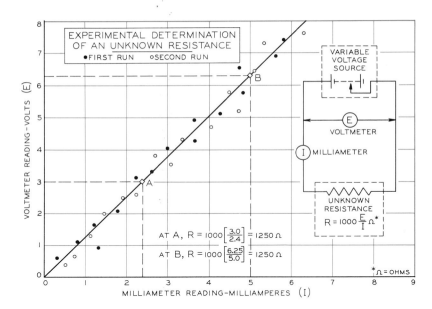

Fig. 23–12 An engineering test chart.

Fig. 23–13 A nomogram.

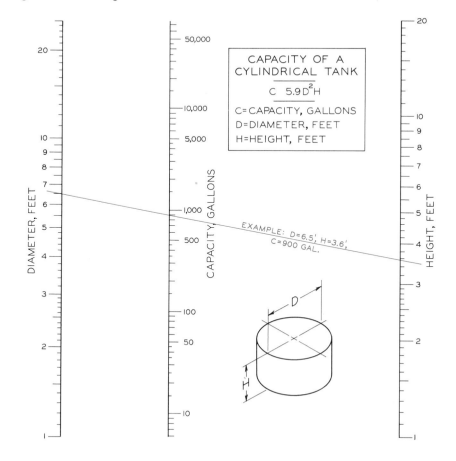

omega.) Values can be taken from two points and inserted in the formula to obtain and check the value of the unknown resistance. Notice that tests were made on two occasions, and the results were plotted on the chart. A straight-line curve was drawn along the center of the path made by the dots.

Nomograms are charts that show the solutions to problems containing three or more variables (kinds of information). Figure 23–13 is an example of this type of chart. A straight-edge from values on the outside scales will cross the inside scale, where the solution to the equation may be read. Nomography is a special division of chart construction which requires more than simple mathematics.

Bar Charts. Bar charts are probably the most familiar and most easily read and understood kind of graphic charts. A bar chart may consist of a single rectangle representing 100 percent (Fig. 23–14E). This chart represents the total number of games played during a season. It is divided to show the proportional part, or percentage, of the total number of games won, lost, and tied.

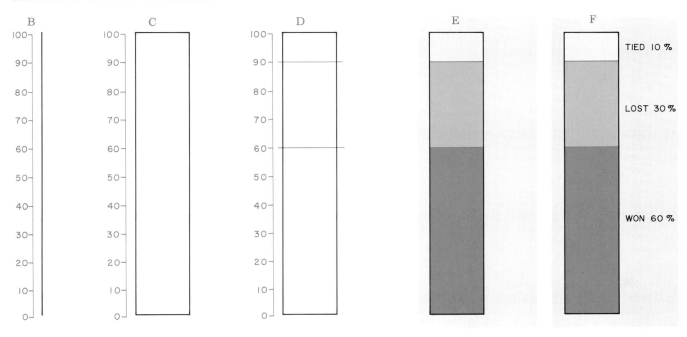

	QUANTITY	PERCENTAGE
GAMES WON	18	60
GAMES LOST	9	30
GAMES TIED	3	10
GAMES PLAYED	30	100

Fig. 23-14 Steps in drawing a one-column bar chart.

To Draw a One-column Bar Chart.
A one-column bar chart is made by following steps A through F.

A. Prepare and list the information to be presented (Fig. 23-14A).

B. Lay off the long side equal to 100 units. For example, 10 in. = 100 tenths with an engineer's scale, or 12½ in. = 100 eighths, etc. (Fig. 23-14B).

C. Lay off a suitable width and complete the rectangle (Fig. 23-14C).

D. Lay off the percentage of the parts and draw lines parallel to the base (Fig. 23-14D).

E. Crosshatch, shade, or color the various parts, as shown in Fig. 23-14E.

F. Letter all necessary information in or near the parts so that it can be read easily (Fig. 23-14F).

To Draw a Multiple-column Bar Chart. A multiple-column bar chart is made by following steps A through D.

A. Prepare and list the information to be presented (Fig. 23-15A).

B. Select a suitable scale and lay off the X and Y axes. Lay off the scale divisions (Fig. 23-15B).

C. Block in the bars using the information gathered in Fig. 23-15A. Allow enough space between bars for all necessary lettering. Make the bars any convenient width for best appearance (Fig. 23-15C).

D. Complete the bar chart by adding shading or color to the bars, lettering, and any other lines and information needed to make the chart easily understood (Fig. 23-15D).

A bar chart with horizontal bars is shown in Fig. 23-16. It gives speed ranges for Caterpillar tractors. Note that the bars do not start at a common line because they show different speed ranges. This is called a *progressive chart*.

A *compound-bar chart* is shown in Fig. 23-17, where the total length of the bars is made up of two parts. The black part is the distance traveled at a given speed before starting to apply the brakes. The red part is the dis-

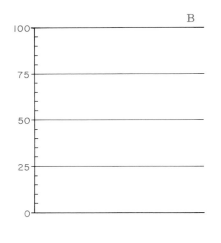

PERSONAL SAVINGS IN DOLLARS	
1968 —	$ 65.00
1969 —	78.00
1970 —	52.00
1971 —	85.00
1972 —	60.00

A

B

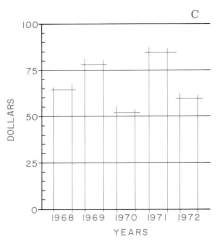

C

D

Fig. 23–15 Steps in drawing a multiple-bar graph.

Fig. 23–16 This horizontal-bar chart is a form of progressive chart. (*Caterpillar Tractor Co.*)

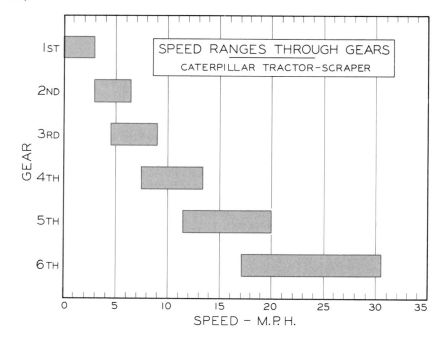

tance traveled in coming to a stop. The gray and the red parts are added together graphically to give the total distance.

The multiple-bar chart shown in Fig. 23–18 has minus values, and these are set off below the X axis by red bars. The same information is given in Fig. 23–19, where the vertical scale is the same as in Fig. 23–18. The dashed line shows the net gain and is found by laying off the minus values down from the total-gain values.

Pie Charts. This is a form of 100-percent chart in which a circle represents 100 percent, and sectors represent parts of the whole (Fig. 23–20).

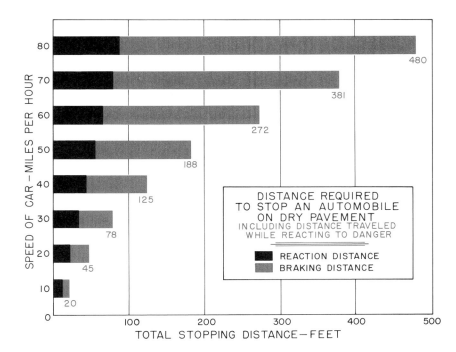

Fig. 23-17 A compound-bar chart in which the total length of each bar is the sum of two parts.

DISTANCE REQUIRED
TO STOP AN AUTOMOBILE
ON DRY PAVEMENT
INCLUDING DISTANCE TRAVELED
WHILE REACTING TO DANGER

■ REACTION DISTANCE
▨ BRAKING DISTANCE

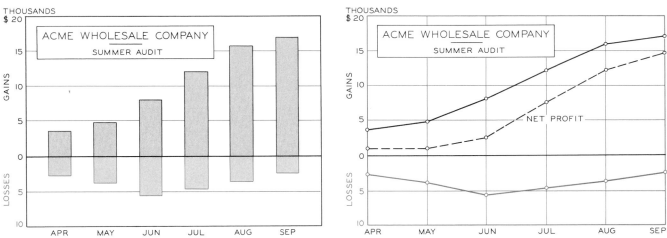

Fig. 23-18 A multiple-bar chart in which the bars have plus and minus values.

Fig. 23-19 A line chart for the same information as that in Fig. 23-18.

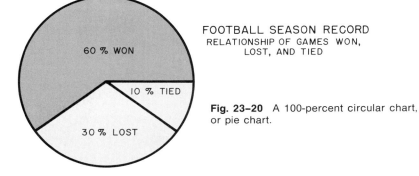

FOOTBALL SEASON RECORD
RELATIONSHIP OF GAMES WON,
LOST, AND TIED

Fig. 23-20 A 100-percent circular chart, or pie chart.

DISTRIBUTION OF CLASS TREASURY		
ITEM	COST	%
DANCE	$ 72.00	40
PARTY	36.00	20
PICNIC	32.40	18
CLASS PLAY	21.60	12
PHOTOGRAPHS	18.00	10
TOTAL	$180.00	100%

A

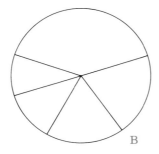

B

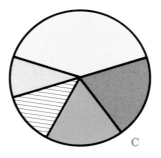

C

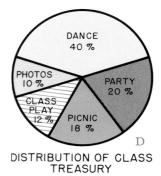

DISTRIBUTION OF CLASS TREASURY

D

Fig. 23–21 Steps in drawing a pie chart.

To Draw a Pie Chart. A pie chart is made by following steps A through D.

A. Prepare and list the information to be presented (Fig. 23–21A).

B. Draw a circle of the desired size. Lay off and draw the radial lines representing the amount or percentage for each part on the circumference of the circle (Fig. 23–21B). If a protractor is used, $3.6° = 1$ percent. Thirty percent is $30 \times 3.6°$, or $108°$, and so on. If a circle is to be divided into a 24-hour day, each hour represents $15°$ on the circle. This can be drawn by using a T-square and triangles.

C. Crosshatch, shade, or color the various parts, as shown in Fig. 23–21C.

D. Complete the pie chart by adding all necessary information (Fig. 23–21D).

Pictorial Charts, or Pictographs. Pictorial charts, or pictographs, are, in effect, bar charts which use pictures or symbols in place of bars. Figure 23–22 illustrates a pictorial graphic chart, or pictograph. Each figure represents 100 people; it could represent 1000 or any assigned number. The figure is a multiple-bar chart. There are four kinds of individuals represented. Adhesive symbols are available in many forms (Fig. 23–23). These make pictorial charts easy to construct.

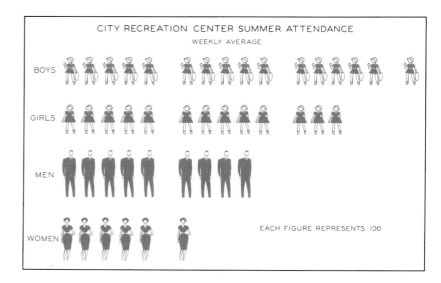

Fig. 23–22 A pictorial graphic chart (also called a pictograph).

Fig. 23–23 Many styles of adhesive symbols are available for use on graphic charts. (*Chart-Pak, Inc.*)

Organization and Flow Charts.
Organization charts are of many kinds but have the features of a flow chart. Figure 23–24 is an example. It shows the path, or flow, of drawings from the top engineer to the shop. It also shows the organization of the drafting department.

A *flow chart* may show the path or series of operations in manufacturing a product or in producing a material, such as the flow chart of steelmaking (Fig. 23–25).

Tape Drafting.
Tape drafting is a convenient method of preparing graphic charts. Adhesive tape comes in many colors, designs, and widths. It is applied from a roll dispenser (Fig. 23–26) and pressed onto the chart in the desired position. Tapes of selected widths provide a quick and simple way of making bar charts.

The Use of Color.
Black-and white charts are often used by scientists and mathematicians for recording data. Newspapers use black-and-white charts and they usually serve the particular purpose rather well. However, the use of color has become quite common in the preparation of charts and diagrams for magazines, books, brochures, and in other publications

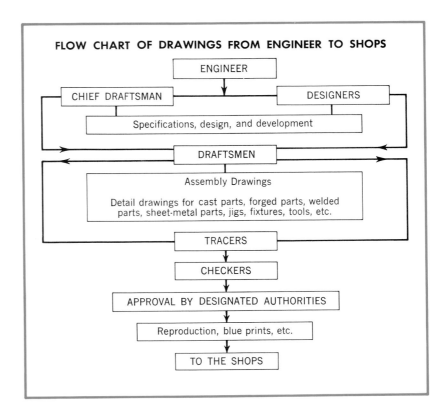

Fig. 23–24 An organization and flow chart.

where it is economically practical. Color is also used extensively in making charts for display purposes.

The use of color adds a great deal to appearance, readability, and em-

phasis (Fig. 23–27). Color may be added in a variety of ways. Colored pencils, felt-tipped pens, water colors, or other similar materials are easy to use and can probably be found in the

Fig. 23–25 A flow chart of steelmaking. (*American Iron and Steel Institute.*)

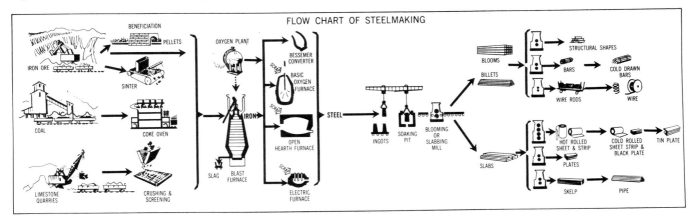

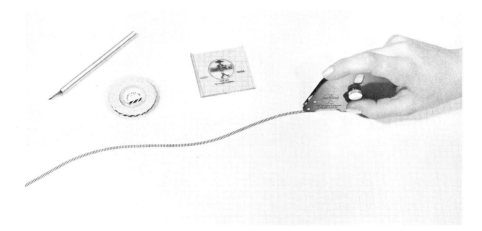

Fig. 23-26 Applying adhesive tape to a graphic chart. (*Chart-Pak, Inc.*)

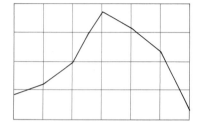

Fig. 23-28 Scales should be selected with care so that the curve plotted from the data will aid the understanding of the information by its appearance.

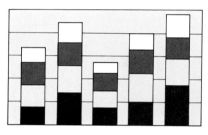 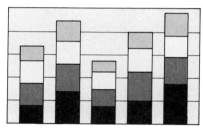

Fig. 23-27 A multiple-bar chart in black and white and the same one in color.

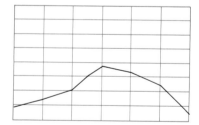 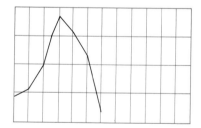 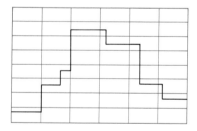

Fig. 23-29 The vertical scale used here is too small. It gives the effect of very little change in values. The movement is slow.

Fig. 23-30 This curve is plotted from the same data used for Figs. 23-28 and 23-29. The horizontal scale is too small (if Fig. 23-28 is a correct picture). The movement is fast.

Fig. 23-31 A step chart shows data which remains constant during regular or irregular intervals. This figure might show time periods during which a price remained constant or was raised or lowered.

drafting room, art room, or at home. Commercially prepared pressure-sensitive materials are available at art and engineering supply stores.

Charts and Elements of Charts. There is an almost endless variety of graphic charts which may be made for visual communication. This chapter has suggested the general character of a few and how they are used.

A variety of charts is shown in Figs. 23-28 to 23-36 to indicate some of the elements of chart making and to sug-

gest the ways in which these elements affect the chart.

Every chart should have a suitable title, well lettered and placed within the area of the chart. Every chart should also have a key to tell the reader what the elements represent.

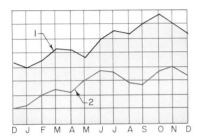

Fig. 23–32 A shaded-surface, or strata, chart uses shaded areas for contrast. This illustration might show the total amount of each of two materials used each month.

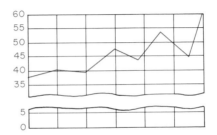

Fig. 23–33 An omission chart may be used for some purposes, as shown, in order to use a larger vertical scale. There are no values below 35, and so a portion of the chart is broken out.

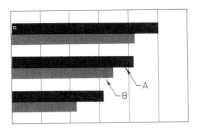

Fig. 23–34 A comparison-bar chart which might be used for two or three values. The illustration might show the amount made (A) and the amount sold (B) of an item by various companies in one year or by one company for several years.

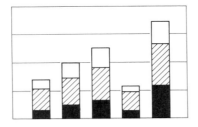

Fig. 23–35 A multiple-bar chart with divided bars. The bars are divided to show the amount of each of three substances which make up the total.

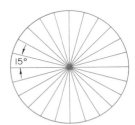

Fig. 23–36 A pie-chart representing 24 hours in a day may be used to show time relationships. Each division is 15° and represents one hour. Fractional parts of an hour may be estimated.

Review

1. What kind of chart is used to show the solution to problems containing three or more kinds of information?

2. Is the curve on a chart always a curved line?

3. Refer to the chart in Fig. 23-8. What was the dollar volume of sales in the Northeast Region in 1968?

4. Refer to the chart in Fig. 23-17. How far does a car travel during reaction time at a speed of 60 mph?

5. In Fig. 23-17, what is the total stopping distance at a speed of 60 mph?

6. Refer to Fig. 23-13. What is the capacity in gallons of a cylindrical tank 10 ft in diameter and 17 ft high?

7. In Fig. 23-13, what is the height of a cylindrical tank that holds 100 gallons and is 2 ft in diameter?

8. In Fig. 23-13, a tank holds 3000 gallons and is 10 ft high. What is its diameter?

9. What is another name for a pictorial graphic chart?

Problems for Chapter 23

Fig. 23–37 Draw a pie chart to show the population distribution in the following four regions of the United States:

Northeast	24.2%
North Central	27.8%
South	31.2%
West	16.8%

Use color or various types of crosshatching for contrast.

Fig. 23–38 Draw a pie chart showing that 67.4% of the population of the United States lives within metropolitan areas, while only 32.6% lives outside metropolitan areas. Use different colors or crosshatch areas for contrast.

Fig. 23–39 Make a one-column bar chart to show how your allowance was spent last month.

Fig. 23–40 Make a multiple bar chart showing a comparison of how you spent your allowance over a period of four months.

Fig. 23–41 Assignment 1: The average cost per pound for beef has varied over a period of 10 years as follows:

Year	Cost Per Pound
1964	$0.65
1965	0.68
1966	0.70
1967	0.76
1968	0.72
1969	0.78
1970	0.80
1971	0.78
1972	0.95
1973	1.35

Make a line chart showing this relationship. Assignment 2: The average cost of pork varied somewhat differently. Add a second line to your chart showing a comparison of the cost of beef to the cost of pork over the same period of time. Use colored pencils or different types of lines for contrast.

1964	$0.60
1965	0.65
1966	0.75
1967	0.72
1968	0.69
1969	0.70
1970	0.80
1971	0.85
1972	0.90
1973	1.10

Assignment 3: Add a line to the chart showing the average cost of poultry for the same ten years.

1964	$0.22
1965	0.26
1966	0.30
1967	0.20
1968	0.23
1969	0.35
1970	0.40
1971	0.33
1972	0.38
1973	0.45

Fig. 23–42 Draw a flow chart showing how to mass produce a project of your choice in the school shop.

Fig. 23–43 Make an organizational chart showing the administrative structure of your school.

Fig. 23–44 Assignment 1: Draw a pie chart or a one-column bar chart showing a breakdown of the average person's income if 24.5% goes to federal taxes, 5.8% goes to state taxes, and 1.3% goes to local taxes. Be sure the figures total 100%. Assignment 2: Compute the dollar value of each category for a gross income of $12,500. Mark each on your chart. Be sure the figures total $12,500.

Fig. 23–45 Assignment 1: Make a pictorial chart (pictograph) showing male and female population in your grade in school. Assignment 2: Make a bar chart showing male and female population in your drawing class.

Fig. 23–46 Draw a pictorial chart showing the enrollment of technical drawing classes in your school. Your instructor can supply the information.

Fig. 23–47 Assignment 1: Make a line graph showing the hourly change in outside temperature for a 12-hour period during any day. Assignment 2: Record similar information for several days and make a multiline chart to show a comparison.

Fig. 23–48 Draw a bar chart to show home consumption of electricity for 1 year as follows:

Month	Kilowatt Hours Used
January	900
February	885
March	800
April	783
May	722
June	600
July	494
August	478
September	525
October	650
November	735
December	820

Fig. 23–49 Plot the batting averages of the players on your favorite baseball team.

Fig. 23–50 The number of cars per mile of road in the United States is growing. Draw a pictorial chart from the data below to show the growth and anticipated increase.

Year	Cars Per Mile of Road
1930	9
1950	15
1960	20
1970	26
1980 (est.)	38

Fig. 23–51 Draw a vertical bar chart showing the following student attendance

for a given week of school. The total school enrollment is 925.

Day	Attendance
Monday	625
Tuesday	715
Wednesday	800
Thursday	775
Friday	695

Fig. 23–52 Assignment 1: Compute your daily calorie intake for 1 week. Make a line chart representing this information. Assignment 2: Use the same information and prepare a horizontal bar chart.

Fig. 23–53 Make a multiline chart representing individual game scores of the top five players on the school basketball team for any given season.

Fig. 23–54 Assignment 1: From the stock-market listings in the newspaper, select any stock and record its daily status for 10 days. Plot the information on a line chart. Assignment 2: Select several stocks and make a multiline chart showing a comparison of growth and decline.

Fig. 23–55 Make a pictorial chart or a pie chart showing a breakdown of the source of each dollar received by the federal government. Use the information given below.

Individual income tax	$0.38
Employment tax	0.26
Corporate income tax	0.14
Borrowing	0.10
Excise tax	0.07
Other (miscellaneous)	0.05

Fig. 23–56 Make a pictorial chart or a pie chart showing a breakdown of the expenditure of each dollar by the federal government. Use the information given below.

National defense	$0.31
Income security	0.27
Interest	0.08
Health	0.07
Commerce, transportation, housing	0.06
Veterans	0.05
Education	0.04
Agriculture	0.03
Other (miscellaneous)	0.09

Fig. 23–57 The information listed below includes five common foods and the number of calories and grams of carbohydrates in a 4-ounce serving of each. Make a bar chart illustrating these facts.

Food	Calories	Carbohydrates
Chocolate ice cream	150	14
Peas	75	14
Pizza	260	29
Milk	85	6
Strawberries	30	6

Fig. 23–58 Assignment 1: The data in the list below represents a percentage breakdown for a family budget. Make a pie chart illustrating this information.

Food	23.1%
Housing	24.0%
Transportation	8.8%
Clothing	10.9%
Medical Care	5.6%
Income Tax	12.5%
Social Security	3.8%
Miscellaneous	11.3%

Assignment 2: Compute the dollar value of each category for a gross income of $15,000. Mark each on your chart. Be sure the figures total $15,000.

Fig. 23–59 Accidents involving children occur in various places. Draw a horizontal bar chart or a pie chart, using the places and percentages given below.

At home	25%
Between home and school	8%
On school grounds	15%
In school buildings	21%
In other places	31%

Electrical and Electronics Drafting

The Electrical Industry. Progress in the development of electric power started with Thomas Alva Edison's generating station in New York City in 1882. Electricity has become one of mankind's practical servants and has revolutionized the manufacturing, communication, and utility industries. The engineer, designer, technician, and draftsman, working along with the scientist, are creating new environments for work and leisure. The consumer can identify the rapid progress of the industry by reflecting on how communication and appliances have changed. A good example is the tiny electronic components used in controlled systems. Microcircuits placed on a single chip of silicon are replacing transistors, and of course a few years ago the transistor replaced the vacuum tube. The electronics industry is one of the biggest manufacturing industries in the nation.

Career Opportunities. Industry offers special opportunities to young men and women who can create the freehand and formal drawings required for the electronics industry. Although electronic drafting is based upon the same basic principles of orthographic projection and dimensioning as all other drafting, it also requires the skill of preparing schematic diagrams and technical illustrations.

Preparation for Opportunities. Industry may provide some students

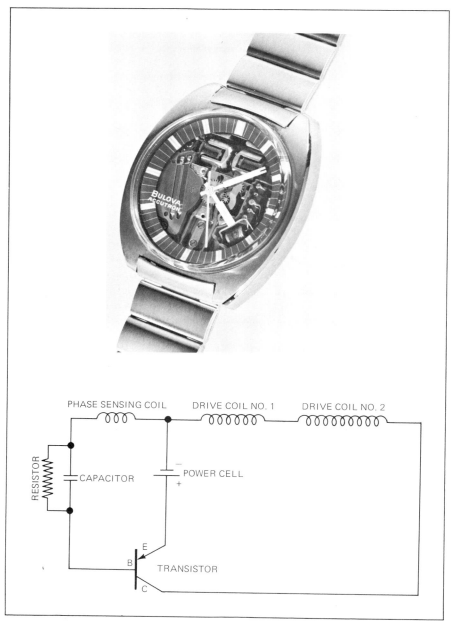

(BULOVA WATCH COMPANY INC.)

with training programs if drafting and electronic aptitudes have been developed through high school course work. A few courses at the technical college or junior college level in electronic drafting and electromechanical systems would help the student who is looking for a technician rating on the design team. Additional career information is available through the following organizations:

IEEE   Institute of Electrical and Electronic Engineers
        345 East 47 Street, New York, N.Y. 10017

EIA   Electronic Industries Association
        11 West 42 Street, New York, N.Y. 10036

ASCET   American Society of Certified Engineering Technicians
        2029 K Street, N.W., Washington, DC 20006

AFTE   American Federation of Technical Engineers (AFL-CIO)
        900 F Street, N.W., Washington, DC 20004

The Electrical or Electronic Draftsman. A person on the engineering design team who serves as an electrical or electronic draftsman must continually develop new skills and expand his technical knowledge. He must exercise good judgment, skill, and originality when working from design sketches or written instructions. The electrical or electronic draftsman may be qualified after training and a few years of experience to serve as a technician on a design team.

Design teams may specialize in developing electrical equipment for use in the generation, transmission, or distribution of electrical power. They

Fig. 24–1 A room specially designed for an electronic environment. (*Motorola, Inc.*)

may also specialize in the manufacture of electronic instruments or appliances. There are career opportunities in areas of equipment manufacturing such as computers or other sophisticated information systems. Future electronic developments promise to be more spectacular than those which serve our needs today. Scientists, along with the electrical engineer, designer, technician, and draftsman, appear to be building an electrical and electronic world.

Electronic Environment. Man has always been involved with the problem of environmental control. Today, electronic devices control the condition of the atmosphere at home and at work. Electronic devices have been developed for regulating the cooling, heating, lighting, and sound systems for various life styles.

The electronic room illustrated in Fig. 24-1 is contained within a 25-ft fiber-glass domed structure. The room features a recessed living and conversation area in the center which contributes to a relaxed atmosphere. It is complemented with an electronic sight and sound system. Mood lighting is programmed to respond to the remote-controlled color television, the cartridge television recorder, and the four-channel sound system.

The floor plan (Fig. 24-2) illustrates the major components of the electronic environment. The center table at 1 is the control center for the components of the room. The rectangle at 2 is a color television set, and the rectangle at 3 is an EVR (electronic video recorder). The rectangles numbered 4 are speaker towers for four-channel sound. They also contain the mood lighting which responds to sound. The sketch (Fig. 24-3) helps to identify the components. Figure 24-4 is an overall view of this electronic environment. The center table rises mechanically to reveal the controls for the sound system, color television, and EVR. The overhead lighting is kinetic (moving) and is modulated to the beat of the music.

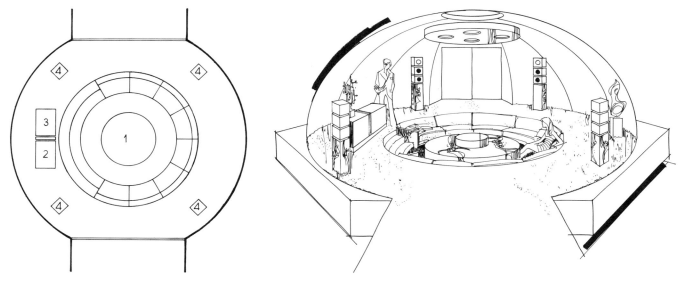

Fig. 24-2 The floor plan of the electronic room. (*Motorola, Inc.*)

Fig. 24-3 A sketch of the electronic room. (*Motorola, Inc.*)

Fig. 24-4 An overall view taken through a fish-eye lens. (*Motorola, Inc.*)

Three of the four sound towers containing speakers and special lighting are visible. The sound towers each contain midrange and tweeter speakers. A down-firing bass speaker is located in the base of each tower.

The electric circuits for this room must be planned as a wiring diagram by the electrical design engineer. The electronic components for the room must be designed by the electronic design team.

Electrical and Electronic Drafting. Students who have had a basic course in electricity will find it of special value in understanding and making electrical or electronic drawings. This chapter will serve as an introduction to electrical and electronic symbols, wiring diagrams, and circuit diagrams as elements of electrical and electronic drafting. For those students who have not had a basic course in electricity, the paragraphs which follow will be a brief introduction to the subject.

Electricity. The source of electrical energy is the tiny atom. In the natural condition, all atoms are made up of several kinds of particles. One of these particles, the electron, is most important in the study of electricity and electronics.

The electrons within an atom rotate about the center, or nucleus, of the atom in definite paths, or orbits (Fig. 24-5). All electrons in all atoms are alike; each possesses what is called a *negative charge* of electricity. Atoms differ from one another in the number of electrons and other particles which they contain. When all the atoms which make up a substance are alike, the substance is called an *element*. Copper, gold, and lead are common elements. When different kinds of atoms are joined together, they form a *compound*. Water, acids, and salt are common compounds. The smallest amount of a compound that retains all the properties of the compound is known as a *molecule*.

Voltage and Current. Under certain conditions, electrons can be made to leave their "parent" atoms. This happens, for example, when a length of wire is connected across the terminals of a battery. The battery produces an electrical pressure called *voltage*. The symbol for voltage is *E*. The voltage causes a steady stream of electrons to flow through the wire. If a light bulb (load) is connected into the wire path, electrons will move through the lamp filament from the battery (power source) (Fig. 24-6). As a result, the energy of the moving electrons is changed into heat energy as the filament becomes white hot. The glow of the filament produces the light.

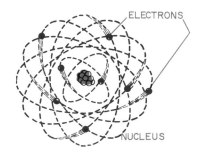

Fig. 24-5 The structure of an atom.

The electron *pathway* formed by the battery, the wire, and the lamp filament is a simple form of electrical circuit. In other circuits, electrical energy is changed into other kinds of energy, such as magnetism, sound, and light.

A direct current (dc) is a flow of electrons through a circuit in one direction only (Fig. 24-7). An alternating current (ac) is a flow of electrons in one direction during a fixed period of time and then in the opposite direction during a like period of time (Fig. 24-8). This is called a *cycle*. The number of times this cycle is repeated in one second is called the *frequency* of the alternating current, such as 60 cycle. Current is measured in amperes, and the symbol is *I*.

Resistance. *Electrons can move through some materials more easily*

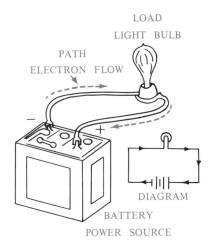

Fig. 24-6 A simple electric circuit.

than through others. Electrical current will flow more easily through a copper wire than through a steel wire of the same size. The steel offers more resistance than copper. Materials with a small resistance to the flow of electrons are called *conductors*. Silver is the best-known conductor, but it is too expensive for general use. Copper and aluminum are good conductors and are the most widely used. Materials through which electrons will not flow easily

Fig. 24-7 Direct current attains magnitude and keeps it as long as the circuit is complete.

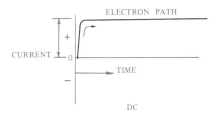

Fig. 24-8 Alternating current builds up from zero to a maximum in a positive direction, falls to zero, and then builds to a maximum in a negative direction and falls back to zero.

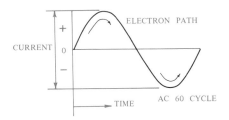

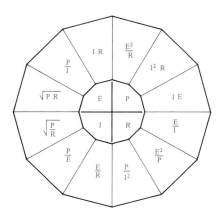

Fig. 24-9 Power studied through Ohm's law.

are called *insulators.* The most commonly used insulating materials are glass, porcelain, plastics, and rubber compounds. Resistance is measured in ohms, and the symbol is R.

Electricity and Electronics.
Electricity has to do with the flow of electrons moving through conductors or wires. Common examples include house wiring systems, generators, and transformers.

Electronics has to do with energy in electrons moving through metallic conductors and through other conductors such as gases, a vacuum, and materials known as *semiconductors.* The most common semiconductor devices are transistors and diodes made of germanium or silicon. Both electricity and electronics are concerned with the flow of electrons through circuits designed to carry energy for a definite purpose.

Basic Electrical Units.
These are volts to measure voltage, ohms to measure resistance, amperes to measure current, and watts to measure power. The value of a unit used to show an electrical quantity is often given by a prefix, such as kilovolts

(kv), where 1 kilovolt equals 1000 volts; 1 milliampere (ma) equals 0.001 ampere; 1 kilowatt (kw) equals 1000 watts; 1 kilohm (kohm) equals 1000 ohms. For other combined forms of units, consult an electrical handbook or text.

Basic Formulas.
There is a definite relation between the values of voltage, current, and resistance in a circuit. This relationship is known as *Ohm's law,* in which E = volts (pressure), I = amperes (current), and R = ohms (resistance). It may be expressed as shown in the 12 formulas of Fig. 24-9. Note the relationship of electrical horsepower to mechanical horsepower:

One electrical horsepower:
  hp = 746 watts
One mechanical horsepower:
  hp = 33,000 foot-pounds per minute
  = work required to raise a weight of 33,000 pounds in 1 minute
  = 550 foot-pounds per second

Graphic Symbols.
These are used on electrical and electronic diagrams to represent the component devices and operations in a circuit. The American National Standards Institute Inc., 1340 Broadway, New York, is the national center for the procedure and development of American Standards by the members, including such organizations as The American Society of Mechanical Engineers (ASME), the Electronic Industries Association (EIA), The Institute of Electrical and Electronic Engineers, Inc. (IEEE), The National Electrical Manufacturers Association (NEMA), and hundreds of others.

Some of the great number of American National Standard symbols for use on electrical and electronic diagrams are shown in Fig. 24-10. A

complete set of symbols is given in American National Standard Graphic Symbols for Electrical and Electronic Diagrams (ANSI Y32.2-1962) and in Military Standards (MIL-STD-15A). Standards are subject to changes and additions; thus, the latest issue should be used.

Graphic symbols are not drawn to scale, but the size of a given symbol should not be changed on a diagram. A variety of templates are available for drawing uniform symbols quickly and easily (Fig. 24-11). Grooved templates similar to those used with scriber guides for lettering are available for drawing electrical symbols.

Basic Electric Circuits.
These include series circuits, parallel circuits, and combinations of series and parallel circuits. These terms are explained in the following paragraphs.

Series circuits. *Series circuits* are those in which the current flows from the source (battery, generator, and so forth) through one resistance (lamp, motor, and so forth) after another, as shown in Figs. 24-12, 24-13, and 24-14.

In Fig. 24-12 a bell (A) is operated from a battery (C) when the circuit is closed by the normally open (NO) type of pushbutton (B).

In Fig. 24-13 a buzzer (A) is operated by the current from the transformer (C). What is item B, and what function does it have in this circuit?

In Fig. 24-14 four lamps (C, D, E, and F) are operated from a generator (A) when the fused switch (B) is closed. All the lights must be on, for if any one is not, the circuit will be open. Remember the Christmas tree light strings that went out completely when just one lamp burned out? This was a series string of lights.

Parallel circuits. Parallel circuits provide for the current to flow through more than one path, as shown in Figs. 24-15 and 24-16.

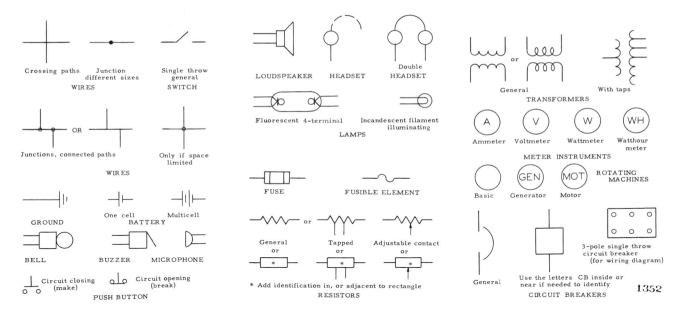

Fig. 24-10 A few electrical symbols from "American National Standard Graphic Symbols for Electrical and Electronic Diagram," ANSI Y32.2–1962. (*With the permission of The Institute of Electrical and Electronics Engineers, Inc.*)

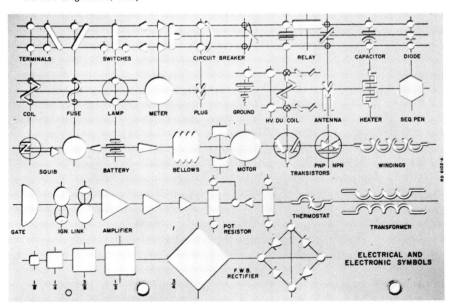

Fig. 24-11 Template for drawing electrical and electronic symbols. (*RapiDesign, Inc.*)

Fig. 24-12 A series-circuit diagram.　　**Fig. 24-13** A series-circuit diagram.

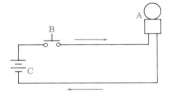

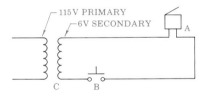

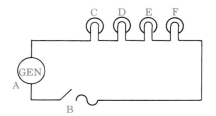

Fig. 24-14 A series-circuit diagram.

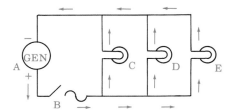

Fig. 24-15 A parallel-circuit diagram.

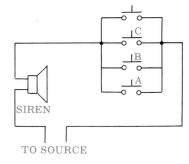

Fig. 24-16 A parallel-circuit diagram.

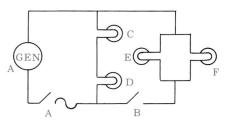

Fig. 24-17 A combination series and parallel circuit.

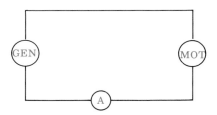

Fig. 24-18 Ammeter connection in circuit.

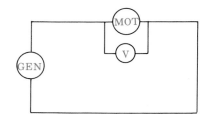

Fig. 24-19 Voltmeter connection in parallel, in circuit.

Fig. 24-20 Ammeter and voltmeter connections.

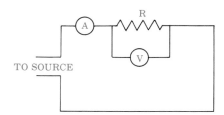

Three separate branches or paths (C, D, and E) with lamps are shown in Fig. 24-15. Each lamp is independent of the others. If one lamp is burned out, the others will continue to operate. With a parallel string of lights on your Christmas tree, the remaining lights continue to burn if some are missing, loose, or burned out.

A siren is shown in Fig. 24-16. It may be activated by any one of the pushbuttons A, B, C, or D, which are all in parallel. An application of this would be in an alarm system to give warning of an attempted holdup in a store. The pushbuttons, connected in parallel, would be located under counters and in the cashier's cage.

Observe that the symbol for the siren is the same as for a loudspeaker (Fig. 24-10) but that here it is accompanied by a note "Siren" to identify it.

Combination circuits. The combination of series and parallel circuits provides many different arrangements, combining both series and parallel connections. In Fig. 24-17, lamps C and D are in series and lamps E and F are in parallel. Both lamps C and D must be on if switch A is closed, since they are in series. When switches A and B are closed, all the lamps (C, D, E, and F) are lighted. Lamps E and F will operate independently. If one fails, the other will remain lighted because they are in parallel; however, because lamps C and D are in series, as we have learned from the example of the Christmas

tree lights, when one fails, the others will not light.

Electrical Instruments. Many kinds of electrical instruments have been developed for measuring purposes. Two principal ones are the ammeter and the voltmeter. The *ammeter* is an instrument which measures electric current in amperes. To measure the amount of current flowing through a resistance, the ammeter is connected directly in series with the resistance which is to be measured (motor, electrical appliance, and so forth), as indicated in Fig. 24-18.

The *voltmeter* is an instrument which measures the electromotive force (pressure) in volts. A voltmeter is connected in parallel with that part of a circuit across which the voltage is to be measured, as in Fig. 24-19.

Figure 24-20 shows both an ammeter and voltmeter connected in a circuit to measure the current flowing through the resistance R and the voltage flowing across the resistance. The amperes and the volts are then measured.

Drafting Practices for Using Graphic Symbols. The small selection of graphic symbols in Fig. 24-21 and the following quotations for using graphic symbols are quoted from American National Standard Graphic Symbols for Electrical and Electronics Diagrams (ANSI Y32.2) by permission of The Institute of Electrical and Electronic Engineers, Inc.

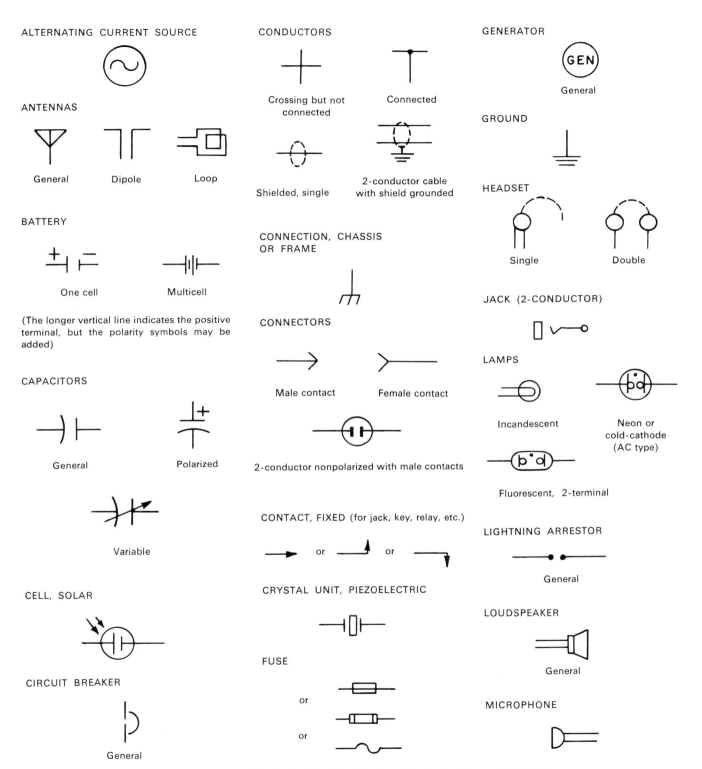

ALTERNATING CURRENT SOURCE

ANTENNAS

General   Dipole   Loop

BATTERY

One cell   Multicell

(The longer vertical line indicates the positive terminal, but the polarity symbols may be added)

CAPACITORS

General   Polarized

Variable

CELL, SOLAR

CIRCUIT BREAKER

General

CONDUCTORS

Crossing but not connected   Connected

Shielded, single   2-conductor cable with shield grounded

CONNECTION, CHASSIS OR FRAME

CONNECTORS

Male contact   Female contact

2-conductor nonpolarized with male contacts

CONTACT, FIXED (for jack, key, relay, etc.)

or   or

CRYSTAL UNIT, PIEZOELECTRIC

FUSE

or

or

GENERATOR

General

GROUND

HEADSET

Single   Double

JACK (2-CONDUCTOR)

LAMPS

Incandescent   Neon or cold-cathode (AC type)

Fluorescent, 2-terminal

LIGHTNING ARRESTOR

General

LOUDSPEAKER

General

MICROPHONE

Fig 24-21 ''American National Standard Graphic Symbols for Electrical and Electronic Diagrams,'' ANSI Y32.2-1962. (*With the permission of the Institute of Electrical and Electronic Engineers, Inc.*)

METER

To indicate a specific type of meter, replace the asterisk by one of the following letters or letter combinations

A	Ammeter
F	Frequency meter
G	Galvanometer
UA	Microammeter
MA	Milliammeter
OHM	Ohmmeter
V	Voltmeter
W	Wattmeter
WH	Watthour meter

MOTOR

General

PLUG (2-CONDUCTOR)

RECTIFIER (SEMICONDUCTOR DIODE OR METALLIC

RESISTOR (general)

or

(When the rectangular symbol is used, always add identification within or adjacent to the rectangle)

RESISTOR (with adjustable contact)

Fig. 24–21 (Cont.)

SWITCH, PUSHBUTTON

Circuit closing (make)   Circuit closing (break)

SWITCHES

Single throw (general)   Double throw (general)

Knife switch (General)   2-pole double throw with terminals shown

THERMISTOR RESISTOR, THERMAL

THERMOCOUPLE, TEMPERATURE MEASURING

THERMOSTAT (with break contact)

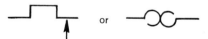

or

TRANSFORMER

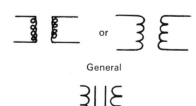

or

General

Magnetic (iron) Core

TRANSISTORS

PNP   NPN

TUBES, ELECTRON

(a) Components or parts of

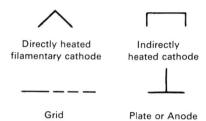

Directly heated filamentary cathode   Indirectly heated cathode

Grid   Plate or Anode

(b) Examples

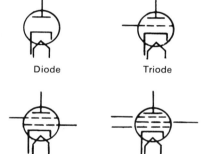

Diode   Triode

Tetrode   Pentode

WINDING, INDUCTOR, or REACTOR (Coil)

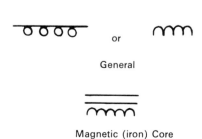

or

General

Magnetic (iron) Core

"Graphic symbols for electrical engineering are a shorthand used to show graphically the functioning or interconnections of a circuit. A graphic symbol represents the function of a part in the circuit. Graphic symbols are used on single-line (one-line) diagrams, on schematic or elementary diagrams or as applicable on connection or wiring diagrams. Graphic symbols are correlated with parts lists, descriptions or instructions by means of designations."

"DRAFTING PRACTICES APPLICABLE TO GRAPHIC SYMBOLS

1. A symbol shall be considered as the aggregate of all its parts.

2. The orientation of a symbol on a drawing, including a mirror image, does not alter the meaning. . . .

3. The width of a line does not affect the meaning of the symbol. In specific cases a wider line may be used for emphasis.

4. The symbols shown in this standard are in their correct relative size. This relationship shall be maintained as nearly as possible on any particular drawing, regardless of the size of the symbol used.

5. A symbol may be drawn to any proportional size that suits a particular drawing, depending on reduction or enlargement anticipated. If essential for purposes of contrast, some symbols may be drawn relatively smaller than the other symbols on a diagram.

6. The arrowhead of a symbol may be closed ——▶ or open ——→ unless otherwise noted in this standard.

7. The standard symbol for a TERMINAL (○) may be added to each point of attachment to connecting lines to any one of the graphic symbols. Such added terminal symbols should not be considered part of the individual graphic symbol unless the

Fig. 24-22 Some electrical and electronic components, with their names and appearance. (*Heath Company, a subsidiary of Daystrom, Inc.*)

terminal symbol is included in the symbol shown in this standard.

8. For simplification of a diagram, parts of a symbol for a device, such as a relay or contactor, may be separated. If this is done provide suitable designations to show proper correlation of the parts.

9. In general, the angle at which a connecting line is brought to a graphic symbol has no particular significance unless otherwise noted in this standard.

10. Associated or future paths and equipment shall be shown by lines composed of short dashes: - - -.

11. Details of type, impedance, rating, etc., may be added, when required,

adjacent to any symbol. If used, abbreviations should be from the American National Standard Abbreviations for Use on Drawings (Z32.13–1950). Letter combinations used as parts of graphic symbols are not abbreviations."

Circuit Components. Figure 24–22 illustrates and names some of the most commonly used electrical and electronic components. The appearance of components should be associated with the symbols by which they are represented on circuit diagrams. Some knowledge of the operation and purpose of the components is most desirable.

Electrical Diagrams. There are many kinds of electrical diagrams to suit the purposes for which they are used. The following definitions adopted as American National Standard are extracted from the *American National Standard Drafting Manual, Electrical Diagrams* (ANSI Y14-15–1966) with the permission of the publisher, The American Society of Mechanical Engineers, 345 East 47 St., New York, N.Y., 10017.

Single-line (one-line) diagram. "A diagram which shows, by means of single lines and graphic symbols, the course of an electric circuit or system of circuits and the component devices or parts used therein."

Schematic, or elementary, diagram. "A diagram which shows, by means of graphic symbols, the electrical connections and functions of a specific circuit arrangement. The schematic diagram facilitates tracing the circuit and its functions without regard to the actual physical size, shape or location of the component device or parts."

Connection or wiring diagram. "A diagram which shows the connections of an installation or its component devices or parts. It may cover internal or external connections, or both, and contains such detail as is needed to make or trace connections that are involved. The connection diagram usually shows general physical arrangement of the component devices or parts."

Interconnection diagram. "A form of connection or wiring diagram which shows only external connections between unit assemblies or equipment. The internal connections of the unit assemblies or equipment are usually omitted."

Line Conventions and Lettering. "The selection of line thickness as well as letter size should take into account size reduction or enlargement when it is felt that legibility will be affected. Line conventions, relative thickness and suggested applications for use on electrical diagrams are shown in [Fig. 24–23].

"A line of medium thickness is recommended for general use on electrical diagrams. A thin line may be used for brackets, leader lines, etc. When emphasis of special features such as main or transmission paths is essential, a line thickness sufficient to provide the desired contrast may be used. Line thickness and lettering used with electrical diagrams shall, in general, conform with American National Standard Y14.2 [latest issue] and local requirements to facilitate microfilming."

Symbols and Layouts. "Graphical symbols may be drawn to any proportional size that suits a particular diagram, provided the selection of size takes into account the anticipated reduction or enlargement. For most electrical diagrams intended for manufacturing purposes, or for ultimate use in a reduced form (2½ to 1 max), it is recommended that symbols be drawn approximately 1½ times the size of those shown in American National Standard Y32.2–1962."

Abbreviations for use with electrical diagrams are given in *American National Standard Abbreviations for Use on Drawings,* ANSI Z32.13 (latest issue).

"Layout of electrical diagrams. The layout of electrical diagrams shall be such that the main features are prominently shown. The parts of the diagram should be spaced to provide an even balance between blank spaces and lines. Sufficient blank area should be provided in the vicinity of symbols to avoid crowding of notes or

LINE APPLICATION	LINE THICKNESS
FOR GENERAL USE	MEDIUM
MECHANICAL CONNECTION, SHIELDING, & FUTURE CIRCUITS LINE	MEDIUM
BRACKET-CONNECTING DASH LINE	MEDIUM
USE OF THESE LINE THICKNESSES OPTIONAL	
BRACKETS, LEADER LINES, ETC.	THIN
BOUNDARY OF MECHANICAL GROUPING	THIN
FOR EMPHASIS	THICK

Fig 24–23 Line conventions for electrical diagrams.

reference information. Large spaces, however, should be avoided, except that space provision may be made for anticipated future circuits if deemed necessary."

Single-Line Diagrams. "The single-line diagram [Fig. 24–24] conveys basic information about the operation of a circuit or a system of circuits, but omits much of the detailed information usually shown on schematic or connection diagrams. This form of presentation lends itself to simplified diagrams of complex circuits and to diagrammatic representation of communication or power systems in which a single line represents a multiconductor communication or power circuit."

In general, the practices established for schematic diagrams apply equally well to the preparation of single-line diagrams.

Schematic Diagrams. "The following sub-paragraphs contain general information for use in the preparation of schematic diagrams.

"Layout. The schematic diagram shall use a layout which follows the circuit, signal or transmission path

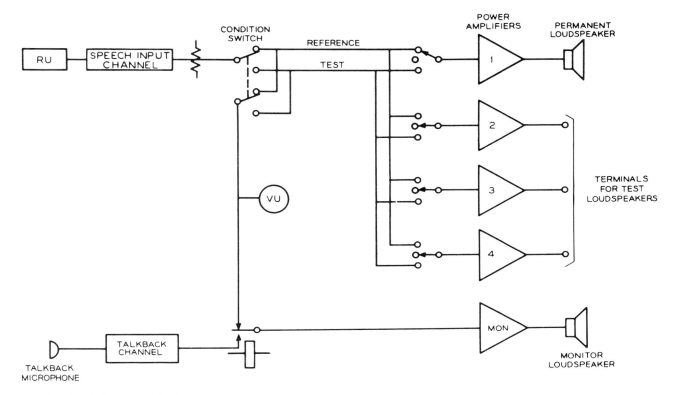

Fig. 24-24 Single-line diagram of an audio system.

either from input to output, source to load or in the order of functional sequence. Long interconnecting lines between parts of the circuit should be avoided.

"Connecting lines. Connecting lines should preferably be drawn horizontally or vertically and with as few bends and crossovers as possible. Connection of four or more lines at one point shall be avoided when it is equally convenient to use an alternative arrangement. When connecting lines are drawn parallel the spacing between lines after reduction shall be a minimum of $\frac{1}{16}$ inch. Parallel lines should be arranged in groups, preferably three, with approximately double spacing between groups of lines. In grouping parallel lines, functional re-

lation of the lines should be considered. Primary power and synchro circuits are examples of the application of this practice.

"Interrupted single lines. For single interrupted lines, the line identification may also serve to indicate destination as shown in [Fig. 24-25] for the power and filament circuit paths. In identification practice for single interrupted lines shall be the same as for grouped and bracketed lines described in the following paragraph.

"Interrupted grouped lines. When interrupted lines are grouped and bracketed, and depending on whether the lines are horizontal or vertical, line identifications shall be indicated as shown in [Fig. 24-26].

Bracket destinations or connections may be indicated either by means of notations outside the brackets as shown in [Fig. 24-26] or as shown in [Fig. 24-27] by means of a dash line. When the dash line is used to connect brackets it shall be drawn so that it will not be mistaken for a continuation of one of the bracketed lines. The dash line shall originate in one bracket and terminate in no more than two brackets." When drawing schematics, carefully observe the above.

Schematic Diagrams for Electronics and Communications. "The following sub-paragraphs contain detailed information which is specifically applicable to schematic dia-

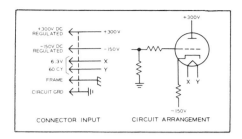

Fig. 24–25 Identification of interrupted lines. At left, a group of lines interrupted on the diagram. At right, single lines interrupted on the diagram.

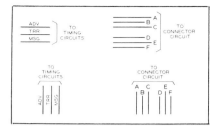

Fig. 24–26 Typical arrangement of line identifications and circuit destinations.

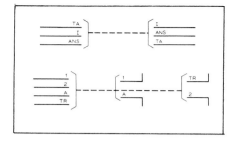

Fig. 24–27 Typical interrupted lines interconnected by dashed lines. The dashed line shows the interrupted paths that are to be connected. Individual line identifications indicate matching connections.

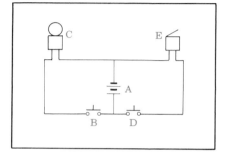

Fig. 24–28 Bell and buzzer circuit.

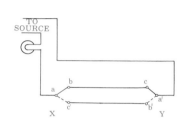

Fig. 24–29 Three-way-switch diagram.

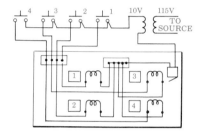

Fig. 24–30 Annunciator diagram.

grams of the type used with electronic and communication equipment. This material is to be used as a supplement to the general standards of [schematic diagrams].

"Layout. In general, schematic diagrams should be arranged so that they can be read functionally from left to right. Complex diagrams should generally be arranged to read from upper left to lower right and may be laid out in two or more layers. Each layer should be read from left to right. The overall result shall be a circuit layout which follows the signal, or transmission path, from input to output, or in the order of functional sequence. Where practical, terminations for external connections should be located at the outer edges of the circuit layout.

"Ground symbols. The ground symbol $\overset{\perp}{=}$ shall be used only when the circuit ground is at a potential level equivalent to that of earth potential. The symbol $\diagup\!/\!\!/\!\!/$ shall be used when an earth potential does not result from connecting to the structure which houses or supports the circuit parts."

Some Electrical Circuits. In Fig. 24–28 the bell (C) and the buzzer (E) are operated independently from the same battery (A) by the pushbuttons (B and D).

In Fig. 24–29 the current is from an outside source. The three-way switches (X and Y) are used so that the light at L may be turned off or on by either of the switches. Switch X might be at the garage and switch Y at the house. Each switch has three terminals. If either switch is opened, the light will be turned off, but the light may be turned on by the switch at the opposite end.

A circuit diagram is shown in Fig. 24–30 for an annunciator (an arrangement for signaling from different places to a station or post). It provides for ringing a buzzer in the annunciator when any of the buttons is pressed. Each button releases or allows a tab to drop down to identify the place where the button is pressed. Trace the circuits which are operated by each of the buttons. The source of the current is from the secondary of a step-down transformer.

Some single-line graphic symbols are shown in Fig. 24–31. A single-line, or one-line, diagram is shown in Fig. 24–32. Such a diagram shows the component parts or devices of a circuit or circuit system using single lines and graphic symbols. The single lines represent two or more conductors. The diagram gives the necessary basic information about the operation of

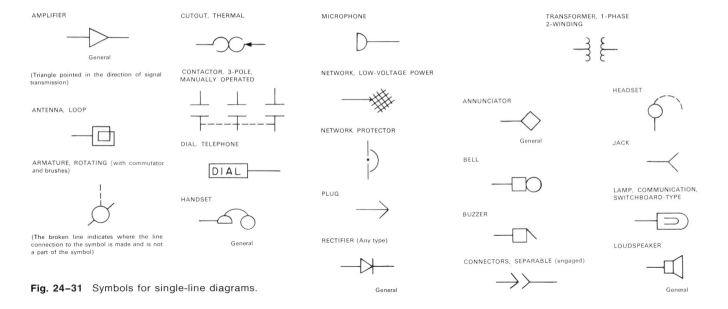

AMPLIFIER

General

(Triangle pointed in the direction of signal transmission)

ANTENNA, LOOP

ARMATURE, ROTATING (with commutator and brushes)

(The broken line indicates where the line connection to the symbol is made and is not a part of the symbol)

CUTOUT, THERMAL

CONTACTOR, 3-POLE, MANUALLY OPERATED

DIAL, TELEPHONE

DIAL

HANDSET

General

MICROPHONE

NETWORK, LOW-VOLTAGE POWER

NETWORK PROTECTOR

PLUG

RECTIFIER (Any type)

General

TRANSFORMER, 1-PHASE 2-WINDING

ANNUNCIATOR

General

BELL

BUZZER

CONNECTORS, SEPARABLE (engaged)

HEADSET

JACK

LAMP, COMMUNICATION, SWITCHBOARD-TYPE

LOUDSPEAKER

General

Fig. 24-31 Symbols for single-line diagrams.

Fig 24-32 A single-line diagram for a power-distribution system.

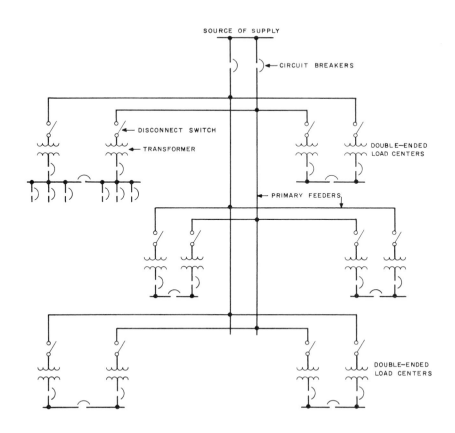

SOURCE OF SUPPLY

CIRCUIT BREAKERS

DISCONNECT SWITCH

TRANSFORMER

DOUBLE-ENDED LOAD CENTERS

PRIMARY FEEDERS

DOUBLE-ENDED LOAD CENTERS

the circuit but not the detailed information of a schematic diagram. A single-line diagram is a simplified representation and is useful in the fields of communications and electrical power transmission (Fig. 24-32). The highest voltage is usually placed at the top or left side of the diagram. The lower-voltage lines are placed in the order of their value below or on the right side of the diagram. Information concerning line location, component ratings, types of equipment, etc., is given at appropriate places on the diagram.

A motor-starter wiring diagram is shown at A in Fig. 24-33, and a schematic, or one-line, diagram is shown at B for the same circuit. A motor starter is required for the following purposes:

1. To give the proper protection against burnouts caused by sustained overloads. This is known as *thermal overload protection,* a protection not afforded by ordinary fuses.

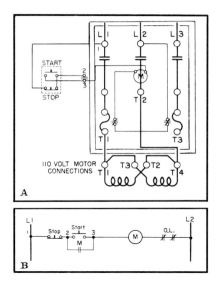

Fig. 24–33 A single-phase starter.

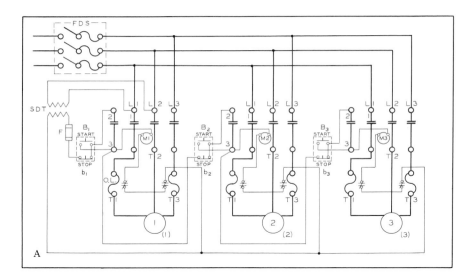

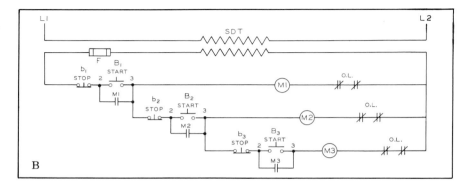

Fig. 24–34 A three-way-conveyor system.

2. To provide for remote control by manual start-stop buttons or automatic devices, such as thermostats, pressurestats, limit switches, and so forth.

3. To furnish provisions for sequence control. This is illustrated in Fig. 24–34.

Figure 24–34 illustrates a wiring diagram at A and a schematic, or one-line, diagram at B for the same circuit. This figure shows the sequence control for a conveyor system consisting of three separate motors. Notice that a thin line (light-value line) is used to indicate the pilot circuit (low voltage) in contrast to a thick line (heavy-value line) to indicate the line-voltage part of the circuit. The low voltage is obtained from the step-down transformer (SDT). A transformer is an electrical device which changes the voltage of alternating current, the type used in most electrical systems. In this case, the step-down transformer changes the current from a higher voltage to a lower voltage.

In Fig. 24–34 at A (upper rectangle) start button B_1 will start the No. 1 conveyor motor at (1), but start button B_2 will not start the No. 2 conveyor motor at (2) unless conveyor motor No. 1 has started. Likewise, start button B_3 starts conveyor motor No. 3 at (3) only after conveyor motor No. 2 has started. The conveyor system may be completely stopped by the fused disconnect switch (FDS) by stop button b_1 or by its overload (OL). The stop button b_2 or the No. 2 motor overload will stop both conveyors Nos. 2 and 3. The stop button b_3, or overload, will only stop the No. 3 con-

veyor. The fuse, F, is for protection of the low-voltage control circuit.

In Fig. 24–34 at B (lower rectangle) the same circuit is shown by a one-line diagram. The fuse F is for the protection of the low-voltage control circuit.

Color Codes. A color-code system is a convenient way of presenting information when preparing circuit diagrams and for use on the actual wiring of the circuit. In electrical and electronic work a color code is used to indicate certain characteristics of components, such as capacitors and resis-

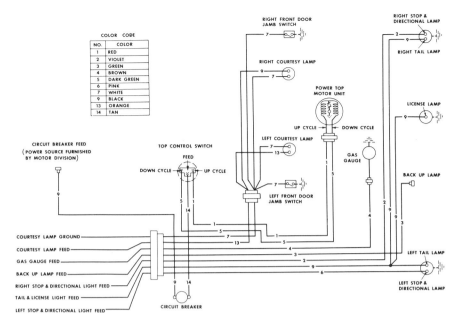

Fig. 24–35 Wire color scheme used on an automobile body wiring circuit. (*Chevrolet Motor Division, General Motors Corporation.*)

EIA Standard Color Code

Color	Abbre-viation	Number
Black	BLK	0
Brown	BRN	1
Red	RED	2
Orange	ORN	3
Yellow	YEL	4
Green	GRN	5
Blue	BLU	6
Violet	VIO	7
Gray	GRA	8
White	WHT	9

Fig. 24–36 EIA standard color code.

tors, to identify wire leads, and to show wire connections. A particular color-code scheme may be a part of the diagram on which it is used (Fig. 24–35).

When using a color code, the Electronic Industries Association (EIA) standards should be consulted, as well as any others which may apply. Note the different code used in Fig. 24–35; it is not the same as the EIA standard in Fig. 24–36.

Color codes are used to give specific information on resistors, capacitors, chassis hooks, and component-lead wire insulation and other purposes. Such uses are covered in published standards and in textbooks.

Block Diagrams. A block diagram (Fig. 24–37) is usually composed of squares or rectangles, or "blocks," joined by single lines and arranged to show the relation between the various component groups or stages in the operation of a circuit. Arrowheads at

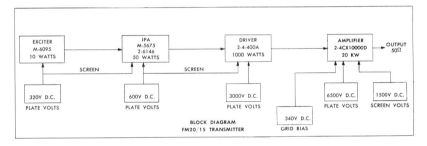

Fig. 24–37 Block diagram of 20,000-watt broadcast transmitter. (*Gates Radio Co., a subsidiary of Harris-Intertype Corporation.*)

the terminal ends of the lines show the direction of the signal path from input to output, reading the diagram from left to right.

Engineers often draw or sketch block diagrams as a first step in designing new equipment. Block diagrams are also used in catalogs, descriptive folders, and advertisements for electrical equipment.

Electrical Layouts for Buildings. The usual architect's indication of

electrical outlets and switch locations is shown in Fig. 24–38. This plan only indicates the location of lights, base plugs, and desired switching arrangements. To provide a satisfactory and adequately wired electrical system upon completion of the structure, it is necessary to have a complete and detailed set of electrical drawings and specifications prepared by someone who knows the engineering requirements. A schedule of the symbols used is shown in Fig. 24–39.

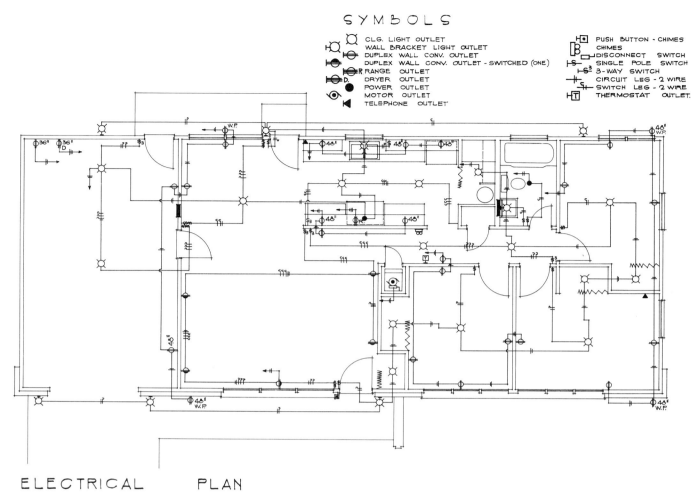

⬚ CLG. LIGHT OUTLET
⬚ WALL BRACKET LIGHT OUTLET
⬚ DUPLEX WALL CONV. OUTLET
⬚ DUPLEX WALL CONV. OUTLET - SWITCHED (ONE)
⬚R RANGE OUTLET
⬚D. DRYER OUTLET
● POWER OUTLET
⬚ MOTOR OUTLET
◀ TELEPHONE OUTLET

⬚ PUSH BUTTON - CHIMES
⬚ CHIMES
⬚ DISCONNECT SWITCH
⬚S SINGLE POLE SWITCH
⬚S³ 3-WAY SWITCH
⬚ CIRCUIT LEG - 2 WIRE
⬚ SWITCH LEG - 2 WIRE
⬚T THERMOSTAT OUTLET.

ELECTRICAL   PLAN

Fig 24-38 Electrical plan for a ranch house.

The Interconnection Diagram. Figure 24-40 is a form of connection, or wiring, diagram which shows the electrical connections between the different assemblies, panels, or units of an electrical or electronics system. Generally, the internal connections within the various units are not shown. The units are identified by name, and they are represented on the diagram by rectangles.

Connection or Wiring Diagrams. Figure 24-41 shows wiring connections in a simplified way so that the connections of the circuit system may be easily followed or traced. Internal or external connections, or both, may be shown. The components are named and are drawn in pictorial form. Auxiliary devices such as terminal blocks, strips, and fuse mountings are shown. Color coding is important for servicing and is indicated on Fig. 24-41.

Such diagrams furnish information needed for manufacture, installation, and maintenance and for use with schematic diagrams.

Printed Circuit Drawings. These consist of accurately drawn layouts of the required pattern. The drawing is made actual size or to an enlarged scale which can be reduced to the desired size by photography. The lines (conductors) on the pattern should be at least $\frac{1}{32}$ in. wide and should be spaced at least $\frac{1}{32}$ in. apart. The circuit layout pattern is transferred to a copper-clad insulating base by photographic or other means. Etching is one process used to remove the copper from all areas of the insulating base except for the required circuits (Fig. 24-42). There are several other different methods of preparing printed circuits.

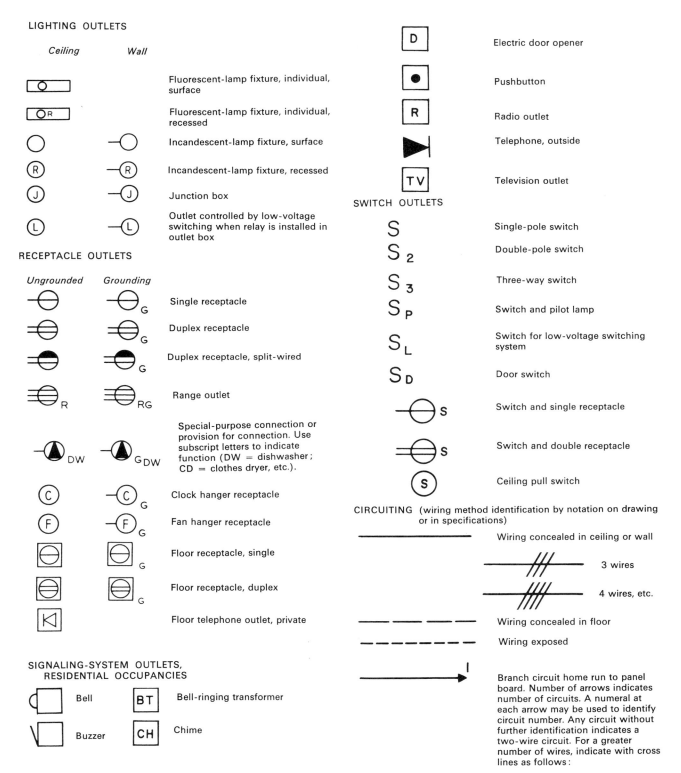

LIGHTING OUTLETS

Ceiling   *Wall*

Fluorescent-lamp fixture, individual, surface

Fluorescent-lamp fixture, individual, recessed

Incandescent-lamp fixture, surface

Incandescent-lamp fixture, recessed

Junction box

Outlet controlled by low-voltage switching when relay is installed in outlet box

RECEPTACLE OUTLETS

Ungrounded   *Grounding*

Single receptacle

Duplex receptacle

Duplex receptacle, split-wired

Range outlet

Special-purpose connection or provision for connection. Use subscript letters to indicate function (DW = dishwasher; CD = clothes dryer, etc.).

Clock hanger receptacle

Fan hanger receptacle

Floor receptacle, single

Floor receptacle, duplex

Floor telephone outlet, private

SIGNALING-SYSTEM OUTLETS, RESIDENTIAL OCCUPANCIES

Bell   BT   Bell-ringing transformer

Buzzer   CH   Chime

D   Electric door opener

●   Pushbutton

R   Radio outlet

Telephone, outside

TV   Television outlet

SWITCH OUTLETS

S   Single-pole switch

S_2 Double-pole switch

S_3 Three-way switch

S_P Switch and pilot lamp

S_L Switch for low-voltage switching system

S_D Door switch

Switch and single receptacle

Switch and double receptacle

Ceiling pull switch

CIRCUITING (wiring method identification by notation on drawing or in specifications)

Wiring concealed in ceiling or wall

3 wires

4 wires, etc.

Wiring concealed in floor

Wiring exposed

Branch circuit home run to panel board. Number of arrows indicates number of circuits. A numeral at each arrow may be used to identify circuit number. Any circuit without further identification indicates a two-wire circuit. For a greater number of wires, indicate with cross lines as follows:

Fig. 24-39 Electrical wiring symbols for architectural layout.

ANTENNA

INDICATOR SYSTEM | SERVO SYSTEM

SYNCHRO TRANSMITTER TXI

D-C SERVO MOTOR

ERROR SIGNAL

SYNCHRO CONTROL TRANSFORMER

CONTROL AMPLIFIER

AMPLIDYNE GENERATOR

SYNCHRO RECEIVER TR

SYNCHRO TRANSMITTER TX2

A-C DRIVE MOTOR

BEARING—INDICATOR DIAL

HAND WHEEL

Fig. 24-40 Interconnection diagram showing the different units of a typical dc servo system used for rotating a search radar system.

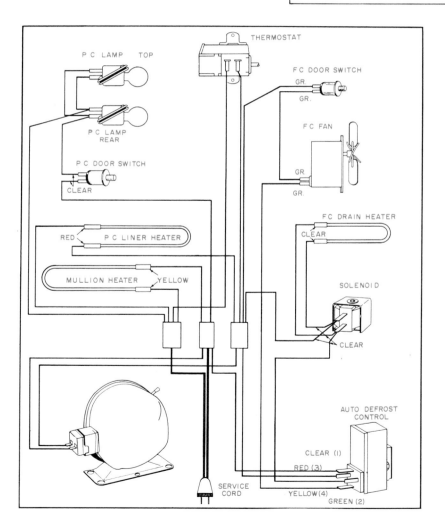

THERMOSTAT

P C LAMP TOP

FC DOOR SWITCH
GR.
GR.

P C LAMP REAR

FC FAN
GR.
GR.

P C DOOR SWITCH
CLEAR

FC DRAIN HEATER
CLEAR

RED   P C LINER HEATER

MULLION HEATER   YELLOW

SOLENOID
CLEAR

AUTO DEFROST CONTROL
CLEAR (I)
RED (3)

SERVICE CORD

YELLOW (4)
GREEN (2)

Fig. 24-41 Connection or wiring diagram for a refrigerator. (*Kelvinator Division, American Motors Corporation.*)

Fig. 24-42 Pressure-sensitive symbols aid the development of printed circuits. (*Keuffel & Esser Co.*)

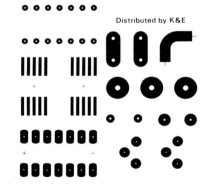

Distributed by K&E

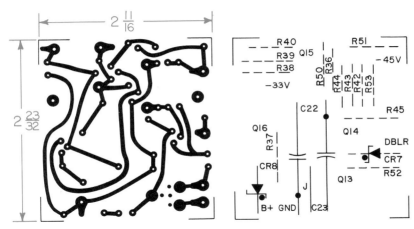

Fig. 24–43 A printed circuit and a component identification overlay. (*Gates Radio Co., a subsidiary of Harris-Intertype Corp.*)

The components may be located on the printed circuit board by the use of symbols or other markings. This information is transferred to the printed circuit diagram from a component identification overlay (Fig. 24–43).

Circuit Diagrams. The circuit diagram for a high-fidelity audio amplifier shown in Fig. 24–44 shows conventional tubes and electrical components. It illustrates one of the important applications of RCA receiving tubes.

The schematic diagrams in Figs. 24–45 and 24–46 represent parts of a storm-detector radar (AN/CP8-9) manufactured by the Raytheon Company. The circuit names are given in the legends for the illustrations (Figs. 24–45 and 24–46).

Fig. 24–44 A circuit diagram of a high-fidelity audio amplifier. (*RCA.*)

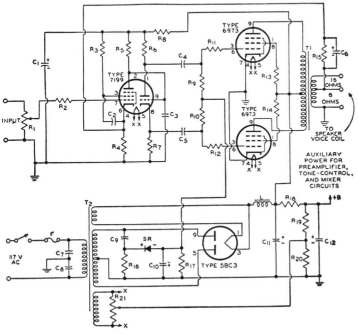

$C_1 = 40$ μf, electrolytic, 450 v.
C_2 C_4 $C_5 = 0.25$ μf, paper, 400 v.
$C_3 = 3.3$ pf, ceramic or mica, 600 v.
$C_6 = 150$ pf, ceramic or mica, 400 v.
C_7 $C_8 = 0.05$ μf, paper, 400 v.
$C_9 = 0.02$ μf, paper, 600 v.
$C_{10} = 100$ μf, electrolytic, 50 v.
$C_{11} = 80$ μf, electrolytic, 450 v.
$C_{12} = 40$ μf, electrolytic, 450 v.
F = Fuse, 3 amperes
L = Choke, 3 h., 160 ma., dc resistance 75 ohms or less
$R_1 =$ Volume control, potentiometer, 1 megohm
$R_2 = 1000$ ohms, 0.5 watt
$R_3 = 0.82$ megohm, 0.5 watt
$R_4 = 820$ ohms, 0.5 watt
$R_5 = 0.22$ megohm, 0.5 watt
R_6 $R_7 = 15000$ ohms ± 5 per cent, 2 watts
$R_8 = 3900$ ohms, 2 watts

R_9 $R_{10} = 0.1$ megohm, 0.5 watt
R_{11} $R_{12} = 1000$ ohms, 0.5 watt
R_{13} $R_{14} = 100$ ohms, 0.5 watt
$R_{15} = 8200$ ohms, 0.5 watt
$R_{16} = 15000$ ohms, 1 watt
$R_{17} = 68000$ ohms, 0.5 watt
$R_{18} = 4700$ ohms, 2 watts
$R_{19} = 0.27$ megohm, 1 watt
$R_{20} = 47000$ ohms, 0.5 watt
$R_{21} =$ Hum balance adjustment, potentiometer, 100 ohms, 0.5 watt
SR = Selenium rectifier, 20 ma., 135 volts rms
$T_1 =$ Output transformer, (having 8-ohm tap for feedback connection) for matching impedance of voice coil to 6600-ohm plate-to-plate tube load; 50 watts; frequency response, 10 to 50000 cps; Stancor A-8056 or equiv.
$T_2 =$ Power transformer, 360-0-360 volts rms, 120 ma.; 6.3 v., 3.5 a; 5v., 3a; Stancor 8410 or equiv.

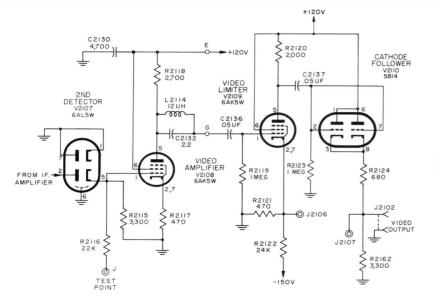

Fig. 24-45 A circuit diagram of a second detector and a video amplifier and limiter. (*Raytheon Co.*)

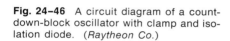

Fig. 24-46 A circuit diagram of a count-down-block oscillator with clamp and isolation diode. (*Raytheon Co.*)

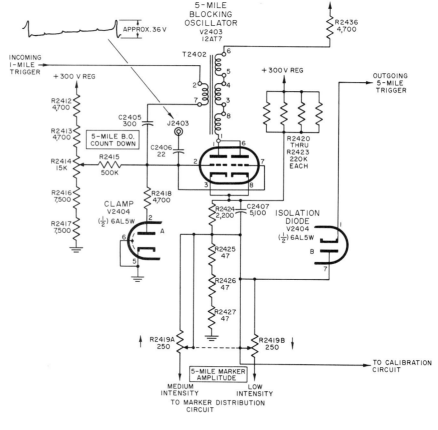

Review

1. Describe a series circuit in technical terms.

2. Illustrate a simple parallel circuit.

3. Explain the difference between a block diagram and a schematic diagram.

4. What is the relationship between electrical energy and mechanical energy?

5. What is the difference between a conductor and a nonconductor? List two of each.

6. What is the significant difference between mechanical drawing and electrical drawing?

Problems for Chapter 24

Figures 24–47 through 24–51 are component diagrams of an AM-FM stereo unit containing integrated circuits. All diagrams should be drawn at least twice the size shown. (*Motorola, Inc.*)

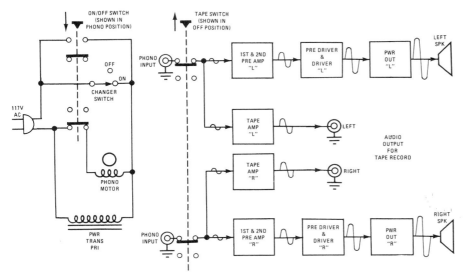

Fig. 24–47 Prepare the complete signal-flow diagram for the phonograph section of the stereo system. Note that only three blocks per channel are shown. Estimate the sizes required, and use a template if one is available.

Fig. 24–48 Draw the tape player signal-flow diagram. This diagram may be drawn as an overlay to Fig. 24–47 as directed by the instructor. Note that when the tape button is depressed, all other functions, AM, FM, and phono, are disabled. Also, the changer switch cannot turn on the phono motor. Since there is no output from the left and right audio-output jacks, no direct tape recording can take place. The tape preamp is grounded to prevent pickup by the tape player.

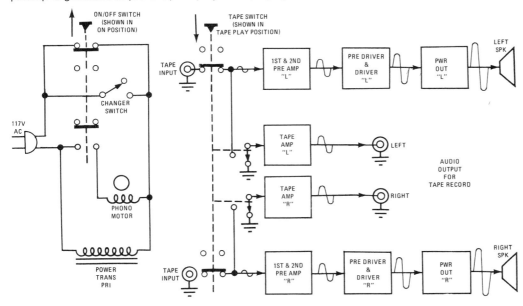

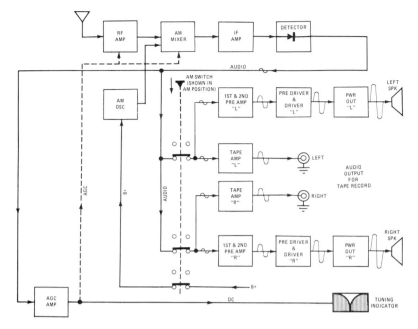

Fig. 24-49 Prepare a block diagram for the complete signal flow of the AM radio. A loop antenna picks up the signal and presents it to the RF amplifier. The mixer receives both the selected AM and oscillator signals. At the AM detector, the audio is recovered and applied to the stereo channels. The stereo channels are connected in parallel to the AM switch, providing monaural operation.

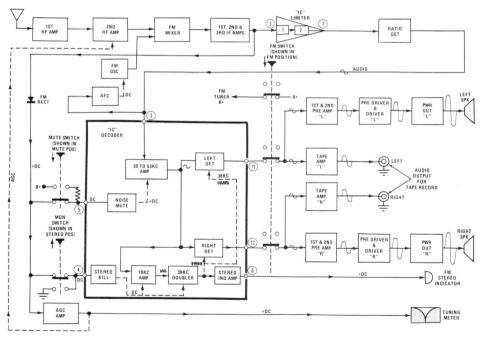

Fig. 24-50 Prepare a block diagram of the stereo FM signal-flow diagram. Two RF amplifiers are used to provide optimum FM selectivity. Selectivity and gain are further improved by three IF amplifier stages. A ratio detector receives the signal and reduces any AM noise that may be present. The audio signal then enters the "IC" FM decoder through pin #3. The primary functions of the decoder are to pass a monaural FM signal into both right and left channels and to separate a stereo FM signal into a right and left channel.

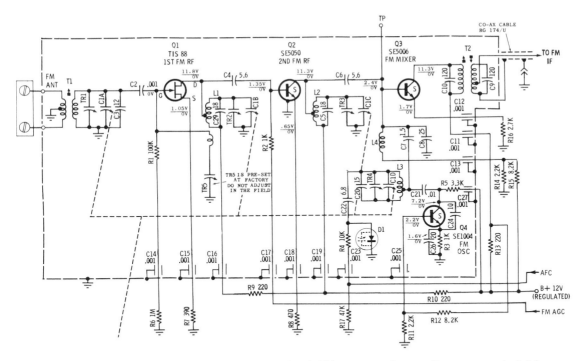

Fig. 24-51 Draw a schematic diagram of the four-circuit FM tuner as shown. Prepare a parts list in alphabetical order of the components shown.

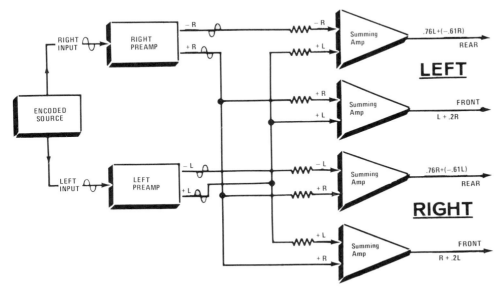

Fig. 24-52 Prepare a diagram illustrating four-channel sound. The four channels are encoded (mixed) into two channels at the recording studio. The resulting two channels have a phase and amplitude relationship which can be decoded into four channels. Develop the diagram approximately twice the size shown. (*Motorola, Inc.*)

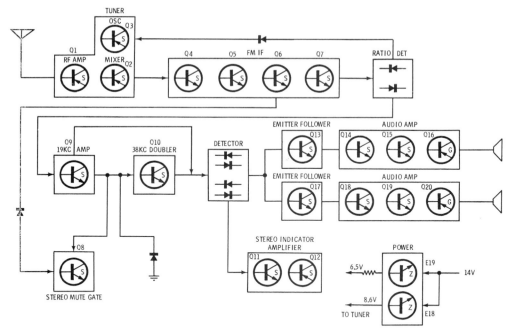

Fig. 24-53 Prepare the block diagram of the solid-state FM stereo auto radio. The main components employed are 14 n-p-n silicon transistors, 4 p-n-p silicon transistors, 2 p-n-p germanium transistors, 9 diodes, and 2 Zener diodes. (*Motorola, Inc.*)

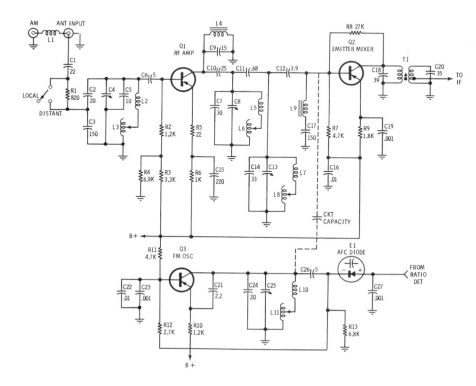

Fig. 24-54 Prepare a schematic diagram of the tuner used in the FM stereo auto radio. Prepare a parts list alphabetically. (*Motorola, Inc.*)

Aerospace Drafting

Chapter 25

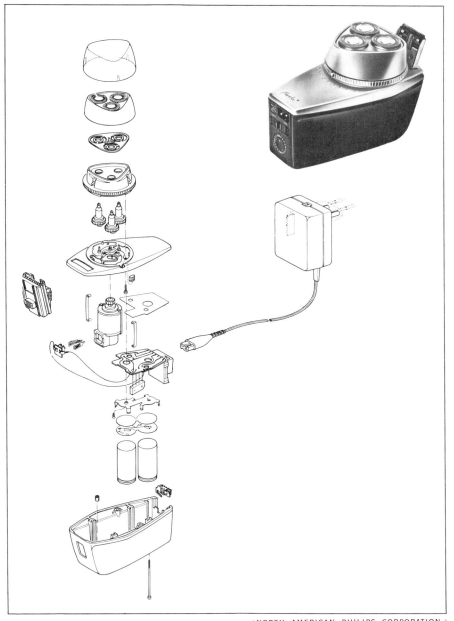

(NORTH AMERICAN PHILIPS CORPORATION.)

The Aerospace Industry. Progress in the development of the airplane, missile, and space vehicle presents challenging opportunities for young men and women interested in research, development, and design and drafting (Fig. 25–1). Aerospace design and drafting deals with all types of vehicles in flight, at all speeds and at all altitudes. The basic aerospace team consists of hundreds of scientists, engineers, designers, draftsmen, and technicians and thousands of skilled craftsmen. Government agencies have cooperated with industry in the research and development of aircraft. The aircraft industry has interests which vary from hovering helicopters, acrobatic biplanes, and supersonic vehicles with intercontinental range to complicated space craft which will rotate 350 million miles to Mars (Fig. 25–2).

Career Opportunities. The aviation industry has expanded with the speed of the jet and the supersonic vehicle. New careers and jobs have been created that did not exist a generation ago (Fig. 25–3). The technology required for aircraft design and drafting begins with formal and industrial training. The starting role is generally as a technician or an engineering aide. Coursework in science and mathematics at the high school and junior college levels may be complemented by a course in technical drawing or engineering graphics. As a member of a design team, the drafting

491

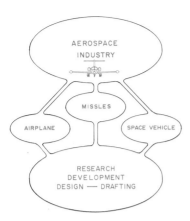

Fig. 25–1 The aerospace industry.

technician may prepare drawings of mechanical or electrical systems from the designer's sketch.

Career Advancement. The drafting technician may advance as he gains experience in a specialized area of development. With some formal education and selected guidance on the job, his title or position may develop to design technician with increased responsibilities. The opportunities for modifying existing designs are apt to be primary assignments for the young designer who has just been promoted. The design team may be called upon to study aircraft tooling, fabrication, and material variations for an improved production and performance. Sources for career guidance are:

Your Career as an Aerospace Engineer
American Institute of Aeronautics and Astronautics
1290 Avenue of the Americas
New York, N.Y. 10010
Engineers Council for Professional Development
345 East 47th Street
New York, N.Y. 10017

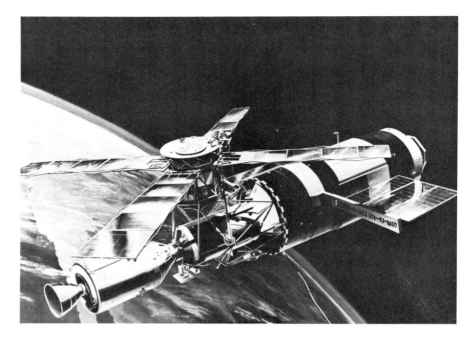

Fig. 25–2 Aircraft industry developed the space lab concept. (*NASA.*)

Many specialists are required in aerospace industries, from the man on the drawing board to the man testing models in the wind tunnel. The large aircraft companies are generally supported by many smaller manufacturing plants across the nation. Perhaps the largest plants are those which assemble the power plants and structural framework. Thousands of components may be produced by the smaller equipment manufacturer.

Testing and Research. The federal government, through various agencies and the National Aeronautics and Space Administration (NASA), cooperates with industry on vehicle testing and research. Aeronautics may be defined as the scientific and engineering studies which deal with design, construction, and operation of aircraft. Experimental research in aircraft explores the travel boundaries of space. Today's leisure aircraft normally cruise at subsonic (Fig. 25–4)

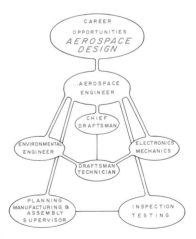

Fig. 25–3 Aerospace industry career opportunities.

Mach numbers (Mach 1 is the speed of sound, about 742 mph). Tomorrow's leisure aircraft may be designed with materials, power plants, and controls which will allow supersonic and hypersonic flight (hypersonic is five times the speed of sound).

Aircraft Materials. Today's high-speed and supersonic aircraft include

Fig. 25-4 Leisure aircraft cruise at subsonic speeds. (*Lear Jet Corp.*)

parts which are made of many different materials. The conventional aluminum, magnesium, and lightweight steel alloys are usually strong enough for the inner, or substructure, parts. The superstructure and skins, which are subject to high temperatures during supersonic flight, must be made of high-temperature-resistant, high-strength materials. The major design thrusts today are a lightweight structure and electronic mechanisms which must sense, compute, and control more quickly and accurately than the human mind.

Major Aircraft Components. Figure 25-5 illustrates the McDonnell Douglas F-4C Phantom jet with 62 basic component parts. The fuselage (or central body) part fourteen contains the operating and passenger compartments. The fuselage structure is composed of three parts or a series of shaped bulkheads and rings (riveted, machined, or welded frames)

and longitudinal (lengthwise) members which together form a rigid structural framework. The outer skin which is a ductile (formable) alloy sheet metal is fastened to the structural framework with rivets or machine screw fasteners.

The wings and airfoil. The wings are composed of ribs which form the "fore and aft" shape of the airfoil. The ribs are attached to spars (beamlike structural members) running inboard and outboard from the fuselage. The wing skins are attached to the ribs and spar structure with metal fasteners. In addition to the wings, other control surfaces, which are called *airfoils,* include ailerons, rudders, stabilizer, flaps, and tabs. A chord line is a straight line between the leading edge and the trailing edge of an airfoil. The general three-view drawing illustrates the sleek form of a vehicle designed for high-speed and high-altitude flying activities (Fig. 25-6).

The landing gear. The landing gear is retracted and extended by hydraulic or electrical drive mechanisms. Hydraulic shock absorbers are used to ease the impact of the landing load.

The power plant. Piston or jet engines are generally located in the lower portion of the fuselage or the lower portion of the wings.

The structural systems are complemented by various other systems such as air conditioning, compartment pressurization, radar, radio (communications), hydraulic, electronic, and plumbing. Many trades and industries are involved in the complex development of aerospace designs of today (Fig. 25-7).

The components and systems prepared for assembly are generally illustrated on a plan which is highly detailed. Many kinds of drawings are required before assembly. They are coordinated by the design-team director. The fundamental principles of orthographic projection and pictorial presentation are essential and basic to understanding assembly.

Aircraft Drafting Practices. The larger aircraft industries provide engineering manuals for their employees so that they may follow the practices best adapted to their company's product. The illustrated practices are prepared carefully for efficient and economical manufacturing procedures. A designer and draftsman must be familiar with his company's manual and with the standards in general use. The SAE (Society of Automotive Engineers) publishes a volume of Aerospace-Automotive Drawing Standards, which includes the best systems for these industries.

Techniques of Undimensioned Drawings. The undimensioned drawing is a method of drafting designed

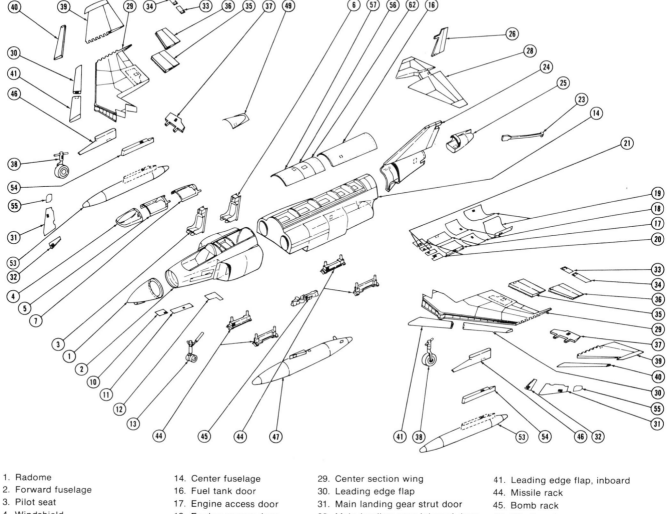

Fig. 25–5 An exploded view of parts of McDonnell Douglas F-4C Phantom. (*McDonnell Douglas Corp.*)

1. Radome
2. Forward fuselage
3. Pilot seat
4. Windshield
5. Forward canopy
6. Radar operation seat
7. Aft canopy
10. Nose landing gear door, forward
11. Nose landing gear door, aft
12. Hydraulic compartment access door
13. Nose landing gear shock strut

14. Center fuselage
16. Fuel tank door
17. Engine access door
18. Engine access door
19. Engine access door
20. Engine access door
21. Auxiliary engine air door
23. Arresting hook
24. Aft fuselage
25. Tail cone
26. Rudder
28. Stabilator

29. Center section wing
30. Leading edge flap
31. Main landing gear strut door
32. Main landing gear inboard door
33. Inboard spoiler
34. Outboard spoiler
35. Flap
36. Aileron
37. Speed brake
38. Main landing gear shock strut
39. Outer wing
40. Leading edge flap, outboard

41. Leading edge flap, inboard
44. Missile rack
45. Bomb rack
46. Missile pylon
47. External center-line fuel tank
49. Data link access door
53. External wing fuel tank
54. External wing fuel tank pylon
55. Landing gear door, outboard
56. Boom IFR receptacle access door
57. Fuel cell access door
62. Fuel cell access door

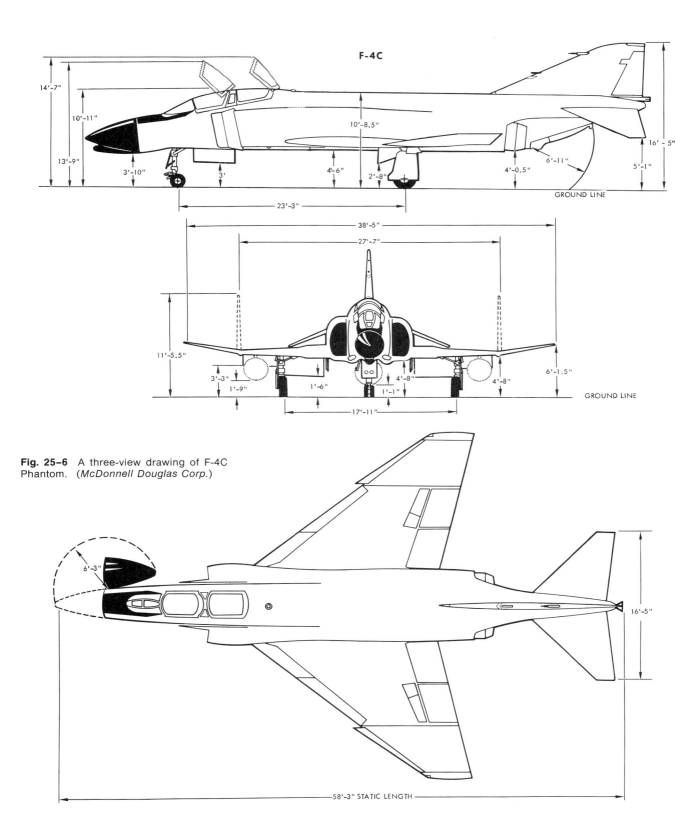

F-4C

14'-7"
10'-11"
13'-9"
10'-8.5"
3'-10"
3'
4'-6"
2'-8"
4'-0.5"
6'-11"
16'-5"
5'-1"
GROUND LINE
23'-3"

38'-5"
27'-7"
11'-5.5"
3'-3"
1'-9"
1'-6"
4'-8"
1'-1"
4'-8"
6'-1.5"
GROUND LINE
17'-11"

Fig. 25-6 A three-view drawing of F-4C Phantom. (*McDonnell Douglas Corp.*)

6'-3"
16'-5"
58'-3" STATIC LENGTH

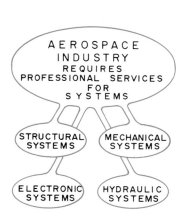

Fig. 25-7 Services needed by the aerospace industry.

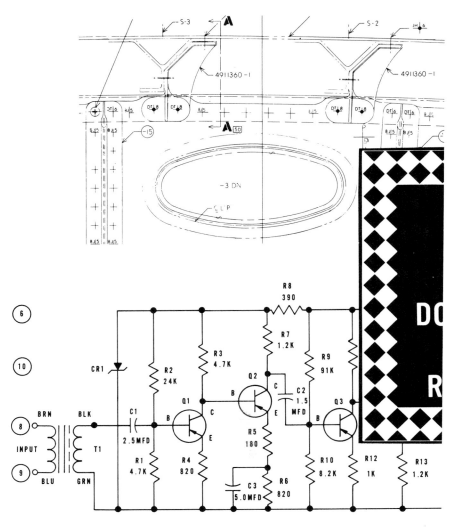

Fig. 25-8 Undimensioned drawing, a method of graphic presentation. (*McDonnell Douglas Corp.*)

to relieve designers and draftsmen from the tedious detail dimensioning required for most finished layouts. This technique reduces the time required to turn finished drawings into completed components ready for assembly. Technical advances in the field of fully automated drafting equipment, with related incremental plotting on three major axes, make undimensioned drafting possible (Fig. 25-8).

The drawings are technically prepared with devices which control graphical accuracy. Some techniques are listed for full size drawings.

1. Lines are drawn on matte-surface polyester film.

2. Drawings are prepared with a technical pen using controlled line widths.

3. Drawing is accurate, and the accuracy is relative to the part being drawn, with ranges of ± 0.005 to ± 0.015 in. on 0.004-in.-thick polyester film and to ± 0.015 in. on 0.002-in.-thick polyester film.

The advantages of undimensioned drawings:

1. Less time required to make finished drawings.

2. Better coordination of drawings. Full-size drawings are a natural proving ground for the fit of parts. If parts do not fit on full-size drawings, they will probably not fit on the final product.

3. Flat-pattern expansion on drawing. A flat pattern provides design teams time for checking out problems.

4. Contact photo creates the master layout. Transfer of the structural form of the drawing can be made with a photograph.

5. Fewer changes due to drawing error. Better coordination with full-sized drawings and the absence of

dimensions which may be wrong or misleading have reduced drawing error by 40 percent.

6. Adapts to automation. In the forthcoming era of automation, the full-size undimensioned drawing will be the ideal instrument which may be digitized directly onto a programed tape for playback of new and configurated drawings. The program will be used for the numerically controlled machine in the automated fabricating areas.

The kinds of typical drawings suitable for the undimensioning technique are:

1. Parts which have flat patterns for development in hydropress forming, power press brake, and stretch brake. An example of flat pattern development is shown in Fig. 25–9.

2. Machined parts which are drawn as undimensioned drawings on a part must be made by profile machine methods (Fig. 25–10).

3. Parts requiring a plaster pattern with a three-dimensional form whose definition is complete within the drawing (Fig. 25–11).

4. Drawings requiring artwork layout. Circuitry and lighting panels are well oriented to undimensioned drawing principles. The superior quality artwork is drawn twice the size and then photoreduced to the proper size for part fabrication because photoreduction makes lines cleaner, sharper, and more accurate (Fig. 25–12).

Fig. 25–9 Undimensioned flat-pattern development. *(McDonnell Douglas Corp.)*

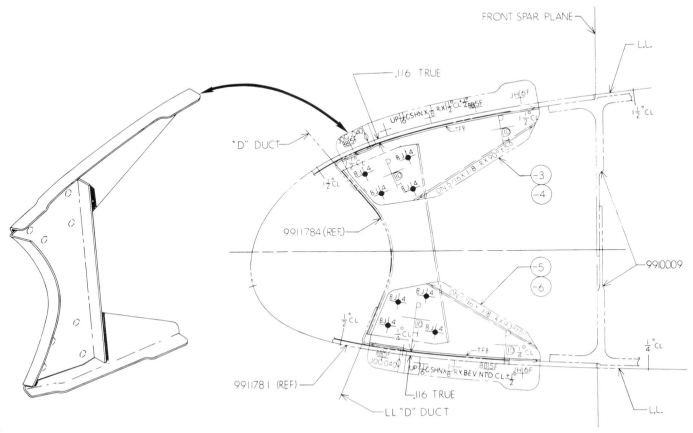

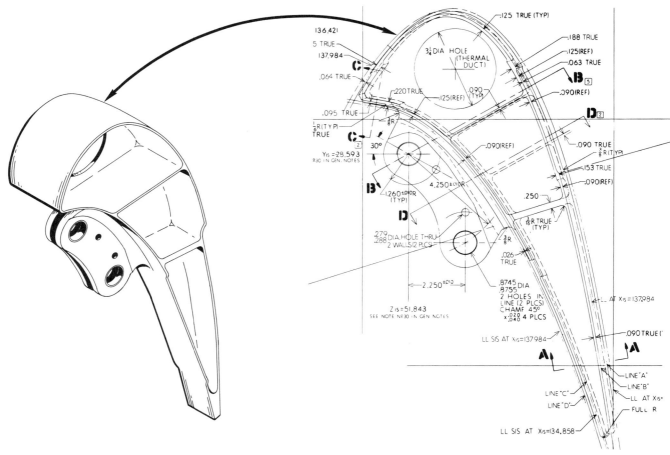

Fig. 25-10 Undimensioned machined parts. (*McDonnell Douglas Corp.*)

Drawings for Large Aircraft. Many types of aircraft drawings are illustrated in the DC-10 study in Figs. 25-13 to 25-21. The assembly drawing is used to describe the inboard profile of the DC-10 McDonnell Douglas transport illustrated in Fig. 25-13. The cutaway pictorial of a wing engine in Fig. 25-14 illustrates the overwhelming size of a power plant as it is contrasted to a human silhouette. Figure 25-15 illustrates the giant transport in a three-view orthographic projection. The wing is developed in a cutaway isometric pictorial with complex detail in Fig. 25-16. The exploded pictorial of the Damper-

Instl-Outbd Aileron in Fig. 25-17 is typical of the way parts are identified and then listed by name in Fig. 25-18. The hydraulic slat control in a pictorial view (Fig. 25-19) is developed in schematic form in Fig. 25-20. Figure 25-21 illustrates the pictorial rendering of a DC-10 galley (kitchen) prepared by the design illustrator. These types of drawings used are essential for complete interpretation of aircraft design, assembly, and flight checkout.

What makes an airplane fly? Figure 25-22 illustrates why an airplane can fly. The wing designs are typical of the high-lift system used on a DC-10 (Fig. 25-23). Any wing pass-

ing through the air at an angle will deflect the air downward, resulting in an equal and opposite upward force. Lift then is caused by the angle of the foil in movement through space. Examine the shape and position of the wing as you attempt to interpret the capacity of an aircraft's mobility.

Smaller Aircraft. Small aircraft have experienced new challenges with the advanced navigational equipment available today. The Beechcraft Bonanza in Fig. 25-24 has been around for a while, and the design features today include a Bendix autopilot and flight director system. This

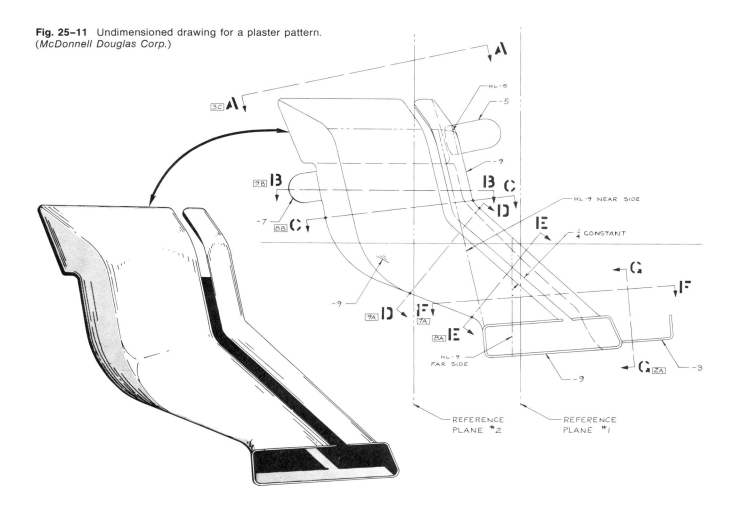

Fig. 25–11 Undimensioned drawing for a plaster pattern. (*McDonnell Douglas Corp.*)

Fig. 25–12 Undimensioned artwork layout. (*McDonnell Douglas Corp.*)

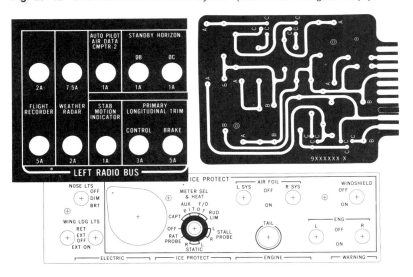

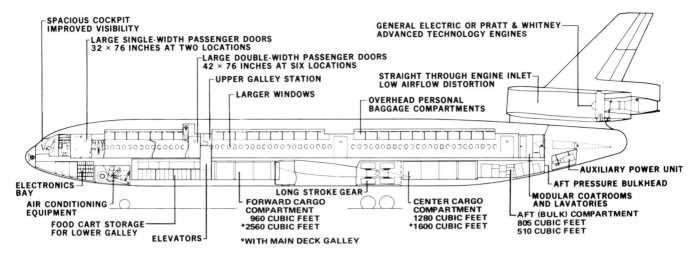

Fig. 25-13 The inboard profile of a commercial jet transport. (*McDonnell Douglas Corp.*)

SPACIOUS COCKPIT
IMPROVED VISIBILITY

LARGE SINGLE-WIDTH PASSENGER DOORS
32 × 76 INCHES AT TWO LOCATIONS

LARGE DOUBLE-WIDTH PASSENGER DOORS
42 × 76 INCHES AT SIX LOCATIONS

UPPER GALLEY STATION

LARGER WINDOWS

GENERAL ELECTRIC OR PRATT & WHITNEY
ADVANCED TECHNOLOGY ENGINES

STRAIGHT THROUGH ENGINE INLET
LOW AIRFLOW DISTORTION

OVERHEAD PERSONAL
BAGGAGE COMPARTMENTS

ELECTRONICS
BAY

AIR CONDITIONING
EQUIPMENT

FOOD CART STORAGE
FOR LOWER GALLEY

ELEVATORS

LONG STROKE GEAR

FORWARD CARGO
COMPARTMENT
960 CUBIC FEET
*2560 CUBIC FEET

*WITH MAIN DECK GALLEY

CENTER CARGO
COMPARTMENT
1280 CUBIC FEET
*1600 CUBIC FEET

AUXILIARY POWER UNIT

AFT PRESSURE BULKHEAD

MODULAR COATROOMS
AND LAVATORIES

AFT (BULK) COMPARTMENT
805 CUBIC FEET
510 CUBIC FEET

Fig. 25-14 Cutaway of wing-mounted jet engine. (*McDonnell Douglas Corp.*)

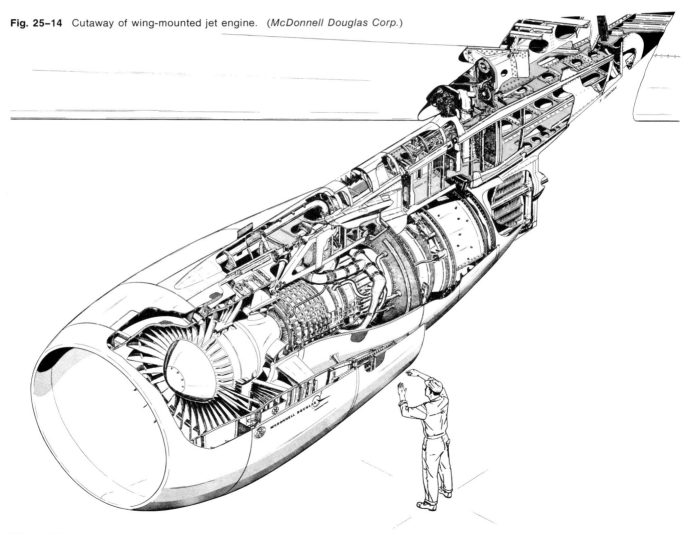

Fig. 25-15 Three-view drawing of a commercial jet transport. (*McDonnell Douglas Corp.*)

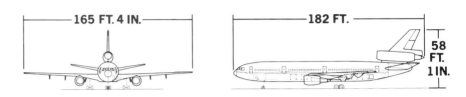

Fig. 25-16 Pictorial drawing of wing section including interior detail. (*McDonnell Douglas Corp.*)

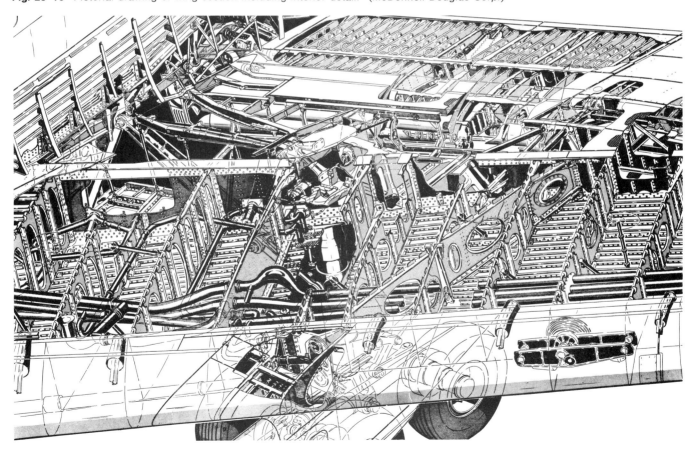

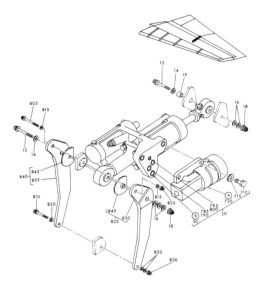

Fig. 25–17 Exploded pictorial drawing of aileron control. (*McDonnell Douglas Corp.*)

Fig. 25–18 Parts list and identification for the drawing in Fig. 25–17. (*McDonnell Douglas Corp.*)

FIG. ITEM	PART NUMBER	1 2 3 4 5 6 7	NOMENCLATURE
1 — 1	NRG6025		DAMPER INSTL–OUTBD AILERON SEE 57-00-05-01 FOR NHA CONFIG AND DET LOCATIONS
12	72214-6D-31		.BOLT-(V56878)
14	MS20002C6		.WASHER
16	MS20002-6		.WASHER
18	RME9868-6		.NUT-(V72962)
19	4931500-6-030		.BUSHING
20	ARG7231-5001		.DAMPER ASSY- SEE 11-65-01-01 FOR DET
30	ARG7231-5003		..CYLINDER ASSY
40	AN814-2DL		...PLUG
50	NAS1612-2		...PACKING
60	ALG7015-1		...NUT-RSVR RETAINING
70	ALG7008-1		...GLAND-RSVR HOUSING
80	ALG7026-1		...WASHER-RSVR
90	ALG7014-1		...SPRING-RSVR HELICAL CPRS
100	NAS1611-112		...PACKING
110	3891431-112		...RING
120	MS21250-04010		...BOLT
130	MS20002C4		...WASHER
140	ALG7010-1		...BUSHING-RSVR CLAMP UP
150	ALG7009-1		...SPRING-SEAL RSVR HELICAL CPRSN
160	ALG7011-1		...RETAINER-RSVR ADAPTOR
170	ALG7012-1		...ADAPTER-RSVR SEAL
180	2922858-19		...PACKING
190	ALG7013-1		...PISTON ASSY-RSVR
200	MS21209F4-20		....INSERT
210	ALG7013-3		....PISTON
220	WC1203-001		...SETSCREW-(V70318)
230	ALG7041-1		...SCREEN-THERMAL RELIEF AN CHECK VALVE
240	ALG7029-1		...RETAINER-THERMAL RELIEF AND CHECK VALVE
250	ALG7037-1		...SPRING-THERMAL RELIEF AN CHECK VALVE HELICAL CPRSN
260	ALG7028-1		...STEM-THERMAL RELIEF AND CHECK VALVE
270	NAS1611-012		...PACKING
280	3891430-012		...RING
285	NAS620A416L		...WASHER
290	ALG7031-501		...VALVE ASSY-THERMAL RELIEF AND CHECK
300	ALG7030-1		....PIN
310	ALG7042-501		....CAP
320	ALG7027-501		....POPPET
330	NAS6204H8		...BOLT
340	AN960-416		...WASHER
			(CONTINUED)

– ITEM NOT ILLUSTRATED

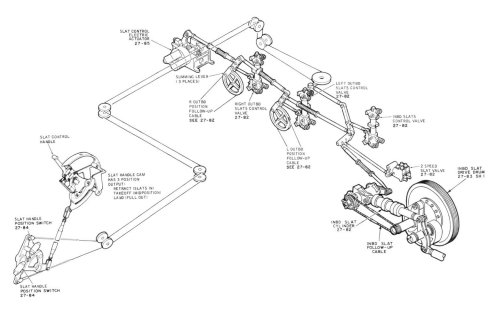

Fig. 25–19 Pictorial of a hydraulic system. (*McDonnell Douglas Corp.*)

Fig. 25–20 Schematic of the hydraulic system in Fig. 25–19. (*McDonnell Douglas Corp.*)

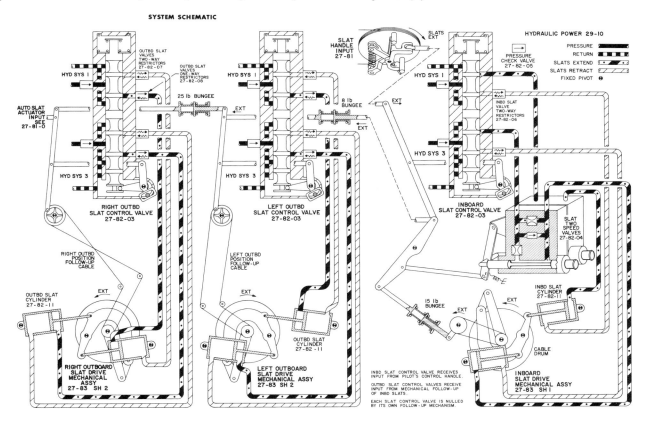

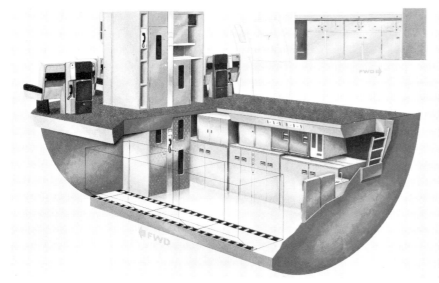

Fig. 25-21 Pictorial rendering of an aircraft interior. (*McDonnell Douglas Corp.*)

Fig. 25-23 Wing designs in flight. (*McDonnell Douglas Corp.*)

Fig. 25-25 Three views of a small aircraft. (*Beech Aircraft Corp.*)

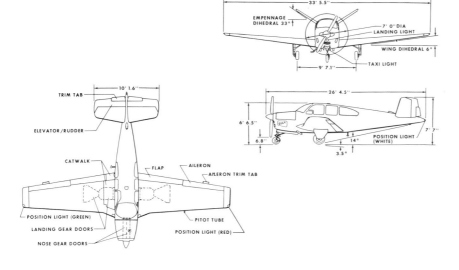

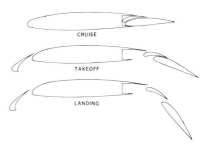

High Lift System

Fig. 25-22 Wing designs. (*McDonnell Douglas Corp.*)

Fig. 25-24 Small aircraft use advanced navigational systems. (*Beech Aircraft Corp.*)

Fig. 25-26 This small aircraft has an all-metal superstructure with fiberglass components. (*Piper Aircraft Corp.*)

Fig. 25-27 Citabria, an aerobatic aircraft. (*Bellanca Aircraft Corp.*)

system is complemented by communications and navigational equipment that allows it to cruise safely from 175 to 200 mph. Figure 25–25 illustrates the three views with pertinent dimensions. This plane has a dihedral stabilizer (V-tail) which has been unique to the design of this small aircraft.

The Piper Aircraft Corporation designed and developed the *Navajo* (Fig, 25–26) as an all-metal superstructure with fiberglass components used where the tough, resilient characteristics of such a material can be used to advantage, such as nose cone, wing, rudder and vertical fin tips, door frame, and windshield channel.

The wing structure of this plane consists of a stepped-down main spar (beam), a front and rear spar, lateral stringers, longitudinal ribs, and stressed skin sheets. The wings are joined together with heavy steel plates, producing in effect a continous main spar from wing tip to wing tip. In addition to the sturdy splice joint, the wing is attached to the fuselage at the front, center, and trailing edge.

Flush riveting is employed forward of the main spar for smoother aerodynamic flow over the wing. The wing root fillet and swept leading edge between fuselage and nacelle (streamlined engine enclosure) fillets smooth the flow of air.

A new adventure in the world of aviation is the rebirth of the biplane and aerobatic aircraft. Citabria is a popular sport model, as illustrated in Fig. 25–27. The tubular steel construction has a Dacron cover for rugged durability. The Bellanca Aircraft Company is only one of the many companies which have participated in designing old styles with new production methods.

New material for small aircraft. A new material has been used in the development of small leisure aircraft in recent years. Windecker Industries, Inc., has produced the plastic plane. The Eagle (Fig. 25–28) is constructed of plastic reinforced with fiberglass. The structure is aerodynamically free of riveted sections or seams formed by lapping skins. The

strength of the new material called Fibaloy has been tested, and the Eagle aircraft structures are gaining strength from a curing process within the plastic. Although the aircraft is made of plastic, it can be considered a chemical airplane because of the molecular action which causes the plastic to harden like steel.

Typical drawings. Some of the typical engineering drawings for a smaller aircraft are illustrated. The torque link from the Aero Commander Model 112 is typical of a dimensioned machine drawing (Fig. 25–29). The forging blank drawing is shown in Fig. 25–30. This drawing

Fig. 25–28 A chemical aircraft made from plastics. (*Windecker Research, Inc.*)

Fig. 25–29 A typical dimensioned drawing for machining. (*North American Rockwell Corp.*)

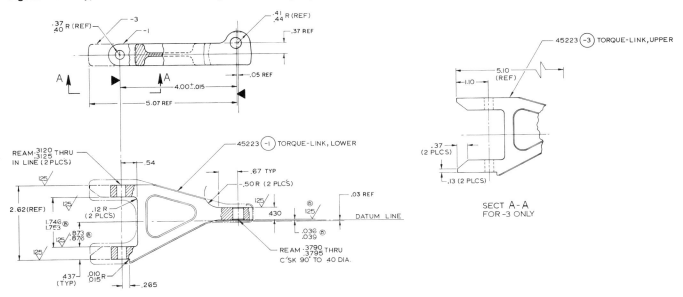

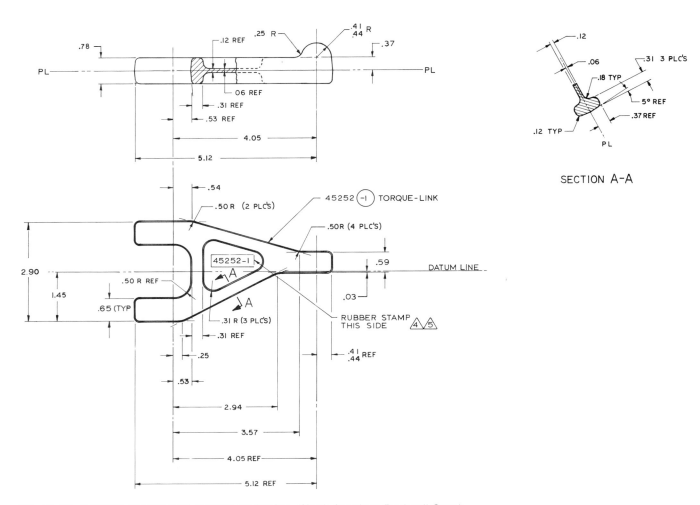

Fig. 25-30 A typical dimensioned drawing for forging. (*North American Rockwell Corp.*)

provides the information required for the forger and inspector of the blank. A forging machine drawing gives information for the machinist, set-up man, inspector, and others who may assemble bushings, bearings, and so forth.

Lofting layouts. Full-size drawings for large projects are made by lofting, a term that comes from the ship loft where the lines of exact shapes of ships are worked out. Lofting is important to aircraft-design layout. Contours of wing sections are accurately developed by lofting. The curves are faired (adjusted or smoothed out) to obtain smooth surfaces, and templates may be made when necessary. The drawing board is generally too small for such work, and so layouts are made on special loft floors where required areas are available. Ribs are detailed with trace chords and a loft line. The drawings of ribs are undimensional. This method is used for sheet aluminum and steel alloy parts wherein the drawing is transmitted (transferred) to the material by a photo process.

Smaller aircraft also include the light jet helicopter. The turbine-powered craft established records for speed, distance, climbing ability, and altitude. The three-view drawing illustrates the structural form, and the overall dimensions are in United States customary and metric measure (Fig. 25-31). The electronic components for navigation and communications (referred to as avionics) provide for the many versatile maneuvers often encountered with this uniquely styled aircraft.

Business Jets. The aircraft manufacturers have designed a few business jets which have unique qualities. The jet illustrated in Fig. 25-32 has a speed

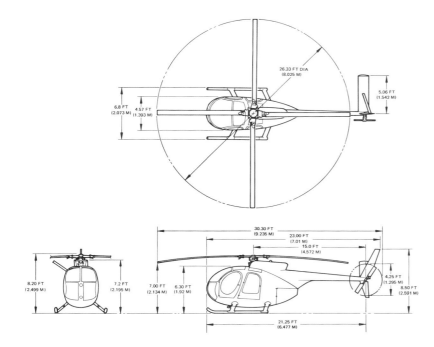

Fig. 25-31 A small jet helicopter. (*Hughes Tool Co.*)

Fig. 25-32 A typical business jet, plan view. (*Cessna Aircraft Co.*)

capability of 508 to 548 mph. The aerodynamic design features an eight-spar fail-safe wing which offers twice the needed strength. A spar is the principal structural member in an airplane wing which runs from tip to tip or from root to tip. The highly polished aluminum skin complements the sleekly designed lines of the high-performance jet.

The three-view drawing of the business jet in Fig. 25-33 illustrates another style for examination. The assembly-line photos show a primary (Figs. 25-34 and 25-35) and final assembly of the fuselage-mounted jet engines. The complex control panel in Fig. 25-36 is the final assembly checkpoint in the cockpit.

One of the design bases for the development of the jet in Fig. 25-32 was a Federal Aviation Regulation, Part 25. This aviation handbook describes testing requirements for airframes and performance expectations. The design team prepares drawings of advanced structures with manufacturing techniques which can produce a lightweight, strong airframe in compliance with the Federal regulation.

Skylab Program. The National Aeronautics and Space Administration (NASA) has prepared designs for Skylab Clusters (Fig. 25-37) which can orbit about the earth. The orbital workshop (Fig. 25-38) is the largest component of the Skylab Cluster. Solar array panels for electrical power extend from the workshop appearing like wings. Attached to the top of the workshop is the airlock module, which is the Skylab nerve center. Above the airlock is a multiple docking adaptor to which an Apollo command/service module is linked. There is an Apollo telescope mount pointed toward the sun, and this is attached to the docking adaptor.

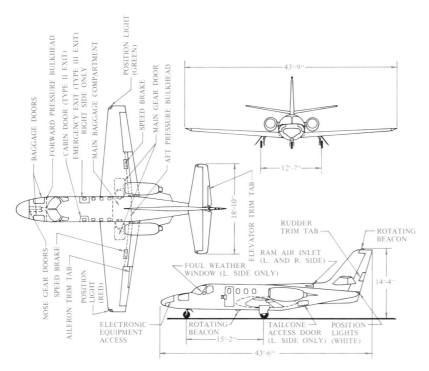

Fig. 25-33 A three-view drawing of a business jet. (*Cessna Aircraft Co.*)

Fig. 25-36 Final assembly checkpoint is the cockpit. (*Cessna Aircraft Co.*)

Fig. 25-34 Primary assembly area. (*Cessna Aircraft Co.*)

Fig. 25-35 Final assembly line. (*Cessna Aircraft Co.*)

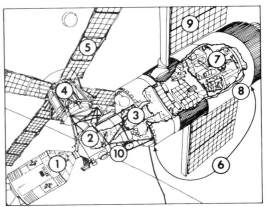

1. Apollo Command & Service Module
2. Multiple Docking Adapter
3. Airlock Module
4. Apollo Telescope Mount (ATM) Solar Experiments
5. ATM Solar Arrays
6. Skylab Saturn Workshop
7. Crew Quarters
8. Micrometeoroid Shield
9. Workshop Solar Arrays
10. Earth Resources Experiments

Fig. 25–37 Skylab clusters which can orbit about the earth. (*NASA*)

Fig. 25–38 Cutaway of the Skylab orbital workshop. (*McDonnell Douglas Corp.*)

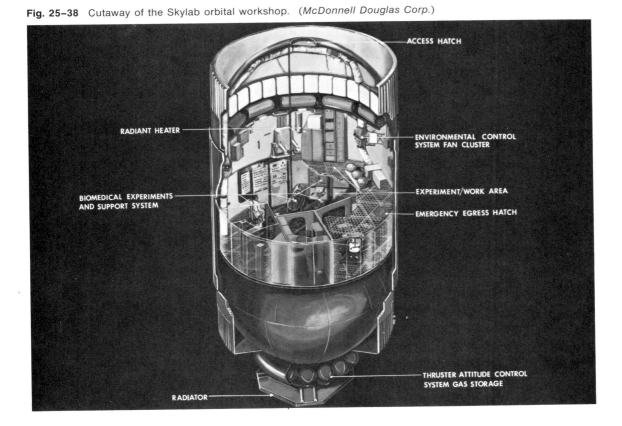

ACCESS HATCH

RADIANT HEATER

ENVIRONMENTAL CONTROL SYSTEM FAN CLUSTER

BIOMEDICAL EXPERIMENTS AND SUPPORT SYSTEM

EXPERIMENT/WORK AREA

EMERGENCY EGRESS HATCH

THRUSTER ATTITUDE CONTROL SYSTEM GAS STORAGE

RADIATOR

The crew area of the workshop is divided into two levels for the various activities. The lower level contains a workroom, sleep compartment, wardroom, and waste management compartment. The upper level is used for storage and experiments.

The designers and draftsmen on the engineering team have detailed many complex details for this Skylab Program. The program is designed to determine the effects of prolonged space flight on man and to provide workshop experiments for engineering and medical research.

Kinds of Aircraft Drawings. A large variety of drawings is prepared for a large variety of aircraft. From helicopters, leisure aircraft, military jets, jet transport, biplanes, to Skylabs, graphic communications are essential. There are drawings for castings, forgings, sheet-metal layout, schematics for controls, lofting for contours, and sketching for study alterations and new designs. Besides the major emphasis on a lightweight frame, sheet-metal forming is important for aerodynamic performance.

Sheet-metal drawings are based upon the principles of intersections and development, as discussed in Chapter 18. Sheet material is used for forming parts of aircraft as well as for the curved skin covering. Many factors enter when the designer and draftsman consider the selection of the proper material. Sketching is used in many ways in aircraft design. Sketched views of small parts, or the overall contour of the aircraft foils, may be examined for use and redesign. Figures 25–39 and 25–40 illustrate the use of sketching to explore interior designs.

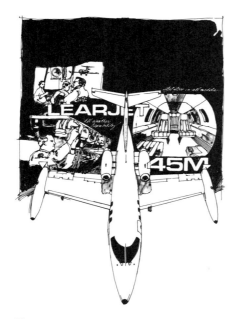

Fig. 25–39. Freehand sketch for design-team evaluation. (*Lear Jet Corp.*)

Fig. 25–40 Freehand sketch for design-team evaluation. (*Lear Jet Corp.*)

Review

1. Name some advantages of the undimensioned drafting practice.

2. Discuss some career opportunities in the aerospace industry.

3. List and discuss career opportunities of professions which are related to the aerospace industry, such as hydraulics engineer, electronics engineer, or interior decorator.

4. Read and discuss the Skylab Cluster concept.

5. List the chronological sequence of aerospace progress.

Problems for Chapter 25

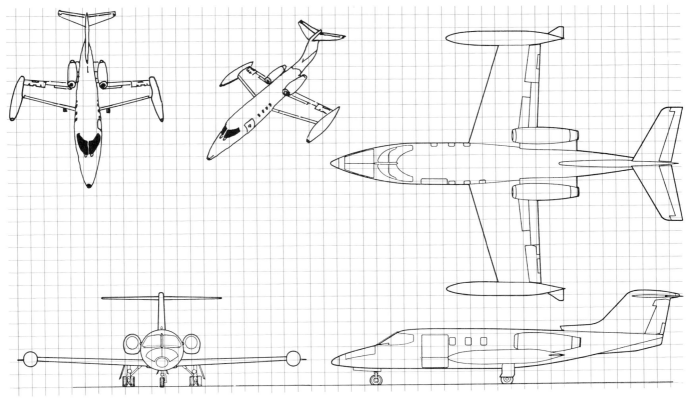

Fig. 25-41 Using grid paper, sketch any two views of the executive jet. Prepare one pictorial view of the jet.

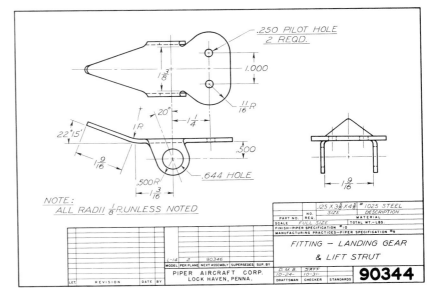

Fig. 25-42 Make a working drawing of the landing gear and lift strut fitting with the following changes. The V-tongue is $1\frac{1}{4}$ instead of $1\frac{9}{16}$. The $1\frac{3}{16}$ location dimension is to be $1\frac{3}{4}$. Note that the $\frac{1}{8}$ label applies to the V-tongue and other radii. List the sequence of operations necessary to make this fitting.

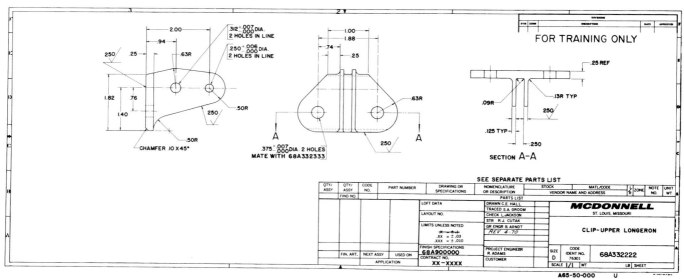

Fig. 25-43 Problem 1. Make a detail drawing of the clip, including the sectional view. Cross hatching may be omitted from the section. Title block and border are optional. Problem 2. Note the zone markings on the border of the illustration. Change order: change 2.00 dimension in zone F-3 to 2.50.

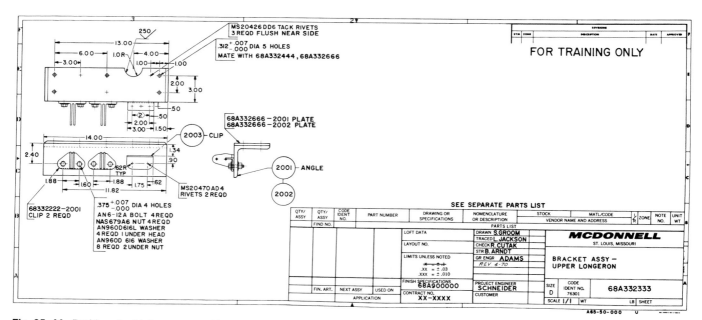

Fig. 25-44 Problem 1. Make an assembly drawing of the bracket assembly, including all notes as shown. Title block and border may be omitted. Examine the details of the clip in Fig. 25-43 before completing this drawing. Problem 2. Make up a parts list for the bracket assembly.

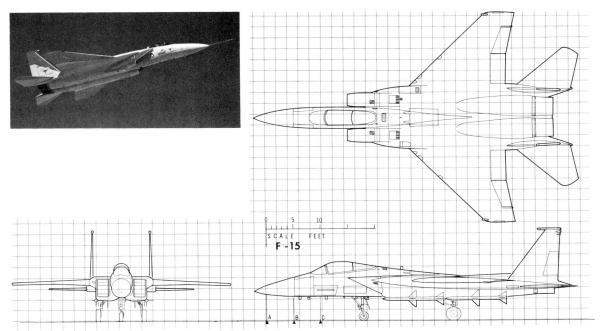

Fig. 25–45 Problem 1. Draw one of the three views of the F-15 aircraft on a grid. Establish the grid 60′ across the longest elevation and 20′ high from the grade line to over the top of the tail. Take dimensions from the scale shown using dividers. The drawing should be at least twice the size shown. Problem 2. Using dividers, locate the contours above the grade line at A, B, and C. Plot the fuselage contours using a grid system.

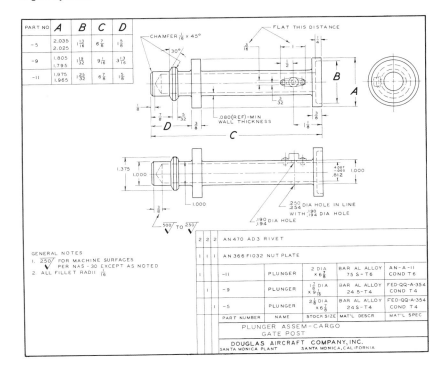

Figure 25–46 Make a working drawing of the Cargo Gate Post Plunger Assembly. Use dimensions for part 5, 9, or 11 as assigned by the instructor. Include the general notes and dimension according to the part no. assigned.

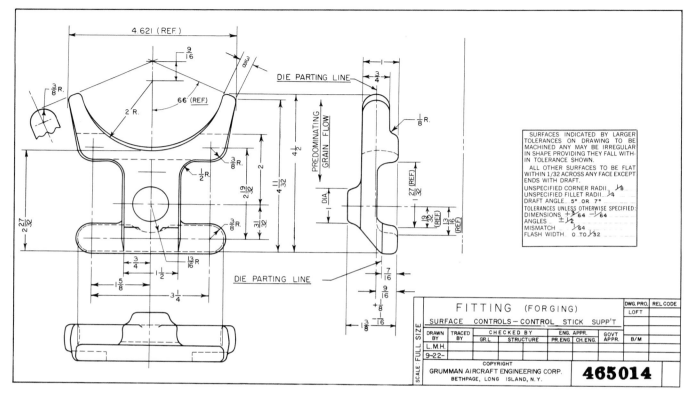

Fig. 25–47 Make a working drawing of the control-stick support fitting. Note that it is a forging drawing, and note the draft angle of forging.

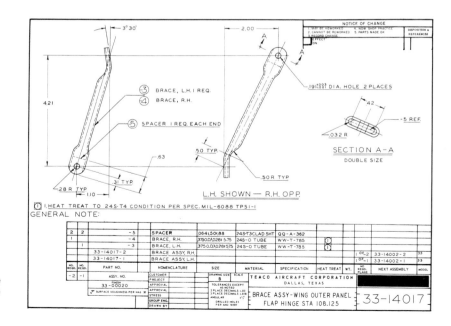

Fig. 25–48 Make a working drawing of the brace assembly. First examine section *A-A* and the size of the material listed in the parts list.

Technical Illustration

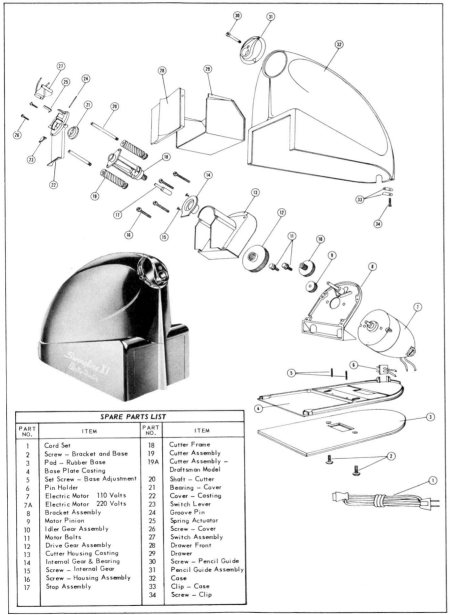

SPARE PARTS LIST			
PART NO.	ITEM	PART NO.	ITEM
1	Cord Set	18	Cutter Frame
2	Screw – Bracket and Base	19	Cutter Assembly
3	Pad – Rubber Base	19A	Cutter Assembly – Draftsman Model
4	Base Plate Casting		
5	Set Screw – Base Adjustment	20	Shaft – Cutter
6	Pin Holder	21	Bearing – Cover
7	Electric Motor 110 Volts	22	Cover – Casting
7A	Electric Motor 220 Volts	23	Switch Lever
8	Bracket Assembly	24	Groove Pin
9	Motor Pinion	25	Spring Actuator
10	Idler Gear Assembly	26	Screw – Cover
11	Motor Bolts	27	Switch Assembly
12	Drive Gear Assembly	28	Drawer Front
13	Cutter Housing Casting	29	Drawer
14	Internal Gear & Bearing	30	Screw – Pencil Guide
15	Screw – Internal Gear	31	Pencil Guide Assembly
16	Screw – Housing Assembly	32	Case
17	Stop Assembly	33	Clip – Case
		34	Screw – Clip

(SWINGLINE DIVISION OF SWINGLINE INC.)

Technical Illustration. Technical illustration has an important place in all phases of engineering and science. Technical illustrations form an essential part of the technical manuals for aircraft, machine tools, automobiles, tractors, air conditioners, and many other things. In technical illustration, pictorial drawings are used to describe parts and the methods for making them. Pictorial drawings show how the parts fit together and the steps that are followed to complete the product on the assembly line. Technical illustrations were probably used to organize and set up the assembly line itself. They are useful for many industrial, engineering, and scientific purposes.

Technical illustration drawings vary from simple sketches to rather extensive shaded drawings. They may be based upon any of the pictorial methods: isometric, perspective, oblique, and so forth. The complete project may be shown, or parts of groups of parts, and the views may be exterior, interior, sectional, cutaway, or phantom (Fig. 26–1). The purpose in all cases is to provide a clear and easily understood description. Previous chapters, in particular Chapter 2, "Sketching," and Chapter 12, "Pictorial Drawing," furnish the basis for making technical illustrations. Drawings for use within a company's plant can sometimes be made by a regular draftsman with artistic talent, but for most purposes the special requirements of such drawings call for work by a professional technical illustrator. Technical illustration has been used for many years for illustrated parts lists, operation and service manuals, process manuals, and similar purposes (Fig. 26-2). The aircraft industry in particular has found production illustration especially valuable. In aircraft construction, pictorial drawings are used when the plane is first designed, at many stages of its manufacture, and as it is completed on the assembly line. When the plane is delivered to the customer, the industry supplies illustrated service, repair, and operation manuals.

The technical illustrator and the commercial artist have several things in common. For example, both must have the ability to draw pictures. The technical illustrator must have a technical or engineering mind to go along with his drawing abilities as draftsman and artist.

Definition. *A technical illustration, in general, means a pictorial drawing made to provide technical information by visible methods.* This is usually accomplished by converting a multiview drawing to a three-dimensional pictorial drawing. It must show shapes and relative positions clearly and concisely. Sufficient shading to bring out the shape may be used, but none for artistic effect, since a technical illustration is for *use;* it is not a work of art. It must, however, be well done or it will be no better than illegible handwriting.

In addition to pictorials, technical illustrations include graphic charts, schematics, flow charts, diagrams, and sometimes circuit layouts. Dimensions are not a part of technical illustrations, as they are not working drawings used for manufacturing.

Tools and Tips. The regular drafting equipment described in Chapters 3

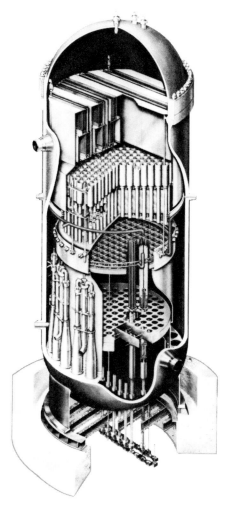

Fig. 26-1 This illustration of a nuclear reactor is a cutaway view showing the relationship of parts and interior details. (*General Electric Company.*)

and 13 provide most of the tools used by the technical illustrator. An H or 2H pencil, with the point kept well sharpened and clean, is the most useful tool. Keep all tools clean. A few other useful items include a crow quill pen, a felt-tip pen, technical pens, masking tape, Scotch tape, X-acto knife, paper stomps, two or three brushes, airbrush, and a reducing glass.

The technical illustrator should become familiar with the use of Craf-

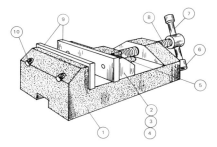

PART NO.	PART NAME	NO. REQD
1	BASE	1
2	MOVABLE JAW	1
3	MOVABLE JAW PLATE	1
4	MACHINE SCREW	1
5	LOCKING PIN	1
6	HANDLE STOP	2
7	HANDLE	1
8	CLAMP SCREW	1
9	JAW FACE	2
10	CAP SCREW	2

Fig. 26-2 An illustrated parts list.

tint, Zip-a-tone, Chart-Pak, and similar press-on section lining, screen tints, etc. He should also learn about the various methods of reproduction and the effect of reduction when his drawing is to be used in a smaller size. Lines must be firm and black. Erasures must be clean. The part of a drawing not being worked on should be kept covered with tracing paper or sheet plastic.

Lettering. Lettering is an important element of technical illustration. In some cases templates or scriber guides can be used, but good freehand lettering is often necessary; it is a required skill for a technical illustrator.

Pictorial Line Drawings. Since nearly all technical illustrations are basically pictorial line drawings, a complete understanding of the various types and their applications is necessary. Chapter 12 describes the various types of pictorial drawing and describes the procedure for drawing each.

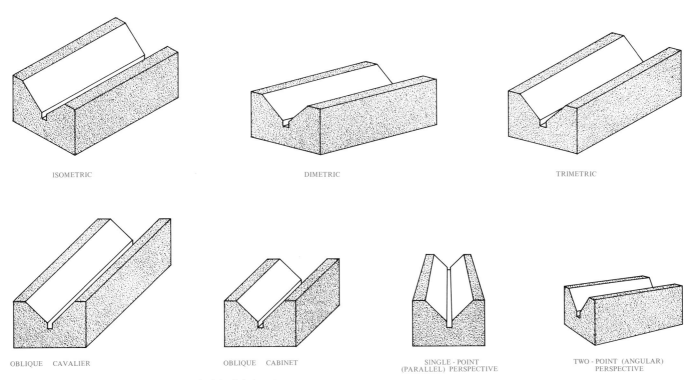

ISOMETRIC   DIMETRIC   TRIMETRIC

OBLIQUE   CAVALIER   OBLIQUE   CABINET   SINGLE - POINT (PARALLEL) PERSPECTIVE   TWO - POINT (ANGULAR) PERSPECTIVE

Fig. 26-3 V-block in various types of pictorial drawing.

While any type of pictorial drawing can be used as the basis for a technical illustration, some types are more suitable than others. This is especially true if the illustration is to be rendered (shaded). Figure 26-3 shows a V-block drawn in various types of pictorial drawing. Notice the difference in the appearance of each. Isometric is the least natural in appearance; perspective is the most natural. This might suggest, then, that all technical illustrations should be drawn in perspective. This is not necessarily true. While perspective is more natural in appearance, it takes more time and is more difficult to draw. Thus, it is more costly.

The shape of the object also helps to determine the type of pictorial drawing to use. Figure 26-4 shows a pipe bracket drawn in isometric and oblique. This shape of object is more

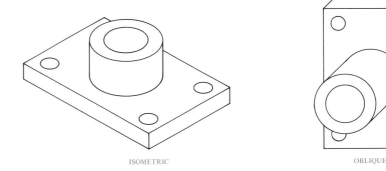

ISOMETRIC   OBLIQUE

Fig. 26-4 The shape of an object helps to determine the most suitable type of pictorial drawing to use.

easily and more quickly drawn in oblique and in many cases will look more natural than in isometric.

If an illustration is to be used only in-plant, the illustrator will usually make the pictorial drawing in isometric or oblique. These are quickest and the least costly to make. If the illus-

tration is to be used in a publication such as a journal, operator's manual, technical publication, etc., dimetric, trimetric, or perspective may be used.

Exploded Views. Perhaps the easiest way to understand an exploded view is to take a single piece and sepa-

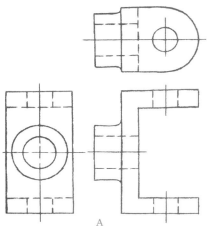
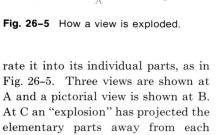
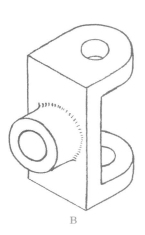
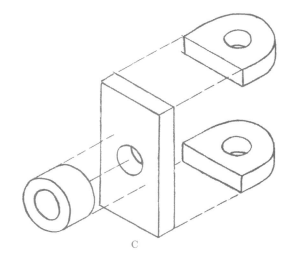

A   B   C

Fig. 26-5 How a view is exploded.

rate it into its individual parts, as in Fig. 26-5. Three views are shown at A and a pictorial view is shown at B. At C an "explosion" has projected the elementary parts away from each other. This illustrates the principle of exploded views.

All such views are based upon the same principle: projecting the parts from the positions they occupy when put together, or just pulling them apart. The exterior of a high-pressure piston pump is shown in Fig. 26-6. An exploded illustration of the pump is shown in Fig. 26-7. Note that all parts are easily identified.

Identification Illustrations. Pictorial drawings are very useful for identifying parts. They help save time when the parts are manufactured or assembled in place and are useful for illustrating operating instruction manuals and spare parts catalogs, and for many other purposes.

Identification illustrations usually take the form of exploded views. If parts are few, they can be identified by names and pointing arrows. The identification illustration in Fig. 26-8 is an example showing numbers for

the parts and a tabulation below with names and quantities.

Rendering. For certain purposes, or where shapes are difficult to read, surface shading or rendering of some kind may be desirable. For most industrial illustrations, accurate descriptions of shapes and positions are more important than fine artistic effects. Desired results can often be obtained without any shading. In general, surface shading should be limited to the least amount necessary to define the shapes illustrated.

Line drawings are used most for both pictorials and schematics. In addition, there are halftone renderings, photographs, and isometric, oblique, and perspective pictorials (Chapter 12). Different ways of rendering technical illustrations include the use of screen tints, pen and ink, wash, stipple, felt-tip pen and ink, smudge, edge emphasis, and other means, which can be observed in the technical illustrations of aircraft companies, automobile manufacturers, machine-tool makers, and even in the directions which come with your TV set.

Fig. 26-6 The exterior of a high-pressure pump. (*Industrial Division, Standard Precision, Inc.*)

Outline Shading. Outline shading may be done mechanically, freehand, or sometimes by a combination of both. The light is generally considered to come from in back of and above the left shoulder of the observer and across the diagonal of the object, as at A in Fig. 26-9. This is a convention used by draftsmen and renderers. At B the upper left and top edges would be in the light and drawn

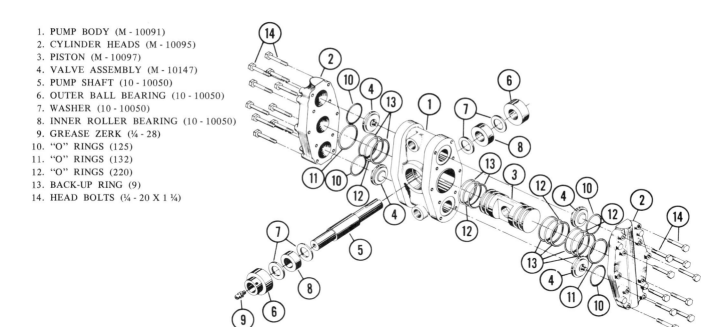

1. PUMP BODY (M - 10091)
2. CYLINDER HEADS (M - 10095)
3. PISTON (M - 10097)
4. VALVE ASSEMBLY (M - 10147)
5. PUMP SHAFT (10 - 10050)
6. OUTER BALL BEARING (10 - 10050)
7. WASHER (10 - 10050)
8. INNER ROLLER BEARING (10 - 10050)
9. GREASE ZERK (¼ - 28)
10. "O" RINGS (125)
11. "O" RINGS (132)
12. "O" RINGS (220)
13. BACK-UP RING (9)
14. HEAD BOLTS (¼ - 20 X 1 ¼)

Fig. 26–7 An exploded view which shows and identifies the parts of a high-pressure piston pump. It is a single-piston, double-ended displacement pump with pressure capabilities up to 1000 psi and a capacity of 2 gpm at 650 psi. (Psi = pounds per square inch; gpm = gallons per minute.) (*Industrial Division, Standard Precision, Inc.*)

Fig. 26–8 An identification illustration. (*The R. K. LeBlond Machine Tool Co.*)

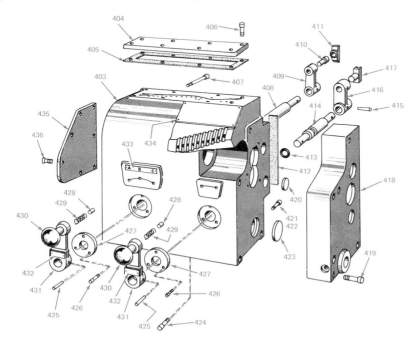

PART NO.	PART NAME	QTY.
403	QUICK CHANGE BOX	1
404	COVER, TOP	1
405	GASKET, COVER	1
406	SCREW, SOCKET HEAD CAP	8
407	SCREW	2
408	SHAFT, SHIFTER	1
409	LINK, SHIFTER	1
410	PIN	1
411	SHOE, SHIFTER	1
412	GASKET (MAKE IN PATTERN SHOP-BOX TO BED)	2
413	"O" RING	2
414	SHAFT, SHIFTER	1
415	PIN, TAPER	2
416	LINK, SHIFTER	1
417	SHOE, SHIFTER	1
418	COVER, SLIP GEAR	1
419	SCREW	4
420	PLUG	2
421	SCREW	3
422	SCREW	1
423	PLUG (NOT USED WITH SCREW REVERSE)	1
424	SCREW	3
425	PIN	2
426	SCREW	6
427	COLLAR	2
428	PLUNGER	2
429	SPRING	2
430	KNOB	2
431	LEVER	2
432	PLATE, FEED-THD.	1
433	PLATE, COMPOUND	1
434	PLATE, ENGLISH INDEX	1
435	COVER	1
436	SCREW	7

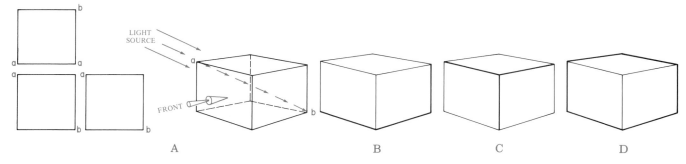

Fig. 26-9 Light source and line-shaded cubes.

with thin lines. The lower right and bottom edges would be shaded and drawn with thick lines. At C the edges meeting in the center are made with thick lines to accent the shape. At D the edges meeting at the center are made with thin lines, and thick lines are used on the other edges to bring out the shape.

An example of the use of a small amount of line shading is shown and described in Figs. 26–10 and 26–11.

Surface Shading. With the light rays coming in the usual conventional direction, as at A in Fig. 26–12, the top and front surfaces would be lighted and the right-hand surface would be shaded, as at B. The front surface can have light shading with heavy shading on the right-hand side, as at C, or solid black may be used on the right-hand side, as at D.

Some Shaded Surfaces. Some shaded surfaces are indicated in Fig. 26–13. An unshaded view is shown at A for comparison. Ruled-surface shading is shown at B, freehand shading at C, stippled shading at D, pressure-sensitive overlay shading at E and F.

Stippling, at D, consists of dots; short, crooked lines; or similar treatment to produce a shaded effect. It is a good method when it is well done, but it requires considerable time.

Fig. 26-10 A maintenance manual illustration. Notice that only the necessary detail is shown and that just enough shading is used to emphasize and give form to the parts. This apparently simple form of shading is effective for many purposes but must be handled carefully. (*Sample supplied courtesy of the Technical Illustrators Association.*)

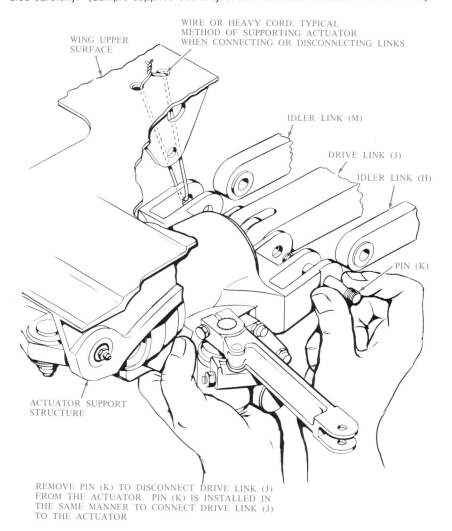

Fig. 26-11 Outline emphasis by a thick black or white line is an effective method of making a shape stand out. (*Rockford Clutch Division, Borg-Warner.*)

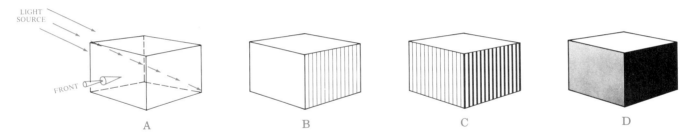

Fig. 26-12 Some methods of rendering the faces of a cube.

Pressure-sensitive overlays (used at E and F) are available in a great variety of patterns and are easily applied (see Chapter 16).

Airbrush Rendering. Airbrush rendering produces illustrations that resemble photographs (Fig. 26-14). The airbrush (Fig. 26-15) is a miniature spray gun used primarily to render illustrations and to retouch photographs. Compressed air is used to spray a solution (usually watercolor) to obtain various shading effects.

Types of airbrushes. Airbrushes may be classified according to the size of their spray pattern and by the type of spray-control mechanism. Care should be taken in the selection of an airbrush so that the size and style match the work to be performed.

The smallest spray pattern can be obtained from an oscillating needle airbrush. This type is capable of spraying very thin lines (hairlines) and small dots. It is the most expensive to buy and is used only by professionals who do highly detailed rendering and retouch work. A slightly larger airbrush, often called the pencil-type, is a general-purpose illustrator's airbrush. It can be adjusted to spray rather thin lines or opened up to spray larger surfaces and backgrounds. This airbrush is most popular for student use. The largest airbrush suitable for use in technical illustrating will spray a pattern large enough to do posters, displays, models, etc. It is often called a poster-type airbrush.

Airbrushes are also classified by the type of spray-control mechanism. That is, an airbrush spray control will be either single action or double action. Oscillating-needle and most pencil-type brushes have double-action mechanisms. Most poster-

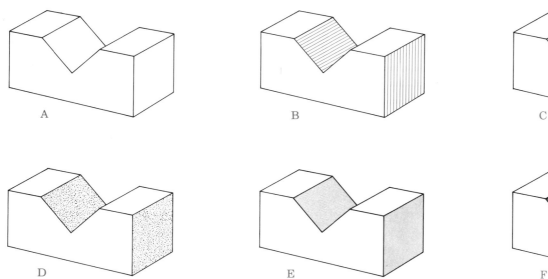

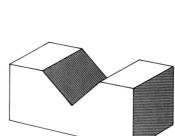

Fig. 26-13 Examples of various kinds of rendering.

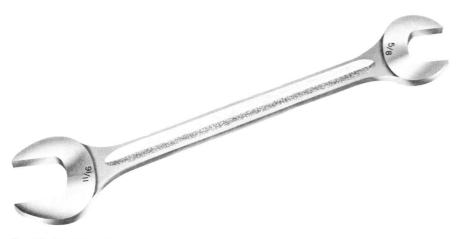

Fig. 26-14 Airbrush rendering.

Fig. 26-15 An airbrush.

type brushes have single-action mechanisms. Single action simply means that when the finger lever is pressed, both color and air are expelled at the same time. Double action means that two motions are necessary. When the lever is pressed, air is released; when the lever is then pulled back, color is released. Much greater control is possible with a double-action airbrush, and it is the kind usually used for rendering.

Air supply. A constant supply of clean air is needed to produce a high-quality spray pattern. An air supply may be obtained from a carbonic gas unit (CO_2) or an air compressor. Up to 32 pounds of pressure is required.

An air transformer (regulator and filter) must be installed between the air supply and the airbrush. The regulated pressure for most airbrush rendering is 32 pounds. Less pressure may be used for special effects.

Supplies and materials. A variety of supplies and materials is needed for airbrush work. Some are common art supplies; others are special and must be obtained through an art or engineering supply house. The following list is in addition to standard drafting supplies and equipment.

Airbrush and hose
Rubber cement
Rubber cement pickup

Razor knife (X-Acto or similar)

White illustration board (hot pressed)

Frisket paper

Watercolor brushes

Medicine dropper

Designer's watercolors (black and white)

Palette

Photo retouch set

Procedure for airbrushing. The following procedure is generally used for airbrushing. Special effects are obtained by experimenting with the equipment and materials.

A. Prepare a line drawing of the desired object and transfer it to the surface of the illustration board. Do not use standard typing carbon paper for transferring the image. Either purchase a special transfer sheet or make one by blackening one side of a sheet of tracing vellum with a soft lead pencil.

B. Cover the image area with frisket paper. This material is available in two forms: prepared and unprepared. Prepared frisket paper has one adhesive side protected with a piece of wax paper. Unprepared frisket paper must be coated with thinned rubber cement (50-50). Prepared frisket paper is more convenient to use and is recommended for the beginner. Cut and remove frisket paper from the area to be airbrushed first (Fig. 26–16). Use the rubber-cement pickup to remove any particles of rubber cement left on the surface. Cover all other areas of the illustration board not covered by the frisket paper.

C. Mix the black watercolor in the palette. Water is placed in the palette cup with the medicine dropper. Squeeze a small amount of black watercolor onto the edge of the palette. Use a watercolor brush to mix the color into the water.

D. Transfer the mixed color from the palette to the color cup on the airbrush using the watercolor brush. Fill the cup about half full.

E. Render the exposed surface as desired.

F. Open a second portion of the frisket and cover the rendered surface. Continue in this way until all surfaces are rendered. Remove frisket. Figure 26–17 shows the finished rendering. Figure 26–18 shows examples of other objects rendered. Highlights may be added using white watercolor if desired.

Photo retouching. Photo retouching is a process used to change details on a photograph. Details may be added, removed, or simply repaired. This process is often necessary in preparing photographs for use in publications or simply for changing the appearance of some detail.

Photo retouching is usually done on a glossy photograph using the same basic procedure outlined above for standard airbrush work. Care must be taken so as not to damage the finish on the photo when cutting frisket paper. The gray tones are obtained by using the tones of gray from

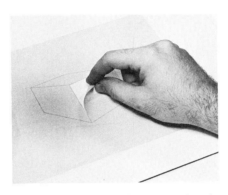

Fig. 26–16 Opening the area to be airbrushed.

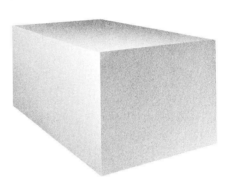

Fig. 26–17 Finished rendering of a cube.

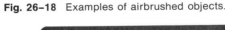

Fig. 26–18 Examples of airbrushed objects.

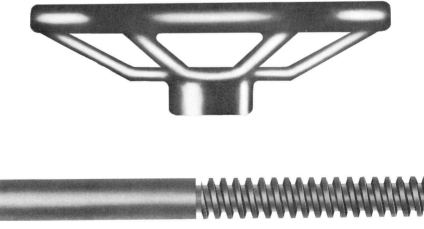

a photo retouching kit. A watercolor brush may also be used for touching up fine details. Figure 26–19 shows a before and after example of a retouched photograph.

Wash Rendering. Wash rendering (also called wash drawing) is a form of watercolor rendering. It is done with watercolor and watercolor brushes. It is commonly used for rendering architectural drawings (Fig. 26–20) and for newspaper advertising of furniture and similar products (Fig. 26–21). This technique is highly specialized and is usually done by a commercial artist. However, some technical illustrators and draftsmen are, at times, required to do this type of illustrating.

Scratchboard. Scratchboard drawing is a form of line rendering. Scratchboard is coated with India ink and a sharp instrument is used to make the lines. This is accomplished by drawing the image on the inked surface and scratching through the ink to expose the desired lines or surfaces (Fig. 26–22).

PHOTOGRAPH BEFORE RETOUCHING

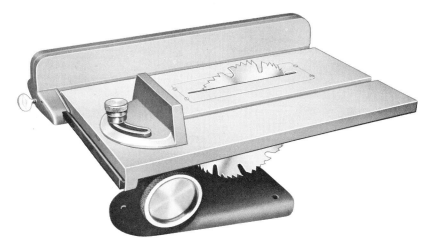

PHOTOGRAPH AFTER RETOUCHING

Fig. 26–19 Before and after retouching.

Fig. 26–20 Wash rendering of an architectural drawing.

Fig. 26–21 Wash rendering of a stereo cabinet.

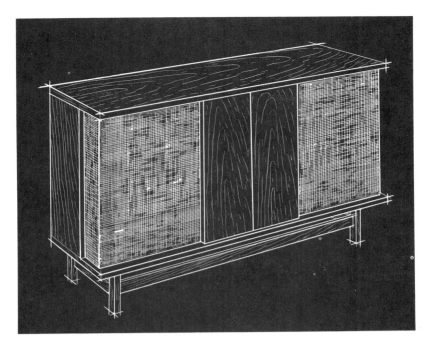

Fig. 26–22   Scratchboard rendering.

Review

1. Another name for surface shading is ——————.

2. Name four shading techniques.

3. Airbrush rendering requires air pressure up to ———————— pounds.

4. Changing details on a photograph is called ——————.

5. Name two major uses for wash rendering.

6. Define technical illustration.

7. Which type of pictorial drawing is most natural in appearance?

8. Which type is the least natural in appearance?

9. When an object is separated into its separate parts and the parts drawn in the correct relative positions, what kind of drawing results?

10. Exploded-view drawings with parts numbered or labeled are called ——————.

Problems for Chapter 26

Fig. 26–23 Pictorial sketching and drawing. Assignment 1. Scale: optional. Most technical illustrations are basically pictorial line drawings. In order to develop a good understanding of the relationship of the various types, make an isometric, oblique-cavalier, oblique-cabinet, single-point perspective, and two-point perspective sketch of the tool support in Fig. 12–61. Assignment 2. Scale: optional. Make instrument drawings of the same tool support (Fig. 12–61) in isometric, oblique-cavalier, oblique cabinet, single-point perspective, and two-point perspective. Compare the sketches with the instrument drawings. Are they similar? Which type of pictorial drawing gives the most natural appearance?

Fig. 26–24 Exploded-view drawings. Make an isometric exploded-view drawing of the letter holder in Fig. 5–64. Scale: optional. Draw your own initials as an overlay $\frac{1}{32}''$ thick. Estimate the height and width.

Fig. 26–25 Surface shading. Scale: optional. Make a pictorial line drawing of the toy boat in Fig. 12–57. Redesign as desired. Refer to Fig. 26–12 on page 521 and render the boat using the technique shown at C or D. Maintain sharp clean lines for contrast.

Fig. 26–26 Pictorial assembly drawing. Scale: optional. Make a pictorial assembly drawing of the level in Fig. 11–25. Number the parts and add a parts list for identification. Do not show sectional views. Redesign as desired. Add outline shading for accent if instructed to do so. Ink tracing is optional.

Fig. 26–27 Identification illustration. Scale: optional. Make an oblique-cavalier drawing of the mini saw horse in Fig. 5–67. Add parts numbers and a parts list to make an identification illustration. Render if required. Trace in ink if required.

Fig. 26–28 Airbrush rendering. Make two-point-perspective line drawings of several basic geometric shapes (solids). Examples: cube, cylinder, sphere, cone, etc. Scale: optional. Transfer each to a piece of hot-pressed illustration board and render them in watercolor, using an airbrush.

Fig. 26–29 Scratchboard rendering. Scale: optional. Make a two-point-perspective drawing of the knife rack in Fig. 5–65. Transfer the line drawing to a piece of scratchboard coated with India ink. Use any sharp instrument to scratch through the ink to expose the desired lines.

Fig. 26–30 Exploded-view drawing. Scale: optional. Make an isometric exploded-view drawing of the tic-tac-toe board in Fig. 5–70. Use outline shading to add contrast.

Fig. 26–31 Identification illustration. Make an isometric exploded-view drawing of the note box in Fig. 5–68. Scale: optional. Add parts numbers and a parts list to make an identification illustration. Your initial should be designed as an inlay attached to a circular disk. Redesign the note box as desired. Render if required.

Fig. 26–32 Wash rendering. Make a one-point-perspective or a two-point-perspective drawing of any piece of wood furniture. Scale: optional. Transfer the line drawing to a piece of cold-pressed illustration board and render it, using watercolor and a watercolor brush. Use a touch of white or black to sharpen the edges. Keep wood grain and other fine detail lines sharp and clean.

Fig. 26–33 Assignment 1: Identification illustration. Scale: optional. Make an isometric assembly drawing of the trammel in Fig. 11–20. Show the full length of the beam and show two complete assemblies of the point, body, and knurled screw on

the beam. Estimate any sizes not given. Redesign as desired. Add parts numbers and a parts list to make an identification illustration. Assignment 2: Render the trammel in pencil or transfer the line drawing to hot-pressed illustration board and render in watercolor with an airbrush. Add the parts numbers and parts list on an overlay sheet.

Fig. 26–34 Photo retouching. Obtain a glossy photograph of a simple machine or machine part. Study the individual features of the object and list all imperfections. Retouch the photograph using a hand watercolor brush or an airbrush, or both. If possible, have two copies of the original photograph so that a *before* and *after* comparison can be made.

Fig. 26–35 Airbrush rendering. Scale: optional. Make a two-point-perspective drawing of the hammer head in Fig. 5–71. Transfer the line drawing to a piece of hot-pressed illustration board and render it in watercolor, using an airbrush. Add a touch of white or black to edges for contrast.

Fig. 26–36 Surface shading. Scale: optional. Refer to the initial letter N in Fig. 12–61. Using the same basic proportions, make a single-point-perspective drawing of one of your initials and render it using the technique shown at C or D in Fig. 26–12 on page 521. This problem may also be used for practice in wash rendering or airbrush rendering. For wash rendering, the line drawing must be transferred to cold-pressed illustration board. For airbrush rendering, transfer the line drawing to hot-pressed illustration board.

Fig. 26–37 Exploded-view drawing. Scale: optional. Make an isometric exploded-view drawing of the garden bench in Fig. 5–69. Use $\frac{1}{2}''$ DIA threaded rods, flat washers, and hex nuts to fasten the parts together. If assigned, add parts numbers and a parts list to make an identification illustration. This problem may also be used for practice in wash rendering.

Fig. 26-38 Surface shading. Scale: optional. Make a two-point-perspective drawing of the V-Block in Fig. 12-61. Refer to Fig. 26-13 on page 522 and render the line drawing, using the technique assigned. Keep all edges crisp and sharp.

Fig. 26-39 Wash rendering. Make a two-point-perspective drawing of a house. Scale: $\frac{1}{4}'' = 1'\text{-}0''$ or $\frac{1}{8}'' = 1'\text{-}0''$. Transfer the line drawing to a piece of cold-pressed illustration board and render it, using watercolor and a watercolor brush.

Fig. 26-40 Scratchboard rendering. Scale: optional. Make a two-point-perspective drawing of any piece of wood furniture. Examples: coffee table, end table, desk, bench, etc. Transfer the line drawing to a piece of scratchboard coated with india ink. Use any sharp instrument to scratch through the ink to expose the desired lines. See Fig. 26-22.

Fig. 26-41 Airbrush rendering. Scale: optional. Make a single-point-perspective drawing of the V-Block in Fig. 12-57. Transfer it to a piece of hot-pressed illustration board and render it in watercolor, using an airbrush. This problem may also be used for practice in outline or surface shading.

Fig. 26-42 Wash rendering. Scale: $\frac{1}{4}'' = 1'\text{-}0''$ or $\frac{1}{8}'' = 1'\text{-}0''$. Make a line drawing of the front elevation of a house similar to the one shown in Fig. 1-14. Transfer the line drawing to a piece of cold-pressed illustration board and render it, using watercolor and a watercolor brush. This problem may also be used for practice in surface shading in pencil or charcoal.

Fig. 26-43 Surface shading. Scale: optional. Make line drawings of geometric solids assigned from Fig. 8-2 on page 156. Render each of the solids, using your choice of surface rendering techniques.

Fig. 26-44 Airbrush rendering. Scale: optional. Make a pictorial sketch or instrument drawing of the racer in Fig. 18-61. Redesign as desired. Transfer the line drawing to a piece of hot-pressed illustration board and render it in watercolor, using an airbrush. Add detail and decorate the body and fins as desired. This problem may also be used for practice in pencil rendering.

Fig. 26-45 Surface shading. Scale: optional. Make a pictorial line sketch or instrument drawing of the edge protector in Fig. 5-54. Add surface shading as assigned by your instructor. Keep all edges sharp and crisp. This problem may also be used for practice in airbrush rendering.

Fig. 26-46 Design and surface shading. Scale: optional. Design and make a pictorial sketch of a "car of the future." Render in pencil.

Fig. 26-47 Airbrush rendering. Scale: optional. Make a pictorial assembly drawing of the coupler in Fig. 11-23. Transfer the drawing to hot-pressed illustration board. Render in watercolor, using an airbrush. Add a touch of white or black to keep the edges crisp and sharp.

Appendixes to the Eighth Edition

Appendix A Lettering

The Use of Lettering. Simple freehand lettering, perfectly legible and quickly made, is an important part of the operation of business, industry, and engineering. Lettering is used on technical drawings to tell the kinds of materials, sizes, distances, and amounts; to identify units; and to give other necessary information.

Various Styles of Lettering. There are various styles of letters, a few of which are suggested in Fig. A–1. Each style is appropriate for a particular use, and care should be taken in its selection. For example, it is not practical to use fancy Roman-style lettering for notes and dimensions on technical drawings. It is time-consuming, more costly to do, and serves no better purpose than the single-stroke styles shown in Fig. A–1.

The standard lettering style most commonly used on working drawings is *single-stroke commercial Gothic* (Fig. A–2). This style is most appropriate because it is easy to read and easy to hand-letter. It consists of uppercase (capital) letters, lowercase (small) letters, and numerals. Nearly all companies now use only uppercase lettering. As a result, emphasis will be placed on uppercase. In addition, letters and numerals may be vertical or inclined (Fig. A–2). The type selected should be used consistently throughout a set of drawings.

Guidelines. In order to make letters the same height and in line, it is necessary to rule (draw) guidelines for the top and the bottom of each line of letters. Guidelines are drawn lightly with a sharp pencil. The clear distance (open space) between lines of letters varies from $\frac{1}{2}$ to $1\frac{1}{2}$ times the height of the letters. A lettering triangle or an Ames lettering instrument may be used for spacing guidelines accurately (Fig. A–3). Guidelines are spaced $\frac{1}{8}$ in. apart for regular letters and numerals on a drawing. Guidelines for fractions are usually twice

Fig. A–1 Various styles of lettering.

VERTICAL SINGLE-STROKE COMMERCIAL GOTHIC

ABCDEFGHIJKLMNOPQRSTUVWXYZ CAPITALS (UPPER CASE)

abcdefghijklmnopqrstuvwxyz LOWER CASE

1234567890 NUMERALS

INCLINED SINGLE-STROKE COMMERCIAL GOTHIC

ABCDEFGHIJKLMNOPQRSTUVWXYZ CAPITALS (UPPER CASE)

abcdefghijklmnopqrstuvwxyz LOWER CASE

1234567890 NUMERALS

Fig. A-2 Single-stroke commercial Gothic letters, vertical and inclined.

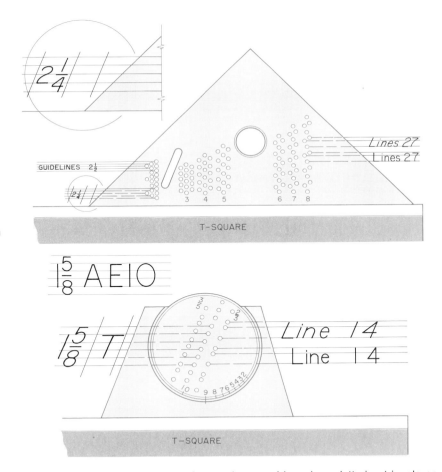

Fig. A-3 Lettering guidelines are to be evenly spaced by using a lettering triangle or an Ames lettering instrument.

the height of whole numbers, as shown in Fig. A-4.

Single-Stroke Vertical Capital Letters and Numerals. The shapes and proportions of letters and numerals are shown in Fig. A-5. These should be studied carefully for a complete understanding of how each is constructed. Each character is shown in a square 6 units high. The squares are divided into unit squares so that the shapes, proportions, and strokes may easily be learned.

Some variations may be used in developing a more individualized lettering technique. Figure A-6 shows some of these variations.

Single-Stroke Inclined Capital Letters and Numerals. Inclined letters and numerals should be slanted at a slope of about 2 in 5. This is $67\frac{1}{2}°$ with the horizontal. Figure A-7 shows various methods for laying out inclined direction lines which may be helpful while developing skill in hand lettering. If used, they should be extremely thin and light so as not to detract from the letters themselves.

Figure A-8 shows single-stroke inclined letters and numerals on an inclined grid. Notice that the only difference between vertical and inclined letters is in the slant of the inclined style.

Fractions. Fractions are always made with a horizontal division line and are usually twice the height of whole numbers. Fraction numbers are about three-fourths the height of the whole number. *They must never touch the division line* (Fig. A-4).

Composition. Composition in lettering means the arrangement and spacing of words and lines with letters of appropriate style and size. Letters in words are not placed at equal dis-

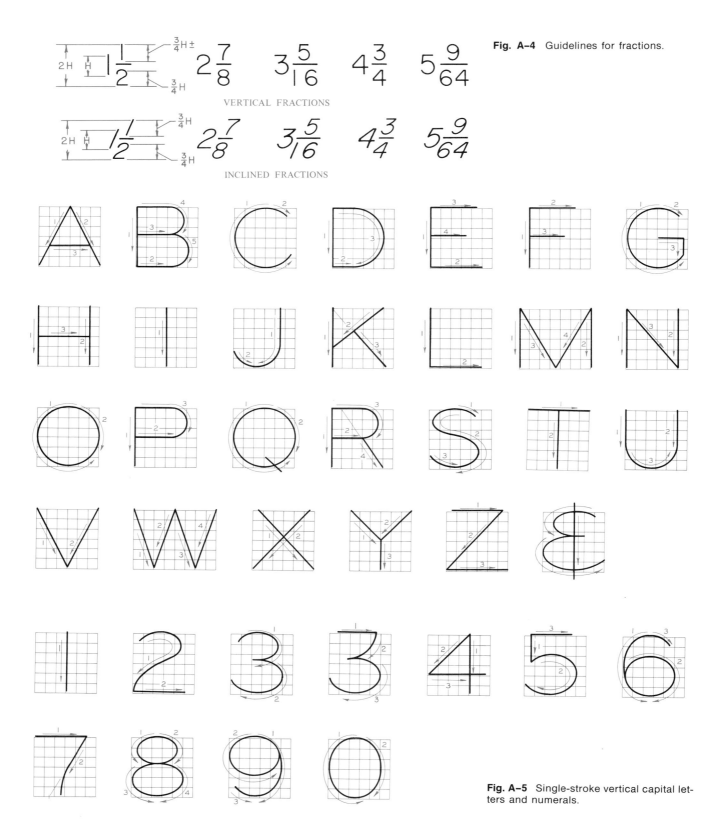

VERTICAL FRACTIONS

INCLINED FRACTIONS

Fig. A–5 Single-stroke vertical capital letters and numerals.

ABCDEFGHIJKLMNOPQRSTUVWXYZ 1234567890
ABCDEFGHIJKLMNOPQRSTUVWXYZ 1234567890
ABCDEFGHIJKLMNOPQRSTUVWXYZ
1234567890

Fig. A-6 Variations in lettering style.

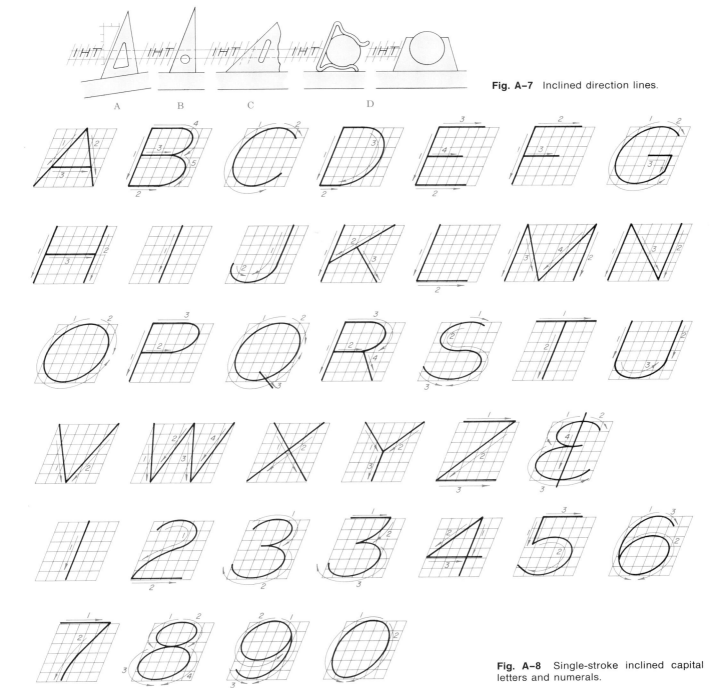

Fig. A-7 Inclined direction lines.

Fig. A-8 Single-stroke inclined capital letters and numerals.

LE TT ERIN G CO MPOSITI O N
INVOLVE S THE   SPA CIN G OF   LETTERS,
W ORD S,   A ND LIN ES AND THE CH OI CE
OF A PPROPR I A TE STYL ES AND SIZ E S.

INCORRECT LETTER, WORD, AND LINE SPACING

LETTERING COMPOSITION
INVOLVES THE SPACING OF LETTERS,
WORDS, AND LINES AND THE CHOICE
OF APPROPRIATE STYLES AND SIZES.

CORRECT LETTER, WORD, AND LINE SPACING

Fig. A–9 Letter and word spacing.

tances from each other but are spaced so that the areas of spaces included between the letters appear to be about equal. The distance between words (called word spacing) should be about equal to the height of the letters. Figure A–9 shows examples of proper and improper letter and word spacing.

Display Lettering. Display lettering is often used for title sheets, display posters, and other applications where large easy-to-read lettering is desired. While any style may be used for display, the draftsman most often uses Gothic, since it is the most legible of all styles. Figure A–10 shows a block-style and a regular-style Gothic.

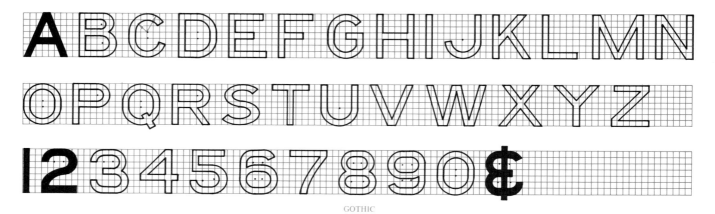

GOTHIC

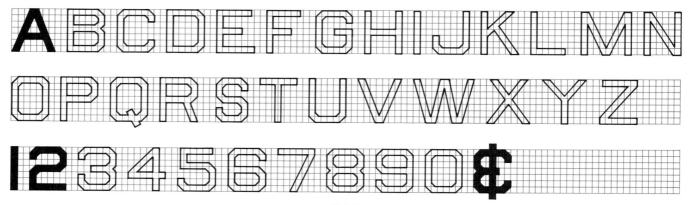

BLOCK

Fig. A–10 Block-style and regular-style Gothic capital letters.

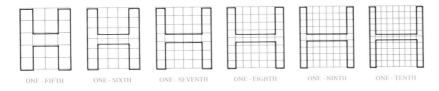

ONE - FIFTH ONE - SIXTH ONE - SEVENTH ONE - EIGHTH ONE - NINTH ONE - TENTH

Fig. A–11 The width of strokes may vary.

The block style is the easiest to make. It has no curved strokes and requires only the use of a T-square and triangle. A light-line grid may be drawn on the sheet and later erased, or a grid sheet may be placed under the drawing sheet if a tracing medium is being used. The individual letters are blocked in in pencil and may then be traced in ink if desired. Letters may simply be outlined or may be filled in, as shown in Fig. A–10. The thickness of the strokes may vary from one-fifth to one-tenth the height of the letter. In Fig. A–10, one-seventh is used. In Fig. A–11, one-fifth through one-tenth are illustrated.

Regular Gothic capital letters are made in much the same way. The only difference is that curved strokes must be made. This may be done using circle templates, compass, or freehand.

The size and proportion of letters may be changed by using the grid method shown in Fig. A–12. Notice that letters can be made larger or smaller, and they can be extended or condensed by using this method.

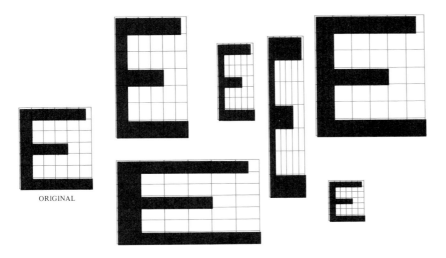

ORIGINAL

Fig. A–12 Grid method for changing the size and proportion of letters.

Appendix B The Metric System

Measurement. For as long as man has been civilized, he has had the need to measure things. One man measuring alone could use anything convenient for a measuring tool. The most convenient things for measuring small distances were his fingers, hands, and feet.

As civilizations grew, a system of standard measurements that everyone could use became desirable. The standards of ancient times were not very precise, however. It has only been in the last few centuries that nations have attempted to establish accurate standard measurements. As the accuracy of man's measuring ability increased, his standards have also become more precise.

U.S. Customary Measurements. The measuring standards used in the United States today are the result of a long history of man's attempt to standardize measurement. The Romans divided a foot into 12 parts, and the name given to that division has become the inch. The Romans also measured 1000 paces, which has become the mile. After the fall of the Roman Empire the use of standards diminished. English standard measurements that emerged slowly in the centuries following the Roman Empire were brought to America by the Colonists. Eventually the United States established its own measuring standards, some of which were interchangeable with the British standards and some of which were not.

Development of the Metric System. The metric system of measurement was developed in France in the late 1700s. Most countries of the world adopted the system by the end of the nineteenth century. By 1960, nearly all countries had made complete use of metrics. In 1965, Britain began a 10-year conversion, and the United States (the last holdout) is now progressing in that direction.

The metric system is a very practical system uniformly divided into units of ten. It was to be a *natural measurement* based on a portion of the earth's surface. The meter was thus originally defined as one ten-millionth of the distance from the North Pole to the equator. Measures of volume and weight were related to linear distance. Such a relationship was never made before. Today the meter, which is the basis of metric measurement, is more precisely defined as a wavelength of the red-orange light of krypton 86.

Because of international trade, worldwide standards are necessary. The problem of repair or replacement parts becomes difficult unless one standard is used from country to country.

In 1960, the International Organization of Weights and Measures, representing countries from all over the world, gave the metric system its formal title of Système International d'Unités, which is abbreviated SI. The basic SI units are shown in Fig. B-1.

Unit	Name	Symbol
Length	Meter	m
Mass	Kilogram	kg
Time	Second	s
Electric current	Ampere	A
Temperature	Kelvin	K
Luminous intensity	Candela	cd

Fig. B-1 The basic SI units.

Using the Metric System. The most common measurements of the metric system are the *meter,* for length; the *gram,* for mass or weight; and the *liter,* for volume.

The division and the multiples of these units are based on the number 10. This is similar to our money system and our number system, which are also based on 10. It is easier to figure with units to the base 10 than to have to divide by 12 (inches) or by 16 (ounces).

Special names are given to certain multiples and divisions of the basic unit. These names become a prefix to the name of the unit. The prefixes apply to any of the unit names— meter, gram, or liter. The prefix names and their corresponding amounts are shown in Fig. B-2.

Prefix	Amount	Fraction	Decimal
Milli	One-thousandth	$\frac{1}{1000}$	0.001
Centi	One-hundredth	$\frac{1}{100}$	0.01
Deci	One-tenth	$\frac{1}{10}$	0.1
Deka	Ten	10	10.0
Hecto	Hundred	100	100.0
Kilo	Thousand	1000	1000.0

Fig. B-2 Metric prefixes and their corresponding amounts.

Adding the prefixes to the word *meter* gives the following:

millimeter (mm) = one-thousandth of a meter

centimeter (cm) = one-hundredth of a meter

decimeter (dm) = one-tenth of a meter

dekameter (dam) = ten meters

hectometer (hm) = one hundred meters

kilometer (km) = one thousand meters

Adding the prefixes to the word *gram* gives the following:

milligram (mg) = one-thousandth of a gram

centigram (cg) = one-hundredth of a gram

decigram (dg) = one-tenth of a gram

dekagram (dag) = ten grams

hectogram (hg) = one hundred grams

kilogram (kg) = one thousand grams

Adding the prefixes to the word *liter* gives the following:

milliliter (ml) = one-thousandth of a liter

centiliter (cl) = one-hundredth of a liter

deciliter (dl) = one-tenth of a liter

dekaliter (dal) = ten liters

hectoliter (hl) = one hundred liters

kiloliter (kl) = one thousand liters

As a basis for comparison, approximate United States Customary and metric equivalents to remember are: one liter is a little larger than a quart; one kilogram is a little greater than two pounds; one meter is a little longer than a yard; and one kilometer is about five-eighths of a mile. Figure B-3 is a list of metric system equivalents.

Figure B-4 shows the actual size as well as the relationship between metric measures of length, volume, and mass.

The Metric System in Drafting. In drafting, the primary unit of measure is length. On drawings, the millimeter is used for dimensions. This is true whether the drawing is prepared with all metric dimensions or whether it is dual dimensioned, with sizes expressed in both metric and United States Customary units. In cases where drawings are not dual dimensioned, a metric equivalent chart (Fig. B-5) may be used to make the conversions. However, not all dimensions will be found on the chart, and it is, therefore, useful to know how to con-

Length

Centimeter	= 0.3937 inch
Meter	= 3.28 feet
Meter	= 1.094 yards
Kilometer	= 0.621 statute mile
Kilometer	= 0.5400 nautical mile
Inch	= 2.54 centimeters
Foot	= 0.3048 meter
Yard	= 0.9144 meter
Statute mile	= 1.61 kilometers
Nautical mile	= 1.852 kilometers

Area

Square centimeter	= 0.155 square inch
Square meter	= 10.76 square feet
Square meter	= 1.196 square yards
Hectare	= 2.47 acres
Square kilometer	= 0.386 square miles
Square inch	= 6.45 square centimeters
Square foot	= 0.0929 square meter
Square yard	= 0.836 square meter
Acre	= 0.405 hectare
Square mile	= 2.59 square kilometers

Volume

Cubic centimeter	= 0.0610 cubic inch
Cubic meter	= 35.3 cubic feet
Cubic meter	= 1.308 cubic yards
Cubic inch	= 16.39 cubic centimeters
Cubic foot	= 0.0283 cubic meter
Cubic yard	= 0.765 cubic meter

Capacity

Milliliter	= 0.0338 U.S. fluid ounce
Liter	= 1.057 U.S. liquid quarts
Liter	= 0.908 U.S. dry quart
U.S. fluid ounce	= 29.57 milliliters
U.S. liquid quart	= 0.946 liter
U.S. dry quart	= 1.101 liters

Mass or Weight

Gram	= 15.43 grains
Gram	= 0.0353 avoirdupois ounce
Kilogram	= 2.205 avoirdupois pounds
Metric ton	= 1.102 short, or net, tons
Grain	= 0.0648 gram
Avoirdupois ounce	= 28.35 grams
Avoirdupois pound	= 0.4536 kilogram
Short, or net, ton	= 0.907 metric ton

Fig. B-3 Metric System Equivalents.

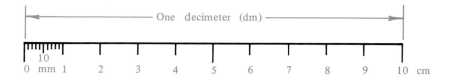

The meter (m), unit of length, is divided into 100 cm

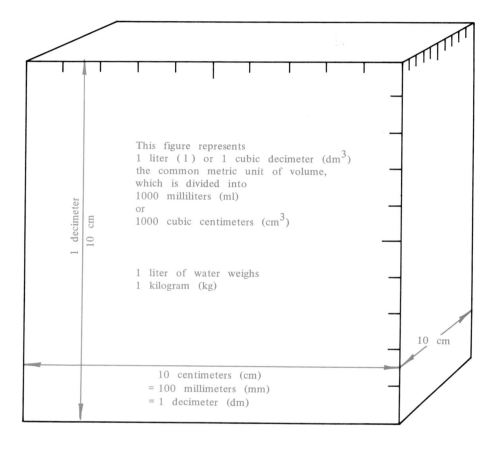

This figure represents
1 liter (1) or 1 cubic decimeter (dm³)
the common metric unit of volume,
which is divided into
1000 milliliters (ml)
or
1000 cubic centimeters (cm³)

1 liter of water weighs
1 kilogram (kg)

10 centimeters (cm)
= 100 millimeters (mm)
= 1 decimeter (dm)

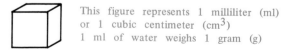

This figure represents 1 milliliter (ml)
or 1 cubic centimeter (cm³)
1 ml of water weighs 1 gram (g)

Fig. B-4 Relationship of metric measures of length, volume, and mass. (*Metric Association.*)

vert mathematically from one system to the other.

Inches to millimeters. One inch is equal to approximately 25.400 millimeters. Multiplying an inch dimension by 25.4 will give an answer in millimeters. To simplify the calculation, the inch dimension should be in decimal form.

Example: How many millimeters equal $2\frac{5}{8}$ inches? Convert $2\frac{5}{8}$ to a decimal, 2.625, then multiply by 25.4:

$$
\begin{array}{r}
2.625 \\
25.4 \\
\hline
10500 \\
13125 \\
5250 \\
\hline
666750
\end{array}
$$

Step off four decimal places to get the result of 66.6750 mm.

Millimeters to inches. One millimeter is equal to approximately 0.0394 inch. Multiplying a millimeter dimension by 0.0394 will give the answer in inches.

Example: How many inches equal 66.675 mm? Multiply 66.675 by 0.0394:

$$
\begin{array}{r}
66.675 \\
.0394 \\
\hline
266700 \\
599075 \\
200025 \\
\hline
26259950
\end{array}
$$

Step off seven decimal places to get the result of 2.6259950, or 2.625 ($2\frac{5}{8}$).

Rounding Off. It is sometimes necessary to round numbers to fewer digits. For example, 25.63220 mm rounded to three decimal places would be 25.632 mm. The same number rounded to two places would be 25.63 mm. The procedure is as follows:

Fig. B-5 Inch-millimeter equivalent chart.

Mm	In.*	Mm	In.	In.	Mm†	In.	Mm
1 =	0.0394	17 =	0.6693	$\frac{1}{32}$ (0.03125) =	0.794	$\frac{17}{32}$ (0.53125) =	13.493
2 =	0.0787	18 =	0.7087	$\frac{1}{16}$ (0.0625) =	1.587	$\frac{9}{16}$ (0.5625) =	14.287
3 =	0.1181	19 =	0.7480	$\frac{3}{32}$ (0.09375) =	2.381	$\frac{19}{32}$ (0.59375) =	15.081
4 =	0.1575	20 =	0.7874	$\frac{1}{8}$ (0.1250) =	3.175	$\frac{5}{8}$ (0.6250) =	15.875
5 =	0.1969	21 =	0.8268	$\frac{5}{32}$ (0.15625) =	3.968	$\frac{21}{32}$ (0.65625) =	16.668
6 =	0.2362	22 =	0.8662	$\frac{3}{16}$ (0.1875) =	4.762	$\frac{11}{16}$ (0.6875) =	17.462
7 =	0.2756	23 =	0.9055	$\frac{7}{32}$ (0.21875) =	5.556	$\frac{23}{32}$ (0.71875) =	18.256
8 =	0.3150	24 =	0.9449	$\frac{1}{4}$ (0.2500) =	6.349	$\frac{3}{4}$ (0.7500) =	19.050
9 =	0.3543	25 =	0.9843	$\frac{9}{32}$ (0.28125) =	7.144	$\frac{25}{32}$ (0.78125) =	19.843
10 =	0.3937	26 =	1.0236	$\frac{5}{16}$ (0.3125) =	7.937	$\frac{13}{16}$ (0.8125) =	20.637
11 =	0.4331	27 =	1.0630	$\frac{11}{32}$ (0.34375) =	8.731	$\frac{27}{32}$ (0.84375) =	21.431
12 =	0.4724	28 =	1.1024	$\frac{3}{8}$ (0.3750) =	9.525	$\frac{7}{8}$ (0.8750) =	22.225
13 =	0.5118	29 =	1.1418	$\frac{13}{32}$ (0.40625) =	10.319	$\frac{29}{32}$ (0.90625) =	23.018
14 =	0.5512	30 =	1.1811	$\frac{7}{16}$ (0.4375) =	11.112	$\frac{15}{16}$ (0.9375) =	23.812
15 =	0.5906	31 =	1.2205	$\frac{15}{32}$ (0.46875) =	11.906	$\frac{31}{32}$ (0.96875) =	24.606
16 =	0.6299	32 =	1.2599	$\frac{1}{2}$ (0.5000) =	12.699	1 (1.0000) =	25.400

* Calculated to *nearest* fourth decimal place.   † Calculated to *nearest* third decimal place.

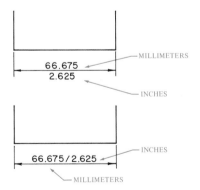

Fig. B-6 Dual dimensioning.

A. When the first number (digit) dropped is less than 5, the last number kept does not change. For example, 15.232 rounded to two decimal places would be 15.23.

B. When the first number dropped is greater than 5, the last number kept should be increased by one. For example, 6.436 rounded to two decimal places would be 6.44.

C. When the first number dropped is 5 followed by at least one number greater than 0, the last number kept should be increased by one. For example, 8.4253 rounded to two decimal places would be 8.43.

D. When the first number dropped is 5 followed by zeros, the last number kept is increased by one if it is an odd number. If it is an even number, no change is made. For example, 3.23500 rounded to two places would be 3.24. However, 3.22500 would be 3.22.

Dual Dimensions. It is standard practice as recommended by ANSI to place millimeters above a solid dimension line and inches below the line (Fig. B-6). When the dimensions are to be placed on one line, the millimeter dimension is first and the inch dimension is second. Examples of metric and dual-dimensioned drawings may be found in Chapter 6, Dimensioning.

Angular Dimensions. Angular dimensions do not need to be converted. Angles dimensioned in degrees, minutes, and seconds are common to both systems.

Appendix C Contemporary French Design House Specifications

In addition to a complete set of house plans developed by an architect, there should also be a set of written specifications. The specifications take care of those details of building a house that cannot be a part of the plans and schedules. Details such as the kinds of materials, kinds of fixtures, quality of workmanship, and the responsibilities of the owner and the contractor should be specified before construction begins. There are abbreviations and trade names throughout. These would not normally appear in a textbook, but are left intact for a sense of realism. The following specifications apply to the house plans in Figs. 20–66 through 20–75.

General Conditions

These general conditions are a part of the specifications and apply to the work of the prime contractor and all subcontractors.

All materials and fixtures throughout shall be of first quality.

Each contractor shall carry ample liability insurance and insurance to protect the owner from any and all claims under the Workman's Compensation Act. The owner shall assume responsibility and carry ample insurance during the progress of the work to protect himself and the contractor from loss through damage by fire, tornado, etc., or loss by theft, vandalism, etc., from time of delivery.

The contractor shall properly barricade his work at all times and erect proper shelters wherever necessary to protect the work from damage.

All work shall be executed in strict accordance with the building ordinance requirements and to the satisfaction of the building commissioner.

Each and every trade shall be responsible for the removal of all rubbish resulting from its operation.

It shall be the duty and responsibility of each contractor and subcontractor to arrange with every other trade so that his own work may proceed without interruption and in conjunction with and in harmony with every other trade. The owner shall assume no responsibility of any sort for the cost of changes or extra work on the account of failure of trades to cooperate during construction.

Figured dimensions take precedence over scaled dimensions.

Where finished surfaces of materials are available in different colors or finishes, the contractor shall submit samples to the owner for approval. The finished work shall match approved samples to a reasonable degree.

The general contractor shall guarantee all work under this contract. Guarantee shall be for a period of one year from date of occupancy, or final payment of contract, whichever shall occur first. Should any defect develop during said period, due to improper workmanship or materials, the same shall be made good by the contractor without expense. This does not include the reasonable amount of settlement or shrinkage, etc., normally appearing in a new house. No guarantee shall be made against shrinkage on panel doors, plaster cracks, or drywall nail popping. Basement shall not be guaranteed against water if finished grade of landscaping is not kept 4″ below first course of brick, which by code must have weep holes. Any additional work involved in soil conditions or change of grade shall be charged accordingly.

The contractor shall leave the building broom-clean upon completion.

Excavating

This contractor shall excavate to firm, bearing soil as per plan.

Garage foundation and crawl space shall be carried to a depth of 4′-0″ below finish grade. All other footings shall be full depth. If any excavation is erroneously carried below specified depth, extend footings of work affected to undisturbed earth without extra charge. As soon as foundations are in place and other trades have completed their work thereon, backfill around all walls and piers to within 6″ of finished grades. Final grade shall be to the finish grades indicated on the building elevations. Finish grades with existing black dirt only. Drive allowance $696.00 for 2900 sq ft of driveway, 8″ of stone machine spread. Tree removal for excavation and 175 yards of granular fill included.

Structural Steel

Provide beams, columns, plates, and erection to complete all structural steel as per plan.

Concrete

The contractor shall provide all footings, foundation walls, floors, walks, etc., as called for in plan.

Concrete shall comply with the ASTM "Standard Specs for Ready Mix Concrete." No mixture shall have less than 5 bags of cement per cu yd of concrete. All concrete used shall be a compressive strength of 3000 lb per sq in. at the end of 28 days and have a maximum water content of $5\frac{1}{2}$ gallons per bag of cement. Garage floor and all outside flat work shall be 5″ thick reinforced with 6″ × 6″ welded-wire fabric. Basement floors shall be 4″ concrete over a minimum of 6″ stone fill. Concrete walls against which earth is to be backfilled shall be damp-proofed with asphaltic compound, dehydrative #10 semi-mastic. All fins on exposed finished concrete shall be removed, all wires removed flush with the surface, and all voids filled with portland cement mortar. Basement steel windows and galvanized window wells shall be provided. Brush coat inside foundation walls with portland cement or equal. Rods shall be placed in foundation at all step downs in level. Crawl space shall have 4″ stone over visqueen.

Masonry

This contractor shall provide labor and materials to complete all masonry as per plan.

Exterior brick shall be used common brick with raked joints, standard mortar. Library and family room fireplaces shall be used brick, raked joints per plan. Door sills shall be stone, window sills shall be stone or wood per plan. Chimney caps shall be concrete poured in place. Sufficient flue size for hot-water heater and furnaces shall be provided. Fireplaces shall have cast iron damper, ash pit and cleanout door, and gas liters. Gas logs by owner. Wing walls stone cap per plan.

Carpentry

This contractor shall provide labor and materials to complete all carpentry as per plan.

All framing lumber shall be standard and better, kiln-dried douglas or white fir. All soffit, fascia, and trim shall be cedar per plan. Siding shall be prefinish RUF X90 panel groove. Screened overhang vents per plan. All exterior doors shall be $1\frac{3}{4}$″ thick per plan. Interior doors shall be $1\frac{3}{8}$″ birch flush doors. All trim shall be ranch-design clear white pine. Vanity cabinets shall be standard lipped doors in birch or oak with marble tops as manufactured by Marblecast Products. All floors shall be plywood or ceramic per plan. Garage doors shall be Barcol or equal, design per plan. Front, inside garage, and side doors shall be keyed alike. All hardware shall be Weiser or equal solid polished brass. Windows shall be Pella standard casements with insulated glass and screens. Shutters shall be per plan. Library and family room panel shall be Bruce pecan or equal $\frac{1}{4}$″ prefinished paneling and $3\frac{5}{8}$″ crown mould. Family room, eating area, and kitchen beams shall be rough sawn cedar per plan.

Kitchen. Eating cabinets shall be birch or oak. Base cabinets shall have formica tops. A $600.00 allowance is included for cabinet design above standard flush.
Approximately 20 ft base cabinets
1 24″-deep upper cabinet over refrigerator

1 lazy susan
1 bread board
1 3′6″ desk unit, formica top
1 door and drawer island side of base cabinet
Refrigerator side panels
Approximately 12 ft upper cabinet
1 bank drawers
2 pull out shelves
1 3′0″ × 7′0″ pantry storage unit

Laundry. Cabinets shall be birch or oak standard flush doors.
1 10′0″ upper cabinet 30″ high
1 base cabinet

Vanity cabinets shall be oak or birch standard flush doors and drawers. Tops shall be cultured marble as manufactured by Marblecast Products.

Powder Room
1 3′-6″ vanity with top
1 louver and flush door medicine cabinet

Bath #1 dressing
1 7′-6″ × 8′-0″ linen cabinet
1 sliding glass tub enclosure for steam unit

Bath #2
1 5′-6″ vanity with top

Bath #3
1 5′-6″ vanity with top
1 laundry chute, lift-top door

Family room
1 oak or birch 10′-0″ × 8′-0″ bookcase, adjustable shelves

Library
Oak or birch standard flush doors, wood top base
Base cabinets 18″ wide, book shelves to ceiling
1 unit 7′-7″, 2 units 5′-0″ each

Bedroom #2
Birch shelf unit 9′ × 8′

Bedroom #3
Birch shelf unit 7′-6″ × 8′-0″

Bedroom #4
Birch shelf unit 3′ × 8′

Bedroom #5
Birch shelf unit 3'–6" × 8'

All cabinet and book shelf work shall be prefinished. Closets shall have 1 rod and 1 shelf unless otherwise noted. A 3-riser stair to library, 4-riser stair in foyer, 10-riser stair and railing to master suite, and 11-riser stair to family room shall all be mill made. Stairs to basement shall be carpenter built.

Roofing

This contractor shall provide and cover roof with ¾–1¼ × 24" heavy cedar shakes laid 10–11" to the weather with 18" 30-lb felt underlayment—shingle starter—tar and gravel roofing and roof vents.

Heating, Sheetmetal, and Air-conditioning

This contractor shall provide and install two G.E. or equal gas-fired furnaces and air-conditioning units. All controls shall be G.E. or equal. This system shall be designed to maintain 72° inside temperature at minus 20° outside. All heating outlets shall be perimeter type; returns shall be high inside wall. Adequate supply and return ducts shall be installed to the basement. Duct work and outlets shall be laid out in such a way as to ensure even, draft-free heating and cooling in all areas. Air conditioner to maintain 20° differential. This contractor shall provide all sheetmetal flashing, saddles, gutters, downspouts, etc. This outside sheetmetal shall be 26-gauge galvanized iron as shown. This contractor shall also provide drier vent and kitchen and powder room exhaust fan ducts and 1 clothes chute; furnish and install April Air humidifiers, 1 #112 and 1 #110; and furnish and install 2 1-cell Honeywell air cleaners.

Sheetrock

This contractor shall install ½" S.W. sheetrock in all areas called for, and ⅝" F.C. in garage and basement-furnace and stair area. This material shall be glued to studs on sidewalls and screwed to ceilings. Backer board shall be used in the family room and library for paneling. All joints shall be taped and sanded by experts in this trade. Garage shall be finished.

Insulation

This contractor shall furnish labor and material to insulate all exposed sidewalls, overhangs, and ceilings. Sidewalls and areas inaccessible to blow shall be insulated with full, thick, foil-faced fiberglass. Ceilings shall be insulated with 6" blown fiberglass. Sound insulate bath #2 tub, master bedroom area.

Resilient Tile

Kitchen, eating area, laundry, lower hall, and powder room. Furnish and install Seranada Corlon by Armstrong direct to floors prepared by us. Supply necessary metals.

Ceramic and Quarry Tile

Foyer. Furnish and install 6' × 6' or 4' × 8' glazed-quarry-tile floor set in cement mortar bed.

Bath #1. Ceramic-tile (3) walls and ceiling in tub area all set in cement mortar bed. Ceramic-tile floor set in cement mortar bed. Install 4" ceramic-tile base set in mastic on rest of walls. Install 4-piece set chrome or tile fixtures.

Powder Room. Install 2-piece set chrome fixtures.

Bath #2 and Lavatory #2. Ceramic tile (3) walls in tub area in a cement mortar bed. Install 4" ceramic-tile base on rest of walls set in mastic. Install 4-piece set chrome or tile fixtures.

Bath #3 and Lavatory #3. Ceramic tile (3) walls in shower stall to soffit set in cement mortar bed. Ceramic tile soffit ceiling and base on rest of walls all set in mastic. Install 4-piece set chrome or tile fixtures. Ceramic-tile floor set in cement mortar bed.

Glass and Glazing

This contractor shall glaze all basement steel sash and entry side lites. Side lites shall be Lozenge glass. All mirrors by owner.

Plumbing

Powder Room
1 Crane Radcliff water closet with seat
1 Crane oval lav., moen trim

Bath #1 dressing
2 Crane oval lavs., moen trim

Bath #1
1 Crane Radcliff water closet with seat
1 Crane Premier bath tub, moen trim, Speakman #2 shower head
1 Thermosal steam unit

Bath #2 Lavatory.
1 Crane oval lav., moen trim

Bath #2
1 Crane Radcliff water closet with seat
1 Crane Premier bath tub, moen trim, Speakman #2 shower head

Bath #3 Lavatory
1 Crane oval lav., moen trim

Bath #3
1 Crane Radcliff water closet with seat
1 fibreglass shower base 60" × 34", moen trim, Speakman #2 showerhead

All Bath Fixtures White

Kitchen. This contractor shall provide all connections for dishwasher, disposal, and kitchen sink.

1 Elkay 33″ × 22″ SS sink per plan
1 BSR-15 SS bar sink and faucet

Laundry. 1 fiberglass laundry sink with cabinet and faucets.

1 laundry box—hot and cold water

Basement. 1 Rheem Fury 75-gal quick-recovery glass-lined water heater. This unit shall carry the manufacturer's guarantee. Provide floor drains in garage and basement as shown. Provide 1 submersible sump pump Swaby or equal with discharge line run through wall and discharged at ground level to splash block.

Run continuous open-joint drain around the outside perimeter of the basement running to the sump pit. This tile shall be covered with 12″ of crushed stone. Risers with metal grilles will be installed to all basement window wells. All piping shall be copper within the building. Sewer service shall be run from the stub in the parkway. Water service shall be 1″ copper from the roundway in the parkway. Gas shall be run from the meter to the heating unit and water heater. Hose outlets per plan on separate hard line. Exposed shut-off valves will be provided at all fixtures. Gas line to fireplace. Sanitary sewer permit included. Water softener by owner.

Electric

This contractor shall provide all labor and material to complete the electric work as per plan. He shall include all switches, outlets, and recessed-type fixtures in his base proposal. He shall also include the following:

200-amp main service with circuit breakers
110-V washer outlet
220-V dryer outlet
220-V range wiring
220-V oven wiring
110-V disposal outlet
Wiring for 2 thermostats and heating controls
1 dishwasher circuit
1 sump-pump circuit
Wiring for 2 furnaces, 2 air-conditioning units, 2 April Airs, 2 air cleaners
110-V range-hood-lite connection
1 water-softener circuit
1 dimmer
8 recess fixtures per plan
Fan exhaust for powder room
Fan lite bath #1, bath #2, and bath #3

All switches shall be silent rocker type. All wiring shall be thin-wall metal conduit. Lighting fixtures and chimes shall be hung by this contractor. This contract includes hanging all available fixtures at the time he trims the building and one later return trip to hang any additional fixtures. Any additional trips required to hang fixtures will be billed to owner on an additional cost basis.

Painting and Decorating

Interior walls shall receive 2 coats of latex-base flat enamel. All interior doors and trim shall be thoroughly cleaned, stained, sealed, rubbed down, and followed by 1 coat of varnish. Exterior painting will consist of all windows back primed. All wood surfaces shall be primed in oil-base paint followed by 1 coat of exterior house paint, Pratt and Lambert or equal. Gutters and exposed sheetmetal will be primed with Galvite or equal. Interior structural steel and basement windows shall receive 1 shop coat and 1 coat of oil-base paint. Any wallpaper will be handled directly by owner—no allowance in contract.

Appliances

All built-in appliances (oven, range and hood, disposal, and dishwasher) shall be installed under this contract.

Allowances in Contract

A $300.00 light fixtures and chime allowance—installed under electric contract.

A $750.00 built-in appliances allowance—installed under this contract.

A $600.00 intercom allowance—including installation to be handled by owner directly.

Architectural Drawings and Foundation Design

All cost for working drawings, blueprinting, engineering and foundation design, and miscellaneous involved therein will be handled by the general contractor at no expense to the owner.

Appendix D Reference Tables

American National Standards

A few standards that are useful for reference are listed below. Standards are subject to revisions and latest issues should be consulted. A catalog with prices is published by the American National Standards Institute, Inc., 1430 Broadway, New York, N.Y. 10018.

Table D-1 Decimal Equivalents of Common Fractions

$\frac{1}{64}$	0.015625	$\frac{17}{64}$	0.265625	$\frac{33}{64}$	0.515625	$\frac{49}{64}$	0.765625
$\frac{1}{32}$	0.03125	$\frac{9}{32}$	0.28125	$\frac{17}{32}$	0.53125	$\frac{25}{32}$	0.78125
$\frac{3}{64}$	0.046875	$\frac{19}{64}$	0.296875	$\frac{35}{64}$	0.546875	$\frac{51}{64}$	0.796875
$\frac{1}{16}$	0.0625	$\frac{5}{16}$	0.3125	$\frac{9}{16}$	0.5625	$\frac{13}{16}$	0.8125
$\frac{5}{64}$	0.078125	$\frac{21}{64}$	0.328125	$\frac{37}{64}$	0.578125	$\frac{53}{64}$	0.828125
$\frac{3}{32}$	0.09375	$\frac{11}{32}$	0.34375	$\frac{19}{32}$	0.59375	$\frac{27}{32}$	0.84375
$\frac{7}{64}$	0.109375	$\frac{23}{64}$	0.359375	$\frac{39}{64}$	0.609375	$\frac{55}{64}$	0.859375
$\frac{1}{8}$	0.1250	$\frac{3}{8}$	0.3750	$\frac{5}{8}$	0.6250	$\frac{7}{8}$	0.8750
$\frac{9}{64}$	0.140625	$\frac{25}{64}$	0.390625	$\frac{41}{64}$	0.640625	$\frac{57}{64}$	0.890625
$\frac{5}{32}$	0.15625	$\frac{13}{32}$	0.40625	$\frac{21}{32}$	0.65625	$\frac{29}{32}$	0.90625
$\frac{11}{64}$	0.171875	$\frac{27}{64}$	0.421875	$\frac{43}{64}$	0.671875	$\frac{59}{64}$	0.921875
$\frac{3}{16}$	0.1875	$\frac{7}{16}$	0.4375	$\frac{11}{16}$	0.6875	$\frac{15}{16}$	0.9375
$\frac{13}{64}$	0.203125	$\frac{29}{64}$	0.453125	$\frac{45}{64}$	0.703125	$\frac{61}{64}$	0.953125
$\frac{7}{32}$	0.21875	$\frac{15}{32}$	0.46875	$\frac{23}{32}$	0.71875	$\frac{31}{32}$	0.96875
$\frac{15}{64}$	0.234375	$\frac{31}{64}$	0.484375	$\frac{47}{64}$	0.734375	$\frac{63}{64}$	0.984375
$\frac{1}{4}$	0.2500	$\frac{1}{2}$	0.5000	$\frac{3}{4}$	0.7500	1	1.0000

Table D-2 American National Standard Unified and American Thread Series

Threads per inch for coarse, fine, extra-fine, 8-thread, 12-thread, and 16-thread, and 16-thread series[b] [tap-drill sizes for approximately 75 per cent depth of thread (not American Standard)]

Nominal size (basic major diam.)	Coarse-thd. series UNC and NC[c] in classes 1A, 1B, 2A, 2B, 3A, 3B, 2, 3		Fine-thd. series UNF and NF[c] in classes 1A, 1B, 2A, 2B, 3A, 3B, 2, 3		Extra-fine thd. series UNEF and NEF[d] in classes 2A, 2B, 2, 3		8-thd. series 8N[c] in classes 2A, 2B, 2, 3		12-thd. series 12UN and 12N[d] in classes 2A, 2B, 2, 3		16-thd. series 16UN and 16N[d] in classes 2A, 2B, 2, 3	
	Thd./in.	Tap drill	Thd./in.	Tap drill	Thd./in.	Tap drill	Thd./in.	Tap drill	Thd./in.	Tap drill	Thd./in.	Tap drill
0(0.060)	....	.....	80	$\frac{3}{64}$								
1(0.073)	64	No. 53	72	No. 53								
2(0.086)	56	No. 50	64	No. 50								
3(0.099)	48	No. 47	56	No. 45								
4(0.112)	40	No. 43	48	No. 42								
5(0.125)	40	No. 38	44	No. 37								
6(0.138)	32	No. 36	40	No. 33								
8(0.164)	32	No. 29	36	No. 29								
10(0.190)	24	No. 25	32	No. 21								
12(0.216)	24	No. 16	28	No. 14	32	No. 13						
$\frac{1}{4}$	20	No. 7	28	No. 3	32	$\frac{7}{32}$						
$\frac{5}{16}$	18	Let. F	24	Let. I	32	$\frac{9}{32}$						
$\frac{3}{8}$	16	$\frac{5}{16}$	24	Let. Q	32	$\frac{11}{32}$						
$\frac{7}{16}$	14	Let. U	20	$\frac{25}{64}$	28	$\frac{13}{32}$						
$\frac{1}{2}$	13	$\frac{27}{64}$	20	$\frac{29}{64}$	28	$\frac{15}{32}$	..	...	12	$\frac{27}{64}$		
$\frac{9}{16}$	12	$\frac{31}{64}$	18	$\frac{33}{64}$	24	$\frac{33}{64}$	..	...	12	$\frac{31}{64}$		
$\frac{5}{8}$	11	$\frac{17}{32}$	18	$\frac{37}{64}$	24	$\frac{37}{64}$	..	...	12	$\frac{35}{64}$		
$\frac{11}{16}$	....	.....	..	.....	24	$\frac{41}{64}$	..	...	12	$\frac{39}{64}$		
$\frac{3}{4}$	10	$\frac{21}{32}$	16	$\frac{11}{16}$	20	$\frac{45}{64}$	..	...	12	$\frac{43}{64}$	16	$\frac{11}{16}$
$\frac{13}{16}$	....	.....	..	.....	20	$\frac{49}{64}$	..	...	12	$\frac{47}{64}$	16	$\frac{3}{4}$
$\frac{7}{8}$	9	$\frac{49}{64}$	14	$\frac{13}{16}$	20	$\frac{53}{64}$	..	...	12	$\frac{51}{64}$	16	$\frac{13}{16}$
$\frac{15}{16}$	....	.....	..	.....	20	$\frac{57}{64}$	..	...	12	$\frac{55}{64}$	16	$\frac{7}{8}$
1	....	.....	14	$\frac{15}{16}$	..	.....	8	$\frac{7}{8}$				
1	8	$\frac{7}{8}$	12	$\frac{59}{64}$	20	$\frac{61}{64}$	..	...	12	$\frac{59}{64}$	16	$\frac{15}{16}$
$1\frac{1}{16}$	....	.....	..	.....	18	1	..	...	12	$\frac{63}{64}$	16	1
$1\frac{1}{8}$	7	$\frac{63}{64}$	12	$1\frac{3}{64}$	18	$1\frac{5}{64}$	8	1	12	$1\frac{3}{64}$	16	$1\frac{1}{16}$
$1\frac{3}{16}$	....	.....	..	.....	18	$1\frac{9}{64}$	..	...	12	$1\frac{7}{64}$	16	$1\frac{1}{8}$
$1\frac{1}{4}$	7	$1\frac{7}{64}$	12	$1\frac{11}{64}$	18	$1\frac{3}{16}$	8	$1\frac{1}{8}$	12	$1\frac{11}{64}$	16	$1\frac{3}{16}$
$1\frac{5}{16}$	....	.....	..	.....	18	$1\frac{17}{64}$	..	...	12	$1\frac{15}{64}$	16	$1\frac{1}{4}$
$1\frac{3}{8}$	6	$1\frac{7}{32}$	12	$1\frac{19}{64}$	18	$1\frac{5}{16}$	8	$1\frac{1}{4}$	12	$1\frac{19}{64}$	16	$1\frac{5}{16}$
$1\frac{7}{16}$	....	.....	..	.....	18	$1\frac{3}{8}$	..	...	12	$1\frac{23}{64}$	16	$1\frac{3}{8}$
$1\frac{1}{2}$	6	$1\frac{11}{32}$	12	$1\frac{27}{64}$	18	$1\frac{7}{16}$	8	$1\frac{3}{8}$	12	$1\frac{27}{64}$	16	$1\frac{7}{16}$
$1\frac{9}{16}$	....	.....	..	.....	18	$1\frac{1}{2}$	..	...	..	.....	16	$1\frac{1}{2}$
$1\frac{5}{8}$	....	.....	..	.....	18	$1\frac{9}{16}$	8	$1\frac{1}{2}$	12	$1\frac{35}{64}$	16	$1\frac{9}{16}$
$1\frac{11}{16}$	....	.....	..	.....	18	$1\frac{5}{8}$	..	...	..	.....	16	$1\frac{5}{8}$

Table D-2 (*Cont.*)

Nominal size (basic major diam.)	Coarse-thd. series UNC and NC[c] in classes 1A, 1B, 2A, 2B, 3A, 3B, 2, 3		Fine-thd. series UNF and NF[c] in classes 1A, 1B, 2A, 2B, 3A, 3B, 2, 3		Extra-fine thd. series UNEF and NEF[d] in classes 2A, 2B, 2, 3		8-thd. series 8N[c] in classes 2A, 2B, 2, 3		12-thd. series 12UN and 12N[d] in classes 2A, 2B, 2, 3		16-thd. series 16UN and 16N[d] in classes 2A, 2B, 2, 3	
	Thd./in.	Tap drill	Thd./in.	Tap drill	Thd./in.	Tap drill	Thd./in.	Tap drill	Thd./in.	Tap drill	Thd./in.	Tap drill
$1\frac{3}{4}$	**5**	$1\frac{9}{16}$	..	.....	**16**	$1\frac{11}{16}$	8[e]	$1\frac{5}{8}$	**12**	$1\frac{43}{64}$	**16**	$1\frac{11}{16}$
$1\frac{13}{16}$	....	.....	..	.....	..	.....			..	.....	**16**	$1\frac{3}{4}$
$1\frac{7}{8}$	....	.....	..	.....	..	.....	8	$1\frac{3}{4}$	**12**	$1\frac{51}{64}$	**16**	$1\frac{13}{16}$
$1\frac{15}{16}$	....	.....	..	.....	..	.....		...	..	.....	**16**	$1\frac{7}{8}$
2	$4\frac{1}{2}$	$1\frac{25}{32}$	..	.....	**16**	$1\frac{15}{16}$	8[e]	$1\frac{7}{8}$	**12**	$1\frac{59}{64}$	**16**	$1\frac{15}{16}$
$2\frac{1}{16}$	....	.....	..	.....	..	.....				.....	**16**	2
$2\frac{1}{8}$	....	.....	..	.....	..	.....	8	2	**12**	$2\frac{3}{64}$	**16**	$2\frac{1}{16}$
$2\frac{3}{16}$	....	.....	..	.....	..	.....				.....	**16**	$2\frac{1}{8}$
$2\frac{1}{4}$	$4\frac{1}{2}$	$2\frac{1}{32}$	..	.....	..	.....	8[e]	$2\frac{1}{8}$	**12**	$2\frac{11}{64}$	**16**	$2\frac{3}{16}$
$2\frac{5}{16}$	....	.....	..	.....	..	.....				.....	**16**	$2\frac{1}{4}$
$2\frac{3}{8}$	....	.....	..	.....	..	.....		...	**12**	$2\frac{19}{64}$	**16**	$2\frac{5}{16}$
$2\frac{7}{16}$	....	.....	..	.....	..	.....					**16**	$2\frac{3}{8}$
$2\frac{1}{2}$	**4**	$2\frac{1}{4}$	..	.....	..	.....	8[e]	$2\frac{3}{8}$	**12**	$2\frac{27}{64}$	**16**	$2\frac{7}{16}$
$2\frac{5}{8}$	....	.....	..	.....	..	.....		...	**12**	$2\frac{35}{64}$	**16**	$2\frac{9}{16}$
$2\frac{3}{4}$	**4**	$2\frac{1}{2}$	..	.....	..	.....	8[e]	$2\frac{5}{8}$	**12**	$2\frac{43}{64}$	**16**	$2\frac{11}{16}$
$2\frac{7}{8}$	....	.....	..	.....	..	.....		...	**12**	$2\frac{51}{64}$	**16**	$2\frac{13}{16}$
3	**4**	$2\frac{3}{4}$	..	.....	..	.....	8[e]	$2\frac{7}{8}$	**12**	$2\frac{59}{64}$	**16**	$2\frac{15}{16}$
$3\frac{1}{8}$	....	.....	..	.....	..	.....		...	**12**	$3\frac{3}{64}$	**16**	$3\frac{1}{16}$
$3\frac{1}{4}$	**4**	3	..	.....	..	.....	8[e]	$3\frac{1}{8}$	**12**	$3\frac{11}{64}$	**16**	$3\frac{3}{16}$
$3\frac{3}{8}$	....	.....	..	.....	..	.....		...	**12**	$3\frac{19}{64}$	**16**	$3\frac{5}{16}$
$3\frac{1}{2}$	**4**	$3\frac{1}{4}$	..	.....	..	.....	8[e]	$3\frac{3}{8}$	**12**	$3\frac{27}{64}$	**16**	$3\frac{7}{16}$
$3\frac{5}{8}$	....	.....	..	.....	..	.....		...	**12**	$3\frac{35}{64}$	**16**	$3\frac{9}{16}$
$3\frac{3}{4}$	**4**	$3\frac{1}{2}$	..	.....	..	.....	8[e]	$3\frac{5}{8}$	**12**	$3\frac{43}{64}$	**16**	$3\frac{11}{16}$
$3\frac{7}{8}$	....	.....	..	.....	..	.....		...	**12**	$3\frac{51}{64}$	**16**	$3\frac{13}{16}$
4	**4**	$3\frac{3}{4}$	..	.....	..	.....	8[e]	$3\frac{7}{8}$	**12**	$3\frac{59}{64}$	**16**	$3\frac{15}{16}$
$4\frac{1}{4}$	....	.....	..	.....	..	.....	8[e]	$4\frac{1}{8}$	**12**	$4\frac{11}{64}$	**16**	$4\frac{3}{16}$
$4\frac{1}{2}$	....	.....	..	.....	..	.....	8[e]	$4\frac{3}{8}$	**12**	$4\frac{27}{64}$	**16**	$4\frac{7}{16}$
$4\frac{3}{4}$	....	.....	..	.....	..	.....	8[e]	$4\frac{5}{8}$	**12**	$4\frac{43}{64}$	**16**	$4\frac{11}{16}$
5	....	.....	..	.....	..	.....	8[e]	$4\frac{7}{8}$	**12**	$4\frac{59}{64}$	**16**	$4\frac{15}{16}$
$5\frac{1}{4}$	....	.....	..	.....	..	.....	8[e]	$5\frac{1}{8}$	**12**	$5\frac{11}{64}$	**16**	$5\frac{3}{16}$
$5\frac{1}{2}$	....	.....	..	.....	..	.....	8[e]	$5\frac{3}{8}$	**12**	$5\frac{27}{64}$	**16**	$5\frac{7}{16}$
$5\frac{3}{4}$	....	.....	..	.....	..	.....	8[e]	$5\frac{5}{8}$	**12**	$5\frac{43}{64}$	**16**	$5\frac{11}{16}$
6	....	.....	..	.....	..	.....	8[e]	$5\frac{7}{8}$	**12**	$5\frac{59}{64}$	**16**	$5\frac{15}{16}$

[a] ANSI B1.1—1960. Dimensions are in inches. [b] Bold type indicates unified combinations.
[c] Limits of size for classes are based on a length of engagement equal to the nominal diameter.
[d] Limits of size for classes are based on length of engagement equal to nine times the pitch.
[e] These sizes, with specified limits of size, based on a length of engagement of 9 threads in classes 2A and 2B, are designated UN.
Note. If a thread is in both the 8-, 12-, or 16-thread series and the coarse, fine, or extra-fine-thread series, the symbols and tolerances of the latter series apply.

Table D-3   Sizes of Numbered and Lettered Drills

No.	Size	No.	Size	No.	Size	Letter	Size
80	0.0135	53	0.0595	26	0.1470	A	0.2340
79	0.0145	52	0.0635	25	0.1495	B	0.2380
78	0.0160	51	0.0670	24	0.1520	C	0.2420
77	0.0180	50	0.0700	23	0.1540	D	0.2460
76	0.0200	49	0.0730	22	0.1570	E	0.2500
75	0.0210	48	0.0760	21	0.1590	F	0.2570
74	0.0225	47	0.0785	20	0.1610	G	0.2610
73	0.0240	46	0.0810	19	0.1660	H	0.2660
72	0.0250	45	0.0820	18	0.1695	I	0.2720
71	0.0260	44	0.0860	17	0.1730	J	0.2770
70	0.0280	43	0.0890	16	0.1770	K	0.2810
69	0.0292	42	0.0935	15	0.1800	L	0.2900
68	0.0310	41	0.0960	14	0.1820	M	0.2950
67	0.0320	40	0.0980	13	0.1850	N	0.3020
66	0.0330	39	0.0995	12	0.1890	O	0.3160
65	0.0350	38	0.1015	11	0.1910	P	0.3230
64	0.0360	37	0.1040	10	0.1935	Q	0.3320
63	0.0370	36	0.1065	9	0.1960	R	0.3390
62	0.0380	35	0.1100	8	0.1990	S	0.3480
61	0.0390	34	0.1110	7	0.2010	T	0.3580
60	0.0400	33	0.1130	6	0.2040	U	0.3680
59	0.0410	32	0.1160	5	0.2055	V	0.3770
58	0.0420	31	0.1200	4	0.2090	W	0.3860
57	0.0430	30	0.1285	3	0.2130	X	0.3970
56	0.0465	29	0.1360	2	0.2210	Y	0.4040
55	0.0520	28	0.1405	1	0.2280	Z	0.4130
54	0.0550	27	0.1440				

Table D-5   American National Standard Regular Hexagon Bolts

Diameter	Flats	Height		
		Unfinished	Semi-finished	Finished
$\frac{1}{4}$	$\frac{7}{16}$	$\frac{11}{64}$	$\frac{5}{32}$	$\frac{5}{32}$
$\frac{5}{16}$	$\frac{1}{2}$	$\frac{7}{32}$	$\frac{13}{64}$	$\frac{13}{64}$
$\frac{3}{8}$	$\frac{9}{16}$	$\frac{1}{4}$	$\frac{15}{64}$	$\frac{15}{64}$
$\frac{7}{16}$	$\frac{5}{8}$	$\frac{19}{64}$	$\frac{9}{32}$	$\frac{9}{32}$
$\frac{1}{2}$	$\frac{3}{4}$	$\frac{11}{32}$	$\frac{5}{16}$	$\frac{5}{16}$
$\frac{9}{16}$	$\frac{13}{16}$	....	....	$\frac{23}{64}$
$\frac{5}{8}$	$\frac{15}{16}$	$\frac{27}{64}$	$\frac{25}{64}$	$\frac{25}{64}$
$\frac{3}{4}$	$1\frac{1}{8}$	$\frac{1}{2}$	$\frac{15}{32}$	$\frac{15}{32}$
$\frac{7}{8}$	$1\frac{5}{16}$	$\frac{37}{64}$	$\frac{35}{64}$	$\frac{35}{64}$
1	$1\frac{1}{2}$	$\frac{43}{64}$	$\frac{39}{64}$	$\frac{39}{64}$
$1\frac{1}{8}$	$1\frac{11}{16}$	$\frac{3}{4}$	$\frac{11}{16}$	$\frac{11}{16}$
$1\frac{1}{4}$	$1\frac{7}{8}$	$\frac{27}{32}$	$\frac{25}{32}$	$\frac{25}{32}$
$1\frac{3}{8}$	$2\frac{1}{16}$	$\frac{29}{32}$	$\frac{27}{32}$	$\frac{27}{32}$
$1\frac{1}{2}$	$2\frac{1}{4}$	1	$\frac{15}{16}$	$\frac{15}{16}$
$1\frac{3}{4}$	$2\frac{5}{8}$	$1\frac{5}{32}$	$1\frac{3}{32}$	$1\frac{3}{32}$
2	3	$1\frac{11}{32}$	$1\frac{7}{32}$	$1\frac{7}{32}$
$2\frac{1}{4}$	$3\frac{3}{8}$	$1\frac{1}{2}$	$1\frac{3}{8}$	$1\frac{3}{8}$
$2\frac{1}{2}$	$3\frac{3}{4}$	$1\frac{21}{32}$	$1\frac{17}{32}$	$1\frac{17}{32}$
$2\frac{3}{4}$	$4\frac{1}{8}$	$1\frac{13}{16}$	$1\frac{11}{16}$	$1\frac{11}{16}$
3	$4\frac{1}{2}$	2	$1\frac{7}{8}$	$1\frac{7}{8}$
$3\frac{1}{4}$	$4\frac{7}{8}$	$2\frac{3}{16}$	2	
$3\frac{1}{2}$	$5\frac{1}{4}$	$2\frac{5}{16}$	$2\frac{1}{8}$	
$3\frac{3}{4}$	$5\frac{5}{8}$	$2\frac{1}{2}$	$2\frac{5}{16}$	
4	6	$2\frac{11}{16}$	$2\frac{1}{2}$	

Table D-4   Acme and Stub Acme Threads*

ANSI preferred diameter-pitch combinations.

Nominal (major) diam.	Threads/in.	Nominal (major) diam.	Threads/in.	Nominal (major) diam.	Threads/in.	Nominal (major) diam.	Threads/in.
$\frac{1}{4}$	16	$\frac{3}{4}$	6	$1\frac{1}{2}$	4	3	2
$\frac{5}{16}$	14	$\frac{7}{8}$	6	$1\frac{3}{4}$	4	$3\frac{1}{2}$	2
$\frac{3}{8}$	12	1	5	2	4	4	2
$\frac{7}{16}$	12	$1\frac{1}{8}$	5	$2\frac{1}{4}$	3	$4\frac{1}{2}$	2
$\frac{1}{2}$	10	$1\frac{1}{4}$	5	$2\frac{1}{2}$	3	5	2
$\frac{5}{8}$	8	$1\frac{3}{8}$	4	$2\frac{3}{4}$	3		

*ANSI B1.5 and B1.8-1952. Diameters in inches.

Table D-6   American National Standard Regular Hexagon Nuts

Diameter	Unfinished		Semifinished		Finished	
	Flats	Thickness	Flats	Thickness	Flats	Thickness
1/4	7/16	7/32	7/16	13/64	7/16	7/32
5/16	9/16	17/64	9/16	1/4	1/2	17/64
3/8	5/8	21/64	5/8	5/16	9/16	21/64
7/16	3/4	3/8	3/4	23/64	11/16	3/8
1/2	13/16	7/16	13/16	27/64	3/4	7/16
9/16	7/8	1/2	7/8	31/64	7/8	31/64
5/8	1	35/64	1	17/32	15/16	35/64
3/4	1 1/8	21/32	1 1/8	41/64	1 1/8	41/64
7/8	1 5/16	49/64	1 5/16	3/4	1 5/16	3/4
1	1 1/2	7/8	1 1/2	55/64	1 1/2	55/64
1 1/8	1 11/16	1	1 11/16	31/32	1 11/16	31/32
1 1/4	1 7/8	1 3/32	1 7/8	1 1/16	1 7/8	1 1/16
1 3/8	2 1/16	1 13/64	2 1/16	1 11/64	2 1/16	1 11/64
1 1/2	2 1/4	1 5/16	2 1/4	1 9/32	2 1/4	1 9/32
1 5/8	...	...	2 7/16	1 25/64		
1 3/4	...	...	2 5/8	1 1/2	2 5/8	1 1/2
1 7/8	...	...	2 13/16	1 39/64		
2	...	...	3	1 23/32	3	1 23/32
2 1/4	...	...	3 3/8	1 59/64	3 3/8	1 59/64
2 1/2	...	...	3 3/4	2 9/64	3 3/4	2 9/64
2 3/4	...	...	4 1/8	2 23/64	4 1/8	2 23/64
3	...	...	4 1/2	2 37/64	4 1/2	2 37/64

Table D-8   American National Standard Regular Square Bolts and Nuts

Diameter	Bolthead		Nut	
	Flats	Height of head	Flats	Thickness of nut
1/4	3/8	11/64	7/16	7/32
5/16	1/2	13/64	9/16	17/64
3/8	9/16	1/4	5/8	21/64
7/16	5/8	19/64	3/4	3/8
1/2	3/4	21/64	13/16	7/16
5/8	15/16	27/64	1	35/64
3/4	1 1/8	1/2	1 1/8	21/32
7/8	1 5/16	19/32	1 5/16	49/64
1	1 1/2	21/32	1 1/2	7/8
1 1/8	1 11/16	3/4	1 11/16	1
1 1/4	1 7/8	27/32	1 7/8	1 3/32
1 3/8	2 1/16	29/32	2 1/16	1 13/64
1 1/2	2 1/4	1	2 1/4	1 5/16
1 5/8	2 7/16	1 3/32		

Table D-7   American National Standard Heavy Hexagon Bolts

Diameter	Flats (Unfinished semifinished finished)	Height of head (Unfinished)	Height of head (Semifinished finished)	Diameter	Flats (Unfinished semifinished finished)	Height of head (Unfinished)	Height of head (Semifinished finished)
1/2	7/8	7/16	13/32	1 5/8	2 9/16	1 9/32	1 7/32
5/8	1 1/16	17/32	1/2	1 3/4	2 3/4	1 3/8	1 5/16
3/4	1 1/4	5/8	19/32	1 7/8	2 15/16	1 15/32	1 13/32
7/8	1 7/16	23/32	11/16	2	3 1/8	1 9/16	1 7/16
1	1 5/8	13/16	3/4	2 1/4	3 1/2	1 3/4	1 5/8
1 1/8	1 13/16	29/32	27/32	2 1/2	3 7/8	1 15/16	1 13/16
1 1/4	2	1	15/16	2 3/4	4 1/4	2 1/8	2
1 3/8	2 3/16	1 3/32	1 1/32	3	4 5/8	2 5/16	2 3/16
1 1/2	2 3/8	1 3/16	1 1/8				

Note: 1 5/8, 1 7/8 not in unfinished or semifinished bolts.

Table D-10 American National Standard Square-head Setscrews and Points

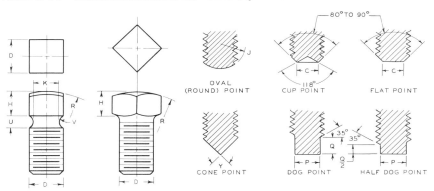

OVAL (ROUND) POINT — CUP POINT — FLAT POINT — CONE POINT — DOG POINT — HALF DOG POINT

D	H nom.	R nom.	K max.	U min.	V max.	C nom.	J nom.	P max.	Q max.
10 (0.190)	$\frac{9}{64}$	$\frac{15}{32}$	0.145	0.083	0.027	$\frac{3}{32}$	0.141	0.127	0.090
12 (0.216)	$\frac{5}{32}$	$\frac{35}{64}$	0.162	0.091	0.029	$\frac{7}{64}$	0.156	0.144	0.110
$\frac{1}{4}$	$\frac{3}{16}$	$\frac{5}{8}$	0.185	0.100	0.032	$\frac{1}{8}$	0.188	0.156	0.125
$\frac{5}{16}$	$\frac{15}{64}$	$\frac{25}{32}$	0.240	0.111	0.036	$\frac{11}{64}$	0.234	0.203	0.156
$\frac{3}{8}$	$\frac{9}{32}$	$\frac{15}{16}$	0.294	0.125	0.041	$\frac{13}{64}$	0.281	0.250	0.188
$\frac{7}{16}$	$\frac{21}{64}$	$1\frac{3}{32}$	0.345	0.143	0.046	$\frac{15}{64}$	0.328	0.297	0.219
$\frac{1}{2}$	$\frac{3}{8}$	$1\frac{1}{4}$	0.400	0.154	0.050	$\frac{9}{32}$	0.375	0.344	0.250
$\frac{9}{16}$	$\frac{27}{64}$	$1\frac{13}{32}$	0.454	0.167	0.054	$\frac{5}{16}$	0.422	0.391	0.281
$\frac{5}{8}$	$\frac{15}{32}$	$1\frac{9}{16}$	0.507	0.182	0.059	$\frac{23}{64}$	0.469	0.469	0.313
$\frac{3}{4}$	$\frac{9}{16}$	$1\frac{7}{8}$	0.620	0.200	0.065	$\frac{7}{16}$	0.563	0.563	0.375
$\frac{7}{8}$	$\frac{21}{32}$	$2\frac{3}{16}$	0.731	0.222	0.072	$\frac{33}{64}$	0.656	0.656	0.438
1	$\frac{3}{4}$	$2\frac{1}{2}$	0.838	0.250	0.081	$\frac{19}{32}$	0.750	0.750	0.500
$1\frac{1}{8}$	$\frac{27}{32}$	$2\frac{13}{16}$	0.939	0.283	0.092	$\frac{43}{64}$	0.844	0.844	0.562
$1\frac{1}{4}$	$\frac{15}{16}$	$3\frac{1}{8}$	1.064	0.283	0.092	$\frac{3}{4}$	0.938	0.938	0.625
$1\frac{3}{8}$	$1\frac{1}{32}$	$3\frac{7}{16}$	1.159	0.333	0.109	$\frac{53}{64}$	1.031	1.031	0.688
$1\frac{1}{2}$	$1\frac{1}{8}$	$3\frac{3}{4}$	1.284	0.333	0.109	$\frac{29}{32}$	1.125	1.125	0.750

Note: Threads may be coarse-, fine-, or 8-thread series, class 2A. Coarse thread normally used on $\frac{1}{4}$ in. and larger. When length equals nominal diameter or less $Y = 118°$. When length exceeds nominal diameter $Y = 90°$.

Table D-9 American National Standard Heavy Nuts, Square and Hexagon

Diameter	Flats — Unfinished semi-finished square and hexagon	Thickness of nut — Unfinished square and hexagon	Semi-finished hexagon
$\frac{1}{4}$	$\frac{1}{2}$	$\frac{1}{4}$	$\frac{15}{64}$
$\frac{5}{16}$	$\frac{9}{16}$	$\frac{5}{16}$	$\frac{19}{64}$
$\frac{3}{8}$	$\frac{11}{16}$	$\frac{3}{8}$	$\frac{23}{64}$
$\frac{7}{16}$	$\frac{3}{4}$	$\frac{7}{16}$	$\frac{27}{64}$
$\frac{1}{2}$	$\frac{7}{8}$	$\frac{1}{2}$	$\frac{31}{64}$
$\frac{9}{16}$	$\frac{15}{16}$	. . . .	$\frac{35}{64}$
$\frac{5}{8}$	$1\frac{1}{16}$	$\frac{5}{8}$	$\frac{39}{64}$
$\frac{3}{4}$	$1\frac{1}{4}$	$\frac{3}{4}$	$\frac{47}{64}$
$\frac{7}{8}$	$1\frac{7}{16}$	$\frac{7}{8}$	$\frac{55}{64}$
1	$1\frac{5}{8}$	1	$\frac{63}{64}$
$1\frac{1}{8}$	$1\frac{13}{16}$	$1\frac{1}{8}$	$1\frac{7}{64}$
$1\frac{1}{4}$	2	$1\frac{1}{4}$	$1\frac{7}{32}$
$1\frac{3}{8}$	$2\frac{3}{16}$	$1\frac{3}{8}$	$1\frac{11}{32}$
$1\frac{1}{2}$	$2\frac{3}{8}$	$1\frac{1}{2}$	$1\frac{15}{32}$
$1\frac{5}{8}$	$2\frac{9}{16}$	. . . .	$1\frac{19}{32}$
$1\frac{3}{4}$	$2\frac{3}{4}$	$1\frac{3}{4}$	$1\frac{23}{32}$
$1\frac{7}{8}$	$2\frac{15}{16}$	. . . .	$1\frac{27}{32}$
2	$3\frac{1}{8}$	2	$1\frac{31}{32}$
$2\frac{1}{4}$	$3\frac{1}{2}$	$2\frac{1}{4}$	$2\frac{13}{64}$
$2\frac{1}{2}$	$3\frac{7}{8}$	$2\frac{1}{2}$	$2\frac{29}{64}$
$2\frac{3}{4}$	$4\frac{1}{4}$	$2\frac{3}{4}$	$2\frac{45}{64}$
3	$4\frac{5}{8}$	3	$2\frac{61}{64}$
$3\frac{1}{4}$	5	$3\frac{1}{4}$	$3\frac{3}{16}$
$3\frac{1}{2}$	$5\frac{3}{8}$	$3\frac{1}{2}$	$3\frac{7}{16}$
$3\frac{3}{4}$	$5\frac{3}{4}$	$3\frac{3}{4}$	$3\frac{11}{16}$
4	$6\frac{1}{8}$	4	$3\frac{15}{16}$

Note: $\frac{9}{16}$, $1\frac{5}{8}$, $1\frac{7}{8}$ not in unfinished nuts.

Table D-11 American National Standard Slotted-head Cap Screws

Fillister Head

Flat Head

Button Head

Nominal diameter D	A max.	B max.	C max.	E max.	F max.	G max.	H average	I max.	J max.	K max.	M max.
1/4	0.375	0.216	0.172	0.075	0.097	0.500	0.140	0.068	0.437	0.191	0.117
5/16	0.437	0.253	0.203	0.084	0.115	0.625	0.177	0.086	0.562	0.245	0.151
3/8	0.562	0.314	0.250	0.094	0.142	0.750	0.210	0.103	0.625	0.273	0.168
7/16	0.625	0.368	0.297	0.094	0.168	0.812	0.210	0.103	0.750	0.328	0.202
1/2	0.750	0.413	0.328	0.106	0.193	0.875	0.210	0.103	0.812	0.354	0.218
9/16	0.812	0.467	0.375	0.118	0.213	1.000	0.244	0.120	0.937	0.409	0.252
5/8	0.875	0.521	0.422	0.133	0.239	1.125	0.281	0.137	1.000	0.437	0.270
3/4	1.000	0.612	0.500	0.149	0.283	1.375	0.352	0.171	1.250	0.546	0.338
7/8	1.125	0.720	0.594	0.167	0.334	1.625	0.423	0.206			
1	1.312	0.803	0.656	0.188	0.371	1.875	0.494	0.240			
1 1/8	....	....	....	0.196	....	2.062	0.529	0.257			
1 1/4	....	....	....	0.211	....	2.312	0.600	0.291			
1 3/8	....	....	....	0.226	....	2.562	0.665	0.326			
1 1/2	....	....	....	0.258	....	2.812	0.742	0.360			

Table D-12 American National Standard Machine Screws
Maximum Dimensions

Nominal width of slot for all heads = M

FLAT HEAD OVAL HEAD

Semi-elliptical
ROUND HEAD FILLISTER HEAD

PAN HEAD

Diameter Nominal	Max.	A	B	C	E	F	G	H	I	J	K	M
0	0.060	0.119	0.035	0.056	0.113	0.053	0.096	0.045	0.059	....	....	0.023
1	0.073	0.146	0.043	0.068	0.138	0.061	0.118	0.053	0.071	....	....	0.026
2	0.086	0.172	0.051	0.080	0.162	0.069	0.140	0.062	0.083	0.167	0.053	0.031
3	0.099	0.199	0.059	0.092	0.187	0.078	0.161	0.070	0.095	0.193	0.060	0.035
4	0.112	0.225	0.067	0.104	0.211	0.086	0.183	0.079	0.107	0.219	0.068	0.039
5	0.125	0.252	0.075	0.116	0.236	0.095	0.205	0.088	0.120	0.245	0.075	0.043
6	0.138	0.279	0.083	0.128	0.260	0.103	0.226	0.096	0.132	0.270	0.082	0.048
8	0.164	0.332	0.100	0.152	0.309	0.120	0.270	0.113	0.156	0.322	0.096	0.054
10	0.190	0.385	0.116	0.176	0.359	0.137	0.313	0.130	0.180	0.373	0.110	0.060
12	0.216	0.438	0.132	0.200	0.408	0.153	0.357	0.148	0.205	0.425	0.125	0.067
1/4	0.250	0.507	0.153	0.232	0.472	0.175	0.414	0.170	0.237	0.492	0.144	0.075
5/16	0.3125	0.635	0.191	0.290	0.590	0.216	0.518	0.211	0.295	0.615	0.178	0.084
3/8	0.375	0.762	0.230	0.347	0.708	0.256	0.622	0.253	0.355	0.740	0.212	0.094
7/16	0.4375	0.812	0.223	0.345	0.750	0.328	0.625	0.265	0.368	....	....	0.094
1/2	0.500	0.875	0.223	0.354	0.813	0.355	0.750	0.297	0.412	....	....	0.106
9/16	0.5625	1.000	0.260	0.410	0.938	0.410	0.812	0.336	0.466	....	....	0.118
5/8	0.625	1.125	0.298	0.467	1.000	0.438	0.875	0.375	0.521	....	....	0.133
3/4	0.750	1.375	0.372	0.578	1.250	0.547	1.000	0.441	0.612	....	....	0.149

Table D-13 American National Standard Hexagon-head Cap Screws

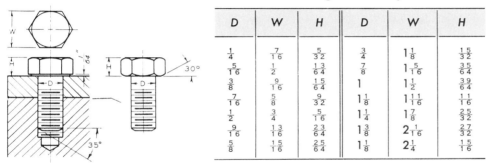

D	W	H	D	W	H
$\frac{1}{4}$	$\frac{7}{16}$	$\frac{5}{32}$	$\frac{3}{4}$	$1\frac{1}{8}$	$\frac{15}{32}$
$\frac{5}{16}$	$\frac{1}{2}$	$\frac{13}{64}$	$\frac{7}{8}$	$1\frac{5}{16}$	$\frac{35}{64}$
$\frac{3}{8}$	$\frac{9}{16}$	$\frac{15}{64}$	1	$1\frac{1}{2}$	$\frac{39}{64}$
$\frac{7}{16}$	$\frac{5}{8}$	$\frac{9}{32}$	$1\frac{1}{8}$	$1\frac{11}{16}$	$\frac{11}{16}$
$\frac{1}{2}$	$\frac{3}{4}$	$\frac{5}{16}$	$1\frac{1}{4}$	$1\frac{7}{8}$	$\frac{25}{32}$
$\frac{9}{16}$	$\frac{13}{16}$	$\frac{23}{64}$	$1\frac{3}{8}$	$2\frac{1}{16}$	$\frac{27}{32}$
$\frac{5}{8}$	$\frac{15}{16}$	$\frac{25}{64}$	$1\frac{1}{8}$	$2\frac{1}{4}$	$\frac{15}{16}$

Note: Bearing surfaces shall be flat and either washer faced or with chamfered corners. Minimum thread length shall be twice the diameter plus $\frac{1}{4}$ in. for lengths up to and including 6 in.; twice the diameter plus $\frac{1}{2}$ in. for lengths over 6 in.

Table D-14 American National Standard Plain Washers

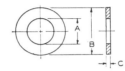

Inside diameter A	Outside diameter B	Thickness, C Gage	Thickness, C Nom.	Inside diameter A	Outside diameter B	Thickness, C Gage	Thickness, C Nom.	Inside diameter A	Outside diameter B	Thickness, C Gage	Thickness, C Nom.	Inside diameter A	Outside diameter B	Thickness, C Gage	Thickness, C Nom.
$\frac{5}{64}$	$\frac{3}{16}$	25	0.020	$\frac{5}{16}$	$\frac{7}{8}$	16	0.065	$\frac{5}{8}$	$2\frac{1}{8}$	10	0.134	$1\frac{3}{8}$	3	8	0.165
$\frac{3}{32}$	$\frac{7}{32}$	25	0.020	$\frac{11}{32}$	$\frac{11}{16}$	16	0.065	$\frac{21}{32}$	$1\frac{5}{16}$	13	0.095	$1\frac{7}{16}$	3	7	0.180
$\frac{3}{32}$	$\frac{1}{4}$	25	0.020	$\frac{3}{8}$	$\frac{3}{4}$	16	0.065	$\frac{11}{16}$	$1\frac{1}{2}$	10	0.134	$1\frac{1}{2}$	$3\frac{1}{4}$	7	0.180
$\frac{1}{8}$	$\frac{1}{4}$	24	0.022	$\frac{3}{8}$	$\frac{7}{8}$	14	0.083	$\frac{11}{16}$	$1\frac{3}{4}$	10	0.134	$1\frac{9}{16}$	$3\frac{1}{4}$	7	0.180
$\frac{1}{8}$	$\frac{5}{16}$	21	0.032	$\frac{3}{8}$	$1\frac{1}{8}$	16	0.065	$\frac{11}{16}$	$2\frac{3}{8}$	8	0.165	$1\frac{5}{8}$	$3\frac{1}{2}$	7	0.180
$\frac{5}{32}$	$\frac{5}{16}$	20	0.035	$\frac{13}{32}$	$\frac{13}{16}$	16	0.065	$\frac{13}{16}$	$1\frac{1}{2}$	10	0.134	$1\frac{11}{16}$	$3\frac{1}{2}$	7	0.180
$\frac{5}{32}$	$\frac{3}{8}$	18	0.049	$\frac{7}{16}$	$\frac{7}{8}$	14	0.083	$\frac{13}{16}$	$1\frac{3}{4}$	9	0.148	$1\frac{3}{4}$	$3\frac{3}{4}$	7	0.180
$\frac{11}{64}$	$\frac{13}{32}$	18	0.049	$\frac{7}{16}$	1	14	0.083	$\frac{13}{16}$	2	9	0.148	$1\frac{13}{16}$	$3\frac{3}{4}$	7	0.180
$\frac{3}{16}$	$\frac{3}{8}$	18	0.049	$\frac{7}{16}$	$1\frac{3}{8}$	14	0.083	$\frac{13}{16}$	$2\frac{7}{8}$	8	0.165	$1\frac{7}{8}$	4	7	0.180
$\frac{3}{16}$	$\frac{7}{16}$	18	0.049	$\frac{15}{32}$	$\frac{59}{64}$	16	0.065	$\frac{15}{16}$	$1\frac{3}{4}$	10	0.134	$1\frac{15}{16}$	4	7	0.180
$\frac{13}{64}$	$\frac{15}{32}$	18	0.049	$\frac{1}{2}$	$1\frac{1}{8}$	14	0.083	$\frac{15}{16}$	2	8	0.165	2	$4\frac{1}{4}$	7	0.180
$\frac{7}{32}$	$\frac{7}{16}$	18	0.049	$\frac{1}{2}$	$1\frac{1}{4}$	14	0.083	$\frac{15}{16}$	$2\frac{1}{4}$	8	0.165	$2\frac{1}{16}$	$4\frac{1}{4}$	7	0.180
$\frac{7}{32}$	$\frac{1}{2}$	18	0.049	$\frac{1}{2}$	$1\frac{5}{8}$	14	0.083	$\frac{15}{16}$	$3\frac{3}{8}$	7	0.180	$2\frac{1}{8}$	$4\frac{1}{2}$	7	0.180
$\frac{15}{64}$	$\frac{17}{32}$	18	0.049	$\frac{17}{32}$	$1\frac{1}{16}$	13	0.095	$1\frac{1}{16}$	2	10	0.134	$2\frac{3}{8}$	$4\frac{3}{4}$	5	0.220
$\frac{1}{4}$	$\frac{1}{2}$	18	0.049	$\frac{9}{16}$	$1\frac{1}{4}$	12	0.109	$1\frac{1}{16}$	$2\frac{1}{4}$	8	0.165	$2\frac{5}{8}$	5	4	0.238
$\frac{1}{4}$	$\frac{9}{16}$	18	0.049	$\frac{9}{16}$	$1\frac{3}{8}$	12	0.109	$1\frac{1}{16}$	$2\frac{1}{2}$	8	0.165	$2\frac{7}{8}$	$5\frac{1}{4}$	3	0.259
$\frac{1}{4}$	$\frac{9}{16}$	16	0.065	$\frac{9}{16}$	$1\frac{7}{8}$	12	0.109	$1\frac{1}{16}$	$3\frac{3}{8}$	4	0.238	$3\frac{1}{8}$	$5\frac{1}{2}$	2	0.284
$\frac{17}{64}$	$\frac{5}{8}$	18	0.049	$\frac{19}{32}$	$1\frac{3}{16}$	13	0.095	$1\frac{3}{16}$	$2\frac{1}{2}$	8	0.165				
$\frac{9}{32}$	$\frac{5}{8}$	16	0.065	$\frac{5}{8}$	$1\frac{3}{8}$	12	0.109	$1\frac{1}{4}$	$2\frac{3}{4}$	8	0.165				
$\frac{5}{16}$	$\frac{3}{4}$	16	0.065	$\frac{5}{8}$	$1\frac{1}{2}$	12	0.109	$1\frac{5}{16}$	$2\frac{3}{4}$	8	0.165				

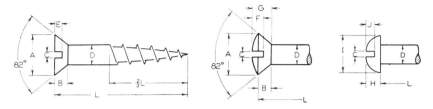

Maximum Dimensions

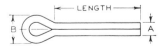

Design of head may vary but outside diameters should be adhered to.

A nominal	B min.	Hole sizes recommended
0.031	1/16	3/64
0.047	3/32	1/16
0.062	1/8	5/64
0.078	5/32	3/32
0.094	3/16	7/64
0.109	7/32	1/8
0.125	1/4	9/64
0.141	9/32	5/32
0.156	5/16	11/64
0.188	3/8	13/64
0.219	7/16	15/64
0.250	1/2	17/64
0.312	5/8	5/16
0.375	3/4	3/8
0.438	7/8	7/16
0.500	1	1/2
0.625	1 1/4	5/8
0.750	1 1/2	3/4

Nominal size	D	A	B	C	E	F	G	H	I	J	Number threads per inch
0	0.060	0.119	0.035	0.023	0.015	0.030	0.056	0.053	0.113	0.039	32
1	0.073	0.146	0.043	0.026	0.019	0.038	0.068	0.061	0.138	0.044	28
2	0.086	0.172	0.051	0.031	0.023	0.045	0.080	0.069	0.162	0.048	26
3	0.099	0.199	0.059	0.035	0.027	0.052	0.092	0.078	0.187	0.053	24
4	0.112	0.225	0.067	0.039	0.030	0.059	0.104	0.086	0.211	0.058	22
5	0.125	0.252	0.075	0.043	0.034	0.067	0.116	0.095	0.236	0.063	20
6	0.138	0.279	0.083	0.048	0.038	0.074	0.128	0.103	0.260	0.068	18
7	0.151	0.305	0.091	0.048	0.041	0.081	0.140	0.111	0.285	0.072	16
8	0.164	0.332	0.100	0.054	0.045	0.088	0.152	0.120	0.309	0.077	15
9	0.177	0.358	0.108	0.054	0.049	0.095	0.164	0.128	0.334	0.082	14
10	0.190	0.385	0.116	0.060	0.053	0.103	0.176	0.137	0.359	0.087	13
12	0.216	0.438	0.132	0.067	0.060	0.117	0.200	0.153	0.408	0.096	11
14	0.242	0.491	0.148	0.075	0.068	0.132	0.224	0.170	0.457	0.106	10
16	0.268	0.544	0.164	0.075	0.075	0.146	0.248	0.187	0.506	0.115	9
18	0.294	0.597	0.180	0.084	0.083	0.160	0.272	0.204	0.555	0.125	8
20	0.320	0.650	0.196	0.084	0.090	0.175	0.296	0.220	0.604	0.134	8
24	0.372	0.756	0.228	0.094	0.105	0.204	0.344	0.254	0.702	0.154	7

Table D-17   American National Standard Taper Pins

TAPER PINS   *Maximum length for which standard reamers are available. Taper ¼ in. per ft.

Size No.	0000000	000000	00000	0000	000	00	0
Size (large end)	0.0625	0.0780	0.0940	0.1090	0.1250	0.1410	0.1560
Maximum length*	0.625	0.750	1.000	1.000	1.000	1.250	1.250

Size No.	1	2	3	4	5	6	7
Size (large end)	0.1720	0.1930	0.2190	0.2500	0.2890	0.3410	0.4090
Maximum length*	1.250	1.500	1.750	2.000	2.250	3.000	3.750

Size No.	8	9	10	11	12	13	14
Size (large end)	0.4920	0.5910	0.7060	0.8600	1.032	1.241	1.523
Maximum length*	4.500	5.250	6.000	(Special sizes. Special lengths.)			

Table D-18   American National Standard Square- and Flat-stock Keys and Shaft Diameters

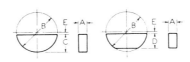

Diameter of shaft D inclusive	Square keys W	Flat keys W × H
$\frac{1}{2}-\frac{9}{16}$	$\frac{1}{8}$	$\frac{1}{8} \times \frac{3}{32}$
$\frac{5}{8}-\frac{7}{8}$	$\frac{3}{16}$	$\frac{3}{16} \times \frac{1}{8}$
$\frac{15}{16}-1\frac{1}{4}$	$\frac{1}{4}$	$\frac{1}{4} \times \frac{3}{16}$
$1\frac{5}{16}-1\frac{3}{8}$	$\frac{5}{16}$	$\frac{5}{16} \times \frac{1}{4}$
$1\frac{7}{16}-1\frac{3}{4}$	$\frac{3}{8}$	$\frac{3}{8} \times \frac{1}{4}$
$1\frac{13}{16}-2\frac{1}{4}$	$\frac{1}{2}$	$\frac{1}{2} \times \frac{3}{8}$
$2\frac{5}{16}-2\frac{3}{4}$	$\frac{5}{8}$	$\frac{5}{8} \times \frac{7}{16}$
$2\frac{7}{8}-3\frac{1}{4}$	$\frac{3}{4}$	$\frac{3}{4} \times \frac{1}{2}$
$3\frac{3}{8}-3\frac{3}{4}$	$\frac{7}{8}$	$\frac{7}{8} \times \frac{5}{8}$
$3\frac{7}{8}-4\frac{1}{2}$	1	$1 \times \frac{3}{4}$
$4\frac{3}{4}-5\frac{1}{2}$	$1\frac{1}{4}$	$1\frac{1}{4} \times \frac{7}{8}$
$5\frac{3}{4}-6$	$1\frac{1}{2}$	$1\frac{1}{2} \times 1$

Table D-19   Woodruff Keys   Dimensions in Inches

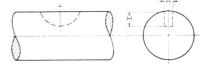

Key No.	Nominal			Maximum		
	A	B	E	C	D	H
204	$\frac{1}{16}$	$\frac{1}{2}$	$\frac{3}{64}$	0.203	0.194	0.1718
304	$\frac{3}{32}$	$\frac{1}{2}$	$\frac{3}{64}$	0.203	0.194	0.1561
305	$\frac{3}{32}$	$\frac{5}{8}$	$\frac{1}{16}$	0.250	0.240	0.2031
404	$\frac{1}{8}$	$\frac{1}{2}$	$\frac{3}{64}$	0.203	0.194	0.1405
405	$\frac{1}{8}$	$\frac{5}{8}$	$\frac{1}{16}$	0.250	0.240	0.1875
406	$\frac{1}{8}$	$\frac{3}{4}$	$\frac{1}{16}$	0.313	0.303	0.2505
505	$\frac{5}{32}$	$\frac{5}{8}$	$\frac{1}{16}$	0.250	0.240	0.1719
506	$\frac{5}{32}$	$\frac{3}{4}$	$\frac{1}{16}$	0.313	0.303	0.2349
507	$\frac{5}{32}$	$\frac{7}{8}$	$\frac{1}{16}$	0.375	0.365	0.2969
606	$\frac{3}{16}$	$\frac{3}{4}$	$\frac{1}{16}$	0.313	0.303	0.2193
607	$\frac{3}{16}$	$\frac{7}{8}$	$\frac{1}{16}$	0.375	0.365	0.2813
608	$\frac{3}{16}$	1	$\frac{1}{16}$	0.438	0.428	0.3443
609	$\frac{3}{16}$	$1\frac{1}{8}$	$\frac{5}{64}$	0.484	0.475	0.3903
807	$\frac{1}{4}$	$\frac{7}{8}$	$\frac{1}{16}$	0.375	0.365	0.2500
808	$\frac{1}{4}$	1	$\frac{1}{16}$	0.438	0.428	0.3130
809	$\frac{1}{4}$	$1\frac{1}{8}$	$\frac{5}{64}$	0.484	0.475	0.3590
810	$\frac{1}{4}$	$1\frac{1}{4}$	$\frac{5}{64}$	0.547	0.537	0.4220
811	$\frac{1}{4}$	$1\frac{3}{8}$	$\frac{3}{32}$	0.594	0.584	0.4690
812	$\frac{1}{4}$	$1\frac{1}{2}$	$\frac{7}{64}$	0.641	0.631	0.5160
1008	$\frac{5}{16}$	1	$\frac{1}{16}$	0.438	0.428	0.2818
1009	$\frac{5}{16}$	$1\frac{1}{8}$	$\frac{5}{64}$	0.484	0.475	0.3278
1010	$\frac{5}{16}$	$1\frac{1}{4}$	$\frac{5}{16}$	0.547	0.537	0.3908
1011	$\frac{5}{16}$	$1\frac{3}{8}$	$\frac{3}{32}$	0.594	0.584	0.4378
1012	$\frac{5}{16}$	$1\frac{1}{2}$	$\frac{7}{64}$	0.641	0.631	0.4848
1210	$\frac{3}{8}$	$1\frac{1}{4}$	$\frac{5}{64}$	0.547	0.537	0.3595
1211	$\frac{3}{8}$	$1\frac{3}{8}$	$\frac{3}{32}$	0.594	0.584	0.4060
1212	$\frac{3}{8}$	$1\frac{1}{2}$	$\frac{7}{64}$	0.641	0.631	0.4535

Note: Nominal dimensions are indicated by the key number in which the last two digits give the diameter (B) in eighths and the ones in front of them give the width (A) in thirty-seconds. For example, No. 809 means $B = \frac{9}{8}$ or $1\frac{1}{8}$ and $A = \frac{8}{32}$ or $\frac{1}{4}$.

Table D-20 Wire and Sheet-metal Gages

Dimensions in Decimal Parts of an Inch

No. of wire gage	American, or Brown & Sharpe	Birming- ham, or Stubs wire	Washburn & Moen or American Steel & Wire Co.	W. & M. steel music wire	New American S. & W. Co. music wire gage	Imperial wire gage	U.S. Standard gage for sheet amd plate iron and steel
00000000	. . . .	. . . .	. . . .	0.0083			
0000000	. . . .	. . . .	. . . .	0.0087			
000000	. . . .	. . . .	. . . .	0.0095	0.004	0.464	0.46875
00000	. . . .	. . . .	. . . .	0.010	0.005	0.432	0.4375
0000	0.460	0.454	0.3938	0.011	0.006	0.400	0.40625
000	0.40964	0.425	0.3625	0.012	0.007	0.372	0.375
00	0.3648	0.380	0.3310	0.0133	0.008	0.348	0.34375
0	0.32486	0.340	0.3065	0.0144	0.009	0.324	0.3125
1	0.2893	0.300	0.2830	0.0156	0.010	0.300	0.28125
2	0.25763	0.284	0.2625	0.0166	0.011	0.276	0.265625
3	0.22942	0.259	0.2437	0.0178	0.012	0.252	0.250
4	0.20431	0.238	0.2253	0.0188	0.013	0.232	0.234375
5	0.18194	0.220	0.2070	0.0202	0.014	0.212	0.21875
6	0.16202	0.203	0.1920	0.0215	0.016	0.192	0.203125
7	0.14428	0.180	0.1770	0.023	0.018	0.176	0.1875
8	0.12849	0.165	0.1620	0.0243	0.020	0.160	0.171875
9	0.11443	0.148	0.1483	0.0256	0.022	0.144	0.15625
10	0.10189	0.134	0.1350	0.027	0.024	0.128	0.140625
11	0.090742	0.120	0.1205	0.0284	0.026	0.116	0.125
12	0.080808	0.109	0.1055	0.0296	0.029	0.104	0.109375
13	0.071961	0.095	0.0915	0.0314	0.031	0.092	0.09375
14	0.064084	0.083	0.0800	0.0326	0.033	0.080	0.078125
15	0.057068	0.072	0.0720	0.0345	0.035	0.072	0.0703125
16	0.05082	0.065	0.0625	0.036	0.037	0.064	0.0625
17	0.045257	0.058	0.0540	0.0377	0.039	0.056	0.05625
18	0.040303	0.049	0.0475	0.0395	0.041	0.048	0.050
19	0.03589	0.042	0.0410	0.0414	0.043	0.040	0.04375
20	0.031961	0.035	0.0348	0.0434	0.045	0.036	0.0375
21	0.028462	0.032	0.03175	0.046	0.047	0.032	0.034375
22	0.025347	0.028	0.0286	0.0483	0.049	0.028	0.03125
23	0.022571	0.025	0.0258	0.051	0.051	0.024	0.028125
24	0.0201	0.022	0.0230	0.055	0.055	0.022	0.025
25	0.0179	0.020	0.0204	0.0586	0.059	0.020	0.021875
26	0.01594	0.018	0.0181	0.0626	0.063	0.018	0.01875
27	0.014195	0.016	0.0173	0.0658	0.067	0.0164	0.0171875
28	0.012641	0.014	0.0162	0.072	0.071	0.0149	0.015625
29	0.011257	0.013	0.0150	0.076	0.075	0.0136	0.0140625
30	0.010025	0.012	0.0140	0.080	0.080	0.0124	0.0125
31	0.008928	0.010	0.0132	. . . .	0.085	0.0116	0.0109375
32	0.00795	0.009	0.0128	. . . .	0.090	0.0108	0.01015625
33	0.00708	0.008	0.0118	. . . .	0.095	0.0100	0.009375
34	0.006304	0.007	0.0104	. . . .	. . . .	0.0092	0.00859375
35	0.005614	0.005	0.0095	. . . .	. . . .	0.0084	0.0078125
36	0.005	0.004	0.0090	. . . .	. . . .	0.0076	0.00703125
37	0.004453	. . . .	. . . .	. . . .	. . . .	0.0068	0.006640625
38	0.003965	. . . .	. . . .	. . . .	. . . .	0.0060	0.00625
39	0.003531	. . . .	. . . .	. . . .	. . . .	0.0052	
40	0.003144	. . . .	. . . .	. . . .	. . . .	0.0048	

Table D-21 American National Standard Welded and Seamless Steel Pipe

Nominal pipe size	Out- side diam.	Nominal wall thickness			Threads per inch
		Stand- ard wall	Extra strong wall	Double extra strong wall	
$\frac{1}{8}$	0.405	0.068	0.095	. . . .	27
$\frac{1}{4}$	0.540	0.088	0.119	. . . .	18
$\frac{3}{8}$	0.675	0.091	0.126	. . . .	18
$\frac{1}{2}$	0.840	0.109	0.147	0.294	14
$\frac{3}{4}$	1.050	0.113	0.154	0.308	14
1	1.315	0.133	0.179	0.358	$11\frac{1}{2}$
$1\frac{1}{4}$	1.660	0.140	0.191	0.382	$11\frac{1}{2}$
$1\frac{1}{2}$	1.900	0.145	0.200	0.400	$11\frac{1}{2}$
2	2.375	0.154	0.218	0.436	$11\frac{1}{2}$
$2\frac{1}{2}$	2.875	0.203	0.276	0.552	8
3	3.500	0.216	0.300	0.600	8
$3\frac{1}{2}$	4.000	0.226	0.318	. . . .	8
4	4.500	0.237	0.337	0.674	8
5	5.563	0.258	0.375	0.750	8
6	6.625	0.280	0.432	0.864	8
8	8.625	0.322	0.500	0.875	8
10	10.750	0.365	0.500	. . . .	8
12	12.750	0.375	0.500	. . . .	8
14	14.000	0.375	0.500	. . . .	8
16	16.000	0.375	0.500	. . . .	8
18	18.000	0.375	0.500	. . . .	8
20	20.000	0.375	0.500	. . . .	8
24	24.000	0.375	0.500	. . . .	8

Note: To find the inside diameter subtract twice the wall thickness from the outside diameter. Schedule numbers have been set up for wall thicknesses for pipe and the American Standard should be consulted for complete information. Standard wall thicknesses are for Schedule 40 up to and including nominal size 10. Extra strong walls are Schedule 80 up to and including size 8, and Schedule 60 for size 10.

Table D-22   Steel-wire Nails

American Steel & Wire Company Gage

Size	Length	Common wire nails and brads		Casing nails		Finishing nails	
		Gage, diam.	No. to pound	Gage, diam.	No. to pound	Gage, diam.	No. to pound
2d	1	15	876	$15\frac{1}{2}$	1010	$16\frac{1}{2}$	1351
3d	$1\frac{1}{4}$	14	568	$14\frac{1}{2}$	635	$15\frac{1}{2}$	807
4d	$1\frac{1}{2}$	$12\frac{1}{2}$	316	14	473	15	584
5d	$1\frac{3}{4}$	$12\frac{1}{2}$	271	14	406	15	500
6d	2	$11\frac{1}{2}$	181	$12\frac{1}{2}$	236	13	309
7d	$2\frac{1}{4}$	$11\frac{1}{2}$	161	$12\frac{1}{2}$	210	13	238
8d	$2\frac{1}{2}$	$10\frac{1}{4}$	106	$11\frac{1}{2}$	145	$12\frac{1}{2}$	189
9d	$2\frac{3}{4}$	$10\frac{1}{4}$	96	$11\frac{1}{2}$	132	$12\frac{1}{2}$	172
10d	3	9	69	$10\frac{1}{2}$	94	$11\frac{1}{2}$	121
12d	$3\frac{1}{4}$	9	64	$10\frac{1}{2}$	87	$11\frac{1}{2}$	113
16d	$3\frac{1}{2}$	8	49	10	71	11	90
20d	4	6	31	9	52	10	62
30d	$4\frac{1}{2}$	5	24	9	46		
40d	5	4	18	8	35		
50d	$5\frac{1}{2}$	3	14				
60d	6	2	11				

Table D-23   American National Standard Large Rivets

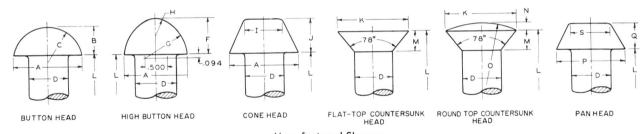

BUTTON HEAD   HIGH BUTTON HEAD   CONE HEAD   FLAT-TOP COUNTERSUNK HEAD   ROUND TOP COUNTERSUNK HEAD   PAN HEAD

Manufactured Shapes

D nominal	A basic	B basic (min.)	C	E basic	F basic	G	H	I basic	J basic (min.)	K basic	M basic (min.)	N	O	P basic	Q basic (min.)	S basic
$\frac{1}{2}$	0.875	0.375	0.443	0.781	0.500	0.656	0.094	0.469	0.438	0.905	0.250	0.095	1.125	0.800	0.381	0.500
$\frac{5}{8}$	1.094	0.469	0.553	0.969	0.594	0.750	0.188	0.586	0.547	1.131	0.312	0.119	1.406	1.000	0.469	0.625
$\frac{3}{4}$	1.312	0.562	0.664	1.156	0.688	0.844	0.282	0.703	0.656	1.358	0.375	0.142	1.688	1.200	0.556	0.750
$\frac{7}{8}$	1.531	0.656	0.775	1.344	0.781	0.937	0.375	0.820	0.766	1.584	0.438	0.166	1.969	1.400	0.643	0.875
1	1.750	0.750	0.885	1.531	0.875	1.031	0.469	0.938	0.875	1.810	0.500	0.190	2.250	1.600	0.731	1.000
$1\frac{1}{8}$	1.969	0.844	0.996	1.719	0.969	1.125	0.563	1.055	0.984	2.036	0.562	0.214	2.531	1.800	0.835	1.125
$1\frac{1}{4}$	2.188	0.938	1.107	1.906	1.062	1.218	0.656	1.172	1.094	2.262	0.625	0.238	2.812	2.000	0.922	1.250
$1\frac{3}{8}$	2.406	1.031	1.217	2.094	1.156	1.312	0.750	1.290	1.203	2.489	0.688	0.261	3.094	2.200	1.009	1.375
$1\frac{1}{2}$	2.625	1.125	1.328	2.281	1.250	1.406	0.844	1.406	1.312	2.715	0.750	0.285	3.375	2.400	1.113	1.500
$1\frac{5}{8}$	2.844	1.219	1.439	2.469	1.344	1.500	0.938	1.524	1.422	2.941	0.812	0.309	3.656	2.600	1.201	1.625
$1\frac{3}{4}$	3.062	1.312	1.549	2.656	1.438	1.594	1.032	1.641	1.531	3.168	0.875	0.332	3.938	2.800	1.288	1.750

Table D-24   American Welding Society Standard Welding Symbols

AMERICAN WELDING SOCIETY ⬡ STANDARD WELDING SYMBOLS

Basic Weld Symbols and Their Location Significance

LOCATION SIGNIFICANCE	FILLET	PLUG OR SLOT	SPOT OR PROJECTION	SEAM	SQUARE	V	BEVEL	GROOVE U	J	FLARE V	FLARE BEVEL	BACK OR BACKING	SURFACING	FLANGE EDGE	FLANGE CORNER
ARROW SIDE				NOT USED											
OTHER SIDE				NOT USED									NOT USED	NOT USED	NOT USED
BOTH SIDES				NOT USED								NOT USED	NOT USED	NOT USED	NOT USED
NO ARROW SIDE OR OTHER SIDE SIGNIFICANCE				NOT USED								NOT USED EXCEPT FOR FLASH OR UPSET		NOT USED	NOT USED

(continued across right columns: ARC SEAM / ARC SPOT, RESISTANCE SEAM / RESISTANCE SPOT, PROJECTION, FLASH OR UPSET)

Supplementary Symbols

WELD ALL AROUND	FIELD WELD	MELT-THRU	CONTOUR FLUSH	CONTOUR CONVEX	CONTOUR CONCAVE

Location of Elements of a Welding Symbol

FINISH SYMBOL
CONTOUR SYMBOL
ROOT OPENING; DEPTH OF FILLING FOR PLUG AND SLOT WELDS
GROOVE ANGLE; INCLUDED ANGLE OF COUNTERSINK FOR PLUG WELDS
LENGTH OF WELD
PITCH (CENTER-TO-CENTER SPACING) OF WELDS
ARROW CONNECTING REFERENCE LINE TO ARROW SIDE OR ARROW-SIDE MEMBER OF JOINT
FIELD WELD SYMBOL
WELD-ALL-AROUND SYMBOL
NUMBER OF SPOT OR PROJECTION WELDS
SIZE; SIZE OR STRENGTH FOR CERTAIN WELDS
SPECIFICATION, PROCESS, OR OTHER REFERENCE
TAIL (MAY BE OMITTED WHEN REFERENCE IS NOT USED)
BASIC WELD SYMBOL OR DETAIL REFERENCE
ELEMENTS IN THIS AREA REMAIN AS SHOWN WHEN TAIL AND ARROW ARE REVERSED

(F) (A) (R) (S) (N) (L - P) (BOTH SIDES) (OTHER SIDE) (ARROW SIDE)

Typical Welding Symbols

BACK OR BACKING WELD SYMBOL
SURFACING WELD SYMBOL INDICATING BUILT-UP SURFACE
DOUBLE FILLET WELDING SYMBOL
CHAIN INTERMITTENT FILLET WELDING SYMBOL
STAGGERED INTERMITTENT FILLET WELDING SYMBOL
SINGLE-V-GROOVE WELDING SYMBOL
SINGLE-V-GROOVE WELDING SYMBOL INDICATING ROOT PENETRATION
DOUBLE-BEVEL-GROOVE WELDING SYMBOL
WELDING SYMBOLS FOR COMBINED WELDS
PLUG WELDING SYMBOL
SLOT WELDING SYMBOL
SPOT WELDING SYMBOL
PROJECTION WELDING SYMBOL
SEAM WELDING SYMBOL
FLASH OR UPSET WELDING SYMBOL
SQUARE-GROOVE WELDING SYMBOL
FLARE-V- AND FLARE-BEVEL-GROOVE WELDING SYMBOL
EDGE- AND CORNER-FLANGE WELD SYMBOLS
MELT-THRU SYMBOL
FLUSH-CONTOUR SYMBOL
CONVEX-CONTOUR SYMBOL

Supplementary Symbols Used with Welding Symbols

WELD-ALL-AROUND SYMBOL
FIELD WELD SYMBOL

Basic Joints — Identification of Arrow Side and Other Side of Joint and Arrow-Side and Other-Side Member of Joint

BUTT JOINT
CORNER JOINT
TEE JOINT
LAP JOINT
EDGE JOINT

DESIGNATION OF WELDING PROCESSES BY LETTERS

DESIGNATION OF CUTTING PROCESSES BY LETTERS
AAC .... Air Carbon-Arc Cutting
AC .... Arc Cutting
OAC .... Oxygen-Arc Cutting
CAC .... Carbon-Arc Cutting
CHC .... Chemical Flux Cutting
MAC .... Metal-Arc Cutting
OC .... Oxygen Cutting
PAC .... Plasma-Arc Cutting
POC .... Metal Powder Cutting

AMERICAN WELDING SOCIETY
2501 N. W. 7th Street, Miami, Fla. 33125

Copyright 1968 by the

AWS A2.1-68

Appendix E Glossary

(*v*) = Verb   (*n*) = Noun

acme (*n*) A screw-thread form.

addendum (*n*) The distance the gear tooth extends above the pitch circle.

allowance (*n*) The minimum clearance between mating parts.

alloy (*n*) Two or more metals combined to form a new metal.

anneal (*v*) To heat slowly to a critical temperature and gradually cool. Used to soften and to remove internal stresses.

babbitt or **babbitt metal** (*n*) A friction bearing metal, invented by Isaac Babbitt. Composed of antimony, tin, and copper.

bearing (*n*) Any part that bears up, or supports, another part. In particular, the support for a revolving shaft.

bevel (*n*) A surface slanted to another surface. Called a *miter* when the angle is 45°.

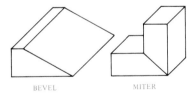

BEVEL   MITER

bolt circle (*n*) A circular center line locating the centers of holes located about a common center point.

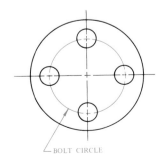

BOLT CIRCLE

bore (*v*) To enlarge or to finish a hole by means of a cutting tool called a *boring bar,* used in a boring mill or lathe.

boss (*n*) A raised surface of circular outline as used on a casting or forging.

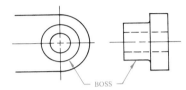

BOSS

brass (*n*) An alloy of copper and zinc or copper with zinc and lead.

braze (*v*) To join two pieces of metal by using hard solder such as brass or zinc.

broach (*v*) To machine and change the forms of holes or outside surfaces to a desired shape, generally other than round. (*n*) A tool with a series of chisel edges used to broach.

bronzes (*n*) Alloys of copper and tin in varying proportions, mostly copper. Sometimes other metals, such as zinc, are added.

buff (*v*) To polish a surface on a fabric wheel using an abrasive material.

burnish (*v*) To smooth or polish by a sliding or rolling motion.

burr (*n*) A rough or jagged edge caused by cutting or punching.

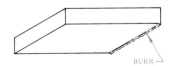

BURR

bushing (*n*) A hollow cylindrical sleeve used as a bearing or as a guide for drills or other tools.

caliper (*n*) A measuring device with two adjustable legs, used for measuring thicknesses or diameters.

cam (*n*) A machine part mounted on a revolving shaft, used for changing rotary motion into an alternating back-and-forth motion. See Chapter 19.

caseharden (carbonize or carburize) (*v*) To heat-treat steel to harden the surface by causing it to absorb carbon by quenching in an oil or lead bath.

casting (*n*) A part formed by pouring molten metal into a hollow form (mold) of the desired shape and allowing it to harden.

center drill (*n*) A special combination drill and countersink used to produce bearing surfaces in the ends of stock to be turned between centers on a lathe.

chamfer (*v*) To bevel an edge. (*n*) An edge which has been beveled.

CHAMFER

circular pitch (*n*) The distance from a point on one gear tooth to the same point on the next tooth measured along the pitch circle.

core (*v*) To form the hollow part of a casting by means of a part made of sand and shaped like the hollow part (called a *core*) and placed in the mold (see **casting**). The core is broken up and removed after the casting is cool.

counterbore (*v*) To enlarge an end of a hole to a desired depth and to cylindrical form. Such an enlargement is called a *counterbore*.

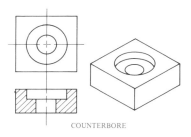

COUNTERBORE

countersink (*v*) To form a conical space at the end of a hole.

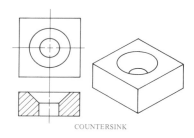

COUNTERSINK

crown (*n*) The contour of the face of a belt pulley, rounded or angular, used to keep the belt in place. The belt tends to climb to the highest place.

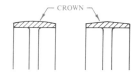

CROWN

dedendum (*n*) The distance the gear tooth extends below the pitch circle.

development (*n*) The pattern of the surface of an object drawn on a flat plane or surface.

diametral pitch (*n*) The number of gear teeth per inch of pitch diameter.

die (*n*) A hardened metal block shaped to form a desired shape by cutting or pressing. Also, a tool used to cut external screw threads.

die casting (*n*) A casting made of molten alloy (or plastic composition) by pouring it into a metal mold or die, generally forced under pressure. Die castings are smooth and accurate.

die stamping (*n*) A piece which has been formed or cut from sheet material, generally sheet metal.

drill (*v*) To make a cylindrical hole using a revolving tool with cutting edges, called a *drill,* generally a twist drill.

dowel (*n*) A cylindrical-shaped pin used for fastening parts together.

draft (*n*) The tapered sides on a foundry pattern to allow it to be easily pulled from the sand.

drop forging (*n*) A piece formed between dies while hot, using pressure or a drop hammer.

exploded view (*n*) Separate parts of a single assembly projected away from each other, or separated to show relationships among the parts of one drawing.

face (*v*) To machine (finish) a flat surface on a lathe with the surface perpendicular to the axis of rotation.

FAO Finished all over.

file (*v*) To smooth, finish, or shape with a file.

fillet (*n*) The rounded-in corner between two surfaces.

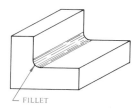

FILLET

fit (*n*) The tightness or looseness between two mating parts.

fixture (*n*) A device used for holding a workpiece during a machining operation.

flange (*n*) A rim extension, as at the end of a pipe or similar construction.

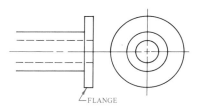

FLANGE

forge (*v*) To give the desired shape or form to hot metal by hammering or pressure.

functional drawing (*n*) A drawing using the fewest number of views and the fewest number of lines to provide the exact information required.

galvanize (*v*) To give a coating of zinc or zinc and lead.

gasket (*n*) A thin piece of material placed between two surfaces to produce a tight joint.

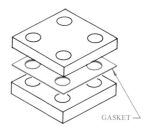

GASKET

gage (*n*) A device for determining whether a specified dimension on an object is within specified limits.

gear (*n*) A toothed wheel used to transmit power or motion from one shaft to another. A machine element used to transmit motion or force.

grind (*v*) To use an abrasive wheel to polish or to finish a surface.

heat-treat (*v*) To change the properties of metal by carefully controlled heating and cooling.

jig (*n*) A special device used to guide a cutting tool; it may also hold the workpiece.

kerf (*n*) A slot or groove made by a cutting tool.

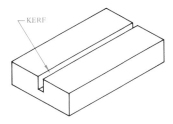

key (*n*) A piece used to fasten a hub to a shaft, or for a similar purpose.

keyway or **keyseat** (*n*) A groove or slot in a shaft or hub into which a key is placed.

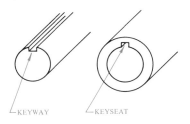

knurl (*v*) To form a series of regular dents to roughen a cylindrical surface so that it can be held or turned by hand.

limit (*n*) A boundary. Indication of only the largest and smallest permissible dimensions. See Chapter 6.

lug (*n*) An "ear" forming a part of, and extending from, a part.

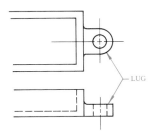

malleable casting (*n*) A casting which has been toughened by annealing.

micrometer caliper (*n*) A measuring device used to determine the exact measurements of diameter thicknesses.

mill (*v*) To machine a part on a milling machine, using a rotating toothed cutter.

neck (*v*) To cut a groove around a cylindrical part, generally at a change in diameter.

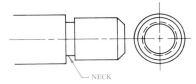

peen (*v*) To stretch or head over material with the peen, or ball, end of a machinist's hammer.

photo drawing (*n*) A drawing prepared from a photograph, or a photograph on which dimensions, changes, or additional parts have been drawn.

pickle (*v*) To clean a metal object by using a weak sulfuric acid bath.

plane (*n*) A flat surface.

plate (*v*) To electrochemically coat a metal object with another metal.

polish (*v*) To smooth with a very fine abrasive.

punch (*v*) To pierce thin metal by pressing a tool of the desired shape through it.

ream (*v*) To make a hole the exact size by finishing with a rotating fluted cutting tool.

round (*n*) The rounded-over corner of two surfaces.

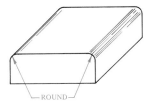

shaft (*n*) Round stock upon which gears, pulleys, or rotating pieces are attached for support or to transmit power.

shear (*v*) To cut material between two blades.

shim (*n*) A thin plate of metal used between two surfaces to adjust the distance between them.

spline (*n*) A long keyway.

spot-face (*v*) To finish a circular spot slightly below a rough surface on a casting to provide a smooth, flat seat for a bolthead or other fastening.

steel casting (*n*) A part made of cast iron to which scrap steel has been added when melted.

tap (*v*) To cut threads in a hole with a threading tool called a tap. (*n*) A hardened screw, fluted to provide cutting edges.

taper pin (*n*) A piece of round stock made with a gradual and uniform decrease in diameter.

technical illustration (*n*) A pictorial drawing rendered to simplify and interpret technical information.

temper (*v*) To reduce the brittleness in hardened steel by heating in various ways, as in a bath of oil, salt, sand, or lead, to a specified temperature, and then cooling.

template or **templet** (*n*) A flat form or pattern of full size, used to lay out a shape and to locate holes or other features.

tolerance (*n*) A specific allowance for variation from a given dimension. The total amount by which a given dimension may vary. The difference between limits. See Chapter 6.

turn (*v*) To machine a piece on a lathe by rotating the piece against a cutting tool (as when forming a cylindrical surface, and so forth).

upset (*v*) To make an enlarged section or shoulder on a rod, bar, or similar piece while forging.

vernier (*n*) A small auxiliary scale used to obtain fractional parts of a major scale.

washer (*n*) A ring of metal used to form a seat for a bolt or nut.

weld (*v*) To join pieces of metal, which have been heated to a fusing temperature, by pressing or hammering them together. See Chapter 17.

Index

8 9 10 KuKu 82 81 80